Advanced Oxidation Catalysts

Advanced Oxidation Catalysts

Guest Editor

Meng Li

Basel • Beijing • Wuhan • Barcelona • Belgrade • Novi Sad • Cluj • Manchester

Guest Editor
Meng Li
School of Environmental
Science and Engineering
Guangzhou University
Guangzhou
China

Editorial Office
MDPI AG
Grosspeteranlage 5
4052 Basel, Switzerland

This is a reprint of the Special Issue, published open access by the journal *Catalysts* (ISSN 2073-4344), freely accessible at: www.mdpi.com/journal/catalysts/special_issues/oxidation_catalysts_2022.

For citation purposes, cite each article independently as indicated on the article page online and using the guide below:

Lastname, A.A.; Lastname, B.B. Article Title. *Journal Name* **Year**, *Volume Number*, Page Range.

ISBN 978-3-7258-3076-3 (Hbk)
ISBN 978-3-7258-3075-6 (PDF)
https://doi.org/10.3390/books978-3-7258-3075-6

© 2025 by the authors. Articles in this book are Open Access and distributed under the Creative Commons Attribution (CC BY) license. The book as a whole is distributed by MDPI under the terms and conditions of the Creative Commons Attribution-NonCommercial-NoDerivs (CC BY-NC-ND) license (https://creativecommons.org/licenses/by-nc-nd/4.0/).

Contents

Preface vii

Meng Li and Chenxi Li
Advanced Oxidation Catalysts—Innovative Approaches and Emerging Trends
Reprinted from: *Catalysts* 2024, 14, 700, https://doi.org/10.3390/catal14100700 1

Asiyat Magomedova, Abdulgalim Isaev, Farid Orudzhev, Dinara Sobola, Rabadanov Murtazali and Alina Rabadanova et al.
Magnetically Separable Mixed-Phase α/γ-Fe_2O_3 Catalyst for Photo-Fenton-like Oxidation of Rhodamine B
Reprinted from: *Catalysts* 2023, 13, 872, https://doi.org/10.3390/catal13050872 6

Aisha Umar, Łukasz Smółka and Marek Gancarz
The Role of Fungal Fuel Cells in Energy Production and the Removal of Pollutants from Wastewater
Reprinted from: *Catalysts* 2023, 13, 687, https://doi.org/10.3390/catal13040687 19

Yuxin Wang, Yinxiu Jin, Minghan Jia, Hu Ruan, Xuefen Tao and Xuefeng Liu et al.
Enhanced Visible-Light Photocatalytic Activities of $CeVO_4$-V_2O_3 Composite: Effect of Ethylene Glycol
Reprinted from: *Catalysts* 2023, 13, 659, https://doi.org/10.3390/catal13040659 46

Yogesh M. Gote, Pankaj S. Sinhmar and Parag R. Gogate
Sonocatalytic Degradation of Chrysoidine R Dye Using Ultrasonically Synthesized $NiFe_2O_4$ Catalyst
Reprinted from: *Catalysts* 2023, 13, 597, https://doi.org/10.3390/catal13030597 59

Barbora Kamenická and Tomáš Weidlich
A Comparison of Different Reagents Applicable for Destroying Halogenated Anionic Textile Dye Mordant Blue 9 in Polluted Aqueous Streams
Reprinted from: *Catalysts* 2023, 13, 460, https://doi.org/10.3390/catal13030460 75

Fahad A. Alharthi, Alanood Sulaiman Ababtain, Hend Khalid Aldubeikl, Hamdah S. Alanazi and Imran Hasan
Synthesis of Novel $Zn_3V_2O_8$/Ag Nanocomposite for Efficient Photocatalytic Hydrogen Production
Reprinted from: *Catalysts* 2023, 13, 455, https://doi.org/10.3390/catal13030455 96

Fahad A. Alharthi, Wedyan Saud Al-Nafaei, Alanoud Abdullah Alshayiqi, Hamdah S. Alanazi and Imran Hasan
Hydrothermal Synthesis of Bimetallic (Zn, Co) Co-Doped Tungstate Nanocomposite with Direct Z-Scheme for Enhanced Photodegradation of Xylenol Orange
Reprinted from: *Catalysts* 2023, 13, 404, https://doi.org/10.3390/catal13020404 109

Zahra Lahootifar, Aziz Habibi-Yangjeh, Shima Rahim Pouran and Alireza Khataee
G-C_3N_4 Dots Decorated with Hetaerolite: Visible-Light Photocatalyst for Degradation of Organic Contaminants
Reprinted from: *Catalysts* 2023, 13, 346, https://doi.org/10.3390/catal13020346 126

Fahad Ahmed Alharthi, Alanoud Abdullah Alshayiqi, Wedyan Saud Al-Nafaei, Adel El Marghany, Hamdah Saleh Alanazi and Imran Hasan
Synthesis of Nanocrystalline Metal Tungstate $NiWO_4/CoWO_4$ Heterojunction for UV-Light-Assisted Degradation of Paracetamol
Reprinted from: *Catalysts* **2023**, *13*, 152, https://doi.org/10.3390/catal13010152 139

María J. Cruz-Carrillo, Rosa M. Melgoza-Alemán, Cecilia Cuevas-Arteaga and José B. Proal-Nájera
Removal of Persistent Acid Pharmaceuticals by a Biological-Photocatalytic Sequential Process: Clofibric Acid, Diclofenac, and Indomethacin
Reprinted from: *Catalysts* **2022**, *12*, 1488, https://doi.org/10.3390/catal12111488 157

Felipe de J. Silerio-Vázquez, Cynthia M. Núñez-Núñez, José B. Proal-Nájera and María T. Alarcón-Herrera
A Systematic Review on Solar Heterogeneous Photocatalytic Water Disinfection: Advances over Time, Operation Trends, and Prospects
Reprinted from: *Catalysts* **2022**, *12*, 1314, https://doi.org/10.3390/catal12111314 173

Tobias Weissenberger, Ralf Zapf, Helmut Pennemann and Gunther Kolb
Effect of the Active Metal on the NO_x Formation during Catalytic Combustion of Ammonia SOFC Off-Gas
Reprinted from: *Catalysts* **2022**, *12*, 1186, https://doi.org/10.3390/catal12101186 195

Umme Habiba, Sadaf Mutahir, Muhammad Asim Khan, Muhammad Humayun, Moamen S. Refat and Khurram Shahzad Munawar
Effective Removal of Refractory Pollutants through Cinnamic Acid-Modified Wheat Husk Biochar: Experimental and DFT-Based Analysis
Reprinted from: *Catalysts* **2022**, *12*, 1063, https://doi.org/10.3390/catal12091063 209

Guoyan Ma, Le Wang, Xiaorong Wang, Lu Li and Hongfei Ma
CO Oxidation over Alumina-Supported Copper Catalysts
Reprinted from: *Catalysts* **2022**, *12*, 1030, https://doi.org/10.3390/catal12091030 226

Svetlana Pavlova, Yulia Ivanova, Sergey Tsybulya, Yurii Chesalov, Anna Nartova and Evgenii Suprun et al.
Sr_2TiO_4 Prepared Using Mechanochemical Activation: Influence of the Initial Compounds' Nature on Formation, Structural and Catalytic Properties in Oxidative Coupling of Methane
Reprinted from: *Catalysts* **2022**, *12*, 929, https://doi.org/10.3390/catal12090929 235

Ermelinda Falletta, Claudia Letizia Bianchi, Franca Morazzoni, Alessandra Polissi, Flavia Di Vincenzo and Ignazio Renato Bellobono
Tungsten Trioxide and Its TiO_2 Mixed Composites for the Photocatalytic Degradation of NO_x and Bacteria (*Escherichia coli*) Inactivation
Reprinted from: *Catalysts* **2022**, *12*, 822, https://doi.org/10.3390/catal12080822 255

Fei Chen, Jiesen Guo, Dezhong Meng, Yuetong Wu, Ruijin Sun and Changchun Zhao
Strong Pyro-Electro-Chemical Coupling of Elbaite/H_2O_2 System for Pyrocatalysis Dye Wastewater
Reprinted from: *Catalysts* **2021**, *11*, 1370, https://doi.org/10.3390/catal11111370 271

Preface

Water contamination has become a major global issue due to fast-paced industrialization, urban development, and agricultural practices. Conventional treatment techniques frequently struggle to effectively manage complex pollutants such as pharmaceuticals, pesticides, and microplastics. This reprint, *Advanced Oxidation Catalysts*, examines the utilization of advanced oxidation catalysts (AOCs), which present a promising approach to enhancing pollutant removal through improved oxidation processes.

The scope of this reprint includes the principles, design, and development of AOCs, along with their application in various wastewater scenarios. It digs into essential catalytic materials, reaction mechanisms, and strategies for optimizing processes, while also tackling practical challenges associated with the scaling up of these technologies. In addition, this text discusses anticipated research directions and the environmental effects of catalytic oxidation processes, providing a thorough overview of how these technologies can promote sustainable water management.

The main goal of this reprint is to deliver both theoretical knowledge and practical skills that can advance the development and deployment of AOCs in wastewater treatment. This initiative is driven by the rising need for innovative and sustainable solutions that can effectively eliminate emerging pollutants, a task that traditional treatments are often unable to resolve. This volume acts as a connection between research and practical applications, providing readers with the necessary scientific and technical knowledge to implement successful catalytic oxidation processes.

This reprint targets a wide range of readers, including researchers, engineers, environmental scientists, and professionals working in water management. Graduate students in fields such as chemical engineering and environmental science will also find the insights valuable. In addition, industry stakeholders and policymakers may find the content to be beneficial for adopting advanced treatment solutions to comply with strict environmental regulations.

The successful completion of this reprint is the result of the collaborative efforts of numerous contributors. We sincerely thank the co-authors for their expertise, commitment, and major research contributions. Our appreciation also extends to the peer reviewers and editors for their constructive feedback and support in enhancing the content. Finally, we are grateful to the institutions and funding organizations that assisted the research that is included in this work.

We hope that this reprint will be a valuable asset for advancing wastewater treatment technologies and inspire continued innovation in the field. By promoting the utilization of AOCs, we aim to play a major role in sustainable water management practices, helping to protect the environment and ensure access to clean water for future generations.

Meng Li
Guest Editor

Editorial

Advanced Oxidation Catalysts—Innovative Approaches and Emerging Trends

Meng Li [1,2,*] and Chenxi Li [1]

[1] School of Environmental Science and Engineering, Guangzhou University, Guangzhou 510006, China; chenxili3927@163.com

[2] Guangdong Provincial Research Center for Environment Pollution Control and Remediation Materials, College of Life Science and Technology, Jinan University, Guangzhou 510632, China

* Correspondence: mengli@gzhu.edu.cn

Citation: Li, M.; Li, C. Advanced Oxidation Catalysts—Innovative Approaches and Emerging Trends. *Catalysts* **2024**, *14*, 700. https://doi.org/10.3390/catal14100700

Received: 29 September 2024
Accepted: 29 September 2024
Published: 8 October 2024

Copyright: © 2024 by the authors. Licensee MDPI, Basel, Switzerland. This article is an open access article distributed under the terms and conditions of the Creative Commons Attribution (CC BY) license (https:// creativecommons.org/licenses/by/ 4.0/).

Novel oxidation catalysts in water treatment and purification have garnered major attention due to the growing demand for clean water resources. As the use of advanced oxidation technologies is recognized as one of the most sustainable methods for breaking down and removing organic pollutants, this Special Issue on advanced oxidation catalysts brings together 17 articles that explore the development, analysis, and use of advanced oxidation catalysts designed for eliminating organic pollutants and disinfecting water to improve its quality. This article emphasizes the most recent advancements and discoveries in this field and summarizes the key results and innovative findings presented in the selected papers.

Lahootifar and his team (Contribution 1) developed a novel photocatalyst composed of hetaerolite-embellished g-C_3N_4 dots. This novel material has increased photocatalytic effectiveness when exposed to visible light, allowing organic pollutants to be easily decomposed. The combination of these compounds causes a synergistic effect that increases the capacity of visible light absorption and extends the time of photo-generated charges, thus increasing the efficiency of pollutant degradation. This work confirms that creating heterojunctions can greatly boost the photocatalytic activity of chemicals, which is promising for their use in treating water pollutants.

An important study was carried out by Alharthi et al., building upon previous research. The authors provide a detailed explanation of the creation of a nanocrystalline composite material that includes metal tungstates, specifically $NiWO_4$ and $CoWO_4$ (Contribution 2). These components work together to form a heterojunction. This complex structure is particularly effective in breaking down paracetamol when exposed to ultraviolet (UV) radiation. When these materials are used in a heterojunction setup, they show an increased rate constant, demonstrating improved photocatalytic activity. This finding emphasizes the necessity of developing heterojunctions to accelerate the breakdown of pharmaceutical pollutants, which pose important environmental and health risks.

Cruz-Carrillo and other researchers focused on decomposing recalcitrant acidic pharmaceuticals, specifically clofibric acid, diclofenac, and indomethacin, through a combined biological and photocatalytic technique (Contribution 3). Their combination of biological processes and photocatalysis achieves almost complete degradation of such pharmaceuticals, where the latter treatment demonstrates excellent efficiency in breaking down any remaining substances. This method has confirmed potential for treating persistent pollutants that are difficult to eliminate using traditional methods.

Magomedova and colleagues focused on the construction of a magnetically divisible α/γ-Fe_2O_3 mixed-phase catalyst for the photo-Fenton-like oxidation process of Rhodamine B (Contribution 4). The magnetic properties of the catalyst make it very easy to separate and reuse, giving it great potential in practical applications. Their study also explored the effect of different factors on degradation efficiency and found that the catalyst performed well under optimal conditions, contributing to the field of sustainable wastewater treatment technology.

The research from Kamenicka and Weidlich investigates the use of different technologies to decompose the halogenated negatively charged textile dye Mordant Blue 9 (Contribution 5). Their study evaluates the efficiency of various oxidants and reducing agents and provides a comparative analysis of their economic feasibility and performance in dye decomposition. This study emphasizes the importance of exploring cost-effective methods to treat dye-contaminated water bodies.

Ma and his colleagues conducted research on the oxidation of carbon monoxide using a copper catalyst supported by alumina (Contribution 6). Their goal was to tackle the issue of toxic gases in industrial emissions. Their study investigated how various factors influence catalytic effectiveness, emphasizing that the characteristics of the substrates involved in mechanochemical activation are important in shaping the properties and reactivity of the resulting catalysts. This research contributes to the larger effort of using advanced catalytic technologies to purify air pollution.

Silerio-Vazquez and colleagues conducted a systematic review of sunlight-catalyzed heterogeneous water disinfection (Contribution 7). This study provides a comprehensive overview of the progress of photocatalytic water treatment technology, with a particular focus on solar energy utilization. The authors systematically analyzed 60 reports from 1043 records, extracting data on the reactor type, photocatalysts, microorganisms, and operational parameters. The results show that titanium dioxide is a widely used photocatalyst, while bismuth oxide and red phosphorus show promising prospects for visible-light-driven disinfection. The review clarifies the importance of further studies to optimize photocatalytic systems in practical water treatment applications.

Umar et al. conducted a thorough exploration of the capabilities of mycelium-based energy conversion and waste purification processes, providing a new view of bioremediation techniques (Contribution 8). Their research clarifies the capability of fungal fuel cells to simultaneously decompose organic pollutants and generate bioelectricity. The researchers investigated the degradation process of pollutants and emphasized the advantages of fungal biomass compared to conventional methods. Their study offers a novel approach to integrating biological processes with advanced oxidation technologies for sustainable wastewater treatment.

An in-depth study on the electrochemical coupling of the Elbaite/H_2O_2 system in the area of dye wastewater catalytic degradation introduces a new method for breaking down pollutants (Contribution 9). This study reveals the combined benefits of using pyrolytic carbon alongside hydrogen peroxide, resulting in the effective degradation of organic dyes. The results indicate that this innovative technique successfully purifies dye-containing wastewater while preventing the creation of external pollutants, positioning it as a potential solution for industrial use.

Falletta et al. carried out an extensive analysis focused on tungsten trioxide (WO_3) and its hybrid composite with titanium dioxide (TiO_2) in the photocatalytic degradation of NO_x and the inactivation of *Escherichia coli* (*E. coli*). This detailed study sought to tackle both air and water pollution simultaneously, demonstrating the important potential of these materials in environmental remediation (Contribution 10). Their research findings suggest that the combined composites exhibit a greater level of photocatalytic effectiveness when compared to WO_3 in its original form, indicating that adding TiO_2 enhances the degradation efficiency. This study emphasizes the promising applications of these composite materials in real-world scenarios.

The research conducted by Pavlova and her team investigated the impact of starting material characteristics on the formation, structural properties, and reactivity of Sr_2TiO_4 created through mechanical activation (Contribution 11). Their findings indicate that the mechanochemical activation process plays a critical role in influencing the catalytic activity of the produced materials. The experimental results demonstrate that the Sr_2TiO_4 catalyst, when activated through mechanical and chemical methods, exhibits enhanced catalytic performance, making it potentially useful for applications in various areas, including methane oxidation coupling.

Gote et al. present a new approach for the sonocatalytic breakdown process, using a $NiFe_2O_4$ catalyst created through ultrasonic irradiation (Contribution 12). The optimal conditions for degrading the azo dye crystal violet R are a hydrogen peroxide loading rate of 75 milligrams per minute, a pH level of 3, an ultrasonic working cycle of 70%, an output power of 120 watts, a reaction time of 160 min, and a catalyst loading rate of 5 g per liter. Under these conditions, the maximum efficiency for sound wave degradation can be achieved, reaching up to 92%. The process achieves a satisfying 83% success rate, demonstrating the effectiveness of ultrasound-enhanced catalytic breakdown for cleaning dye-contaminated water.

The research conducted by Wang and his colleagues examined how the photocatalytic activity of $CeVO_4$-V_2O_3 composites can be improved when exposed to visible light (Contribution 13). A detailed analysis of the effect of ethylene glycol on photocatalytic performance shows that the composite material exhibits increased photocatalytic reactivity under normal light conditions. This finding emphasizes the need to modify the structure and components of photocatalysts to enhance their ability to capture solar energy for water purification.

Alharthi and other researchers (Contribution 14) have conducted intensive research on the synthesis of $Zn_3V_2O_8$ and Ag-nanoparticle composite materials, and their results have attracted much attention. A large transition reducing the optical band gap from 2.33 eV to 2.19 eV was achieved by integrating Ag nanoparticles into a ZnV matrix. The widened band gap improves the light absorption efficiency and thus promotes the photocatalytic hydrogen generation process. The experimental results show that the $Zn_3V_2O_8$/Ag nanocomposite has outstanding performance in terms of its hydrogen production rate, reaching a high efficiency of 37.52 $\mu mol\ g^{-1}\ h^{-1}$, showing its great potential as an economical and environmentally friendly photocatalyst in the field of hydrogen preparation.

The chosen papers often emphasize the significance of developing catalysts with enhanced properties through synthesis. One study describes the use of the hydrothermal method to create a nanocomposite that features bimetallic (Zn, Co) concurrent doping within a tungstate matrix, designed with a Z-scheme architecture (Contribution 15). This material demonstrates outstanding photocatalytic efficiency in breaking down Xylenol Orange (XO), greatly exceeding the performance of pure $ZnWO_4$ and $CoWO_4$, suggesting that co-doping can improve photocatalytic efficiency. The experimental findings confirm the essential role of superoxide anion radicals ($O_2^{\bullet -}$) and light-generated electrons (e^-) in the degradation process. This research emphasizes the nanocomposite's potential as an eco-friendly and cost-effective material for use in industrial applications, especially for treating wastewater contaminated with XO pollutants.

A thorough evaluation of the research conducted by Weissenberger and colleagues has been completed. This study primarily aimed to explore the catalytic effects of active metals in the ammonia combustion process, with a particular emphasis on their role in NO_x formation (Contribution 16). Additionally, it analyzed the catalytic properties of noble and transition metals supported by alumina, specifically for the treatment of emissions from solid oxide fuel cells that arise from ammonia combustion. The tested catalysts can fully convert the hydrogen and ammonia present in the exhaust gases, and the efficiency of selectively generating NO_x improves as the reaction temperature rises. Under low-temperature conditions, the production of NO_x is effectively regulated, demonstrating the potential for optimizing catalysts to minimize NOx emissions in burner systems.

Overall, the articles in this Special Issue display the variety and depth of the catalytic processes applied for environmental remediation (Contributions 1–17). These papers emphasize the promise of innovative materials and integrated strategies in tackling water and air pollution. The guest editors express their sincere gratitude to the contributors for their valuable insights and to the evaluators for their comprehensive reviews. It is hoped that the findings presented in this article will inspire further research and innovations in advanced oxidation catalysts, eventually contributing to a cleaner and safer ecological environment.

Author Contributions: Writing—original draft preparation, C.L.; writing—review and editing, M.L. All authors have read and agreed to the published version of the manuscript.

Acknowledgments: As the Guest Editors of this Special Issue, entitled "Advanced Oxidation Catalysts", we extend our sincere gratitude to all the authors who contributed their invaluable research, which has been instrumental in the success of this compilation. The authors wish to thank the Guangzhou University provincial key college student innovation training program (S202411078005) and Guangzhou University Challenge Cup project (2024TZBNAJ0711) for their support.

Conflicts of Interest: The authors declare no conflicts of interest.

List of Contributions:

1. Lahootifar, Z.; Habibi-Yangjeh, A.; Rahim Pouran, S.; Khataee, A. G-C_3N_4 Dots Decorated with Hetaerolite: Visible-Light Photocatalyst for Degradation of Organic Contaminants. *Catalysts* **2023**, *13*, 346. https://doi.org/10.3390/catal13020346.
2. Alharthi, F.; Alshayiqi, A.; Al-Nafaei, W.; El Marghany, A.; Alanazi, H.; Hasan, I. Synthesis of Nanocrystalline Metal Tungstate $NiWO_4$/$CoWO_4$ Heterojunction for UV-Light-Assisted Degradation of Paracetamol. *Catalysts* **2023**, *13*, 152. https://doi.org/10.3390/catal13010152.
3. Cruz-Carrillo, M.; Melgoza-Alemán, R.; Cuevas-Arteaga, C.; Proal-Nájera, J. Removal of Persistent Acid Pharmaceuticals by a Biological-Photocatalytic Sequential Process: Clofibric Acid, Diclofenac, and Indomethacin. *Catalysts* **2022**, *12*, 1488. https://doi.org/10.3390/catal12111488.
4. Magomedova, A.; Isaev, A.; Orudzhev, F.; Sobola, D.; Murtazali, R.; Rabadanova, A.; Shabanov, N.; Zhu, M.; Emirov, R.; Gadzhimagomedov, S.; et al. Magnetically Separable Mixed-Phase α/γ-Fe_2O_3 Catalyst for Photo-Fenton-like Oxidation of Rhodamine B. *Catalysts* **2023**, *13*, 872. https://doi.org/10.3390/catal13050872.
5. Chen, F.; Guo, J.; Meng, D.; Wu, Y.; Sun, R.; Zhao, C. Strong Pyro-Electro-Chemical Coupling of Elbaite/H_2O_2 System for Pyrocatalysis Dye Wastewater. *Catalysts* **2021**, *11*, 1370. https://doi.org/10.3390/catal11111370.
6. Falletta, E.; Bianchi, C.; Morazzoni, F.; Polissi, A.; Di Vincenzo, F.; Bellobono, I. Tungsten Trioxide and Its TiO_2 Mixed Composites for the Photocatalytic Degradation of NO_x and Bacteria (*Escherichia coli*) Inactivation. *Catalysts* **2022**, *12*, 822. https://doi.org/10.3390/catal12080822.
7. Pavlova, S.; Ivanova, Y.; Tsybulya, S.; Chesalov, Y.; Nartova, A.; Suprun, E.; Isupova, L. Sr_2TiO_4 Prepared Using Mechanochemical Activation: Influence of the Initial Compounds Nature on Formation, Structural and Catalytic Properties in Oxidative Coupling of Methane. *Catalysts* **2022**, *12*, 929. https://doi.org/10.3390/catal12090929.
8. Gote, Y.; Sinhmar, P.; Gogate, P. Sonocatalytic Degradation of Chrysoidine R Dye Using Ultrasonically Synthesized $NiFe_2O_4$ Catalyst. *Catalysts* **2023**, *13*, 597. https://doi.org/10.3390/catal13030597.
9. Wang, Y.; Jin, Y.; Jia, M.; Ruan, H.; Tao, X.; Liu, X.; Lu, G.; Zhang, X. Enhanced Visible-Light Photocatalytic Activities of $CeVO_4$-V_2O_3 Composite: Effect of Ethylene Glycol. *Catalysts* **2023**, *13*, 659. https://doi.org/10.3390/catal13040659.
10. Alharthi, F.; Ababtain, A.; Aldubeikl, H.; Alanazi, H.; Hasan, I. Synthesis of Novel $Zn_3V_2O_8$/Ag Nanocomposite for Efficient Photocatalytic Hydrogen Production. *Catalysts* **2023**, *13*, 455. https://doi.org/10.3390/catal13030455.
11. Alharthi, F.; Al-Nafaei, W.; Alshayiqi, A.; Alanazi, H.; Hasan, I. Hydrothermal Synthesis of Bimetallic (Zn, Co) Co-Doped Tungstate Nanocomposite with Direct Z-Scheme for Enhanced Photodegradation of Xylenol Orange. *Catalysts* **2023**, *13*, 404. https://doi.org/10.3390/catal13020404.
12. Weissenberger, T.; Zapf, R.; Pennemann, H.; Kolb, G. Effect of the Active Metal on the NO_x Formation during Catalytic Combustion of Ammonia SOFC Off-Gas. *Catalysts* **2022**, *12*, 1186. https://doi.org/10.3390/catal12101186.
13. Ma, G.; Wang, L.; Wang, X.; Li, L.; Ma, H. CO Oxidation over Alumina-Supported Copper Catalysts. *Catalysts* **2022**, *12*, 1030. https://doi.org/10.3390/catal12091030.
14. Habiba, U.; Mutahir, S.; Khan, M.; Humayun, M.; Refat, M.; Munawar, K. Effective Removal of Refractory Pollutants through Cinnamic Acid-Modified Wheat Husk Biochar: Experimental and DFT-Based Analysis. *Catalysts* **2022**, *12*, 1063. https://doi.org/10.3390/catal12091063.

15. Silerio-Vázquez, F.; Núñez-Núñez, C.; Proal-Nájera, J.; Alarcón-Herrera, M. A Systematic Review on Solar Heterogeneous Photocatalytic Water Disinfection: Advances over Time, Operation Trends, and Prospects. *Catalysts* **2022**, *12*, 1314. https://doi.org/10.3390/catal12111314.
16. Kamenická, B.; Weidlich, T. A Comparison of Different Reagents Applicable for Destroying Halogenated Anionic Textile Dye Mordant Blue 9 in Polluted Aqueous Streams. *Catalysts* **2023**, *13*, 460. https://doi.org/10.3390/catal13030460.
17. Umar, A.; Smółka, Ł.; Gancarz, M. The Role of Fungal Fuel Cells in Energy Production and the Removal of Pollutants from Wastewater. *Catalysts* **2023**, *13*, 687. https://doi.org/10.3390/catal13040687.

Disclaimer/Publisher's Note: The statements, opinions and data contained in all publications are solely those of the individual author(s) and contributor(s) and not of MDPI and/or the editor(s). MDPI and/or the editor(s) disclaim responsibility for any injury to people or property resulting from any ideas, methods, instructions or products referred to in the content.

Communication

Magnetically Separable Mixed-Phase α/γ-Fe$_2$O$_3$ Catalyst for Photo-Fenton-like Oxidation of Rhodamine B

Asiyat Magomedova [1], Abdulgalim Isaev [1,*], Farid Orudzhev [1], Dinara Sobola [1,2,3], Rabadanov Murtazali [1], Alina Rabadanova [1], Nabi S. Shabanov [4], Mingshan Zhu [5], Ruslan Emirov [1], Sultanakhmed Gadzhimagomedov [1], Nariman Alikhanov [1] and Kaviyarasu Kasinathan [6,7]

[1] Department of Inorganic Chemistry and Chemical Ecology, Dagestan State University, St. M. Gadjieva 43-a, 367015 Makhachkala, Russia; asiyat_magomedova1996@mail.ru (A.M.); farid-stkha@mail.ru (F.O.); sobola@vut.cz (D.S.); rab_mur@mail.ru (R.M.); aderron@mail.ru (R.E.); darkusch@mail.ru (S.G.); alihanov.nariman@mail.ru (N.A.)

[2] Department of Physics, Faculty of Electrical Engineering and Communication, Brno University of Technology, Technická 2848/8, 61600 Brno, Czech Republic

[3] Central European Institute of Technology BUT, Purkyňova 123, 61200 Brno, Czech Republic

[4] Dagestan Federal Research Centre of the Russian Academy of Sciences, Analytical Center for Collective Use, M. Gadjieva 45, 367001 Makhachkala, Russia; shabanov.nabi@yandex.ru

[5] School of Environment, Jinan University, Guangzhou 511443, China; zhumingshan@jnu.edu.cn

[6] UNESCO-UNISA Africa Chair in Nanoscience's/Nanotechnology Laboratories, College of Graduate Studies, University of South Africa (UNISA), Muckleneuk Ridge, Pretoria P.O. Box 392, South Africa; kasinathankaviyarasu@gmail.com

[7] Nanosciences African Network (NANOAFNET), Materials Research Group (MRG), iThemba LABS-National Research Foundation (NRF), 1 Old Faure Road Western Cape Province, Cape Town P.O. Box 722, South Africa

* Correspondence: abdul-77@yandex.ru; Tel.: +7-963-427-97-78

Citation: Magomedova, A.; Isaev, A.; Orudzhev, F.; Sobola, D.; Murtazali, R.; Rabadanova, A.; Shabanov, N.S.; Zhu, M.; Emirov, R.; Gadzhimagomedov, S.; et al. Magnetically Separable Mixed-Phase α/γ-Fe$_2$O$_3$ Catalyst for Photo-Fenton-like Oxidation of Rhodamine B. *Catalysts* **2023**, *13*, 872. https://doi.org/10.3390/catal 13050872

Academic Editors: Meng Li and Marta Pazos Currás

Received: 20 March 2023
Revised: 22 April 2023
Accepted: 8 May 2023
Published: 11 May 2023

Copyright: © 2023 by the authors. Licensee MDPI, Basel, Switzerland. This article is an open access article distributed under the terms and conditions of the Creative Commons Attribution (CC BY) license (https:// creativecommons.org/licenses/by/ 4.0/).

Abstract: Iron oxides are widely used as catalysts for photo-Fenton-like processes for dye oxidation. In this study, we report on the synthesis of an α/γ-Fe$_2$O$_3$ mixed-phase catalyst with magnetic properties for efficient separation. The catalyst was synthesized using glycine–nitrate precursors. The synthesized α/γ-Fe$_2$O$_3$ samples were characterized using scanning electron microscopy, X-ray diffraction spectroscopy (XRD), Raman shift spectroscopy, X-ray photoelectron spectroscopy (XPS), and vibrating sample magnetometer (VSM). The diffraction peaks were indexed with two phases, α-Fe$_2$O$_3$ as the main phase (79.6 wt.%) and γ-Fe$_2$O$_3$ as the secondary phase (20.4 wt.%), determined using the Rietveld refinement method. The presence of Fe^{2+} was attributed to oxygen vacancies. The mixed-phase α/γ-Fe$_2$O$_3$ catalyst exhibited remarkable photo-Fenton-like degradation performance for Rhodamine B (RhB) in neutral pH. The effects of operating parameters, including H$_2$O$_2$ concentration, catalyst concentration, and RhB concentration, on the degradation efficiency were investigated. The removal rates of color were 99.2% after 12 min at optimal conditions of photo-Fenton-like oxidation of RhB. The sample exhibited a high saturation magnetization of 28.6 emu/g. Additionally, the α/γ-Fe$_2$O$_3$ mixed-phase catalyst showed long-term stability during recycle experiments, with only a 5% decrease in activity.

Keywords: iron oxides; α/γ-Fe$_2$O$_3$ mixed-phase catalyst; magnetically separable; Rhodamine B; photo-Fenton-like

1. Introduction

Advanced oxidation processes are currently being utilized as an effective method for treating industrial wastewater that contains non-biodegradable organic compounds [1]. One of the most extensively studied oxidation processes is the Fenton process, which utilizes hydrogen peroxide and ferrous ions and exhibits an effective ability to destroy a wide range of contaminants. However, this process can only be effectively used with acidic or neutral pH wastewater [2]. The Fenton process is a homogeneous catalytic system, where the catalyst (Fe^{2+}) is added in the form of a soluble salt and is removed from the reactor with the outgoing stream of purified water, due to the impossibility of its separation. This problem is further exacerbated by the fact that upon further neutralization of the purified solution, Fe^{3+} precipitates, necessitating the separation of the resulting precipitate, which collectively leads to a decrease in the efficiency of the process [3]. In this regard, the use of the heterogeneous Fenton-like process in the presence of iron compounds in the form of a precipitate on which the hydroxyl radical is generated from hydrogen peroxide is more promising [4]. Heterogeneous Fenton-like reactions can also efficiently degrade organic compounds in wastewater [5,6]. Solid catalysts can be reused after separation [7,8]. The use of the heterogeneous Fenton-like process overcomes some of the disadvantages of the homogeneous Fenton reaction, such as reduced reactivity due to catalyst consumption and the need to adjust pH [9].

At the same time, it is known that in the dark processes of homogeneous and heterogeneous Fenton, Fe^{3+} ions accumulate in the system and the reaction rate decreases significantly with time and stops after the complete consumption of Fe^{2+} ions [10]. The combination of Fenton and Fenton-like processes with simultaneous exposure to UV/visible radiation (λ < 600 nm) can solve this problem due to the photoreduction reaction of Fe^{3+} ions to Fe^{2+} [11].

Iron oxides are among the widely used materials as catalysts for Fenton-like processes for the oxidation of organic compounds [12–16]. Among the most studied and promising iron oxides and hydroxides for use as catalysts in Fenton-like processes, magnetite (Fe_3O_4) [17–24], goethite (α-FeOOH) [25,26], maghemite (γ-Fe_2O_3) [12,27–31], and hematite (α-Fe_2O_3) [32–34] can be noted. Various physicochemical characteristics of these oxides make them more or less favorable for oxidative reactions. These solid catalysts have good potential to degrade bio-oxidation-resistant contaminants [35,36]. Among these crystal structures, hematite is the most stable state of iron oxide under environmental conditions [37]. The prevailing use of α-Fe_2O_3 is due to its excellent physical and chemical properties, which can manifest themselves in samples with different morphology [38–40], particle size [41], and also in composite structures [42–44], which is especially important for creation of efficient heterogeneous catalysts in Fenton-like oxidation processes [45,46]. Maghemite also finds use as a catalyst in Fenton-like processes for the oxidation of organic compounds [47,48]. Unlike α-Fe_2O_3, γ-Fe_2O_3 is magnetic and can be easily reduced using a magnet. At the same time, γ-Fe_2O_3 retains catalytic activity for many cycles of use [49].

Various methods are used to synthesize iron oxides [50]. Specifically, the solution combustion method, using glycine as a fuel, was employed to synthesize α-Fe_2O_3 and γ-Fe_2O_3 [51–53]. This method results in the formation of ultra-small α-Fe_2O_3 nanoparticles (less than 5 nm) that exhibit superparamagnetism in the temperature range of 70–300 K [54]. Additionally, biphasic α/γ-Fe_2O_3 nanoparticles have been reported, which demonstrate high sensitivity to detecting volatile organic compounds such as acetone [55] and ethanol [56]. Biphasic α/γ-Fe_2O_3 exhibits significantly higher sensitivity than α-Fe_2O_3 and γ-Fe_2O_3 alone [57]. The work [58] reports higher photocatalytic activity of heterophase α/γ-Fe_2O_3 during methylene blue oxidation due to a decrease in the rate of electron-hole pair recombination. Therefore, a new approach to enhance metal oxide catalyst performance is to integrate different crystalline forms of the same metal oxide into a single structure. Currently, there is no information available on the use of biphasic α/γ-Fe_2O_3 in Fenton-like processes for organic compound oxidation. Based on this, this paper investigates the production of biphasic iron oxide and its use as a magnetically separable heterogeneous catalyst in the Fenton-like process for RhB oxidation.

2. Results

2.1. Catalyst Characterizations

Iron oxides (Fe_2O_3) with various ratios of fuel and oxidizer (φ) were synthesized using the conventional solution combustion method. Glycine is widely used in the so-called glycine–nitrate synthesis of metal oxide nanoparticles by combustion [56,57].

The ongoing combustion reaction can be written according to Equation (1).

$$6Fe(NO_3)_3 + 10C_2H_5NO_2 \rightarrow 3Fe_2O_3 + 20CO_2 + 14N_2 + 25H_2O \tag{1}$$

The coefficients of the expected reaction were placed based on the theory of combustion [59]. The fuel–oxidizer ratio of 0.4 ($\varphi < 1$) was chosen so that the amount of oxidizer was in excess and there was no need for atmospheric oxygen. Additionally, it was shown that in several systems, solution combustion synthesis of reactive solutions with an excess of fuel ($\varphi > 1$) leads to the formation of pure metals [60].

The morphology of the synthesized powders was studied using SEM. The images at various magnifications are shown in Figure 1.

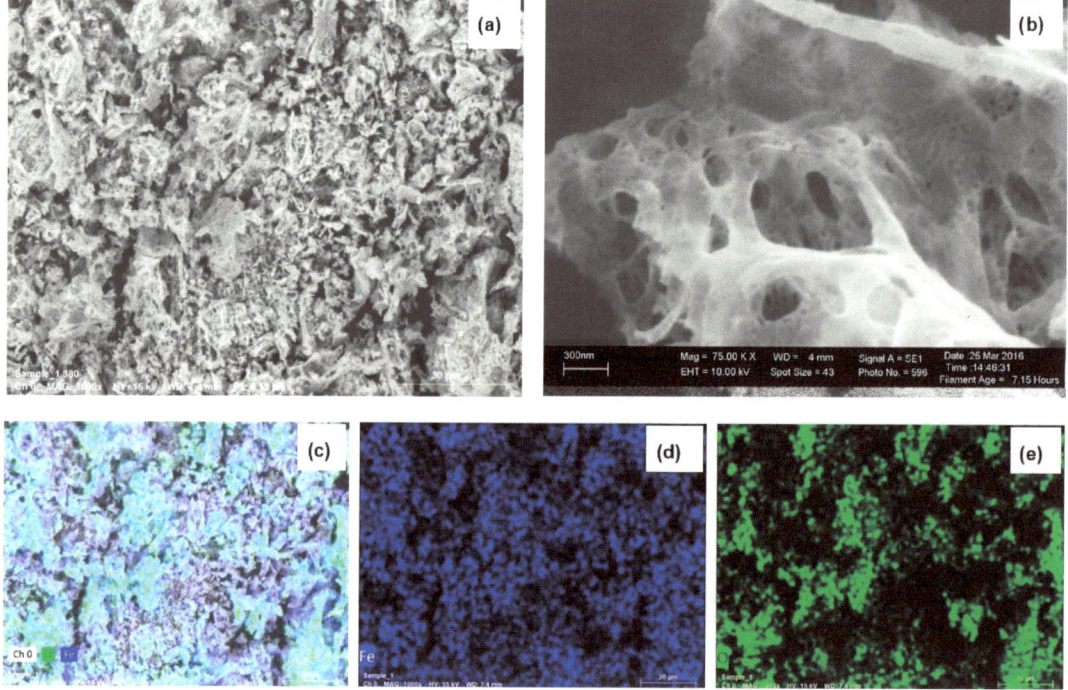

Figure 1. SEM images of catalyst samples at various magnifications (**a**,**b**). EDS elemental mapping of the selected area (**c**–**e**).

In Figure 1a, at low magnifications, it can be seen that the powder has a loose flaky texture characteristic of the combustion method with a large number of pores. At high magnifications in Figure 1b, one can see that the powder is in the form of large submicron agglomerates of a bone-like structure, sintered at high temperatures during synthesis. In this case, the grain boundaries are quite clearly traced. It can be seen that, along with large micron-sized pores, there is a large number of nanopores. From the EDS images (Figure 1c–e), it is clearly seen that the atoms of iron and oxygen are uniformly distributed over the surface under study. EDX spectra are presented in Supplementary Materials (Figure S1). Analysis of the atomic percentage of the elements Fe and O showed that the ratio is close to stoichiometric, Fe (47 wt.%) O (53 wt.%) with a slight oxygen deficiency.

It is known from the literature that varying the fuel-oxidizer ratio affects the phase of the synthesized iron oxide [61]. Therefore, the crystal structure of the powder was investigated by XRD and Raman methods. The data are presented in Figure 2a,b.

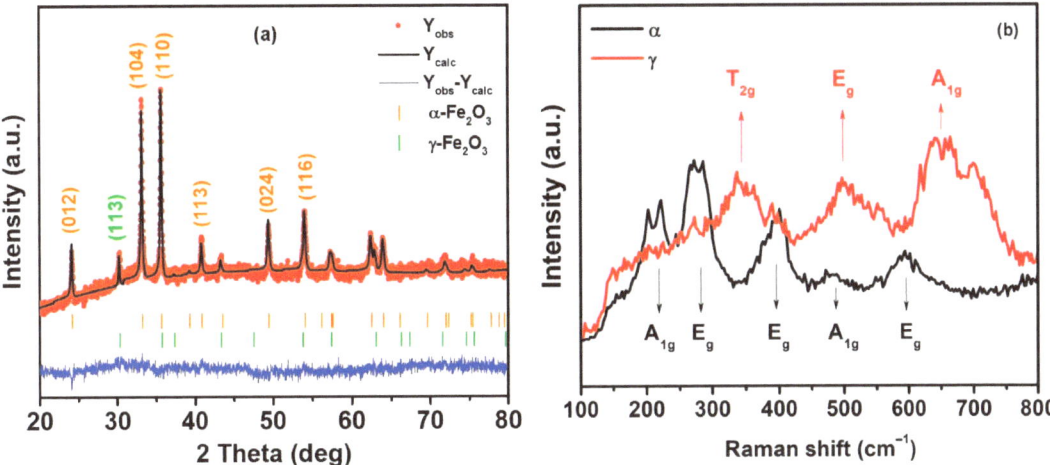

Figure 2. (a) Rietveld refinement graphs of Fe_2O_3. The red circles indicate the experimental data, the black line is the fitting value, the blue line is the difference, and the orange and green ticks are the Bragg reflections of the α-phase and γ-phase, respectively. (b) Raman spectra in different areas.

Figure 2a shows the XRD spectrum of the sample with the structure refined by the Rietveld method. The spectrum is well described by two phases, namely α-Fe_2O_3 with Hexagonal structure and space group R-3c (Ref. Code 98-006-6756) and γ-Fe_2O_3 with Cubic structure and space group Fd-3m (Ref. Code 98-006-6756). Quantitative phase evaluation carried out using the Rietveld method showed that α-Fe_2O_3 exists as the main phase (79.6 wt.%) and γ-Fe_2O_3 is present in an amount (20.4 wt.%). The presence of background noise indicates a high proportion of amorphous Fe_2O_3 in the sample. The Rietveld reliability factors displayed in Table 1 show that the quality of the fit is appreciable.

Table 1. Results of phase quantification and Rietveld refinement of Fe_2O_3.

			Sample	
Phases			α-Fe_2O_3	γ-Fe_2O_3
wt.%			79.6	20.4
Space group			R-3c (No. 167)	F d-3 m (No. 227)
Crystal system			Hexagonal	Cubic
Lattice parameters	a (Å)		5.0339	8.3421
	c (Å)		13.742	-
Cell volume	V (Å^3)		301.57	580.53
Rietveld reliability factors	R_{exp}			1.5628
	R_w			1.7091
	R_p			1.3556
	GoF			1.1961
Crystallite size	L_{Vol-IB} (nm)		47.4	45.7

The resulting fitted D-V function was then used for the calculation of volume-weighted mean crystallite size (L_{Vol-IB}) via the Scherrer equation. The average crystallite size of α-phase was found to be 47.4 nm, while that of the γ-phase was 45.7 nm.

Iron oxide polymorphs of the α- and γ-phases are also distinguishable by Raman spectroscopy. The Raman spectra from two different parts of the sample are shown in Figure 2b. The black line shows two classes of Raman active modes of hematite in the range from 200 to 800 cm^{-1}. The existence of characteristic A 1g bands at 221 and 491 cm^{-1} and Eg bands at 239, 287, 401, and 605 cm^{-1}, respectively, is attributed to the main hematite bands. Low-frequency modes (200–300 cm^{-1}) were attributed to vibrations of the Fe atom, and bands from 400 to 650 cm^{-1} were attributed to vibrations of the O atom [62–65]. The red line in Figure 2b shows three Raman active phonon modes at 365 cm^{-1} (T2g), 511 cm^{-1} (Eg), and 700 cm^{-1} (A1g), characteristic of maghemite. The spectrum is in good agreement with the data for maghemite previously published in the literature [66–68].

The surface states play a key role for heterogeneous photo-Fenton-like catalysis; the surface was investigated by XPS. The obtained results are presented in Figure 3.

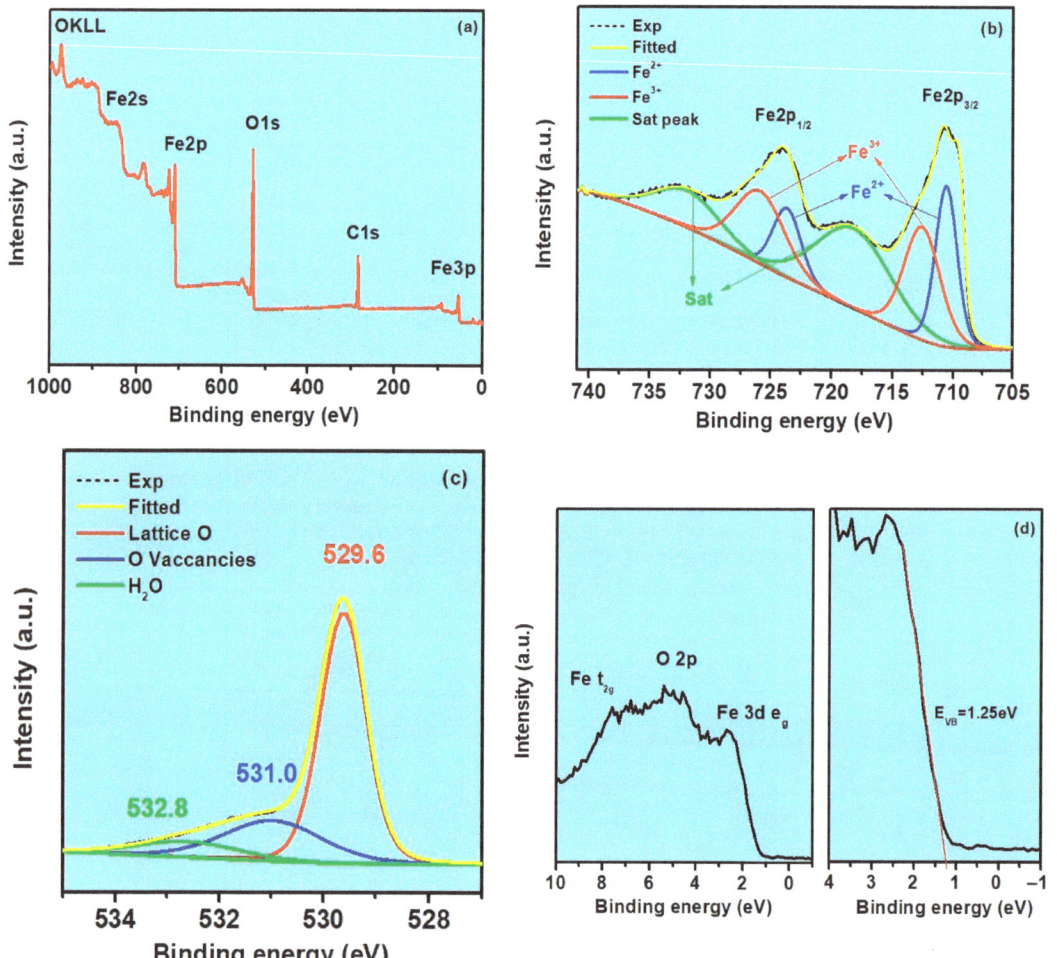

Figure 3. XPS spectra of as-prepared mixed-phase α/γ-Fe$_2$O$_3$ catalyst: full spectra (**a**); Fe 2p (**b**); O 1s (**c**); and Fe 3p (**d**) core-level spectra.

With a wide panoramic scan in Figure 3a, peaks of C 1s, O 1s, and Fe 2p were detected, which indicates the absence of impurities. Peak C 1s comes from random carbon [67,69,70]. The high-resolution spectrum of Fe 2p after deconvolution with approximation of Gaussian peaks is shown in Figure 3b. It can be seen that the spectrum is well described by the superposition of six peaks. There were two peaks at 726.1 and 712.5 eV, which are typical characteristic peaks of Fe^{3+} in 2p 1/2 and 2p 3/2 orbitals [71]. In addition, two deconvoluted peaks at 723.9 and 710.5 eV correspond to Fe^{2+}, which can be due both to the presence of magnetite in the structure, which is quite difficult to distinguish from maghemite by XRD and Raman methods, and to the formation of oxygen vacancies in Fe_2O_3. Generation of oxygen vacancies in the crystal lattice leaves two electrons per missing oxygen atom, which leads to the reduction of Fe^{3+} to Fe^{2+} [72]. The generation of oxygen vacancies is common for the high-temperature combustion method [73]. The two deconvolution peaks at 732.1 and 718.3 eV are attributed to the presence of their satellite vibrational peaks (labeled "Sat."). From a comparison of the integral areas of the Fe^{3+} peaks in Fe^{2+}, it was found that their ratio is 60:40%. Data are presented in Supplementary Materials (Figures S2 and S3).

To confirm the presence of oxygen vacancies, the spectrum of the O 1s level was studied. Figure 3c shows the O 1s spectra after deconvolution with approximation of Gaussian peaks. The spectrum is well described by the superposition of three components centered at 529.6, 531.0, and 532.8 eV, respectively. The peak at 529.6 was a typical lattice oxygen peak, and that at 531.0 eV could be attributed to the low-coordinated oxygen species adsorbed onto the oxygen vacancies. The peak at 532.8 eV was assigned to the hydroxyl species of surface-adsorbed H_2O molecules [74].

The XPS results indicate the co-presence of Fe^{2+} and Fe^{3+} and that the presence of Fe^{2+} is not associated with the presence of magnetite in the structure, confirming the results of XRD and Raman. It is important that the Fe^{2+}/Fe^{3+} redox pair formed on the surface can accelerate the charge transfer in Fe_2O_3, since Fe^{3+} is reduced to Fe^{2+} during heterogeneous Fenton-like catalysis [75].

The XPS valence band (VB) region analysis is a powerful tool for understanding the electronic structure of a material. Figure 3d shows the XPS (VB) spectrum in the binding energy range 0–10 eV. The VB spectrum is the result of hybridization of Fe3d and O 2p atomic orbitals [76] and can apparently be described by three bands, which is consistent with previously published results [77] and corresponds to the states of Fe 3d eg strongly hybridized with O 2p and non-bonding O 2p, and the C characteristic is dominated by bond states of the O 2p and Fe t2g orbitals. The inset to Figure 3d shows that the valence band maximum (VBM) is 1.25 eV below the Fermi level.

2.2. Catalytic Activity in Fenton-like Process

The catalytic activity of two-phase α/γ-Fe_2O_3 was studied by oxidation of the dye RhB under various conditions. Figure 4 shows a typical change in the absorption spectra of RhB during treatment for 12 min.

Changes in the catalytic activity of sample α/γ-Fe_2O_3 in the form of kinetic curves of the RhB oxidation are shown in Figure 4a. When using the heterogeneous Fenton-like system using α/γ-Fe_2O_3 catalysts, RhB slowly decomposes and was 4% after 12-min treatment with α-Fe_2O_3. The use of UV-visible light irradiation leads to a significant acceleration of the oxidation of RhB. Irradiation with light has a dual effect on a heterogeneous system: the oxidation of the dye directly by hydrogen peroxide upon irradiation with light and the acceleration of the formation of hydroxyl radicals (HO$^\bullet$) as a result of the decomposition of H_2O_2 in the presence of a catalyst [78,79].

$$H_2O_2 + h\nu \rightarrow 2HO^\bullet \tag{2}$$

$$\equiv FeIII + H_2O + h\nu \rightarrow \equiv FeII + HO^\bullet + H^+ \tag{3}$$

$$\equiv FeII + H_2O_2 \rightarrow \equiv FeIII - OH + HO^- \tag{4}$$

The effect of H_2O_2 concentration, catalyst dose, and RhB concentration on degradation is also shown in Figure 4. Increasing the H_2O_2 concentration improved the decomposition activity. It has been shown in the literature that only increasing the concentration of H_2O_2 up to 15 mM led to a decrease in the efficiency of decomposition due to the unfavorable consumption of excess H_2O_2 due to the effect of scavenging free radicals [80,81]. The highest performance was achieved using a catalyst dosage of 0.2 g/L. The decrease in activity with an excess of catalyst is associated with blocking the penetration of light and active sites on the catalyst surface. A study of the effect of dye concentration on degradation efficiency (Figure 4d) demonstrated that the lower the initial concentration, the higher the efficiency. At a concentration of 1 mg/L, 99.2% of the dye decomposes in 12 min. The decrease in activity with increasing dye concentration may be due to the formation of a larger number of intermediate products that can occupy active sites on the catalyst surface. As pH is an important parameter for the photo-Fenton process, additional studies were conducted to investigate its influence. The data is presented in the Supplementary Materials (Figure S4), which show that the pH of the medium does not affect the reaction progress.

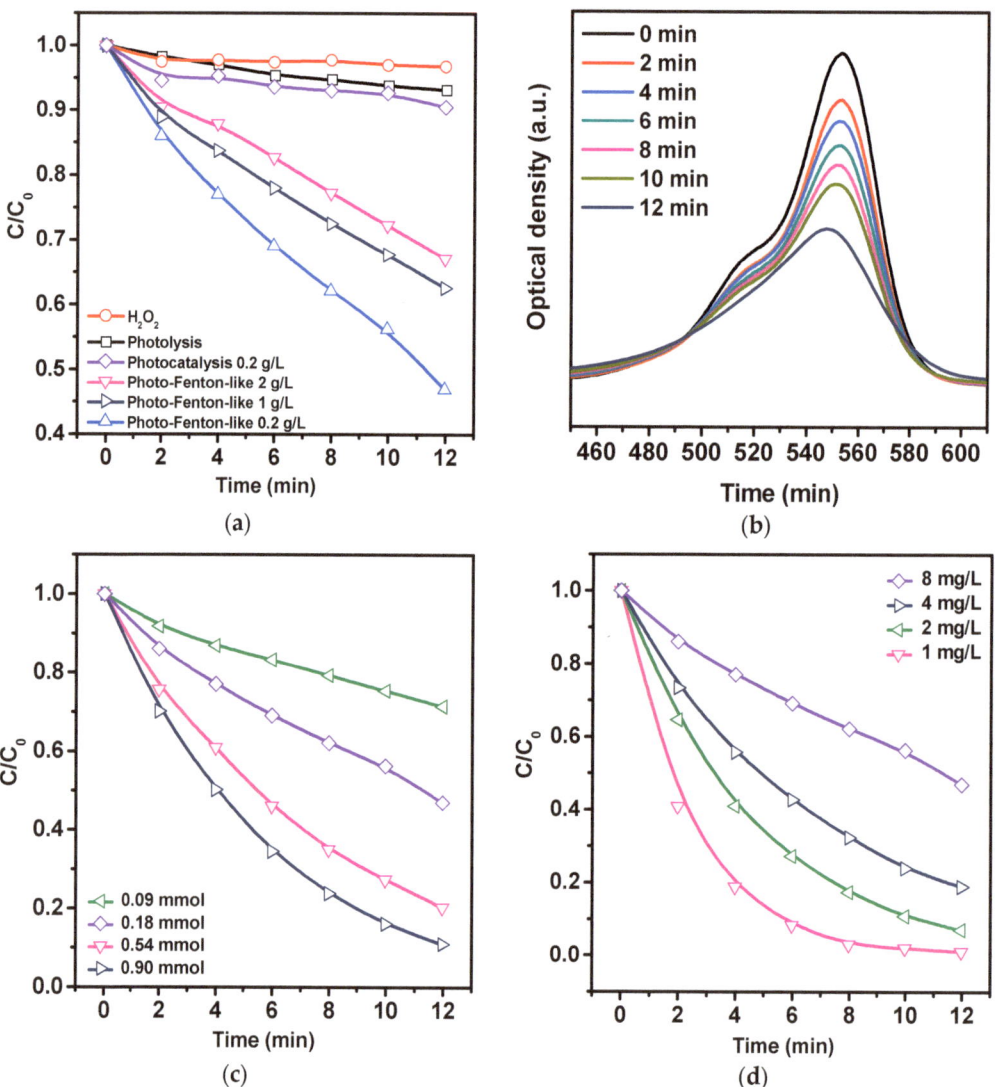

Figure 4. Kinetic curves of RhB degradation: (**a**) effect of catalyst concentration (C_{RhB} = 8 mg/L; $C(H_2O_2)$ = 0.18 mM; t =12 min), (**b**) absorption spectra of RhB during oxidation in the photo-Fenton-like process (C_{RhB} = 8 mg/L; $C(H_2O_2)$ = 0.18 mM; 0.2 g/L α/γ-Fe_2O_3; t =12 min), (**c**) H_2O_2 concentration (C_{RhB} = 8 mg/L; 0.2 g/L α/γ-Fe_2O_3; t =12 min) on the photo-Fenton-like degradation, (**d**) RhB concentration ($C(H_2O_2)$ = 0.18 mM; 0.2 g/L α/γ-Fe_2O_3; t =12 min).

Figure 5 shows the results of catalyst recycling and magnetic properties. The separation of the spent catalyst was carried out by magnetic separation. Figure 5a shows that the sample exhibits long-term stability. After five repeated uses, the activity of the catalyst decreased by 5%. However, it is also important to investigate the leaching of iron ions into the solution in the photo-Fenton-like process. After each cycle, we determined the content of iron ions in the solution using the colorimetric method with nitroso-R-salt. The results showed that the concentration of Fe^{2+} after the process was 330 µg/L.

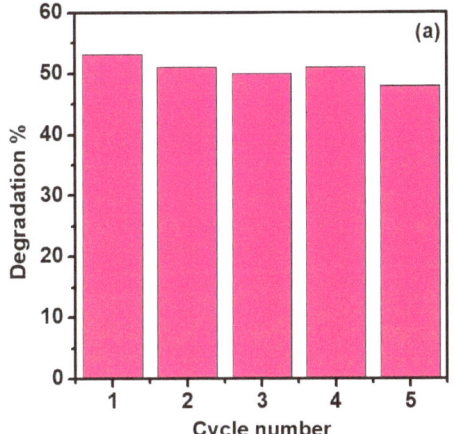
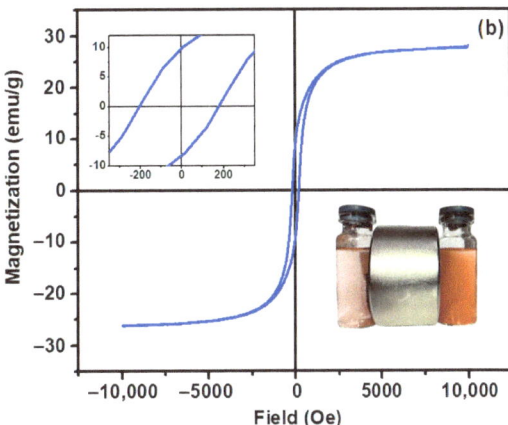

Figure 5. (a) Long-term catalyst stability results (C_{RhB} = 8 mg/L; $C(H_2O_2)$ = 0.18 mM; t = 12 min); (b) magnetic hysteresis loop for mixed α/γ-Fe_2O_3.

The magnetic hysteresis loop (MH) measured at room temperature is shown in Figure 5b. The sample is ferromagnetic at room temperature.

The loop has an obvious hysteresis loop, and the coercive force (Hc) is 383.2 Oe, the magnetization vector (Ms) is 28.6 emu/g, and the remanence intensity (Mr) is 9.7 emu/g, as shown in Figure 5b. The inset to Figure 5b shows a photograph of the magnetic separation process. For clarity, a sample is presented consisting only of the α-Fe_2O_3 phase.

3. Materials and Methods

3.1. Synthesis Procedure

Synthesis of mixed α/γ-Fe_2O_3 was carried out using the combustion of glycine–nitrate precursors [59,60]. An aqueous solution of iron (III) nitrate was used as the starting material for the preparation of the two-phase α/γ-Fe_2O_3 catalyst. The precursor was prepared by mixing glycine and $Fe(NO_3)_3$ in an aqueous solution. The resulting solution was evaporated to a gel state on an electric heater with an operating temperature up to about 180 °C. During further heating, the reaction mixture ignited and iron(III) oxide powder was formed. Combustion was fast and self-sustaining, with a flame temperature of 1100 to 1450 °C. The synthesized samples were annealed at 400 °C for 1 h.

3.2. Characterizations

Characterization of the obtained heterostructures was performed using scanning electron microscopy (SEM) with the Aspex ExPress VP (FEI Company, Hillsboro, OR, USA). X-ray diffraction (XRD) studies were done using an Empyrean PANalytical X-ray diffractometer (Almelo, The Netherlands) in the radiation of a copper anode with a nickel filter, with radiation wavelength λ(CuKα) = 0.154051 nm. Data processing was performed using the High Score Plus application program, included in the instrument software, and the diffraction database ICSD (PDF-2). The surface composition was carried out by an AXIS SupraTM X-ray photoelectron spectrometer (XPS) (Kratos Analytical Ltd., Manchester, UK).

The data were processed by CasaXPS v.2.3.23 software (Casa Software Ltd., Wilmslow, UK). Raman spectra were examined by a Laser Raman 3D scanning confocal microscope (Ntegra Spectra, Moscow, Russia) using a green laser (532 nm) with a spot size of 1 μm and a resolution of 0.5 cm^{-1}.

3.3. Fenton-like Oxidation of the Rhodamine B

The catalytic activity of the samples in Fenton-like process were evaluated using the degradation of RhB in an aqueous solution (8 mg/L). The experiments were carried out in a 50 mL glass beaker. The 250 W high-pressure mercury lamp (Phillips, Amsterdam, The Netherlands) was used as a source of UV-visible light at photo-Fenton-like process investigation. The oxidant (H_2O_2) was added to the Rhodamine B solution with α/γ-Fe_2O_3 suspension. The light source was placed above the reactor at a distance of 10 cm. The RhB concentration was measured using an SF-2000 spectrophotometer (Saint-Petersburg, Russia) from the characteristic absorption peak at a wavelength of 553 nm. After the measurement, the solution was poured back into the reactor and the process continued. The concentration of iron in the solution after the process was determined by photometric method using nitroso-R-salt [82].

4. Conclusions

Heterogeneous photo-Fenton-like degradation of RhB with high efficiency has been demonstrated over a mixed-phase α/γ-Fe_2O_3 catalyst. α/γ-Fe_2O_3 was prepared by a combustion of glycine–nitrate precursors with fuel–oxidizer ratio of 0.4 ($\varphi < 1$). At the same time, a powder with a composition of 80%α/20%γ-Fe_2O_3 was synthesized, with crystal sizes of 47.4 and 45.7 nm, respectively. XPS analysis showed that Fe^{2+} ions, up to 40%, were present on the surface along with Fe^{3+} ions, due to the presence of oxygen vacancies. Optimization of photo-Fenton-like degradation of RhB showed that reducing the dye concentration from 8 to 1 mg/L, increasing the H_2O_2 concentration from 0.09 to 0.90 mmol, and reducing the mass loading from 2 to 0.2 g/L leads to an increase in catalytic activity. At optimal efficiency, 99.2% degradation is achieved in 12 min of the process. It has been shown that the pH of the medium does not affect the catalytic activity of α/γ-Fe_2O_3. The 80% α-Fe_2O_3 and 20% γ-Fe_2O_3 mixed-phase catalyst showed no obvious decrease in degradation performance over five consecutive cycles. The results show that the mixed-phase α/γ-Fe_2O_3 catalyst is a very promising catalyst that is magnetically separable and a suitable candidate for practical applications of dye containing wastewater treatment.

Supplementary Materials: The following supporting information can be downloaded at: https://www.mdpi.com/article/10.3390/catal13050872/s1, Figure S1: EDX spectra of α/γ-Fe_2O_3; Figure S2: Fe 2p level XPS spectra; Figure S3: O 1s level XPS spectra; Figure S4: Dependence of the catalytic activity of the photo-Fenton-like process on pH (C_{RhB} = 8 mg/L; $C(H_2O_2)$ = 0.18 mM; t =12 min).

Author Contributions: Conceptualization, A.M. and A.I.; methodology, D.S., N.S.S., R.E., S.G. and N.A.; formal analysis, M.Z.; investigation, A.M., A.R., D.S., N.S.S., R.E., S.G. and N.A.; writing, A.I., F.O., R.M. and M.Z.; review and editing, A.I., M.Z. and K.K.; visualization, A.M., A.I., F.O. and R.E.; supervision, A.I. and M.Z.; project administration, A.M. and A.I.; funding acquisition, A.I., F.O. and R.M. All authors have read and agreed to the published version of the manuscript.

Funding: This research was funded by Russian Fund of Basic (RFBR) according to the research project No. 20-33-90220\20 and by the Russian Science Foundation (RSF) under project No. 22-73-10091 in part related to the synthesis of nanoparticles and XRD, Raman, XPS, VSM investigation.

Data Availability Statement: The data presented in this study are available on request from the corresponding author.

Acknowledgments: We acknowledge CzechNanoLab Research Infrastructure supported by MEYS CR (LM2023051) and FEKT-S-23-8228.

Conflicts of Interest: The authors declare no conflict of interest.

References

1. Isaev, A.B.; Magomedova, A.G. Advanced Oxidation Processes Based Emerging Technologies for Dye Wastewater Treatment. *Moscow Univ. Chem. Bull.* **2022**, *77*, 181–196. [CrossRef]
2. Boczkaj, G.; Fernandes, A. Wastewater treatment by means of advanced oxidation processes at basic pH conditions: A review. *Chem. Eng. J.* **2017**, *320*, 608–633. [CrossRef]
3. Oturan, M.A.; Aaron, J.J. Advanced oxidation processes in water/wastewater treatment: Principles and applications. A review. *Crit. Rev. Environ. Sci. Technol.* **2014**, *44*, 2577–2641. [CrossRef]
4. Pliego, G.; Zazo, J.A.; Garcia-Muñoz, P.; Munoz, M.; Casas, J.A.; Rodriguez, J.J. Trends in the Intensification of the Fenton Process for Wastewater Treatment: An Overview. *Crit. Rev. Environ. Sci. Technol.* **2015**, *45*, 2611–2692. [CrossRef]
5. Ramirez, J.H.; Maldonado-Hódar, F.J.; Pérez-Cadenas, A.F.; Moreno-Castilla, C.; Costa, C.A.; Madeira, L.M. Azo-dye Orange II degradation by heterogeneous Fenton-like reaction using carbon-Fe catalysts. *Appl. Catal. B Environ.* **2007**, *75*, 312–323. [CrossRef]
6. Palas, B.; Ersöz, G.; Atalay, S. Heterogeneous photo Fenton-like oxidation of Procion Red MX-5B using walnut shell based green catalysts. *J. Photochem. Photobiol. A Chem.* **2016**, *324*, 165–174. [CrossRef]
7. Li, X.; Li, J.; Shi, W.; Bao, J.; Yang, X. A fenton-like nanocatalyst based on easily separated magnetic nanorings for oxidation and degradation of dye pollutant. *Materials* **2020**, *13*, 332. [CrossRef]
8. Fang, Z.; Zhang, K.; Liu, J.; Fan, J.Y.; Zhao, Z.W. Fenton-like oxidation of azo dye in aqueous solution using magnetic Fe_3O_4-MnO_2 nanocomposites as catalysts. *Water Sci. Eng.* **2017**, *10*, 326–333. [CrossRef]
9. He, J.; Yang, X.; Men, B.; Wang, D. Interfacial mechanisms of heterogeneous Fenton reactions catalyzed by iron-based materials: A review. *J. Environ. Sci.* **2016**, *39*, 97–109. [CrossRef]
10. Pouran, S.R.; Abdul Aziz, A.R.; Daud, W.M.A.W. Review on the main advances in photo-Fenton oxidation system for recalcitrant wastewaters. *J. Ind. Eng. Chem.* **2015**, *21*, 53–69. [CrossRef]
11. Nishanth, T.; Dionysiou, D.D.; Pillai, S.C. Heterogeneous Fenton catalysts: A review of recent advances. *J. Hazard. Mater.* **2021**, *404*, 124082.
12. Ghasemi, E.; Ziyadi, H.; Afshar, A.M.; Sillanpää, M. Iron oxide nanofibers: A new magnetic catalyst for azo dyes degradation in aqueous solution. *Chem. Eng. J.* **2015**, *264*, 146–151. [CrossRef]
13. Baldrian, P.; Merhautová, V.; Gabriel, J.; Nerud, F.; Stopka, P.; Hrubý, M.; Beneš, M.J. Decolorization of synthetic dyes by hydrogen peroxide with heterogeneous catalysis by mixed iron oxides. *Appl. Catal. B Environ.* **2006**, *66*, 258–264. [CrossRef]
14. Isaev, A.B.; Shabanov, N.S.; Orudzhev, F.F.; Giraev, K.M.; Emirov, R.M. Electrochemical synthesis and photocatalytic properties of α-Fe_2O_3. *J. Nanosci. Nanotechnol.* **2017**, *17*, 4498–4503. [CrossRef]
15. Isaev, A.B.; Aliev, Z.M.; Adamadzieva, N.K.; Alieva, N.A.; Magomedova, G.A. The photocatalytic oxidation of azo dyes on Fe_2O_3 nanoparticles under oxygen pressure. *Nanotechnol. Russ.* **2009**, *4*, 475–479. [CrossRef]
16. Tariq, M.; Muhammad, M.; Khan, J.; Raziq, A.; Uddin, M.K.; Niaz, A.; Ahmed, S.S.; Rahim, A. Removal of Rhodamine B dye from aqueous solutions using photo-Fenton processes and novel Ni-Cu@ MWCNTs photocatalyst. *J. Mol. Liquids* **2020**, *312*, 113399. [CrossRef]
17. Zhu, X.; Zhang, L.; Zou, G.; Chen, Q.; Guo, Y.; Liang, S.; Hu, L.; North, M.; Xie, H. Carboxylcellulose hydrogel confined-Fe_3O_4 nanoparticles catalyst for Fenton-like degradation of Rhodamine B. *Int. J. Biol. Macromol.* **2021**, *180*, 792–803. [CrossRef]
18. Song, S.; Wang, Y.; Shen, H.; Zhang, J.; Mo, H.; Xie, J.; Zhou, N.-L.; Shen, J. Ultrasmall Graphene Oxide Modified with Fe_3O_4 Nanoparticles as a Fenton-Like Agent for Methylene Blue Degradation. *ACS Appl. Nano Mater.* **2019**, *2*, 7074–7084. [CrossRef]
19. Arshad, A.; Iqbal, J.; Ahmad, I.; Israr, M. Graphene/Fe_3O_4 nanocomposite: Interplay between photo-Fenton type reaction, and carbon purity for the removal of methyl orange. *Ceram. Int.* **2018**, *44*, 2643–2648. [CrossRef]
20. Chai, F.; Li, K.; Song, C.; Guo, X. Synthesis of magnetic porous Fe_3O_4/C/Cu_2O composite as an excellent photo-Fenton catalyst under neutral condition. *J. Colloid Interface Sci.* **2016**, *475*, 119–125. [CrossRef]
21. Jiang, Y.; Xie, Q.; Zhang, Y.; Geng, C.; Yu, B.; Chi, J. Preparation of magnetically separable mesoporous activated carbons from brown coal with Fe_3O_4. *Int. J. Min. Sci. Technol.* **2019**, *29*, 513–519. [CrossRef]
22. Li, Q.; Kong, H.; Li, P.; Shao, J.; He, Y. Photo-Fenton degradation of amoxicillin via magnetic TiO_2-graphene oxide-Fe_3O_4 composite with a submerged magnetic separation membrane photocatalytic reactor (SMSMPR). *J. Hazard. Mater.* **2019**, *373*, 437–446. [CrossRef]
23. Hesas, R.H.; Baei, M.S.; Rostami, H.; Gardy, J.; Hassanpour, A. An investigation on the capability of magnetically separable Fe_3O_4/mordenite zeolite for refinery oily wastewater purification. *J. Environ. Manag.* **2019**, *241*, 525–534. [CrossRef]
24. Baptisttella, A.M.; Araujo, C.M.; da Silva, M.P.; Nascimento, G.F.; Costa, G.R.; do Nascimento, B.F.; Ghislandi, M.G.; Motta Sobrinho, M.A. Magnetic Fe_3O_4-graphene oxide nanocomposite–synthesis and practical application for the heterogeneous photo-Fenton degradation of different dyes in water. *Sep. Sci. Technol.* **2021**, *56*, 425–438. [CrossRef]
25. Yin, R.; Sun, J.; Xiang, Y.; Shang, C. Recycling and reuse of rusted iron particles containing core-shell Fe-FeOOH for ibuprofen removal: Adsorption and persulfate-based advanced oxidation. *J. Clean. Prod.* **2018**, *178*, 441–448. [CrossRef]
26. Bai, Z.; Yang, Q.; Wang, J. Catalytic ozonation of sulfamethazine antibiotics using $Ce_{0.1}Fe_{0.9}OOH$: Catalyst preparation and performance. *Chemosphere* **2016**, *161*, 174–180. [CrossRef]
27. Chen, S.; Wu, Y.; Li, G.; Wu, J.; Meng, G.; Guo, X.; Liu, Z. A novel strategy for preparation of an effective and stable heterogeneous photo-Fenton catalyst for the degradation of dye. *Appl. Clay Sci.* **2017**, *136*, 103–111. [CrossRef]

28. Ribeiro, R.S.; Rodrigues, R.O.; Silva, A.M.; Tavares, P.B.; Carvalho, A.M.; Figueiredo, J.L.; Faria, J.L.; Gomes, H.T. Hybrid magnetic graphitic nanocomposites towards catalytic wet peroxide oxidation of the liquid effluent from a mechanical biological treatment plant for municipal solid waste. *Appl. Catal. B Environ.* **2017**, *219*, 645–657. [CrossRef]
29. Ma, Y.; Wang, B.; Wang, Q.; Xing, S. Facile synthesis of α-FeOOH/γ-Fe$_2$O$_3$ by a pH gradient method and the role of γ-Fe$_2$O$_3$ in H$_2$O$_2$ activation under visible light irradiation. *Chem. Eng. J.* **2018**, *354*, 75–84. [CrossRef]
30. Ai, C.; Wu, S.; Li, L.; Lei, Y.; Shao, X. Novel magnetically separable γ-Fe$_2$O$_3$/Ag/AgCl/g-C$_3$N$_4$ composite for enhanced disinfection under visible light. *Colloids Surf. A Physicochem. Eng. Asp.* **2019**, *583*, 123981. [CrossRef]
31. Wang, Y.; Fan, X.; Wang, S.; Zhang, G.; Zhang, F. Magnetically separable γ-Fe$_2$O$_3$/TiO$_2$ nanotubes for photodegradation of aqueous methyl orange. *Mater. Res. Bull.* **2013**, *48*, 785–789. [CrossRef]
32. Ding, M.; Chen, W.; Xu, H.; Shen, Z.; Lin, T.; Hu, K.; Lu, C.; Xie, Z. Novel A-Fe$_2$O$_3$/MXene nanocomposite as heterogeneous activator of peroxymonosulfate for the degradation of salicylic acid. *J. Hazard. Mater.* **2020**, *382*, 121064. [CrossRef] [PubMed]
33. Yang, S.; Huang, Z.; Wu, P.; Li, Y.; Dong, X.; Li, C.; Zhu, N.; Duan, X.; Dionysiou, D.D. Rapid removal of tetrabromobisphenol A by A-Fe$_2$O$_3$-x@Graphene@Montmorillonite catalyst with oxygen vacancies through peroxymonosulfate activation: Role of halogen and A-hydroxyalkyl radicals. *Appl. Catal. B Environ.* **2020**, *260*, 118129. [CrossRef]
34. Zheng, H.; Bao, J.; Huang, Y.; Xiang, L.; Faheem; Ren, B.; Du, J.; Nadagouda, M.N.; Dionysiou, D.D. Efficient degradation of atrazine with porous sulfurized Fe$_2$O$_3$ as catalyst for peroxymonosulfate activation. *Appl. Catal. B Environ.* **2019**, *259*, 118056. [CrossRef]
35. Rahim Pouran, S.; Abdul Raman, A.A.; Wan Daud, W.M.A. Review on the application of modified iron oxides as heterogeneous catalysts in Fenton reactions. *J. Clean. Prod.* **2014**, *64*, 24–35. [CrossRef]
36. Liu, Y.; Yu, C.; Dai, W.; Gao, X.; Qian, H.; Hu, Y.; Hu, X. One-pot solvothermal synthesis of multi-shelled α-Fe$_2$O$_3$ hollow spheres with enhanced visible-light photocatalytic activity. *J. Alloys Compd.* **2013**, *551*, 440–443. [CrossRef]
37. Santhosh, C.; Malathi, A.; Dhaneshvar, E.; Bhatnagar, A.; Grace, A.N.; Madhavan, J. Iron Oxide Nanomaterials for Water Purification. In *Nanoscale Materials in Water Purification*; Elsevier: Amsterdam, The Netherlands, 2019; pp. 431–446.
38. Xu, W.; Xue, W.; Huang, H.; Wang, J.; Zhong, C.; Mei, D. Morphology controlled synthesis of α-Fe$_2$O$_{3-x}$ with benzimidazole-modified Fe-MOFs for enhanced photo-Fenton-like catalysis. *Appl. Catal. B Environ.* **2021**, *291*, 120129. [CrossRef]
39. Xiao, C.; Li, J.; Zhang, G. Synthesis of stable burger-like α-Fe$_2$O$_3$ catalysts: Formation mechanism and excellent photo-Fenton catalytic performance. *J. Clean. Prod.* **2018**, *180*, 550–559. [CrossRef]
40. Huang, M.; Qin, M.; Chen, P.; Jia, B.; Chen, Z.; Li, R.; Liu, Z.; Qu, X. Facile preparation of network-like porous hematite (α-Fe$_2$O$_3$) nanosheets via a novel combustion-based route. *Ceram. Int.* **2016**, *42*, 10380–10388. [CrossRef]
41. Cheng, X.L.; Jiang, J.; Jin, C.Y.; Lin, C.C.; Zeng, Y.; Zhang, Q.H. Cauliflower-like α-Fe$_2$O$_3$ microstructures: Toluene–water interface-assisted synthesis, characterization, and applications in wastewater treatment and visible-light photocatalysis. *Chem. Eng. J.* **2014**, *236*, 139–148. [CrossRef]
42. Khurram, R.; Wang, Z.; Ehsan, M.F. α-Fe$_2$O$_3$-based nanocomposites: Synthesis, characterization, and photocatalytic response towards wastewater treatment. *Environ. Sci. Pollut. Res.* **2021**, *28*, 17697–17711. [CrossRef] [PubMed]
43. Pang, Y.L.; Lim, S.; Ong, H.C.; Chong, W.T. Synthesis, characteristics and sonocatalytic activities of calcined γ-Fe$_2$O$_3$ and TiO$_2$ nanotubes/γ-Fe$_2$O$_3$ magnetic catalysts in the degradation of Orange G. *Ultrason. Sonochem.* **2016**, *29*, 317–327. [CrossRef] [PubMed]
44. Liu, Y.; Jin, W.; Zhao, Y.; Zhang, G.; Zhang, W. Enhanced catalytic degradation of methylene blue by α-Fe$_2$O$_3$/graphene oxide via heterogeneous photo-Fenton reactions. *Appl. Catal. B Environ.* **2017**, *206*, 642–652. [CrossRef]
45. Xiang, H.; Ren, G.; Yang, X.; Xu, D.; Zhang, Z.; Wang, X. A low-cost solvent-free method to synthesize α-Fe$_2$O$_3$ nanoparticles with applications to degrade methyl orange in photo-Fenton system. *Ecotoxicol. Environ. Saf.* **2020**, *200*, 110744. [CrossRef]
46. Li, M.; Zhang, C. γ-Fe$_2$O$_3$ nanoparticle-facilitated bisphenol A degradation by white rot fungus. *Sci. Bull.* **2016**, *61*, 468–472. [CrossRef]
47. Mao, G.Y.; Bu, F.X.; Wang, W.; Jiang, D.M.; Zhao, Z.J.; Zhang, Q.H.; Jiang, J.S. Synthesis and characterization of γ-Fe$_2$O$_3$/C nanocomposite as an efficient catalyst for the degradation of methylene blue. *Desalin. Water Treat.* **2016**, *57*, 9226–9236. [CrossRef]
48. Wang, F.; Yu, X.; Ge, M.; Wu, S.; Guan, J.; Tang, J.; Wu, X.; Ritchie, R.O. Facile self-assembly synthesis of γ-Fe$_2$O$_3$/graphene oxide for enhanced photo-Fenton reaction. *Environ. Pollut.* **2019**, *248*, 229–237. [CrossRef]
49. Cao, S.; Kang, F.; Li, P.; Chen, R.; Liu, H.; Wei, Y. Photoassisted hetero-Fenton degradation mechanism of Acid Blue 74 by a γ-Fe$_2$O$_3$ catalyst. *RSC Adv.* **2015**, *5*, 66231–66238. [CrossRef]
50. Zhan, J.; Li, M.; Zhang, X.; An, Y.; Sun, W.; Peng, A.; Zhou, H. Aerosol-assisted submicron γ-Fe$_2$O$_3$/C spheres as a promising heterogeneous Fenton-like catalyst for soil and groundwater remediation: Transport, adsorption and catalytic ability. *Chin. Chem. Lett.* **2020**, *31*, 715–720. [CrossRef]
51. Rajoba, S.J.; Badabade, A.R.; Pingale, P.C.; Kale, R.D. Structural, Morphological, and Vibrational Properties of Porous α-Fe$_2$O$_3$ Nanoparticles Prepared by Combustion method. *Macromol. Symp.* **2021**, *400*, 2100033. [CrossRef]
52. Apte, S.K.; Naik, S.D.; Sonawane, R.S.; Kale, B.B.; Baeg, J.O. Synthesis of nanosize-necked structure α- and γ-Fe$_2$O$_3$ and its photocatalytic activity. *J. Am. Ceram. Soc.* **2007**, *90*, 412–414. [CrossRef]
53. de Andrade, M.B.; Guerra, A.C.S.; dos Santos, T.R.T.; Cusioli, L.F.; de Souza Antônio, R.; Bergamasco, R. Simplified synthesis of new GO-α-γ-Fe$_2$O$_3$-Sh adsorbent material composed of graphene oxide decorated with iron oxide nanoparticles applied for removing diuron from aqueous medium. *J. Environ. Chem. Eng.* **2020**, *8*, 103903. [CrossRef]

54. Manukyan, K.V.; Chen, Y.S.; Rouvimov, S.; Li, P.; Li, X.; Dong, S.; Liu, X.; Furdyna, J.K.; Orlov, A.O.; Bernstein, G.H.; et al. Ultrasmall α-Fe_2O_3 Superparamagnetic Nanoparticles with High Magnetization Prepared by Template-Assisted Combustion Process. *J. Phys. Chem. C* **2014**, *118*, 16264–16271. [CrossRef]
55. Tian, R.; Gao, Z.; Lang, R.; Li, N.; Gu, H.; Chen, G.; Guan, H.; Comini, E.; Dong, C. Ru-functionalized Ni-doped dual phases of α/γ-Fe_2O_3 nanosheets for an optimized acetone detection. *J. Nanostruct. Chem.* **2022**. [CrossRef]
56. Yan, S.; Zan, G.; Wu, Q. An ultrahigh-sensitivity and selective sensing material for ethanol: α-/γ-Fe_2O_3 mixed-phase mesoporous nanofibers. *Nano Res.* **2015**, *8*, 3673–3686. [CrossRef]
57. Huang, D.; Li, H.; Wang, Y.; Wang, X.; Cai, L.; Fan, W.; Chen, Y.; Wang, W.; Song, Y.; Han, G.; et al. Assembling a high-performance acetone sensor based on MOFs-derived porous bi-phase α-/γ-Fe_2O_3 nanoparticles combined with $Ti_3C_2T_x$ nanosheets. *Chem. Eng. J.* **2022**, *428*, 131377. [CrossRef]
58. Ghasemifard, M.; Heidari, G.; Ghamari, M.; Fathi, E.; Izi, M. Synthesis of Porous Network-Like α-Fe_2O_3 and α/γ-Fe_2O_3 Nanoparticles and Investigation of Their Photocatalytic Properties. *Nanotechnol. Russ.* **2019**, *14*, 353–361. [CrossRef]
59. Orudzhev, F.F.; Alikhanov, N.M.-R.; Ramazanov, S.M.; Sobola, D.S.; Murtazali, R.K.; Ismailov, E.H.; Gasimov, R.D.; Aliev, A.S.; Ţălu, Ş. Morphotropic Phase Boundary Enhanced Photocatalysis in Sm Doped $BiFeO_3$. *Molecules* **2022**, *27*, 7029. [CrossRef]
60. Alikhanov, N.M.-R.; Rabadanov, M.K.; Orudzhev, F.F.; Gadzhimagomedov, S.K.; Emirov, R.M.; Sadykov, S.A.; Kallaev, S.N.; Ramazanov, S.M.; Abdulvakhidov, K.G.; Sobola, D. Size-dependent structural parameters, optical, and magnetic properties of facile synthesized pure-phase $BiFeO_3$. *J. Mater. Sci. Mater. Electron.* **2021**, *32*, 13323–13335. [CrossRef]
61. Kashyap, S.J.; Sankannavar, R.; Madhu, G.M. Iron oxide (Fe_2O_3) synthesized via solution-combustion technique with varying fuel-to-oxidizer ratio: FT-IR, XRD, optical and dielectric characterization. *Mater. Chem. Phys.* **2022**, *286*, 126118. [CrossRef]
62. Chernyshova, I.V.; Hochella, M.F.; Madden, A.S. Size-dependent structural transformations of hematite nanoparticles. 1. Phase transition. *Phys. Chem. Chem. Phys.* **2007**, *9*, 1736–1750. [CrossRef]
63. El Mendili, Y.; Bardeau, J.F.; Randrianantoandro, N.; Gourbil, A.; Greneche, J.M.; Mercier, A.M.; Grasset, F. New evidences of in situ laser irradiation effects on γ-Fe_2O_3 nanoparticles: A Raman spectroscopic study. *J. Raman Spectrosc.* **2011**, *42*, 239–242. [CrossRef]
64. Wang, L.; Lu, X.; Han, C.; Lu, R.; Yang, S.; Song, X. Electrospun hollow cage-like α-Fe_2O_3 microspheres: Synthesis, formation mechanism, and morphology-preserved conversion to Fe nanostructures. *CrystEngComm* **2014**, *16*, 10618–10623. [CrossRef]
65. Jain, S.; Shah, J.; Negi, N.S.; Sharma, C.; Kotnala, R.K. Significance of interface barrier at electrode of hematite hydroelectric cell for generating ecopower by water splitting. *Int. J. Energy Res.* **2019**, *43*, 4743–4755. [CrossRef]
66. Jubb, A.M.; Heather, C.A. Vibrational spectroscopic characterization of hematite, maghemite, and magnetite thin films produced by vapor deposition. *ACS Appl. Mater. Interfaces* **2010**, *2*, 2804–2812. [CrossRef]
67. Bahari, A. Characteristics of Fe_3O_4, α-Fe_2O_3, and γ-Fe_2O_3 Nanoparticles as Suitable Candidates in the Field of Nanomedicine. *J. Supercond. Nov. Magn.* **2017**, *30*, 2165–2174. [CrossRef]
68. Han, Q.; Liu, Z.; Xu, Y.; Chen, Z.; Wang, T.; Zhang, H. Growth and properties of single-crystalline γ-Fe_2O_3 nanowires. *J. Phys. Chem. C* **2007**, *111*, 5034–5038. [CrossRef]
69. Zhong, Y.; Ma, Y.; Guo, Q.; Liu, J.; Wang, Y.; Yang, M.; Xia, H. Controllable Synthesis of TiO_2@Fe_2O_3 Core-Shell Nanotube Arrays with Double-Wall Coating as Superb Lithium-Ion Battery Anodes. *Sci. Rep.* **2017**, *7*, 40827. [CrossRef]
70. Sun, Y.P.; Li, X.Q.; Cao, J.; Zhang, W.X.; Wang, H.P. Characterization of zero-valent iron nanoparticles. *Adv. Colloid Interface Sci.* **2006**, *120*, 47–56. [CrossRef] [PubMed]
71. Yamashita, T.; Hayes, P. Analysis of XPS spectra of Fe^{2+} and Fe^{3+} ions in oxide materials. *Appl. Surf. Sci.* **2008**, *254*, 2441–2449. [CrossRef]
72. Zhang, C.; Liu, S.; Chen, T.; Li, Z.; Hao, J. Oxygen vacancy-engineered Fe_2O_3 nanocubes via a task-specific ionic liquid for electrocatalytic N_2 fixation. *Chem. Commun.* **2019**, *55*, 7370–7373. [CrossRef] [PubMed]
73. Pandey, J.; Sethi, A.; Uma, S.; Nagarajan, R. Catalytic Application of Oxygen Vacancies Induced by Bi^{3+} Incorporation in ThO_2 Samples Obtained by Solution Combustion Synthesis. *ACS Omega* **2018**, *3*, 7171–7181. [CrossRef] [PubMed]
74. Liang, X.; Wang, L.; Wen, T.; Liu, H.; Zhang, J.; Liu, Z.; Zhu, C.; Long, C. Mesoporous poorly crystalline α-Fe_2O_3 with abundant oxygen vacancies and acid sites for ozone decomposition. *Sci. Total Environ.* **2022**, *804*, 150161. [CrossRef] [PubMed]
75. Sang, Y.; Cao, X.; Ding, G.; Guo, Z.; Xue, Y.; Li, G.; Yu, R. Constructing oxygen vacancy-enriched Fe_2O_3@NiO heterojunctions for highly efficient electrocatalytic alkaline water splitting. *Crystengcomm* **2021**, *24*, 199–207. [CrossRef]
76. Flak, D.; Chen, Q.; Mun, B.S.; Liu, Z.; Rękas, M.; Braun, A. In situ ambient pressure XPS observation of surface chemistry and electronic structure of α-Fe_2O_3 and γ-Fe_2O_3 nanoparticles. *Appl. Surf. Sci.* **2018**, *455*, 1019–1028. [CrossRef]
77. Tian, C.M.; Li, W.W.; Lin, Y.M.; Yang, Z.Z.; Wang, L.; Du, Y.G.; Xiao, H.Y.; Qiao, L.; Zhang, J.Y.; Chen, L.; et al. Electronic Structure, Optical Properties, and Photoelectrochemical Activity of Sn-Doped Fe_2O_3 Thin Films. *J. Phys. Chem. C* **2020**, *124*, 12548–12558. [CrossRef]
78. Dai, H.; Xu, S.; Chen, J.; Miao, X.; Zhu, J. Oxalate enhanced degradation of Orange II in heterogeneous UV-Fenton system catalyzed by Fe_3O_4@γ-Fe_2O_3 composite. *Chemosphere* **2018**, *199*, 147–153. [CrossRef]
79. Chen, J.; Zhu, L. UV-Fenton discolouration and mineralization of Orange II over hydroxyl-Fe-pillared bentonite. *J. Photochem. Photobiol. A Chem.* **2007**, *188*, 56–64. [CrossRef]

80. Jiang, J.; Gao, J.; Li, T.; Chen, Y.; Wu, Q.; Xie, T.; Lin, Y.; Dong, S. Visible-light-driven photo-Fenton reaction with α-Fe_2O_3/BiOI at near neutral pH: Boosted photogenerated charge separation, optimum operating parameters and mechanism insight. *J. Colloid Interface Sci.* **2019**, *554*, 531–543. [CrossRef]
81. Wang, Y.; Hongying, Z.h.a.o.; Zhao, G. Iron-copper bimetallic nanoparticles embedded within ordered mesoporous carbon as effective and stable heterogeneous Fenton catalyst for the degradation of organic contaminants. *Appl. Catal. B Environ.* **2015**, *164*, 396–406. [CrossRef]
82. Griffing, M.; Mellon, M.G. Colorimetric determination of iron with nitroso-R-salt. *Anal. Chem.* **1947**, *19*, 1014–1016. [CrossRef]

Disclaimer/Publisher's Note: The statements, opinions and data contained in all publications are solely those of the individual author(s) and contributor(s) and not of MDPI and/or the editor(s). MDPI and/or the editor(s) disclaim responsibility for any injury to people or property resulting from any ideas, methods, instructions or products referred to in the content.

The Role of Fungal Fuel Cells in Energy Production and the Removal of Pollutants from Wastewater

Aisha Umar [1,*], Łukasz Smółka [2] and Marek Gancarz [2,3,*]

[1] Institute of the Botany, University of the Punjab, Lahore 54590, Pakistan
[2] Faculty of Production and Power Engineering, University of Agriculture in Krakow, Balicka 116B, 30-149 Krakow, Poland
[3] Institute of Agrophysics, Polish Academy of Sciences, Doświadczalna 4, 20-290 Lublin, Poland
* Correspondence: ash.dr88@gmail.com (A.U.); m.gancarz@urk.edu.pl (M.G.)

Abstract: Pure water, i.e., a sign of life, continuously circulates and is contaminated by different discharges. This emerging environmental problem has been attracting the attention of scientists searching for methods for the treatment of wastewater contaminated by multiple recalcitrant compounds. Various physical and chemical methods are used to degrade contaminants from water bodies. Traditional methods have certain limitations and complexities for bioenergy production, which motivates the search for new ways of sustainable bioenergy production and wastewater treatment. Biological strategies have opened new avenues to the treatment of wastewater using oxidoreductase enzymes for the degradation of pollutants. Fungal-based fuel cells (FFCs), with their catalysts, have gained considerable attention among scientists worldwide. They are a new, ecofriendly, and alternative approach to nonchemical methods due to easy handling. FFCs are efficiently used in wastewater treatment and the production of electricity for power generation. This article also highlights the construction of fungal catalytic cells and the enzymatic performance of different fungal species in energy production and the treatment of wastewater.

Keywords: fungi; metals; treatment; energy; catalyst

1. Introduction

The rapid industrial and global population growth has polluted water and depleted the resources of fossil fuels to fulfill the excessive demand for energy production. The quality of water is deteriorating due to the continuous mixing of undesirable chemicals [1]. The need for water quality improvement and preservation is continuously growing day by day due to agricultural, civilization, and industrial activities, leading to environmental and global changes. Wastewater is defined as a combination of liquid, water with wastes from residential areas, commercial sites, institutions, and industrial establishments together with ground, surface, and storm water [2].

Nonpoint sources contaminate valuable water resources. Organic pollutants are hazardous and toxic; hence, chemical processes are most suitable to remediate and eliminate the inorganic matter, dyes, and recalcitrant matter. Various techniques (biological, physical, and chemical) are used to treat organic-compounds polluted the wastewater. Traditional methods have certain limitations for bioenergy production, e.g., large spaces, high capital cost, and complexities linked with the production process. The demands for sustainable bioenergy production have been increasing in the world as an alternative to nonchemical methods for power generation. Biological degradation involves the use of microorganisms (fungi, algae, bacteria, and enzymes), which utilize the maximum land area, exhibit very high sensitivity toward toxic agents, and require a long consumption time [3].

The exploration of novel and efficient approaches have attracted the attention of environmental scientists to cleanup and remediation of the contaminated water bodies. The fungal potential to generate bioelectricity from biodegradable wastewater reduces the cost of conversion [4]. Biotic sources exploit different species of fungi for bioenergy generation. However, very little data are available on the use of "fungal-mediated electrochemical system" for energy production. Minimal resources, higher prices of fossil fuels, and increasing global warming issues have motivated the scientists to design alternative "renewable" energy sources, e.g., fungal cell factories.

The fungal fuel cell is a device that uses fungi as catalysts to generate electricity by oxidizing the inorganic compounds of biomass [5]. A few researchers believe that this technology is not only used for the production of electricity. It also depends on the ability of the electrode associated with the fungi to degrade the toxics and waste materials [6]. Biomass/organic material is a sustainable alternative approach to address this issue. Fungal fuel cells (FFCs) provide electricity directly through the "biodegradation" of raw materials by fungal cells [7].

It is proposed that fungal species are used for energy generation, taking advantage of their potential as "novel cell factories". *Saccharomyces* or *Pichia* fungi are used in these cells [8]. Fungal cells have nine times higher potential to generate energy accumulated in sewage sludge than conventional methods [9]. Fungal species have a strong potential to generate power using the presence of complex enzymatic systems. These species can rapidly grow on waste materials and degrade these materials within a shorter time for the production of "bioenergy". The use of fungi in "bioremediation" is a promising technique [10].

This approach is also called an "Oxidative Biocatalyst" approach. The efficiency of this strategy can be maximized by using different fungal growth and environmental parameters with redox mediator systems [11]. The fungal approach is the best method for energy production and wastewater treatment in a cost-effective manner. Integrated physical, chemical, and biological wastewater treatment are discussed in detail with current challenges in terms of achieving good treatment efficiencies to meet the discharge standards during wastewater treatment and bioelectricity generation.

Objectives

- To construct bioelectrochemical devices, where the fungus (catalyst) is used for the oxidation of inorganic matter for electricity generation.
- To generate power/energy from wastewater (substrate) using fungal electrochemical technology (FET) in an ecofriendly manner.
- Compared to conventional methods, biodegradation, using cost-effective and economical technologies, is greatly preferred with improved outcomes. The goal of this review provides a design for a much more cost-effective system with a principle of wastes removal using fungal fuel cells. Studies on wastewater treatment in FFCs with electricity generation are also presented. Thus, biodegradation using FFCs is considered to be a highly economical, ecofriendly, and more prominent way to solve these problems.

1.1. Oleaginous Fungi

Oleaginous microorganisms have potential for biodiesel formulation and production. These are used as an alternative renewable energy sources. Oleaginous fungi have numerous advantages, e.g., lower land requirements, short cultivation time, and maximum production of fatty acids (oils) [12]. A few oleaginous species metabolize xylose and assist in lipid production from "lignocellulosic hydrolysates" [13]. These fungal species become more oleaginous, when different organic substrates (glucose and sucrose) are added to their growth medium. Each species has particular abilities to utilize organic substrates and enhance the lipid yield. It is noticed that in a fungal consortium, less productive species always follow a more productive species during co-metabolism. This way of combination is yielding more biomass than single cultures.

The genera *Mucor* and *Aspergillus* have been recognized to store up to 80% of oils (in cells) in specific conditions [14]. Strains that have high lipid contents and metabolize TAG (triacylglycerides) usually preferred to formulate and generate biofuels efficiently.

Zygomycetes are a class of excellent oleaginous fungal species, providing palmitic and oleic acids that are used for biodiesel formation. Additionally, anaerobic fungi are an arsenal of extracellular multienzyme complexes. These fungi are involved in the breakdown of various biomasses for biogas generation. Zygomycetes, such as *Mortierella isabelline* has reported to have a 60–70% lipid content [15].

Oleaginous yeast (*Rhodotorula mucilaginosa* SML) has been using for the treatment of food industry effluents. The overall yeast lipid content for the effluent treatment was 67.95 *w/w*% of dry cell biomass. The extracted yeast oil was used for transesterification and showed a 98% conversion of oil to methanol. The fatty acid composition was compatible with petroleum diesel, making it applicable for alternative biofuel production. Thus, this strategy proved efficient in the removal of contaminants of industrial wastes suggested as a new sustainable source for biodiesel production [16].

The biofilm of *Wickerhamomyces anomalus* (yeast) on the anodic electrode of a single-chamber fuel cell fed with zinc and copper electrodes and pineapple waste (substrate) is used for fuel production. Current (4.95667 ± 0.54 mA) and voltage peaks (0.99 ± 0.03 V) were generated for 16 and 20 days, respectively. The maximum power density of 513.99 ± 6.54 mW/m^2 at a current density of 6.123 A/m^2 was generated [17].

1.2. Hydrolytic and Lignolytic Fungi

Hydrolytic and ligninolytic fungi are suitable candidates for the production of biofuels or bioethanol. A few basidiomycetes have been reporting to secrete extracellular enzymes that degrade the waste materials [18]. Fungal peroxidases (manganese-dependent peroxidase and lignin peroxidase) degrade the lignin, hemicellulose, and polyaromatic phenols [19].

Fungal cells are known for the generation of bioelectricity, good-quality biofuel production, and wastewater treatment. The best-known biofuel-producing fungal species are *Rhodosporidium toruloides, Cryptococcus* sp., *Yarrowia lipolytica, Penicillium* sp., *Aspergillus* sp., and *Trichoderma reesei*. Species that have the potential to produce biodiesel or electricity generation transfer the e$^-$ via cytochrome C. These include *Candida* sp., *Colletotrichum* sp., *Saccharomyces cerevisiae, Penicillium* sp., *Alternaria* sp., *Rhizopus* sp., and *Aspergillus* sp. Cells constructed from these species are called "Fungal-based FCs" [20].

Energy-generating fungal biocatalysts increase the electron transmission rate through extensive networking of fungal hyphae and produce stable electricity, which contributes to "external electrochemical operations". Due to this unique property, fungi, and yeasts are preferred over bacterial cells for wastewater treatment and electricity generation [21].

1.3. Effects of Environmental Factors on Fungal Growth and Metabolism

- pH

A few fungal species grow in a broad pH range, while some species grow in a narrower pH range. The optimum growth of fungal species appears to correspond to a specific pH value [22]. The fungal ability to grow at a pH >7 is required during industrial production. The substrate with a pH below 7.00 inhibits the growth of contaminants (bacteria) without affecting the yield. A slight increase in pH of FFCs, inhibits fungal growth and metabolism. Fungal catalyst formation (oxidoreductase) and catalytic action are highly stable at an acidic pH (3–6). A low pH induces mobility and unfolding of the enzyme proteins.

- Temperature

Temperature plays an important role in fungal growth, metabolism, and electricity generation using fungal fuel cells. The temperature of system facilitates the cells metabolism and their enzymatic reactions. In wood-rotting fungi, oxidoreductase is produced in an optimum temperature range (25–30 °C), which depends on mesophilic and thermophilic fungal species [23]. The enzyme system of mesophilic basidiomycetes is thermostable at elevated temperatures. Optimum temperature is also favorable for the efficient maintenance of fungal systems in fuel cells during their metabolic mechanisms. A slight decrease or increase in temperature leads to denaturation and inactivation of the cell components, which consequently stops the work of fuel cells with no power generation.

- Ionic strength

Higher ionic conductivity also influences the work of fungal fuel cell. Ionic conductivity is directly proportional to power generation due to minimum internal resistance. High ionic conductivity increases the power output of FCs. Protons and electrons can easily move from one compartment to another for the completion of a circuit.

- Salinity

About 90% fungal species can tolerate at 3 to 6% salt stress. Halotolerant fungal species are better adapted to the salty environment [24]. Marine fungi with a dark cell wall can tolerate higher salinity than moniliaceous fungi [25]. The habitats of marine fungal species have a strong influence on their adaptation to salt and metabolic functioning.

Hyperosmotic stress in fungi is linked with the inhibition of cell wall extension and cellular expansion, resulting a reduction in their growth [26]. Excess in everything is bad. Maximum ions can alter protein, membrane integrity, and nucleic metabolism, which may change the enzymatic activity and catalytic performance during fungal growth and functioning of fuel cells [27]. Organic osmotica (compounds) are called compatible solutes, as these solutes can store high concentrations of salts without interfering with cell metabolism. Polyols (mannitol, arabitol, and glycerol) and non-reducing saccharides (trehalose) are soluble carbohydrates found in basidiomycetes and ascomycetes. These solutes help the fungi to grow efficiently in a salt-stress environment. A maximum salt range destroys the fungal product yield as well.

1.4. Enzymatic Treatment by Biocatalytic Fungal Species

Catalysts are needed to accelerate the maximum biodiesel production. Biofuels are categorized into bioethanol, biohydrogen, and biodiesel (Figure 1). During the enzymatic treatment of wastewater (substrate), low-chain carbon compounds are produced, which utilized during microbial oxidation. The basic oxidoreductases, such as laccases, lignin peroxidases, and manganese peroxidases, are extracellular enzymes. In contrast, glucose oxidase and cellobiose dehydrogenase are auxiliary enzymes isolated from white-rot fungi.

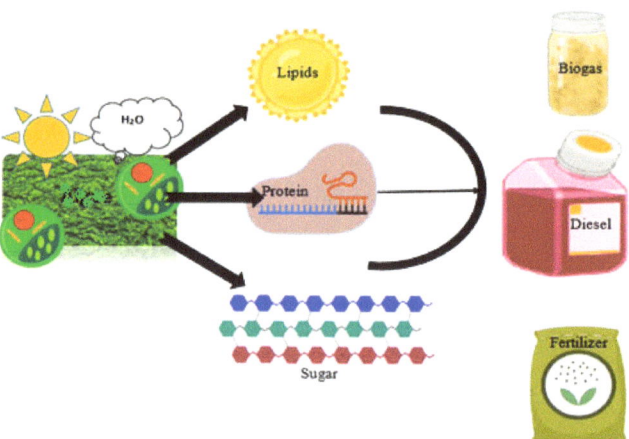

Figure 1. Fungal species utilize sunlight and water for the metabolic production of lipids, protein, and sugar used for the manufacturing of biodiesel, biogas, and biofertilizers.

Biocatalysts are exoelectrogens that oxidize organic material and deliver electrons from anode to cathode to generate the electricity. Exoelectrogens are now investigated for the development of FFCs, which potentially convert the diverse organic substances (activated sludge of waste water) into electricity, ethanol, and H_2 [28]. *Sporotrichum pruinosum* (white-rot fungus) efficiently degrades pollutants, organic biomass, and chemical substances with the use of an extracellular enzyme [29].

Biocatalyst is deposited onto the carbon anode as floating biomass in yeast-based FFCs [30]. Yeast-based FFCs have the following advantages: (i) degradation of very complex substrates (starch and cellulose-based substrates) into simple organic molecules; (ii) survival in an anaerobic environment [31]; and (iii) simple and easy production, rapid development, and sensitivity of the strains. Except for yeast, other fungal species are also exploited as biocatalysts for both wastewater treatment and electrochemical approaches.

Scientists are motivated toward the development of such mediator systems as fungal-based FCs [21]. Pure *Saccharomyces cerevisiae* (yeast) is a model organisms used as biocatalysts in FFCs. Christwardana et al. [30] indicated the significance of yeast cells in MFCs due to their unique features and sustainability. *S. cerevisiae* has been extensively studied and characterized as a biocatalyst in biological fuel cells [32–34]. This species is nonpathogenic to non-target organisms (humans), has a high growth rate, easy to culture in anaerobic environments, and grows very well at room temperature [35]; hence, it can be used for the effective treatment of wastewater [36,37]. In addition to the low cost and rapid multiplication, the species remains active and survives even in a dried environment for a longer period [38]. Carbon-neutral fuel, referred to as bioethanol, is produced from plant waste and bacterial/algal biomass [39]. It is also produced from yeast and fungi in anaerobic conditions [40], especially *S. cerevisiae*, which is considered to have great potential in the production of bioethanol.

Candida melibiosica, Kluyveromyces marxianus, Blastobotrys adeninivorans, Pichia anomala, P. polymorpha, and *Saccharomyces cerevisiae* yeasts are used as biocatalysts in FFCs with/without an external mediator. *Kluyveromyces marxianus* is a promising yeast species producing maximum power at higher temperatures, when grown in natural (organic) substrates. Other fungal species, e.g., *Saccharomyces cerevisiae* [33], *Candida melibiosica* [36], *Blastobotrys adeninivorans* [37], *Hansenula polymorpha* [41], and *Pichia anomala* [42] all are a potential source for catalysts in FFCs.

Exogenous mediators, such as methylene blue (MB) and neutral red (NR), are used to increase the transport of electrons between anodes and microbes. The yeast cell surface-displayed dehydrogenases include cellobiose dehydrogenase (CDH) and pyranose dehydrogenase (PDH) [43]. Both CDH- and PDH-based biocatalysts are used in the anodic compartment of FFCs.

1.5. Structure of Fungal-Mediated Fuel Cells

Protons and electrons are generated through the oxidation of organic matter in the aqueous solution of anode compartment, when fungi used as catalyst. The external circuit is used to transmit the electrons toward the cathode, while the proton exchange membrane (PEM) facilitates proton diffusion [44–46]. At the cathode, e^- and protons are used for the reduction reaction and eventually change oxygen to water [47]. Potter [48] was the first person, who liberated electrical energy from yeast cells in 1911. Fungi can transfer electrons to the anode electrode in three possible ways: (1) direct contact; (2) pili/conductive wires; and (3) redox mediators/electron shuttle [49].

The advantages of FFCs include sustainable nonchemical character, minimum sludge generation, optimum temperature, wide range of substrates, low power consumption, and good performance [50]. This is a promising alternative technique used to explore the fungal potential in the conversion of organic substrates into electricity. The performance of FFCs depends on multiple factors, e.g., configuration of the cell, choice of the substrate, anodic material, biocatalyst, electro-catalyst (at the cathode), and environmental conditions. Prasad et al. [42] observed that fungi are more active than bacteria in MFCs.

Fungal fuel cells (FFCs) are operated in a closed-system mode on the principle of oxidation–reduction reaction through a series of electrochemical and microbial pathways. The anaerobic environment is maintained in the anodic compartment [51]. The e^- and protons in the anode chamber are produced through oxidation of the substrate by fungi, and oxygen reduced by a terminal electron acceptor in the cathodic chamber.

Fungi used at the anode to transport e^- via redox-active fungal protein and synthetic mediators. Anodic fungal cells oxidize the substrates and produce electrons and protons. The e^- is absorbed by the electrode, and protons flow toward the cathode via a PEM. The protons are transferred through the PEM to the cathode. Subsequently, the electron and proton combination produce a molecule of water (at the cathode) for the completion of the bioelectrochemical reaction. At the cathode, the fungal enzyme catalyzes the reaction and the final electron receiver is O_2 and H_2O produced during this reaction. Additionally, airtight compartments, sufficient space within the chamber, an outlet, and inlets are certain prerequisites in the arrangement of the PEM and electrodes into a system (Figure 2).

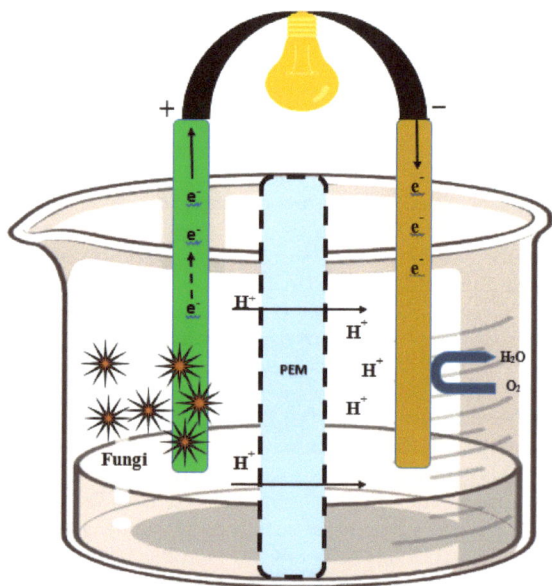

Figure 2. Structure of the fungal fuel cell. The left side of the cell is the fungal catalyst electrode (anodic chamber), and protons are transferred from the proton exchange membrane to the cathodic chamber (where catalytic oxidation takes place).

A membrane separator called a proton exchange membrane (PEM), divides the cell into two distinct cathodic and anodic chambers. An extracellular microorganism with the ability to transfer electrons is called an exoelectrogen (Biocatalyst) [52]. Membrane separators (PEM, salt bridge, anion and cation exchange membrane, microfiltration membrane, glass fiber) have such features as low permeability and high conductivity for optimum FFC performance [53].

The membrane (Nafion), cathode (Platinum), and anode (carbon cloth and carbon paper) materials are expensive and fragile. Fungal FCs with a low-cost electrode, high power output, and membrane materials with good scalability should develop to treat different effluents (de-sizing, bleaching, dyeing, and printing effluents).

Potent microbes improve the electron transfer and facilitate the degradation of biological substrates, e.g., *Shewanella oneidensis* or the hyphal networks of *T. versicolor* facilitate electron transfer onto the anode efficiently. Nearly 30 days are required for the formation of homogeneous bacterial and fungal biofilms on the electrode [54].

1.6. Electron Transfer (ET) Mechanism

There are two types of e⁻ transfer mechanisms. Fungal consortia demonstrated a better ET mechanism than single species.

1. Direct ET: two types via outer cytochrome and nanowire.
2. Indirect ET: (mediated/mediator electron transfer) reactive diffusible redox mediators (RMs) enhance the reaction rate and increase the range of degraded substrates.

1.7. Types of Electrodes

The electrode material influences the performance of FFC, which has a direct impact on the kinetics of electrodes [55].

Anode: A carbon-based anode is cost-effective and noncorrosive (modifier carbon paper, carbon felt, carbon cloth, carbon nanotubes, graphene, stainless steel, titanium, and gold) [56]. These materials improve the characteristics of anodic surface material and provide an appropriate platform for fungal biofilm formation with an active catalyst. The anode quality enhances the high surface area, chemical/electrical stability, and biocompatibility [57]. The anodic electrode increases the efficiency of FFCs. This serves as a driving feature for power generation. Thus, the anodic material seems to be a suitable strategy to enhance performance.

Reaction in the anodic chamber:

$$C_6H_{12}O_6 + 6H_2O \rightarrow 6CO_2 + 24H^+ + 24e^- \tag{1}$$

Iron and iron oxide nanoparticles, graphite, carbon cloth, and carbon felt are effective anode catalysts improving the efficiency of a fungal fuel cell (FFC) for industrial wastewater treatment. Wastewater is used nowadays as an energy source. This is a promising approach to meet the increasing energy needs in place of fossil fuels [58]. Biocatalysts provide clean, sustainable, and renewable energy sources by utilizing exoelectrogenic organisms [59].

There are several advantages of FFCs; however, their utilization is still limited due to the high cost of their components and their low power output [60]. This limitation can overcome by using an appropriate anode surface morphology (large surface area, superhydrophilicity, high electrical conductivity, excellent chemical stability, high porosity, biocompatibility, and chemistry) with improved electron transfer process [6,61]. The hydrophobic nature of anode, negatively affects the microbial adhesion and enhances the interfacial resistance for e⁻ transfer due to insufficient adhesion on anode surface. This minimize the power and current density [62], which can be overcome with the use of a carbonaceous anode and modification methods (chemical function group treatment, physical treatment, acid heat treatment, and transition metal coating techniques) [63].

Cathode and Cathodic compartment: Types of substrate act as a cathode in FFCs (biodegradable waste such as brewery and sewage wastewater, rich in organics such as glucose, sucrose, lignocellulose, acetate, biomass materials, etc). The reduction of oxygen takes place in this compartment. This is a key interaction in energy conversion and biological respiration [64]. Electrons from the anode are received by cathode through an external circuit and protons are transported via the PEM. This is essential in the reduction reaction between electrons and protons resulting in H_2O formation. The cathode affects the total cell voltage output and has a high redox potential.

Reaction in the cathodic chamber

$$6O_2 + 24H^+ + 24e^- \rightarrow 12H_2O \tag{2}$$

The biocathode is an alternative low-cost, sustainable, stable, and nonchemical option used currently in FFCs. Due to certain biological components in the cathode, the term 'biocathode' is used. The fungus is embedded in an oxygenated cathodic chamber and establishes a mutual configuration of a dual compartment-based yeast fuel cell.

1.8. Reactor Configuration

The reactor configuration influences the performance of biological fuel cells. There are two construction designs: (1) single [65] and (2) dual-chambers [66] to evaluate the in vitro performance. According to the mode of aeration, FFCs are classified into different configurations (Figure 3):

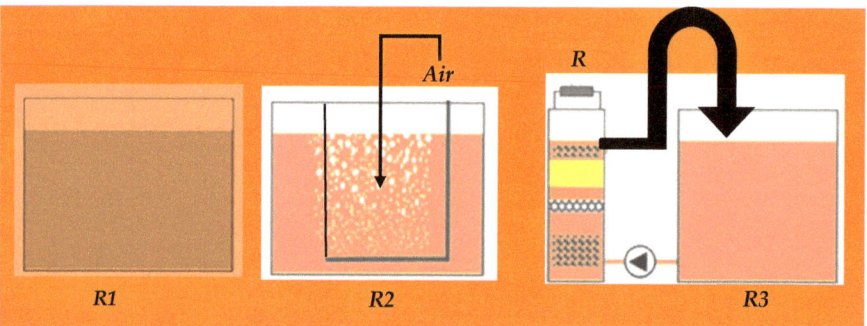

Figure 3. Reactors for the measurement of wastewater parameters; without-air reactor (R1)—the reactor 15 L and the air is interface in wastewater sewage; with-air reactor (R2)—15 L with constantly aerated wastewater; and membrane less FFC (R3)—reduces COD to 90%.

(1) Aqueous cathode: in this cell, water is bubbled with air for the supply of dissolved O_2 to the electrode [67]; (2) air cathode: to minimize the cost and maximize the energy output, the air cathode is designed. Carbon electrodes generate energy in the absence of a PEM. The power density (494 mW/m^2) is much better in this type of cell than in the aqueous cathode [68]; (3) downflow: this membrane less fuel cell is constructed with downflow feeding to generate electricity from wastewater. Water is fed directly onto the cathode, which is horizontally installed in the upper part of the FFC. Oxygen is utilized readily from the air and concentration of dissolved oxygen in the wastewater has little effect on the power generation. The maximum power density of 37.4 mW/m^2 is generated by this type of cell and mostly used in brewery wastewater treatment [69]; (4) upflow: upflow reactors have advantages in retaining the maximum cell density and mass transfer efficiency. In this type of cell, the recirculation rate can improve the upflow rate. At a recirculation rate of 4.8 RV/h, a power density of 356 ± 24 mW/m^2 is produced from this cell [70]; (5) miniature: a low-cost mini tubular fuel cell is developed for the treatment of groundwater contaminated with benzene and for the monitoring of wells. An increase in the length and density and a decrease in size of char particles at the anode effectively reduce the internal resistance. This type of cell removes 95% of benzene and generates a power density of 38 mW/m^2 [71]; (6) stacked: this easy-to-operate FFC in septic tanks comprises a common base and multiple pluggable units, which are connected in series or parallel for electricity generation during waste treatment. Three parallel-connected units produce a power density of 142 ± 6.71 mW/m^2 [72]; (7) large scale: in this cell, multiple operational conditions can be tested (different flow rates, application of external resistors, and poised anodic potentials). This results in the highest COD removal efficiency (94.6 ± 1.0%) at an applied resistance of 10 Ω across each circuit. Results of eight stages of operation (325 days total) indicate that this fuel cell can sustain treatment rates over a long-term period and are robust enough to sustain performance even after system perturbations [73]; (8) tubular: two ceramic stacks, mullite (m-stack) and terracotta (t-stack) are developed to produce energy. Each stack contains 12 identical fuel cells, which are arranged in cascades and tested under different electrical configurations. The m-stack and the t-stack are found to produce a maximum power of 800 µW and 520 µW, respectively [74]; and (9) salt bridge: a salt bridge is used instead of membrane system. The low power output (2.2 mW/m^2) is directly attributed to the higher internal resistance of the salt bridge (19920 ± 50 Ω)

compared to the membrane system (1286 ± 1 Ω). Oxygen diffusion from the cathode to the anode chamber is a factor in power generation [75].

2. Methods for Degradation

2.1. Physical Methods

Physical methods for wastewater treatment remove substances with the use of naturally occurring forces like gravity, electrical attraction, and van der Waal forces. In the mechanism of physical treatment, no change is found in chemical structure of the target substances, while in some cases, the physical state will be changed, as in vaporization, and isolated or scattered substances often caused to agglomerate.

The following methods are easy to use and cost-effective/inexpensive, but have numerous disadvantages.

- **Adsorption:** The method is easy to use and cost-effective, and ensures the regeneration of adsorbents and disposal of generated sludge. Activated lignin and coal are applied as surfaces for adsorption used for degradation [76].
- **Coagulation, flocculation, and sedimentation:** Coagulation, flocculation, and sedimentation techniques are efficient approaches to remove pollutants; however, both tend to be selective toward specific types of contaminants [77].
- **Reverse osmosis and filtration:** These are effective but expensive methods for wastewater treatment. They generate secondary waste during their performances (drawback). Filtration is an integral component of drinking water and wastewater treatment applications, which include ultrafiltration, microfiltration, nanofiltration, and reverse osmosis. These techniques remove the color from wastewater. Each membrane process is best suited for a particular water treatment function [78].

2.2. Chemical Methods

The conversion or removal of contaminants is achieved by the addition of chemicals or chemical reactions. Chemical treatments include precipitation, adsorption, and disinfection. These processes are activated by adding aluminum, calcium, and ferric ions, etc.

The following techniques are promising and effective in the degradation of organic compounds [79].

- Advanced oxidation processes (AOPs): These degrade various organic compounds. This approach generates reactive OH^- radicals for subsequent reactions with organic pollutants resulting in the degradation of pollutants into smaller intermediates. This is a costly process and demands a continuous input of expensive, reactive, and corrosive chemicals with large amounts of energy.
- Electrochemical destruction: Direct electroreduction has lost its popularity as a means of destruction of dyes in an aqueous solution because it offers very poor decontamination of wastewater compared to other electrochemical treatments.
- NaOCl: Wastewater is treated with sodium hypochlorite, allowed to stand for 1 d in the dark, and then neutralized with sodium thiosulfate [80]. The neutralized sample is used for the determination of hypochlorite treatment effects during the wastewater cleanliness.

2.3. Biological Methods

A modern society without the utilization of chemicals in pulp, leather, pharmaceutical, and paper industries is not possible. However, the consumption of chemicals contaminates the environment and causes harmful effects [81]. Chemical and physical methods include electrochemical methods applicable for wastewater decolorization [82]. These methods are quite expensive, have low removal efficiencies, produce toxic intermediates, and exhibit high specificity for dyes [83]. There are environmental friendlier and potentially less expensive methods for the removal of Ops (organic pollutants) from contaminated water. Numerous microorganisms, e.g., bacteria, yeasts, and fungi, have the potential to decolorize different types of organic compounds [84]. The modification of living cells of yeast by polypyrrole (PPy) was evaluated. A microbial fuel cell using yeast modified by a solution containing 0.05 M pyrrole generated maximal power of 47.12 mW/m^2, which is 8.32 mW/m^2 higher than the system, which not based on yeast [85]. The yeast-based FC technology showed great potential to harness energy (bioelectricity and biohydrogen) from xylose. Herein, the yeast strain (*Cystobasidium slooffiae* JSUX1) facilitated the reduction and assembly of graphene oxide (GO) nanosheets reduced to 3D rGO hydrogels on the carbon felt (CF) anode surface. This fuel cell enhanced, by two times, the bioelectricity and biohydrogen production from xylose [86].

The performance of a fuel cell is estimated in terms of pollutant removal and electricity generation. Pollutant removal can measure in terms of organic removal, also called a change in equivalent COD (chemical oxygen demand) between the effluent and the influent. Wastewater pollution from numerous sources is removed by using FFCs, in which organisms decompose organic compounds and convert this chemical energy into electrical energy [2]. The most important advantages of this method are the low concentrations of reagents needed for mild conditions and the degradation of a wide range of substrates. A disadvantage of this method is the high cost of enzymes, which ameliorated by the use of recombinant DNA technology [87].

The biodegradable organic matter ranges from pure compounds (acetate, cysteine, glucose, and ethanol) to mixtures comprising organic compounds (liquid municipal waste, leachate landfills, animal waste, liquid waste of industrial and agricultural origin) [88]. Biological treatment removes these biodegradable organic contaminants. Biological processes degrade dyes contained in wastewater through decolorization carried out by fungal strains [89]. These processes are slow; however, their efficiency is satisfactory. Enzymatic decolorization is now used for the decolorization of dye effluents. There are several problems, e.g., the cost, stability, and product inhibition of enzymes [89]. Anaerobic treatments degrade a wide variety of synthetic dyes [90]. However, successfully aerobic conditions are applied to decolorize the dyes. A few industries use biological treatment to dispose of biodegradable materials, e.g., food processing dyes, dairy wastes, paper, plastics, brewery wastes, and petrochemicals [91,92]. The fungal method is more economical in decolorization via adsorption (living or dead), microbial biomass formation, and bioremediation systems. Fungal organisms degrade and accumulate different pollutants [93].

3. Metabolism of Fungi

Fungi, which are ubiquitous organisms, play a vital role in ecosystems. They can grow on different substrates and function for an indefinite time period. A pivotal parameter determining the cell potential is a metabolic pathway of the microorganism. The performance of FFCs is influenced by fungal metabolism and growth. Fungi have a complex cellular organization and use two pathways for electron transfer. The substrate (glucose) oxidation results in the production of two molecules of NADH/glucose (glycolysis), while mediators interact with the component of ETC (keeps ETC functioning and produces electrons from TCA) [94]. Both metabolic pathways are essential for the removal of waste from the substrates by providing electrons. Any disturbance in these pathways disturb the system, resulting in lower power generation.

The consumption of non-renewable energy creates many problems such as the "availability of fossil fuel stocks for future releases of a huge quantity of toxic gases or particles", which have influence at the global level and stimulate changes in the climate. Fungal biofuel cells (bioelectrochemical system) utilize the living cell for the production of bioelectricity. This cell can drive electricity or other energy generation currents by the use of living cell interaction. Fungal fuel cells and enzymatic biofuel cells can improve sustainable energy production with an efficient conversion system compared to chemical fuels [95].

The diverse group of yeast, molds, and filamentous fungi can remediate various industrial wastewaters. Mycodegradation destroys wood, paper, textile, plastic, and leather materials. The mycelia of several species facilitate degradation. Fungi degrade pesticides, dyes, polychlorinated biphenyls, hydrocarbons, and phenolic and chlorinated compounds with the use of different enzymes (laccases, manganese peroxidase, and lignin peroxidases) [96]. *Irpex lacteus* and *Pleurotus ostreatus* degrade PAH from contaminated industrial soil [97]. Many fungi (*Fusarium oxysporum, Mucor alternans, Tricoderma viride,* and *Phanerochaete chrysosporium*) can degrade DDT.

Numerous white-rot fungi (*Phanerochaete chrysosporium, Pleurotus ostreatus, Trametes versicolor, Irpex lacteus,* and *Lentinula edodes*) can degrade various toxic compounds through their numerous reductive and oxidative mechanisms. Endosulfan is oxidized to endosulfan sulfate through the catalyze-based mechanism of *Tricoderma harzianum*. Fungi are suitable for biotreating oil-based sediments and PAH-contaminated cuttings [98]. *Phanerochaete chrysosporium* fungi oxidize pyrene, benzo[a]pyrene, anthracene, and fluorine into quinines using MnP and LiP [99].

Fungal Chitosan: Fungal chitin is an economically attractive pollutant-adsorbing material next to cellulose [100]. Chitosan or glucosamine is a derivative of chitin found in the cell wall of a few fungi (Mucorales). Reactive Red 2 contaminated water is decolorized by macro fungi [101]. Chitosan beads are used for the treatment of aqueous solutions containing perfluorooctane sulfonate (PFOS) [102]. The biosorption efficiency of fungal biomass can be increased by modification processes. Chitosan poly vermiculite hydrogel adsorbents remove methylene blue from aqueous solutions [103].

4. Role of Fungal Enzymes and Modifications in FFCs

Extracellular ligninolytic oxidative enzymes, e.g., valuable extracellular oxidoreductases (laccases, lignin peroxide, and manganese peroxide) help the fungi to degrade dyes and xenobiotic compounds [104]. Intracellular enzymes are recovered from the cell wall of fungal mycelia [105]. Multiple enzyme systems are successfully used for the efficient breakdown of diverse types of organic pollutants by oxidation or degradation into smaller intermediates.

Oxidoreductase is renowned for its ability to degrade numerous types of organic pollutants [106]. Reactive diffusible redox mediators (RMs) based on oxidoreductase dramatically increase the reaction rate and a broad range of substrates are degraded by these enzymes [107]. The advantages of enzymatic degradation include the low concentration of reagents in mild conditions, with their ability to break down a wide range of substrates. The disadvantage of high cost of enzymes can be improved by using recombinant DNA technology.

Lactate dehydrogenase and ferricyanide reductase are redox enzymes [42]. Fungi exhibit a similar mechanism of electron transfer to that in bacteria. Mediator-less FFCs and fungal electrogenic efficiency are examined by Sayed and Abdelkareem [21], and studies on fungal-mediated electron transport have received attention due to the presence of fungal redox proteins (lactic acid dehydrogenase or ferricyanide reductase).

Wood degraders (white-rot fungi) secrete many extracellular enzymes such as laccase a multi copper oxidative enzyme. Fungal laccase is a 4Cu-containing oxidoreductase biocatalyst that can transfer electrons and has a higher capacity for redox reactions (for organic and aromatic compounds) using an enzyme-mediated system. Laccase catalyzes the oxidation reactions and increases the chances of smooth biological degradation mechanisms. Laccase accepts e^- from atm. O_2 (electron acceptor) catalyzes the $1e^-$ oxidation reaction of phenolic compounds, which facilitates the catabolism of organic compounds [108]. Consequently, at the same time, laccase act as a cathode catalyst in a fungal fuel cell (FFC). White-rot fungi metabolize the laccase to accumulate the nutrients in the soil by lignin degeneration or organic components of the ecosystem. They are suitable and cost-effective for the sustainable development of power generation from FFCs through the in situ elimination of laccase.

Fungi play a dual role in FFCs [109], e.g., at the anode, fungi facilitate e^- transfer via their respiratory proteins or chemical mediators. At the cathode, they reduce the terminal electron acceptors (oxygen). Recent investigations have shown the direct electron transfer through cytochrome C [110]. The in situ laccase of white-rot fungi shown to increase the efficiency of FFCs. This involved the oxidization of aromatic amines and phenolic mixtures using atmospheric oxygen (terminal electron acceptor) [111]. Laccase is extensively manipulated to degrade the organic pollutants, phenolic compounds, triclosan, bisphenol A [112], synthetic and natural hormones such as estrone (E1), 17b-estradiol (E2), estriol (E3), and aromatic dyes [113]. Lignin is degraded as laccase is generated. In the treatment of rubber-processing wastewater, sludge is used at the anode (as a substrate), while laccase is deposited on the cathode under optimal systems to generate electricity. *Ganoderma lucidum* (strain BCRC 36123), *Pleurotus ostreatus*, and *T. versicolor* are well-known laccase-producing fungal species with high efficiency in the production of energy because of an incapacitated layer of fungal enzymes (at the cathode) [114]. Laccase also returns the nutrients by degradation of plant lignin in soil and produces large amount of power in dual-chamber rather than single-chamber FFCs [115]. Laccase used on the biocathode to minimizes the cost of FFC manufacturing.

Laccase-producing *Ganoderma lucidum* BCRC 36123 was planted on the cathode surface of a single-chamber FFC to degrade a dye synergistically with a community of anaerobic microbes in the anode chamber (Figure 4). The laccase activity (1063 ± 26 U/L) of white-rot fungi (*Ceriporiopsis subvermispora, Pycnoporus cinnabarinus, Phanerochaete chrysosporium*, and *Trametes pubescens*) efficiently removed 71–77% of COD and 87–92% of phenolic compounds at a pH of 5.0 [116] (Strong 2010). Fungal laccase hydrolyzed winery wastewater and effectively removed the phenolic compounds, COD, and colors [117].

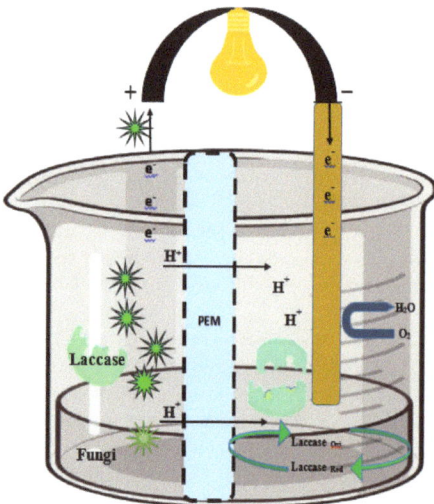

Figure 4. Extracellular fungal laccase catalyst in the anodic chamber, which enzymatically oxidizes wastes during wastewater treatment in the cathodic chamber with reduction of O_2 into water. The movement of e^- from the anode to cathode lit the bulb during the completion of circuit.

Peroxidases are other oxidoreductase enzymes found in fungi. They efficiently oxidize a wide range of substrates [118]. These are effective and extensively used in dye degradation [113]. Only a few studies have reported the use of peroxidases for PPCP degradation [119]. Pollutant-degrading peroxidases are heme peroxidases and non-heme peroxidases. The heme-based peroxidases are further classified into four superfamilies: peroxidase cyclooxygenase, peroxidase–catalase, peroxidase–peroxygenase, and peroxidase–chlorite dismutase. The peroxidase–catalase superfamily is the most abundant in fungi and further subdivided into three additional families, e.g., F1 (intracellular bacterial catalase), F2 (secretory fungal peroxidases such as lignin and manganese), and F3 (secreted plant peroxidases) [120].

4.1. Modification in Yeast-Based Cells

The yeast-based fuel cell (YBFC) is a novel technique used to purify different types of wastewater and convert chemical energy into electrical energy at the cost of active biocatalysts [121]. The performance of YBFCs is influenced by the performance of the electrodes. The electrode performance of yeast-based cells are investigated by coating carbon paper with a thin layer of gold or cobalt (of thickness 5 or 30 nm). The electrode performance is assessed by measuring the electrode half-cell potential efficiency during degradation of chemical substances, pollutants, and organic biomass on the budget of extracellular enzymes [29].

4.2. Factors Affecting FFC Performance

Certain biological (external resistance, substrate type with concentration, and choice of inoculum) and physical factors (reactor configuration, electrode material, and separator) efficiently increase the cell functioning [122], e.g.:

- Using a fungal biocatalyst.
- The type of fuel for the FFC.
- The chemical energy of the substrate (converted into electrical energy).
- The use of mediators (ABTS 2,2′-Azino-bis (3-ethylbenzthiazoline-6-sulfonic acid)) that are effective in e^- transfer from the electrode to laccase.
- An airtight anodic chamber in a dual-chamber.
- Cathode chamber is filled with laccase secreted by white-rot fungi and sufficient nutrient growth medium for the optimum growth of fungal cells.

5. Mechanism of FFCs in Wastewater Treatment

Different types of FCs are operated and reported in wastewater treatment with the generation of biohydrogen, biogas, and other biofuels/energy. Later on, biogas is converted into electric power. Recent developments and researches on designing of fungal fuel cell and its application emphasize the bioenergy generation for future.

Environmental scientists try to devise the methods of degradation or removal of pollutants from the water supply. Water bodies are rich in organic pollutants. These pollutants have diverse chemical structures, e.g., dyes present in textile waste streams are classified into anthraquinone, basic, acidic, and azo dyes [123]. Organic aromatic compounds are carcinogenic in nature and pose serious health risks for humans and aquatic organisms [124]. Organic pollutants are suitable substrates for the growth of bacteria that reduce the oxygen level in water bodies and increase their turbidity with color to decrease the photosynthetic growth of water biota [125]. Human consumption and improper disposal of personal care products contribute to their release into ground and surface water [126]. Efficient degradation depends on the proper binding and orientation of organic compounds in the active sites of peroxidase enzymes. Hence, the substrate structure is a very important parameter in the degradation of pollutants.

Biological approaches are environmentally friendlier, less expensive, and widely used for the handling and removal of harmful substances from wastewater [127,128]. Fungal enzymes (peroxidases and laccases) metabolize and degrade pollutants into less-toxic forms [129]. Diverse fungal enzymes (exocellular and endocellular) in fungus-based FCs are used for waste biodegradation activity. Peroxidases and laccases degrade many types of organic pollutants. Mostly, studies are focused on the treatment of "pure (neat) pollutants" or "simulated wastewater". This presents a major challenge and weakness in the field of enzyme-based remediation of organic pollutants. Only a few studies use enzyme-based systems for real wastewater samples (complex and complicated) rather than neat solutions.

In FFCs, the anaerobic anode and aerobic cathode chambers are separated by a proton exchanger membrane. The anode chamber comprises both respiring and fermenting microbial communities, which enhances the versatility of FCs in the degradation of pollutants [130]. In the fungal fuel cell, electrochemically active fungal species can oxidize different organic compounds present in wastewater in the anode chamber and produce electrons and protons, which are transported toward the cathodic chamber for reduction of O_2 into water (Figure 5). A membrane separates both the anode and cathode compartments. An external resistor set between the anode and the cathode harvests the electricity easily.

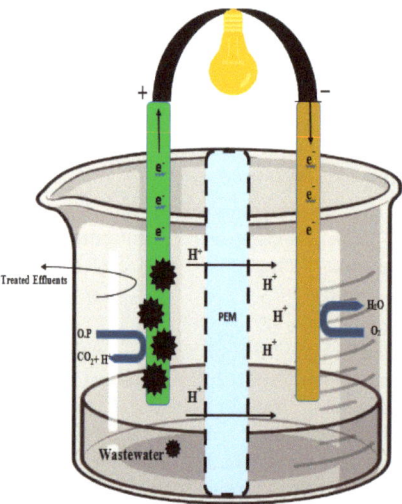

Figure 5. Wastewater with effluents and organic pollutants is placed in anodic regions, where fungal oxidoreductase enzymes are deposited on the electrode. Carbon, electrons, and protons are generated from the effluents by oxidoreductase action and transferred to the cathode where the reduction of O_2 into water takes place. The external resistor between the anode and the cathode harvests the electricity easily.

First, wastewater enters the anode chamber, where fermenting microbes convert the large organic molecules into smaller fermented products (lactate), which are further oxidized by anaerobic (respiring) bacteria to release CO_2, protons, and e^-. When e^- does not find a suitable acceptor in the anolyte, electrons pass to the anode interface and transfer through an external wire to the cathode. Here, they bind O_2 and protons and convert them into water molecules. This is a final step of the circuit [92]. Electrons are released during organic matter degradation and travel toward anodes, eventually being oxidized by electron acceptors (cathode) [131]. The electron donating substrates are used in the anode chamber. A few electron acceptors are characterized by fast kinetics, high redox potential, low cost, and by their importance in environmental sustainability [132]. Oxygen is one of the most convenient electron acceptor utilize commonly in FFCs. The oxygen availability, with its high oxidation potential, produces a clean product, i.e., water [133]. These fuel cells not only generate electricity, also recover and eliminate the compounds from wastewater [134].

The concentration of cations is higher than that of protons, and these cations accumulate in the cathode chamber, resulting an increase in pH of chamber and a decrease in pH of the anodic chamber. This reduces the efficiency of fungal activity and thermodynamics of the fuel cell [135]. Hence, modification of the pore size of membranes can ensure the transfer of only protons and no other cations. Different approaches, e.g., cation and anion exchange membranes (EM), have been suggested to solve the problem of the pH gradient on both sides of the cell membrane [136]. Ionic liquid membranes can improve the fuel cells and ensure the selective transport of only protons and no cations via the membrane. This improvement enhances the efficiency of fungal activity in the anode chamber (not affected) due to absence of transport of cations present in textile wastewater across the membrane. The energy obtained in this process (wastewater treatment) is five times greater than energy consumed during the treatment [137].

The conventional biological process is not sufficient in wastewater treatment due to the inhibition of "biocatalytic activity". The deposition of catalysts on electrodes has gained huge interest, as it accelerate the reaction kinetics (at electrodes). Cathode catalysts enhance the rates of the reduction reaction, while anode catalysts enhance the oxidation rate.

Enzymes secreted by fungi and algae, either individually or in symbiotic association, catalyze the toxic pollutants into a less harmful form. Valuable products are generated in wastewater contaminated by fungal mycelia and fungal interventions such as different enzymes and lipids during the treatment of distillery wastewater to minimize contaminants [138].

COD removal takes place at the cathode and anode, which is a major concern in the treatment of wastewater. A fungal-based study on the treatment of water and sewage wastewater for energy generation was conducted by Fernandez de Dios et al. [139]. The biodegradation involved efficient species remediating harmful toxicants for energy production. Agro-industries release organic pollutants dissolved in wastewater [115], as do fermentation, pulp, and paper industries. Many countries have implemented stringent regulations and norms for the disposal of agro-based industrial wastewater. Such norms are quite challenging, as they insist on the steady adoption of cost-effective and innovative technologies.

The catalytic platinum and gold increase the price of electrodes. To reduce electricity production and wastewater treatment costs, catalysts should not comprise valuable metals [140]. Therefore, fungi play a role as anode catalysts to adjust the appropriate cathode catalysts.

5.1. Role of the Cathode in Wastewater Treatment

The cathode must be good, robust, mechanically strong, and highly conductive with the best catalytic properties [55]. Cathode materials require high power generation with Coulombic efficiency. Therefore, the main challenge to minimize the expenses is the target aspect of fuel cell technology. Such electrode is successfully applied for the treatment of wastewater [141]. Cathodes are abiotic and biocathodic. Biotic types can be aerobic or anaerobic [142]. Mediators/or catalysts are required in abiotic cathodes for oxygen reduction reactions. Maximum expense and poor stability causes the catalyst poisoning (Pt-based), which vanquish by engaging the biotic cathodes (where microorganisms assist in the cathodic reactions) [143]. Biocathodes have an advantage over abiotic cathodes, a few interesting benefits are cost reduction and the omission of costly mediators or expensive metal catalysts (platinum) [57].

Role of the Anode in Wastewater Treatment

The selection of an anode is also critical in the performance of FFCs. It plays an important role in e^- transfer from microbes to the electron acceptor. Conventional cells comprise bioanodes and abiotic cathodes.

5.2. Integrated Treatment Processes

Water for dilution is required in bioelectrochemical processes to reduce the organic matter in fungal pretreated effluents, which is a major disadvantage of FCs. Effluent dilution is possible by using two repeated fungal cocultivation processes followed by treatment in FFCs. Hence, ecofriendly treatment processes with lesser environmental and economical footprints should focus on the achievement of highest treatment efficiency. Different enzymes and other derivatives of fermentation broth make the fungal fermentation technology profitable for industrial implementation. Catalytic thermolysis is an effective process for the destruction of color and organic compounds in complex wastewater [144].

Coupling processes can improve the efficiency of distillery wastewater treatment with improved COD and color removal. Single-stage anaerobic digestion is insufficient for the higher removal of organic matter. Therefore, combined single- or multistage aerobic and chemical oxidation treatment is required [145]. Sequential anaerobic–aerobic treatment of fungal fuel cell (FFC) is a better modification [146]. The treatment of malt whiskey distillery wastewater removed 52% of COD and BOD in a UASB reactor (first stage), 70% COD and BOD in batch aerobic degradation (second stage), and >99% of COD and BOD, when combined with thermophilic fungi (third stage) [147].

Trametes pubescens pretreatment followed by anaerobic digestion was helpful in the removal of COD (53.3%) and polyphenolic compounds (72.5%) from wine distillery wastewater. Similarly, winery effluent treated with fungi supplemented with C and nutrient media followed by anaerobic digestion for 2 days; this procedure reduced the COD level (99.5%) [148]. A study consisting of a combination of fungal pretreatment, submerged MBR, and the secondary digestion of wine distillery wastewater showed 86% decrease in COD concentration [149]. Fungal biodegradation carried out with exo-enzymes released by *Aspergillus* sp. combined with grain-based distillery stillage removed 94% of total COD (first time), while higher efficiency (99%) was achieved by support of fungal degradation processes (second time) [105].

5.3. Role of Yeast Cells in Wastewater Treatment

Fungal degraders are widely used in the treatment of wastewater due to their consumption of substrates. The yeast wastewater is a dual-stage (anaerobic) biological wastewater [150]. Yeast cells used in FFCs represent two types: oxidizing and fermenting [151]. Fermenting yeasts utilize 6C sugar and ferment into CO_2 and alcohol. Yeast cells metabolize inorganic compounds and hazardous materials. The yeast cell wall is very thick and cell membrane can separate from the cell wall, which is difficult to obtain. Therefore, yeasts are processed in a transplasma membrane ET system also called "plasma membrane (PM) oxidoreductase". The PM–NADH–oxidoreductase system located throughout the membrane is also involved in transportation of e^- using NADH and NADPH to the external electron acceptor/anode. Yeast fuel cells play an active role in e^- transfer during ATP synthesis along with NAD^+ into NADH reduction. The oxidation of glucose is the main source of energy production by yeast cell factories [33].

Yeasts efficiently treat wastewater. Yeasts and fungi shown decolorization of various organic compounds efficiently [152]. Yeasts are used to remove dyes and heavy metals from wastewater [153]. Yeast cells are well-known producers of different lipids, enzymes, and glycolipids suitable for wastewater treatment, as wastewater contains maximum concentrations of heavy metals, ion, organic matter, and domestic sewages also. The yeast consortium quickly degrades high concentrations of organic matter [154,155]. Biodegradable organic wastewater, ferricyanide potassium, acid orange, and acetate are also used as anolytes or anolyte feed.

Algae, in combination with fungi, are effective in the generation of electricity in MFCs (microbial fuel cells) using molasses substrate. The transport of protons from yeasts to algae proceeds with the help of microporous tubes rooted in activated bleaching earth, called an ion exchange medium, for the production of electricity. A strain of *Galactomyces reessii* reported to utilize synthetic wastewater and rubber industry sludge for the generation of electricity [156]. The unique ability of yeast species to transfer electrons extracellularly helps to advance the fuel cell technologies and to purify wastewater as a biocatalyst. Natural and genetically modified yeasts with improved enzyme action are useful for the degradation of toxic substances and exhibit the capacity for the generation of electricity. *Candida melibiosica* has shown the high phytase activity to remediate phytates and a maximum potential for electricity generation. This opens up novel avenues for nonchemical and more sustainable ecofriendly approaches in the purification of phosphate-polluted wastewater as well as many other xenobiotic contaminants [157].

A double-chambered fuel cell has generated electricity due to the ability of *Pichia fermentans* to utilize wheat straw hydrolyzate. This hydrolyzate was prepared by degrading the wheat straw in the presence of *Phlebia brevispora, Phanerochaete chrysospo-*

rium, and *Phlebia floridensis* (white-rot fungi). The maximum electrochemical response of 20.13 ± 0.052 mWm^{-2} and 20.42 ± 0.071 mWm^{-2} was recorded in hydrolysate using *P. floridensis* and *P. chrysosporium*, respectively [158].

The electricity was produced from yeast wastewater in membraneless (ML) fuel cells with a Cu–Ag cathode and a power of 6.38 mW and a cell voltage of 1.09 V were obtained. This research proved a feasible way to obtain the maximum bioelectricity from fuel cells (fed with yeast wastewater) [159].

5.4. Pharmaceutical Wastewater Treatment

Pharmaceutical industries release pharmaceutical products in water on daily basis [160], the treatment of which is a challenging task. Wastewater treatment plants are the point sources of pharmaceutically active compounds; however, these are not designed for the elimination of active compounds [160]. Long-term exposure to compounds in the environment may result in chronic and acute damage, behavioral changes, reproductive impairment, inhibition of cell proliferation, and accumulation in tissues [161]. Unchanged antibiotics from pharmaceutical wastewater reaching the sewage system may result in the emergence of antibiotic resistance in fungi and aquatic organisms. They may also change the microbial community structure. Different treatment techniques for pharmaceutical wastewater require a large area, costly chemicals, treatment methods, and high energy input. In turn, the low biodegradability [162] and generation of toxic byproducts have stimulated scientists to construct FFCs as an efficient alternative not requiring costly chemicals and capable of generating energy without toxic byproducts.

5.5. Heavy Metal-Loaded Wastewater

Water, which faces various types of pollution and degradation, is a precious commodity. It only seems inexhaustible, whereas the whole world in the short or long term will face the problem of its scarcity. This urge makes wastewater one of the most valuable resource for energy and water. Conventional remediation treatments are neither environmentally friendly nor economical. A new process is required to overcome these issues, where the conservation and recovery of energy will become possible. Fungal fuel cells are emerging as a promising technique for the mitigation of pollution.

FFCs provide a strategy to treat wastewater and remove or recover heavy metals. These bioelectrochemical systems utilize fungal catalytic activity in the form of biofilms to oxidize inorganic compounds with the production of electric current. It is a sustainable bioremediation method for energy production (Figure 6).

5.6. Agro–Industrial Wastewater Management

Agro-industries demand large volume of water during manufacturing processes and generate wastewater. This agricultural wastewater has a high content of N, P, and organic compounds can also be treated using FFC technology [163]. The wastewater is loaded with organic matter, proteins, oils, grease, and sugars, which increase the COD of ground and surface water. This causes severe environmental pollution, because such wastewater not biodegradable easily due to the addition of bacterial end products and high chemical stability. Industrial wastewater is especially useful in FFCs because of its constant composition.

5.7. Biodegradation of Distillery Wastewater

The bioremediation of distillery wastewater using biological and natural pathways leads to the degradation of organic pollutants. Via aerobic degradation, the white-rot fungus *Phanerochaete chrysosporium* can remove organic matter (COD and BOD) effectively [164]. Fungal biodegradation is an inexpensive technique where wastewater dilution is not required. A pseudo-second-order rate is used for the kinetic model of fungal biodegradation of distillery wastewater [165]. *Aspergillus awamori* removed 39.3% of COD from grain-based distillation [166]. A mixed consortium of six fungal species, i.e., *Penicillium pinophilum*,

Aspergillus flavus, *Alternaria gaisen*, *Fusarium verticillioides*, *Pleurotus florida*, and *Aspergillus niger* are used to remove 65.4% of COD in distillery wastewater treatment [167].

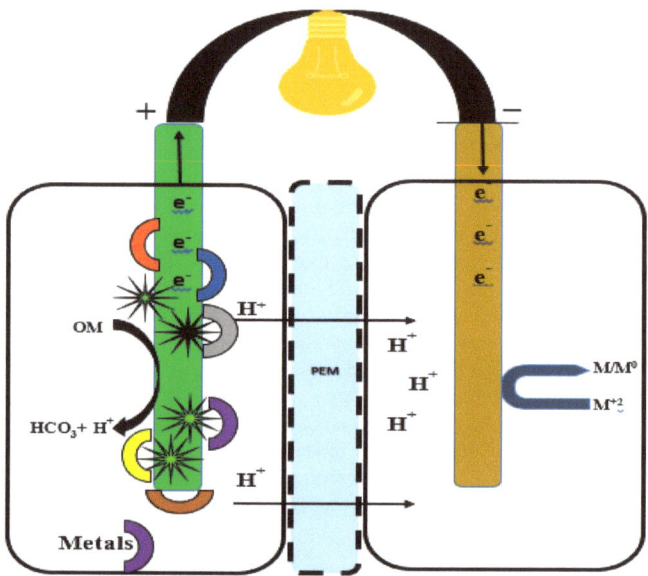

Figure 6. A double-chambered fungal fuel cell for the treatment of heavy metals. Heavy metal-contaminated water is placed in the anodic cell (aerobic compartment), where fungi enzymatically release carbon, e^- and H^+ transfer via PEM and external wire to the cathodic cell (anaerobic compartment), where the reduction of the metal from M^{+2} to M/M^0 takes places.

Batch aerobic treatment with the use of yeast and fungal species is very effective in the remediation of distillery wastewater. This treatment remove >85% of COD in a single phase, while a minimum cultivation time of 7–10 days is required for a higher removal of organics [164].

Distillery fermentation produces a large volume of wastewater called stillage. This wastewater has a lower pH, unconvertible organic fractions, and a maximum % of dissolved inorganic matter. Dark brown molasses stillage comprises higher chemical oxygen demand, biochemical oxygen demand, and inorganic impurities. Grain stillage is characterized by a lower chemical oxygen demand, acidic pH (3.4–4.1), and exhibits significant potential for pollution upon discharge. Physical, chemical, and biological integrated treatments of distillery stillage are discussed in detail with current challenges. Distillery stillage is used as a substrate in fungal fuel cells, encouraging observations in terms of achieving high treatment efficiency to meet the discharge standards, bioelectricity generation, and value-added products recovery.

Xylanase (17.3 U/mL) production from the phyllosphere yeast (*Pseudozyma antarctica*) removed 63% of DOC (dissolved organic compound) from wastewater by utilizing bioethanol distillery wastewater (substrate) [168]. This is an expensive source for the production of biomaterials. Oleaginous yeasts (*Rhodosporidium toruloides* and *Chlorella pyrenoidosa*) successfully degraded distillery wastewater effluent with a 43.65 ± 1.74% lipid content and a 3.54 ± 0.04 g/L lipid yield associated with an 86.11 ± 0.41% COD reduction [169]. Maximum mycelial biomass of *Calvatia gigantea* (2.75 g/100 mL/4.5 days) was obtained in optimized conditions when the fungus was cultivated on raw distillery wastewater [170]. *Candida utilis* biomass utilized *shochu* wastewater and removed 62.9% of DOC [171].

5.8. Degradation of Ethanol Distillery Wastewater

Pollutants from ethanol production vary significantly in wastewater. Winery wastewater comprises low concentration of organic matters and high amount of polyphenolics and nutrients [172]. The fermentation of beet molasses and sugarcane releases aldehydes and ketones, which impart flavor. The wastewater from this process was characterized by a high level of biodegradable organic matter [173], maximum nutrient contents (sulfate, chloride, calcium, magnesium, and potassium), and a low pH [174]. This type of wastewater constitutes severe water and land pollution.

COD and phenol is removed from winery wastewater using yeast-mediated fuel cells. In this study, electrochemical properties are monitored. The results indicated the laccase activity of yeast strain ET-KK. A maximal current and power density of 139.17 ± 1.44 mA/m^2 and 38.74 ± 0.80 mW/m^2 is generated. The COD and phenol removal from winery wastewater was $79.14 \pm 0.92\%$ and $85.04 \pm 0.07\%$, respectively [175].

5.9. Degradation of Dye Wastewater

Trametes versicolor is a well-known white-rot fungus used for the treatment of dye wastewater [83]. Fungal culture adjusts its metabolism through its modification in environmental conditions. Intra- and extracellular enzymes are helpful in fungal metabolic activities and the degradation of different dyes from textile wastewater. The best examples are lignin, laccase, peroxidase, and manganese [176] from white-rot fungi (*Coriolopsis* sp., *Pleurotus eryngii*, and *Penicillium simplicissimum*) [177].

In bioanode–biocathode FFCs, the dye wastewater acts as anolyte and catholyte. The maximum COD removal and decolorization are observed in the first 12 h, because a higher substrate consumption at cathode and anode occurs during the initial hours. The maximal cell potential are recorded to be 706 mV in a fungal FC, with power densities of 276.9 mWm^{-2}. The reactor was also tested for the biodegradation of RR 195 dye from wastewater along with bioelectricity production. The overall COD and color removal efficiency was 72%, and 95% [178].

Fungal fuel cells are an emerging technique that effectively treats wastewater with simultaneous electricity production. In a study on the decolorization and degradation of azo dyes (remazol brilliant blue, mordant blue 9, acid red1, and orange G), wheat straw hydrolyzate substrate are used in FFC. The hydrolysate was prepared by the degradation of wheat straw by *P. floridensis, P. chrysosporium,* and *P. brevispora,* while *Pichia fermentans* (yeast) used as a biocatalyst. Dye decolorization was carried out in a fungus yeast-mediated single-chambered FC. The maximum power density was recorded (34.99 mW m^{-2}) on the 21st day. The best response to dye decolorization was observed in MB9 (96%) with *P. floridensis* followed by RBB (90–95%), AR1 (38%), and OG (76%) [179].

Mechanism: FCs transform azo dyes into less colorful compounds but fail to degrade and mineralize them completely. The decolorization of wastewater containing azo dyes and other types of dye takes place through the movement of anode and cathode electrons via external circuit. Azo dyes in the catholyte act as e$^-$ acceptors and decolorize dyes via reductive cathode reactions. The dye-reducing reactions progress better in anaerobic conditions, as O$_2$ competes for e$^-$ from the cathode, and reaction rate heavily depends on the pH of the catholyte [90].

Two-in-one systems: ***First stage*** (reductive transformation): Azo dye wastewater is placed as an anolyte (anode chamber), where dye molecules are reduced by bacteria (anaerobic) to form aromatic amines. ***Second Stage:*** This water is then shifted to an aerobic bioreactor, where it is further degraded into smaller compounds [180]. This two-in-one system is very effective; however, it is structurally complex and expensive to operate and construct. The proton exchanger membrane prevents the movement of pollutants and keeps transformed products in their respective chambers.

White-rot fungi are the only group of organisms that completely degrade azo dyes. Laccase produce, when fungi degrade lignin (a highly heterogeneous aromatic polymer) that is abundant in the natural habitats of white-rot fungi [181]. The laccase catalyst has replaced the noble metals in catalyzing the reduction of O_2 [182]. Single-chamber FCs have no cathode chamber and their cathodes are directly exposed to air for maximum oxygen availability [183]. A feasible test in growing a laccase-producing white-rot fungus on the cathode surface of a single-chamber FC is the possibility of proton exchanger membrane, which, when replaced by a layer of polyvinyl alcohol-hydrogel, allows the pollutant to diffuse from the anode to cathode chamber [184].

5.10. Applications of FFCs

- The management of the environment
- The generation of bioelectricity

6. Future Prospective

In future, fungal fuel cell technique may become the part of energy system at the expense of waste products for electricity generation from wastewater, where organic contents are usually in a much smaller range of 260–450 mWm^{-2} [185]. The absolute goal in near future is to design a cell that capable of producing the highest amount of energy on the expenditure of lowest possible energy input.

7. Conclusions

Conventional methods can treat loaded wastewater, while biological methods such as FFCs represent a sustainable technique in lowering the wastewater impact on the environment. This remediation system is profitable, ecofriendly, and cheaper than other conventional methods. FFCs offer an enduring solution for the degradation of wastewater contaminants into non/less-toxic forms with simultaneous generation of electricity by using fungal oxidoreductase. The fungal enzymes clean up the contaminated wastewater with maintaining of their regular metabolism. The "Industrial Revolution" forces the valorization of chemical wastes. Fungal cells are suitable candidates for the valorization of wastes in water and the simultaneous production of electrical energy.

Author Contributions: Conceptualization and methodology, A.U.; software, M.G.; validation, A.U. and M.G.; formal analysis, A.U.; investigation, A.U.; resources, A.U. and M.G.; data curation, A.U.; writing—original draft preparation, A.U. and M.G.; writing—review and editing, A.U., Ł.S. and M.G.; visualization, A.U.; supervision, M.G.; project administration, A.U.; funding acquisition, A.U. and M.G. All authors have read and agreed to the published version of the manuscript.

Funding: This research received no external funding.

Data Availability Statement: The datasets used and/or analyzed during the current study are available from the corresponding author on reasonable request.

Conflicts of Interest: The authors declare no conflict of interest.

References

1. Lvovich, M.I. *Water Resources and Their Future*; Litho Crafters Inc.: Eastpointe, MI, USA, 1979.
2. Adeel, S.; Abrar, S.; Kiran, S.; Farooq, T.; Gulzar, T.; Jamal, M. *Sustainable Application of Natural Dyes in Cosmetic Industry. Handbook of Renewable Materials for Coloration and Finishing*; John Wiley & Sons, Inc.: Hoboken, NJ, USA, 2018; pp. 189–211.
3. Hanafi, M.F.; Sapawe, N. A review on the current techniques and technologies of organic pollutants removal from water/wastewater. *Mater. Today Proc.* **2022**, *31*, A158–A165. [CrossRef]
4. Narita, J.; Okano, K.; Tateno, T.; Tanino, T.; Sewaki, T.; Sung, M.H.; Fukuda, H.; Kondo, A. Display of active enzymes on the cell surface of *Escherichia coli* using PgsA anchor protein and their application to bioconversion. *Appl. Microbiol. Biotechnol.* **2006**, *70*, 564–572. [CrossRef]
5. Lovley, D.R. Microbial fuel cells: Novel microbial physiologies and engineering approaches. *Curr. Opin. Biotechnol.* **2006**, *17*, 327–332. [CrossRef]
6. Franks, A.E.; Nevin, K.P. Microbial fuel cells. A currentreview. *Energies* **2010**, *3*, 899–919. [CrossRef]
7. Rabaey, K.; Verstraete, W. Microbial fuel cells: Novel biotechnology for energy generation. *Trends Biotechnol.* **2005**, *23*, 291–298. [CrossRef] [PubMed]
8. Mitra, P.; Hill, G.A. Continuous microbial fuel cell using a photoautotrophic cathode and a fermentative anode. *Can. J. Chem. Eng.* **2012**, *90*, 1006–1010. [CrossRef]

9. Flimban, S.G.A.; Ismail, I.M.I.; Kim, T.; Oh, S.-E. Overview of recent advancements in the microbial fuel cell from fundamentals to applications: Design, major elements, and scalability. *Energies* **2019**, *12*, 3390. [CrossRef]
10. Wang, Y.; Liu, P.; Zhang, G.; Yang, Q.; Lu, J.; Xia, T.; Peng, L.; Wang, Y. Cascading of engineered bioenergy plants and fungi sustainable for low-cost bioethanol and high-value biomaterials under green-like biomass processing. *Renew. Sustain. Energy Rev.* **2021**, *137*, 110586. [CrossRef]
11. Vernet, G.; Hobisch, M.; Kara, S. Process intensification in oxidative biocatalysis. *Curr. Opin. Green Sustain. Chem.* **2022**, *38*, 100692. [CrossRef]
12. Ratledge, C. Fatty acid biosynthesis in microorganisms being used for single cell oil production. *Biochimie* **2004**, *86*, 807–815. [CrossRef]
13. Kurosawa, K.; Wewetze, S.J.; Sinskey, A.J. Engineering xylose metabolism in triacylglycerol producing *Rhodococcus opacus* for lignocellulosic fuel production. *Biotechnol. Biofuels* **2013**, *6*, 134–147. [CrossRef] [PubMed]
14. Dhanasekaran, D.; Sundaresan, M.; Suresh, A.; Thajuddin, N.; Thangaraj, R.; Vinothini, G. Oleaginous microorganisms for biofuel development. *Environ. Sci. Eng.* **2017**, *12*, 243–263.
15. Fakas, S.; Papanikolaou, S.; Batsos, A.; Galiotou-Panayotou, M.; Mallouchos, A.; Aggelis, G. Evaluating renewable carbon sources as substrates for single cell oil production by *Cunninghamella echinulata* and *Mortierella isabellina*. *Bio. Bioenergy* **2009**, *33*, 573–580. [CrossRef]
16. Sundaramahalingam, M.A.; Sivashanmugam, P. Concomitant strategy of wastewater treatment and biodiesel production using innate yeast cell (*Rhodotorula mucilaginosa*) from food industry sewerage and its energy system analysis. *Renew Energ.* **2023**, *208*, 52–62. [CrossRef]
17. Rojas-Flores, S.; Nazario-Naveda, R.; Benites, S.M.; Gallozzo-Cardenas, M.; Delfín-Narciso, D.; Díaz, F. Use of Pineapple Waste as Fuel in Microbial Fuel Cell for the Generation of Bioelectricity. *Molecules* **2022**, *27*, 7389. [CrossRef]
18. Beopoulos, A.; Nicaud, J.M. Yeast: A new oil producer? *Ol. Corps Gras. Lipides* **2012**, *19*, 22–28. [CrossRef]
19. Hofrichter, M. Review: Lignin conversion by manganese peroxidase (MnP). *Enzym. Microb. Technol. Recent Adv. Lignin Biodegrad.* **2002**, *30*, 454–466. [CrossRef]
20. Sekrecka-Belniak, A.; Toczyłowska-Mamińska, R. Fungi-based microbial fuel cells. *Energies* **2018**, *19*, 2827. [CrossRef]
21. Sayed, T.; Abdelkareem, M.A. Yeast as a Biocatalyst in Microbial Fuel Cell. *Old Yeasts-New Quest.* **2017**, *317*, 41–65.
22. Hallsworth, J.E.; Magan, N. Culture age, temperature, and pH affect the polyol and trehalose contents of fungal propagules. *Appl. Environ. Microbiol.* **1996**, *62*, 2435–2442. [CrossRef]
23. Deska, M.; Kończak, B. Immobilized fungal laccase as" green catalyst" for the decolourization process–State of the art. *Process Biochem.* **2019**, *84*, 112–123. [CrossRef]
24. Huang, J.; Lu, C.; Qian, X.; Huang, Y.; Zheng, Z.; Shen, Y. Effect of salinity on the growth, biological activity and secondary metabolites of some marine fungi. *Acta Oceanol. Sin.* **2011**, *30*, 118. [CrossRef]
25. Cantrell, S.A.; Casillas-Martínez, L.; Molina, M. Characterization of fungi from hypersaline environments of solar salterns using morphological and molecular techniques. *Mycol. Res.* **2006**, *110*, 962–970. [CrossRef] [PubMed]
26. Hasegawa, P.M.; Bressan, R.A.; Zhu, J.K.; Bohnert, H.J. Plant cellular and molecular responses to high salinity. *Annu. Rev. Plant Physiol. Plant Mol. Biol.* **2000**, *51*, 463–499. [CrossRef] [PubMed]
27. Mansour, M.M.F.; Salama, K.H. Cellular basis of salinity tolerance in plants. *Environ. Exp. Bot.* **2004**, *52*, 113–122. [CrossRef]
28. Patil, S.A.; Hagerhall, C.; Gorton, L. Electron transfer mechanisms between microorganisms and electrodes in bio electrochemical systems. *Bioanal. Rev.* **2012**, *4*, 159–192. [CrossRef]
29. Bugg, T.D.H.; Ahmad, M.; Hardiman, E.M.; Rahmanpour, R. Pathways for degradation of lignin in bacteria and fungi. *Nat. Prod. Rep.* **2011**, *28*, 1883–1896. [CrossRef] [PubMed]
30. Christwardana, M.; Frattini, D.; Accardo, G.; Yoon, S.P.; Kwon, Y. Effects of methylene blue and methyl red mediators on performance of yeast based microbial fuel cells adopting polyethylenimine coated carbon felt as anode. *J. Power Sources* **2018**, *396*, 1–11. [CrossRef]
31. Mao, L.; Verwoerd, W.S. Selection of organisms for systems biology study of microbial electricity generation: A review. *Int. J. Energy Environ. Eng.* **2013**, *4*, 17. [CrossRef]
32. Ganguli, R.; Dunn, B.S. Kinetics of anode reactions for a yeast-catalysed microbial fuel cell. *Fuel Cells* **2009**, *9*, 44–52. [CrossRef]
33. Gunawardena, A.; Fernando, S.; To, F. Performance of a yeast-mediated biological fuel cell. *Int. J. Mol. Sci.* **2008**, *9*, 1893–1907. [CrossRef] [PubMed]
34. Raghavulu, S.V.; Goud, R.K.; Sarma, P.N.; Mohan, S.V. *Saccharomyces cerevisiae* as anodic biocatalyst for power generation in biofuel cell: Influence of redox condition and substrate load. *Bioresour. Technol.* **2011**, *102*, 2751–2757. [CrossRef] [PubMed]
35. Schaetzle, O.; Barriere, F.; Baronian, K. Bacteria and yeasts as catalysts in microbial fuel cells: Electron transfer from microorganisms to electrodes for green electricity. *Energy Environ. Sci.* **2008**, *1*, 607–620. [CrossRef]
36. Hubenova, Y.; Mitov, M. Extracellular electron transfer in yeast-based biofuel cells: A review. *Bioelectrochemistry* **2015**, *106*, 177–185. [CrossRef] [PubMed]
37. Haslett, N.D.; Rawson, F.J.; Barriere, F.; Kunze, G.; Pasco, N.; Gooneratne, R.; Baronian, K.H.R. Characterisation of yeast microbial fuel cell with the yeast *Arxula adeninivorans* as the biocatalyst. *Biosens. Bioelectron.* **2011**, *26*, 3742–3747. [CrossRef]
38. He, L.; Du, P.; Chen, Y.; Lu, H.; Cheng, X.; Chang, B.; Wang, Z. Advances in microbial fuel cells for wastewater treatment. *Renew. Sustain. Energy Rev.* **2017**, *71*, 388–403. [CrossRef]

39. Hanaki, K.; Portugal-Pereira, J. The Effect of Biofuel Production on Greenhouse Gas Emission Reductions. In *Biofuels and Sustainability. Science for Sustainable Societies*; Takeuchi, K., Shiroyama, H., Saito, O., Matsuura, M., Eds.; Springer: Tokyo, Japan, 2018; pp. 53–71. [CrossRef]
40. Robak, K.; Balcerek, M. Review of second generation bioethanol production from residual biomass. *Food Technol. Biotech.* **2018**, *56*, 174. [CrossRef]
41. Shkil, H.; Schulte, A.; Guschin, D.A.; Schuhmann, W. Electron transfer between genetically modified hansenula polymorpha yeast cells and electrode surfaces via complex modified redox polymers. *Chem. Phys. Chem.* **2011**, *12*, 806–813. [CrossRef]
42. Prasad, D.; Arun, S.; Murugesan, M.; Padmanaban, S.; Satyanarayanan, R.S.; Berchmans, S.; Yegnaraman, V. Direct electron transfer with yeast cells and construction of a mediatorless microbial fuel cell. *Biosens. Bioelectron.* **2007**, *22*, 2604–2610. [CrossRef]
43. Gal, I.; Schlesinger, O.; Amir, L.; Alfonta, L. Yeast surface display of dehydrogenases in microbial fuel-cells. *Bioelectrochemistry.* **2016**, *112*, 53–60. [CrossRef]
44. Mathuriya, A.S.; Yakhmi, J.V. Microbial fuel cells to recover heavy metals. *Environ. Chem. Lett.* **2014**, *12*, 483–494. [CrossRef]
45. Miskan, M.; Ismail, M.; Ghasemi, M.; Md Jahim, J.; Nordin, D.; Abu Bakar, M.H. Characterization of membrane biofouling and its effect on the performance of microbial fuel cell. *Int. J. Hydrog. Energy* **2016**, *41*, 543–552. [CrossRef]
46. Darvishi, Y.; Hassan-Beygi, S.R.; Zarafshan, P.; Hooshyari, K.; Malaga-Toboła, U.; Gancarz, M. Numerical Modeling and Evaluation of PEM Used for Fuel Cell Vehicles. *Materials* **2021**, *14*, 7907. [CrossRef]
47. Woo, S.; Lee, S.; Taning, A.Z.; Yang, T.H.; Park, S.H.; Yim, S.D. Current understanding of catalyst/ionomer interfacial structure and phenomena affecting the oxygen reduction reaction in cathode catalyst layers of proton exchange membrane fuel cells. *Curr. Opin. Electrochem.* **2020**, *21*, 289–296. [CrossRef]
48. Potter, M.C. Electrical effects accompanying the decomposition of organic compounds. *Proc. R. Soc. B Biol. Sci.* **1911**, *84*, 260–276. [CrossRef]
49. Kumar, R.; Singh, L.; Zularisam, A.W. Exoelectrogens: Recent advances in molecular drivers involved in extracellular electron transfer and strategies used to improve it for microbial fuel cell applications. *Renew. Sustain. Energy Rev.* **2016**, *56*, 1322–1336. [CrossRef]
50. Sayed, E.T.; Tsujiguchi, T.; Nakagawa, N. Catalytic activity of baker's yeast in a mediatorless microbial fuel cell. *Bioelectrochemistry* **2012**, *86*, 97–101. [CrossRef]
51. Slate, A.J.; Whitehead, K.A.; Brownson, D.A.C.; Banks, C.E. Microbial fuel cells: An overview of current technology. *Renew. Sustain. Energy Rev.* **2019**, *101*, 60–81. [CrossRef]
52. Ho, C.H.; Yi, J.; Wang, X. Biocatalytic continuous manufacturing of diabetes drug: Plantwide process modeling, optimization, and environmental and economic analysis. *ACS Sustain. Chem. Engineer.* **2018**, *7*, 1038–1051. [CrossRef]
53. Kim, I.S.; Chae, K.J.; Choi, M.J.; Verstraete, W.; Kim, I.S.; Chae, K.J.; Choi, M.J.; Verstraete, W. Microbial fuel cells: Recent advances, bacterial communities and application beyond electricity generation. *Environ. Eng. Res.* **2008**, *13*, 51–65. [CrossRef]
54. Nimje, V.R.; Chen, C.Y.; Chen, H.R.; Chen, C.C.; Huang, Y.M.; Tseng, M.J.; Cheng, K.C.; Chang, Y.F. Comparative bioelectricity production from various wastewaters in microbial fuel cells using mixed cultures and a pure strain of *Shewanella oneidensis*. *Bioresour. Technol.* **2012**, *104*, 315–323. [CrossRef] [PubMed]
55. Mustakeem, M. Electrode materials for microbial fuel cells: Nanomaterial approach. *Mater. Renew. Sustain. Energy* **2015**, *4*, 1–11. [CrossRef]
56. Richter, H.; McCarthy, K.; Nevin, K.P.; Johnson, J.P.; Rotello, V.M.; Lovley, D.R. Electricity generation by geobacter sulfurreducens attached to gold electrodes. *Langmuir* **2008**, *24*, 4376–4379. [CrossRef]
57. Watanabe, K. Recent developments in microbial fuel cell technologies for sustainable bioenergy. *J. Biosci. Bioeng.* **2008**, *106*, 528–536. [CrossRef]
58. Wu, D.; Yi, X.; Tang, R.; Feng, C.; Wei, C. Single microbial fuel cell reactor for coking wastewater treatment: Simultaneous carbon and nitrogen removal with zero alkaline consumption. *Sci. Total Environ.* **2018**, *621*, 497–506. [CrossRef]
59. Chae, K.J.; Choi, M.J.; Lee, J.W.; Kim, K.Y.; Kim, I.S. Effect of different substrates on the performance, bacterial diversity, and bacterial viability in microbial fuel cells. *Bioresour. Technol.* **2009**, *100*, 3518–3525. [CrossRef]
60. Zhang, P.; Liu, J.; Qu, Y.; Feng, Y. Enhanced *Shewanella oneidensis* MR-1 anode performance by adding fumarate in microbial fuel cell. *Chem. Engineer. J.* **2017**, *328*, 697–702. [CrossRef]
61. Santoro, C.; Babanova, S.; Artyushkova, K.; Cornejo, J.A.; Ista, L.; Bretschger, O.; Marsili, E.; Atanassov, P.; Schuler, A.J. Influence of anode surface chemistry on microbial fuel cell operation. *Bioelectrochemistry* **2015**, *106*, 141–149. [CrossRef]
62. Xu, H.; Quan, X.; Xiao, Z.; Chen, L. Effect of anodes decoration with metal and metal oxides nanoparticles on pharmaceutically active compounds removal and power generation in microbial fuel cells. *Chem. Engineer. J.* **2018**, *335*, 539–547. [CrossRef]
63. Mohanakrishna, G.; Abu-Reesh, I.M.; Kondaveeti, S.; Al-Raoush, R.I.; He, Z. Enhanced treatment of petroleum refinery wastewater by short-term applied voltage in single chamber microbial fuel cell. *Bioresour. Technol.* **2018**, *253*, 16–21. [CrossRef] [PubMed]
64. Jadhav, D.A.; Ghadge, A.N.; Ghangrekar, M.M. Simultaneous organic matter removal and disinfection of wastewater with enhanced power generation in microbial fuel cell. *Bioresour. Technol.* **2014**, *163*, 328–334. [CrossRef] [PubMed]
65. Huang, Q.; Jiang, F.; Wang, L.; Yang, C. Design of photobioreactors for mass cultivation of photosynthetic organisms. *Engineering* **2017**, *3*, 318–329. [CrossRef]
66. Erable, B.; Bergel, A. First air-tolerant effective stainless steel microbial anode obtained from a natural marine biofilm. *Bioresour. Technol.* **2009**, *100*, 3302–3307. [CrossRef] [PubMed]

67. Utomo, H.D.; Yu, L.S.; Yi, D.C.Z.; Jun, O.J. Recycling solid waste and bioenergy generation in MFC dual-chamber model. *Energy Procedia* **2017**, *143*, 424429. [CrossRef]
68. Liu, H.; Logan, B.E. Electricity generation using an air-cathode single chamber microbial fuel cell in the presence and absence of a proton exchange membrane. *Environ. Sci. Technol.* **2004**, *38*, 40404046. [CrossRef] [PubMed]
69. Zhu, F.; Wang, W.; Zhang, X.; Tao, G. Electricity generation in a membrane-less microbial fuel cell with down-flow feeding onto the cathode. *Bioresour. Technol.* **2011**, *102*, 73247328. [CrossRef] [PubMed]
70. Lay, C.H.; Kokko, M.E.; Puhakka, J.A. Power generation in fed-batch and continuous up-flow microbial fuel cell from synthetic wastewater. *Energy* **2015**, *91*, 235241. [CrossRef]
71. Chang, S.H.; Wu, C.H.; Wang, R.C.; Lin, C.W. Electricity production and benzene removal from groundwater using low-cost mini tubular microbial fuel cells in a monitoring well. *J. Environ. Manag.* **2017**, *193*, 551557. [CrossRef]
72. Yazdi, H.; Alzate-Gaviria, L.; Ren, Z.J. Pluggable microbial fuel cell stacks for septic wastewater treatment and electricity production. *Bioresour. Technol.* **2015**, *180*, 258263. [CrossRef]
73. Lu, M.; Chen, S.; Babanova, S.; Phadke, S.; Salvacion, M.; Mirhosseini, A.; Chan, S.; Carpenter, K.; Cortese, R.; Bretschger, O. Long-term performance of a 20-L continuous flow microbial fuel cell for treatment of brewery wastewater. *J. Power Sources* **2017**, *356*, 274287. [CrossRef]
74. Tremouli, A.; Greenman, J.; Ieropoulos, I. Investigation of ceramic MFC stacks for urine energy extraction. *Bioelectrochemistry* **2018**, *123*, 1925. [CrossRef]
75. Min, B.; Cheng, S.; Logan, B.E. Electricity generation using membrane and salt bridge microbial fuel cells. *Water Res.* **2005**, *39*, 16751686. [CrossRef] [PubMed]
76. Ali, I. New generation adsorbents for water treatment. *Chem. Rev.* **2012**, *112*, 5073–5091. [CrossRef] [PubMed]
77. Sudoh, R.; Islam, M.S.; Sazawa, K.; Okazaki, T.; Hata, N.; Taguchi, S.; Kuramitz, H. Removal of dissolved humic acid from water by coagulation method using polyaluminum chloride (PAC) with calcium carbonate as neutralizer and coagulant aid. *J. Environ. Chem. Eng.* **2015**, *3*, 770–774. [CrossRef]
78. Gautami, G.; Khanam, S. Selection of optimum configuration for multiple effect evaporator system. *Desalination* **2012**, *288*, 16–23. [CrossRef]
79. Serpone, N.; Artemev, Y.M.; Ryabchuk, V.K.; Emeline, A.V.; Horikoshi, S. Lightdriven advanced oxidation processes in the disposal of emerging pharmaceutical contaminants in aqueous media: A brief review. *Curr. Opin. Green Sustain. Chem.* **2017**, *6*, 18–33. [CrossRef]
80. Brillas, E.; Martínez-Huitle, C.A. Decontamination of wastewaters containing synthetic organic dyes by electrochemical methods. An updated review. *Appl. Catal. B Environ.* **2015**, *166*, 603–643. [CrossRef]
81. Harrison, R.M. (Ed.) *Pollution: Causes, Effects and Control*, 3rd ed.; Royal Society of Chemistry: Cambridge, UK, 2001.
82. Robinson, T.; Chandran, B.; Nigam, P. Studies on the production of enzymes by white rot-fungi for the decolorisation of textile dyes. *Enzym. Microb. Technol.* **2001**, *29*, 575–579. [CrossRef]
83. Amaral, P.F.F.; Fernandes, D.L.A.; Tavares, A.P.M.; Xavier, A.B.M.R.; Cammarota, M.C.; Coutinho, J.A.P.; Coelho, M.A.Z. Decolorization of dyes from textile wastewater by *Trametes versicolor*. *Environ. Technol.* **2004**, *25*, 1313–1320. [CrossRef]
84. Qu, J.; Xu, Y.; Ai, G.M.; Liu, Y.; Liu, Z.P. Novel *Chryseobacterium* sp. PYR2 degrades various organochlorine pesticides (OCPs) and achieves enhancing removal and complete degradation of DDT in highly contaminated soil. *J. Environ. Manag.* **2015**, *161*, 350–357. [CrossRef]
85. Zinovicius, A.; Rozene, J.; Merkelis, T.; Bruzaite, I.; Ramanavicius, A.; Morkvenaite-Vilkonciene, I. Evaluation of a yeast–polypyrrole biocomposite used in microbial fuel cells. *Sensors* **2022**, *22*, 327. [CrossRef] [PubMed]
86. Moradian, J.M.; Mi, J.L.; Dai, X.; Sun, G.F.; Du, J.; Ye, X.M.; Yong, Y.C. Yeast-induced formation of graphene hydrogels anode for efficient xylose-fueled microbial fuel cells. *Chemosphere* **2022**, *291*, 132963. [CrossRef] [PubMed]
87. Kumar, V.; Sehgal, R.; Gupta, R. Microbes and wastewater treatment. In *Development in Wastewater Treatment Research and Processes*; Elsevier: Amsterdam, The Netherlands, 2023; pp. 239–255.
88. Pant, D.; Van Bogaert, G.; Diels, L.; Vanbroekhoven, K. A review of the substrates used in microbial fuel cells (MFCs) for sustainable energy production. *Bioresour. Technol.* **2010**, *101*, 1533–1543. [CrossRef] [PubMed]
89. Husain, Q. Peroxidase mediated decolorization and remediation of wastewater containing industrial dyes: A review. *Rev. Environ. Sci. Biotechnol.* **2010**, *9*, 117–140. [CrossRef]
90. Holkar, C.R.; Jadhav, A.J.; Pinjari, D.V.; Mahamuni, N.M.; Pandit, A.B. A critical review on textile wastewater treatments: Possible approaches. *J. Environ. Manag.* **2016**, *182*, 351–366. [CrossRef] [PubMed]
91. Abdullah, N.A.; Ramli, S.; Mamat, N.H.; Khan, S.; Gomes, C. Chemical and biosensor technologies for wastewater quality management. *Int. J. Adv. Res. Publ.* **2017**, *1*, 1–10.
92. Hamza, R.A.; Iorhemen, O.T.; Tay, J.H. Advances in biological systems for the treatment of high-strength wastewater. *J. Water Process Engineer.* **2016**, *10*, 128–142. [CrossRef]
93. Arfin, T.; Sonawane, K.; Saidankar, P.; Sharma, S. Role of microbes in the bioremediation of toxic dyes. *Integ. Green Chem. Sustain. Eng.* **2019**, *26*, 443–472.
94. Behera, B.C. Citric acid from *Aspergillus niger*: A comprehensive overview. *Crit. Rev. Microbiol.* **2020**, *46*, 727–749. [CrossRef]
95. Rojas-Flores, S.; Cabanillas-Chirinos, L.; Nazario-Naveda, R.; Gallozzo-Cardenas, M.; Diaz, F.; Delfin-Narciso, D.; Rojas-Villacorta, W. Use of Tangerine Waste as Fuel for the Generation of Electric Current. *Sustainability* **2023**, *15*, 3559. [CrossRef]

96. Bhattacharya, S.; Angayarkanni, J.; Das, A.; Palaniswamy, M. Mycoremediation of Benzo [a] pyrene by *Pleurotus ostreatus* isolated from Wayanad district in Kerala, India. *Int. J. Pharm. Bio. Sci.* **2012**, *2*, 84–93.
97. Bhatt, M.; Cajthaml, T.; Šašek, V. Mycoremediation of PAH-contaminated soil. *Folia Microbiol.* **2002**, *47*, 255–258. [CrossRef] [PubMed]
98. Okparanma, R.N.; Ayotamuno, J.M.; Davis, D.D.; Allagoa, M. Mycoremediation of polycyclic aromatic hydrocarbons (PAH)-contaminated oil-based drill-cuttings. *Afr. J. Biotech.* **2013**, *10*, 5149–5156.
99. Peng, R.H.; Xiong, A.S.; Xue, Y.; Fu, X.Y.; Gao, F.; Zhao, W.; Tian, Y.S.; Yao, Q.H. Microbial biodegradation of polyaromatic hydrocarbons. *FEMS Microb. Rev.* **2008**, *32*, 927–955. [CrossRef] [PubMed]
100. Roorer, G.L.; Hsien, T.Y.; Way, J.D. Synthesis of porous-magnetic chitosan beads for removal of cadmium ions from wastewater. *Ind. Eng. Chem. Res.* **1993**, *32*, 2170–2178. [CrossRef]
101. Akar, T.; Divriklioglu, M. Biosorption applications of modified fungal biomass for decolorization of reactive red 2 contaminated solutions, batch and dynamic flow mode studies. *Bioresour. Technol.* **2010**, *101*, 7271–7277. [CrossRef] [PubMed]
102. Zhang, Q.; Deng, S.; Yu, G.; Huang, J. Removal of perfluorooctane sulfonate from aqueous solution by crosslinked chitosan beads, sorption kinetics and uptake mechanism. *Bioresour. Technol.* **2011**, *102*, 2265–2271. [CrossRef]
103. Liu, Y.; Zheng, Y.; Wang, A. Enhanced adsorption of methylene blue from aqueous solution by chitosan-g-poly, (acrylic acid)/vermiculite hydrogel composites. *J. Environ. Sci.* **2010**, *22*, 486–493. [CrossRef] [PubMed]
104. Wesenberg, D.; Kyriakides, I.; Agathos, S.N. White-rot fungi and their enzymes for the treatment of industrial dye effluents. Biotechnol. Adv. VI Int. Symp. *Environ. Biotechnol.* **2003**, *22*, 161–187. [CrossRef]
105. Ghosh Ray, S.; Ghangrekar, M.M. Comprehensive review on treatment of high-strength distillery wastewater in advanced physico-chemical and biological degradation pathways. *Int. J. Environ. Sci. Tech.* **2019**, *16*, 527–546. [CrossRef]
106. Rauf, M.A.; Salman Ashraf, S. Survey of recent trends in biochemically assisted degradation of dyes. *Chem. Eng. J.* **2012**, *209*, 520–530. [CrossRef]
107. Adelaja, O.; Keshavarz, T.; Kyazze, G. The effect of salinity, redox mediators and temperature on anaerobic biodegradation of petroleum hydrocarbons in microbial fuel cells. *J. Hazard Mater.* **2015**, *283*, 211–217. [CrossRef] [PubMed]
108. Shleev, S.; Tkac, J.; Christenson, A.; Ruzgas, T.; Yaropolov, A.I.; Whittaker, J.W.; Gorton, L. Direct electron transfer between copper-containing proteins and electrodes. 20th Anniversary of Biosensors and Bioelectronics. *Biosens. Bioelectron.* **2005**, *20*, 2517–2554. [CrossRef] [PubMed]
109. Shabani, M.; Pontié, M.; Younesi, H.; Nacef, M.; Rahimpour, A.; Rahimnejad, M.; Bouchenak Khelladi, R.M. Biodegradation of acetaminophen and its main by-product 4-aminophenol by *Trichoderma harzianum* versus mixed biofilm of *Trichoderma harzianum/Pseudomonas fluorescens* in a fungal microbial fuel cell. *J. Appl. Electrochem.* **2021**, *51*, 581–596. [CrossRef]
110. Wilkinson, S.; Klar, J.; Applegarth, S. Optimizing biofuel cell performance using a targeted mixed mediator combination. *Electroanalysis* **2006**, *18*, 2001–2007. [CrossRef]
111. Leonowicz, A.; Cho, N.; Luterek, J.; Wilkolazka, A.; Wojtas-Wasilewska, M.; Matuszewska, A.; Hofrichter, M.; Wesenberg, D.; Rogalski, J. Fungal laccase: Properties and activity on lignin. *J. Basic Microbiol.* **2001**, *41*, 185–227. [CrossRef]
112. Daâssi, D.; Prieto, A.; Zouari-Mechichi, H.; Martínez, M.J.; Nasri, M.; Mechichi, T. Degradation of bisphenol A by different fungal laccases and identification of its degradation products. *Int. Biodeterior. Biodegr.* **2016**, *110*, 181–188. [CrossRef]
113. Singh, R.L.; Singh, P.K.; Singh, R.P. Enzymatic decolorization and degradation of azo dyes—A review. *Int. Biodeterio. Biodegr.* **2015**, *104*, 21–31. [CrossRef]
114. Mani, P.; Keshavarz, T.; Chandra, T.S.; Kyazze, G. Decolourisation of Acid orange 7 in a microbial fuel cell with a laccase-based biocathode: Influence of mitigating pH changes in the cathode chamber. *Enzym. Microb. Technol.* **2017**, *96*, 170–176. [CrossRef]
115. Wu, C.; Liu, X.W.; Li, W.W.; Sheng, G.P.; Zang, G.L.; Cheng, Y.Y.; Shen, N.; Yang, Y.P.; Yu, H.Q. A white rot fungus is used as a biocathode to improve electricity production of a microbial fuel cell. *Appl. Energy* **2012**, *98*, 594–596. [CrossRef]
116. Strong, P.J. Fungal remediation of Amarula distillery wastewater. *World J. Microbiol. Biotechnol.* **2010**, *26*, 133–144. [CrossRef]
117. Strong, P.J.; Burgess, J.E. Fungal and enzymatic remediation of a wine lees and five wine-related distillery wastewaters. *Bioresour. Technol.* **2008**, *99*, 6134–6142. [CrossRef] [PubMed]
118. Dunford, H.B.; Stillman, J.S. On the function and mechanism of action of peroxidases. *Coord. Chem. Rev.* **1976**, *19*, 187–251. [CrossRef]
119. Wang, J.; Majima, N.; Hirai, H.; Kawagishi, H. Effective removal of endocrine-disrupting compounds by lignin peroxidase from the white-rot fungus *Phanerochaete sordida* YK-624. *Curr. Microbiol.* **2012**, *64*, 300–303. [CrossRef] [PubMed]
120. Zamocky, M.; Jakopitsch, C.; Furtmüller, P.G.; Dunand, C.; Obinger, C. The peroxidase–cyclooxygenase superfamily: Reconstructed evolution of critical enzymes of the innate immune system. *Proteins* **2008**, *72*, 589–605. [CrossRef] [PubMed]
121. Abubackar, H.N.; Biryol, I.; Ayol, A. Yeast industry wastewater treatment with microbial fuel cells: Effect of electrode materials and reactor configurations. *Int. J. Hydrog. Energy* **2023**, *48*, 12424–12432. [CrossRef]
122. Bhagchandanii, D.D.; Babu, R.P.; Khanna, N.; Pandit, S.; Jadhav, D.A.; Khilari, S.; Prasad, R. A Comprehensive Understanding of Electro-Fermentation. *Fermentation* **2020**, *6*, 92. [CrossRef]
123. Langhals, H. Color chemistry. In *Synthesis, Properties and Applications of Organic Dyes and Pigments*, 3rd ed.; Heinrich, Z., Ed.; VCH: Weinheim, Germany, 2004; pp. 5291–5292.
124. Martins, M.; Santos, J.M.; Diniz, M.S.; Ferreira, A.M.; Costa, M.H.; Costa, P.M. Effects of carcinogenic versus non-carcinogenic AHR-active PAHs and their mixtures: Lessons from ecological relevance. *Environ. Res.* **2015**, *138*, 101–111. [CrossRef]

125. Pachauri, R.K.; Sridharan, P.V. (Eds.) *Looking Back to Think Ahead: GREEN India 2047 (Growth with Resource Enhancement of Environment and Nature)*; TERI: New Delhi, India, 1998.
126. Burkina, V.; Zlabek, V.; Zamaratskaia, G. Effects of pharmaceuticals present in aquatic environment on Phase I metabolism in fish. *Environ. Toxicol. Pharmacol.* **2015**, *40*, 430–444. [CrossRef]
127. Chaalal, O.; Zekri, A.Y.; Soliman, A.M. A novel technique for the removal of strontium from water using thermophilic bacteria in a membrane reactor. *J. Ind. Eng. Chem.* **2015**, *21*, 822–827. [CrossRef]
128. Liu, H.; Guo, S.; Jiao, K.; Hou, J.; Xie, H.; Xu, H. Bioremediation of soils contaminated with heavy metals and 2,4,5-trichlorophenol by fruiting body of *Clitocybe maxima*. *J. Hazard Mater.* **2015**, *294*, 121–127. [CrossRef] [PubMed]
129. Martinez-Alcala, I.; Guillén-Navarro, J.M.; Fernández-López, C. Pharmaceutical biological degradation, sorption and mass balance determination in a conventional activated-sludge wastewater treatment plant from Murcia, Spain. *Chem. Eng. J.* **2017**, *316*, 332–340. [CrossRef]
130. Li, J.; Zhang, Y.; Peng, J.; Wu, X.; Gao, S.; Mao, L. The effect of dissolved organic matter on soybean peroxidase-mediated removal of triclosan in water. *Chemosphere.* **2017**, *172*, 399–407. [CrossRef]
131. Momoh, O.L.Y. A novel electron acceptor for microbial fuel cells: Nature of circuit connection on internal resistance. *J. Biochem. Technol.* **2011**, *2*, 216–220.
132. Lu, M.; Li, F.Y. Cathode reactions and applications in microbial fuel cells: A review. *Crit. Rev. Environ. Sci. Technol.* **2012**, *42*, 2504–2525. [CrossRef]
133. Logan, B.E.; Regan, J.M. Microbial fuel cells-challenges and applications. *Environ. Sci. Technol.* **2006**, *40*, 5172–5180. [CrossRef] [PubMed]
134. He, C.S.; Mu, Z.X.; Yang, H.Y.; Wang, Y.Z.; Mu, Y.; Yu, H.Q. Electron acceptors for energy generation in microbial fuel cells fed with wastewaters: A mini-review. *Chemosphere* **2015**, *140*, 12–17. [CrossRef]
135. Hernandez-Fernandez, F.J.; Perez de los Ríos, A.; Mateo-Ramírez, F.; Juarez, M.D.; Lozano-Blanco, L.J.; Godínez, C. New application of supported ionic liquids membranes as proton exchange membranes in microbial fuel cell for wastewater treatment. *Chem. Eng. J.* **2015**, *279*, 115–119. [CrossRef]
136. Leong, J.X.; Daud, W.R.W.; Ghasemi, M.; Liew, K.B.; Ismail, M. Ion exchange membranes as separators in microbial fuel cells for bioenergy conversion: A comprehensive review. *Renew. Sustain. Energy Rev.* **2013**, *28*, 575–587. [CrossRef]
137. Xie, S.; Lawlor, P.G.; Frost, J.P.; Hu, Z.; Zhan, X. Effect of pig manure to grass silage ratio on methane production in batch anaerobic co-digestion of concentrated pig manure and grass silage. *Bioresour. Technol.* **2011**, *102*, 5728–5733. [CrossRef]
138. Mohana, S.; Acharya, B.K.; Madamwar, D. Bioremediation concepts for treatment of distillery effluent. In *Biotechnology for Environmental Management and Resource Recovery*; Springer: New Delhi, India, 2013; pp. 261–278.
139. Fernandez de Dios, M.A.; del Campo, A.G.; Fernandez, F.J.; Rodrigo, M.; Pazos, M.; Sanroman, M.A. Bacterial–fungal interactions enhance power generation in microbial fuel cells and drive dye de colorization by an ex situ and in situ electroFenton process. *Bioresour. Technol.* **2013**, *148*, 39–46. [CrossRef] [PubMed]
140. Liew, K.B.; Daud, W.R.W.; Ghasemia, M.; Leong, J.X.; Lim, S.S.; Ismail, M. Non-Pt catalyst as oxygen reduction reaction in microbial fuel cells: A review. *Int. J. Hydrog. Energy* **2014**, *39*, 4870–4883. [CrossRef]
141. Rahimnejad, M.; Adhami, A.; Darvari, S.; Zirepour, A.; Oh, S.E. Microbial fuel cell as new technology for bioelectricity generation: A review. *Alexandr. Eng. J.* **2015**, *54*, 745–756. [CrossRef]
142. Modestra, J.A.; Chiranjeevi, P.; Mohan, S.V. Cathodic material effect on electron acceptance towards bioelectricity generation and wastewater treatment. *Renew. Energy* **2016**, *98*, 178–187. [CrossRef]
143. Kalathil, S.; Patil, S.A.; Pant, D. Microbial fuel cells: Electrode materials. In *Encyclopedia of Interfacial Chemistry: Surface Science and Electrochemistry*; Wandelt, K., Vadgama, P., Eds.; Elsevier: Amsterdam, The Netherlands, 2017; pp. 309–318. [CrossRef]
144. Kazemi, N.; Tavakoli, O.; Seif, S.; Nahangi, M. High-strength distillery wastewater treatment using catalytic sub-and supercritical water. *J. Supercrit. Fluids* **2015**, *97*, 74–80. [CrossRef]
145. Latif, M.A.; Ghufran, R.; Wahid, Z.A.; Ahmad, A. Integrated application of upflow anaerobic sludge blanket reactor for the treatment of wastewaters. *Water Res.* **2011**, *45*, 4683–4699. [CrossRef] [PubMed]
146. Anupama, S.; Pradeep, N.V.; Hampannavar, U.S. Anaerobic followed by aerobic treatment approaches for Spentwash using MFC and RBC. *Sugar Tech.* **2013**, *13*, 197–202. [CrossRef]
147. Apollo, S.; Onyango, M.S.; Ochieng, A. An integrated anaerobic digestion and UV photocatalytic treatment of distillery wastewater. *J. Hazard Mater.* **2013**, *261*, 435–442. [CrossRef]
148. Melamane, X.; Tandlich, R.; Burgess, J. Anaerobic digestion of fungally pre-treated wine distillery wastewater. *Afr. J. Biotechnol.* **2007**, *6*, 1990–1993.
149. Melamane, X.L.; Tandlich, R.; Burgess, J.E. Submerged membrane bioreactor and secondary digestion in the treatment of wine distillery wastewater. Part II: The effect of fungal pre-treatment on wine distillery wastewater digestion. *Fresenius Environ. Bull.* **2007**, *16*, 162–167.
150. Włodarczyk, B.; Włodarczyk, P.P. Analysis of the potential of an increase in yeast output resulting from the application of additional process wastewater in the evaporator station. *Appl. Sci.* **2019**, *9*, 2282. [CrossRef]
151. Rozene, J.; Morkvenaite-Vilkonciene, I.; Bruzaite, I.; Dzedzickis, A.; Ramanavicius, A. Yeast-based microbial biofuel cell-mediated by 9,10- phenantrenequinone. *Electrochim. Acta* **2021**, *373*, 137918. [CrossRef]

152. Que, Y.; Xie, Y.; Li, T.; Yu, C.; Tu, S.; Yao, C. An N-heterocyclic carbene-catalyzed oxidative γ-aminoalkylation of saturated carboxylic acids through in situ activation strategy: Access to δ-lactam. *Org. Lett.* **2015**, *17*, 6234–6237.143. [CrossRef] [PubMed]
153. Tamjidi, S.; Moghadas, B.K.; Esmaeili, H.; Khoo, F.S.; Gholami, G.; Ghasemi, M. Improving the surface properties of adsorbents by surfactants and their role in the removal of toxic metals from wastewater: A review study. *Process Saf. Environ. Prot.* **2021**, *148*, 775–795. [CrossRef]
154. Chigusa, K.; Hasegawa, T.; Yamamoto, N.; Watanabe, Y. Treatment of wastewater from oil manufacturing plant by yeasts. *Water Sci. Technol.* **1996**, *34*, 51–58. [CrossRef]
155. Zheng, S.; Yang, M.; Lv, W.; Liu, F. Study on sludge expansion during treatment of salad oil manufacturing wastewater by yeast. *Environ. Technol.* **2001**, *22*, 533–542. [CrossRef]
156. Chaijak, P.; Lertworapreecha, M.; Sukkasem, C. Phenol removal from palm oil mill effluent using *Galactomyces reessii* termite-associated yeast. *Pol. J. Environ. Stud.* **2018**, *27*, 39–44. [CrossRef]
157. Hubenova, Y.; Georgiev, D.; Mitov, M. Enhanced phytate dephosphorylation by using *Candida melibiosica* yeast-based biofuel cell. *Biotechnol. Lett.* **2014**, *36*, 1993–1997. [CrossRef]
158. Pal, M.; Shrivastava, A.; Sharma, R.K. Electroactive biofilm development on carbon fiber anode by *Pichia fermentans* in a wheat straw hydrolysate based microbial fuel cell. *Biomass Bioenerg.* **2023**, *168*, 106682. [CrossRef]
159. Włodarczyk, B.; Włodarczyk, P.P. Electricity Production from Yeast Wastewater in Membrane-Less Microbial Fuel Cell with Cu-Ag Cathode. *Energies* **2023**, *16*, 2734. [CrossRef]
160. Cecconet, D.; Molognoni, D.; Callegari, A.; Capodaglio, A.G. Biological combination processes for efficient removal of pharmaceutically active compounds from wastewater: A review and future perspectives. *J. Environ. Chem. Eng.* **2017**, *5*, 35903603. [CrossRef]
161. Patneedi, C.B.; Prasadu, K.D. Impact of pharmaceutical wastes on human life and environment. *Rasayan J. Chem.* **2015**, *8*, 6770.
162. Ismail, Z.Z.; Habeeb, A.A. Experimental and modeling study of simultaneous power generation and pharmaceutical wastewater treatment in microbial fuel cell based on mobilized biofilm bearers. *Renew. Energy* **2017**, *101*, 12561265. [CrossRef]
163. Zub, S.; Kurisso, T.; Menert, A.; Blonskaja, V. Combined biological treatment of high-sulphate wastewater from yeast production. *Water Environ. J.* **2008**, *22*, 274–286. [CrossRef]
164. Hossain, S.M. Aerobic treatment of distillery wastewater using *Phanerochaete chrysosporium*. *Indian J. Environ. Prot.* **2007**, *27*, 362–366.
165. Mullai, P.; Sathian, S.; Sabarathinam, P.L. Kinetic modelling of distillery wastewater biodegradation using *Paecilomyces variotii*. *Chem. Eng. World* **2007**, *42*, 98–106.
166. Ray, S.G.; Ghangrekar, M.M. Enhancing organic matter removal, biopolymer recovery and electricity generation from distillery wastewater by combining fungal fermentation and microbial fuel cell. *Bioresour. Technol.* **2015**, *176*, 8–14.
167. Ravikumar, R.; Vasanthi, N.S.; Saravanan, K. Single factorial experimental design for decolorizing anaerobically treated distillery spent wash using *Cladosporium Cladosporioides*. *Int. J. Environ. Sci. Technol.* **2011**, *8*, 97–106. [CrossRef]
168. Watanabe, T.; Suzuki, K.; Sato, I.; Morita, T.; Koike, H.; Shinozaki, Y.; Ueda, H.; Koitabashi, M.; Kitamoto, H.K. Simultaneous bioethanol distillery wastewater treatment and xylanase production by the phyllosphere yeast *Pseudozyma antarctica* GB-4(0). *AMB Express* **2015**, *5*, 121. [CrossRef]
169. Ling, J.; Tian, Y.; de Toledo, R.A.; Shim, H. Cost reduction for the lipid production from distillery and domestic mixed wastewater by *Rhodosporidium toruloides* via the reutilization of spent seed culture medium. *Energy* **2016**, *136*, 135–141. [CrossRef]
170. Zhu, W.; Guo, C.; Luo, F.; Zhang, C.; Wang, T.; Wei, Q. Optimization of *Calvatia gigantea* mycelia production from distillery wastewater. *J. Inst. Brew.* **2015**, *121*, 78–86. [CrossRef]
171. Watanabe, T.; Iefuji, H.; Kitamoto, H.K. Treatment of *Candida utilis* biomass production from shochu wastewater; the effects of maintaining a low pH on DOC removal and feeding cultivation on biomass production. *SpringerPlus* **2013**, *2*, 514. [CrossRef] [PubMed]
172. Melamane, X.L.; Strong, P.J.; Burgess, J.E. Treatment of wine distillery wastewater: A review with emphasis on anaerobic membrane reactors. *S. Afr. J. Enol. Vitic.* **2007**, *28*, 25. [CrossRef]
173. Sheehan, G.J.; Greenfield, P.F. Utilisation, treatment and disposal of distillery wastewater. *Water Res.* **1980**, *14*, 257–277. [CrossRef]
174. Tang, Y.Q.; Fujimura, Y.; Shigematsu, T.; Morimura, S.; Kida, K. Anaerobic treatment performance and microbial population of thermophilic upflow anaerobic filter reactor treating awamori distillery wastewater. *J. Biosci. Bioeng.* **2007**, *104*, 281–287. [CrossRef] [PubMed]
175. Kongthale, G.; Sotha, S.; Michu, P.; Madloh, A.; Wetchapan, P.; Chaijak, P. Electricity Production and Phenol Removal of Winery Wastewater by Constructed Wetland—Microbial Fuel Cell Integrated with Ethanol Tolerant Yeast. *Biointerface Res. Appl. Chem.* **2023**, *13*, 157.
176. Chen, S.H.; Yien Ting, A.S. Biodecolorization and biodegradation potential of recalcitrant triphenylmethane dyes by *Coriolopsis* sp. isolated from compost. *J. Environ. Manag.* **2015**, *150*, 274–280. [CrossRef]
177. Chen, S.H.; Yien Ting, A.S. Biosorption and biodegradation potential of triphenylmethane dyes by newly discovered *Penicillium simplicissimum* isolated from indoor wastewater sample. *Int. Biodeterior. Biodegr.* **2015**, *103*, 1–7. [CrossRef]
178. Raqba, R.; Rafaqat, S.; Ali, N.; Munis, M.F.H. Biodegradation of Reactive Red 195 azo dye and Chlorpyrifos organophosphate along with simultaneous bioelectricity generation through bacterial and fungal based biocathode in microbial fuel cell. *J. Water Process Eng.* **2022**, *50*, 103177. [CrossRef]

179. Pal, M.; Shrivastava, A.; Sharma, R.K. Wheat straw-based microbial electrochemical reactor for azo dye decolorization and simultaneous bioenergy generation. *J. Environ. Manag.* **2022**, *323*, 116253. [CrossRef]
180. Rawat, D.; Sharma, R.S.; Karmakar, S.; Arora, L.S.; Mishra, V. Ecotoxic potential of a presumably non-toxic azo dye. *Ecotoxicol. Environ. Saf.* **2018**, *148*, 528–537. [CrossRef] [PubMed]
181. Kadir, W.N.A.; Lam, M.K.; Uemura, Y.; Lim, J.W.; Lee, K.T. Harvesting and pre-treatment of microalgae cultivated in wastewater for biodiesel production: A review. *Energy Conv. Manag.* **2018**, *171*, 1416–1429. [CrossRef]
182. Mani, A.; Hameed, S.A.S. Improved bacterial-fungal consortium as an alternative approach for enhanced decolourization and degradation of azo dyes: A review. *Nat. Environ. Pollut. Technol.* **2019**, *18*, 49–64.
183. Crini, G.; Lichtfouse, E.; Wilson, L.D.; Morin-Crini, N. Conventional and non-conventional adsorbents for wastewater treatment. *Environ. Chem. Lett.* **2019**, *17*, 195–213. [CrossRef]
184. Srivastava, R.K.; Boddula, R.; Pothu, R. Microbial fuel cells: Technologically advanced devices and approach for sustainable/renewable energy development. *Energ. Conver. Manag. X.* **2022**, *13*, 100160. [CrossRef]
185. Cusick, R.D.; Kim, Y.; Logan, B.E. Energy capture from thermolytic solutions in microbial reverse-electrodialysis cells. *Science* **2012**, *335*, 1474–1477. [CrossRef] [PubMed]

Disclaimer/Publisher's Note: The statements, opinions and data contained in all publications are solely those of the individual author(s) and contributor(s) and not of MDPI and/or the editor(s). MDPI and/or the editor(s) disclaim responsibility for any injury to people or property resulting from any ideas, methods, instructions or products referred to in the content.

Article

Enhanced Visible-Light Photocatalytic Activities of CeVO$_4$-V$_2$O$_3$ Composite: Effect of Ethylene Glycol

Yuxin Wang [1,*], Yinxiu Jin [1], Minghan Jia [2], Hu Ruan [3], Xuefen Tao [1], Xuefeng Liu [1], Guang Lu [4,*] and Xiaodong Zhang [5,*]

1. Institute of Applied Biotechnology, Taizhou Vocation & Technical College, Taizhou 318000, China
2. Zhejiang Jiuzhou Pharmaceutical Co., Ltd., Taizhou 318000, China
3. Taizhou Institute of Environmental Science Design and Research Co., Ltd., Taizhou 318000, China
4. School of Environmental and Safety Engineering, Liaoning Shihua University, Fushun 113001, China
5. School of Environment and Architecture, University of Shanghai for Science and Technology, Shanghai 200093, China
* Correspondence: wyx790914@aliyun.com (Y.W.); luguang20121101@126.com (G.L.); fatzhxd@126.com (X.Z.)

Abstract: CeVO$_4$-V$_2$O$_3$ composites were prepared by simple hydrothermal method, and the effects of ethylene glycol(EG) on the products were studied by XRD, N$_2$ adsorption–desorption, SEM, EDS, XPS, PL and UV-vis spectra. The characterization reveals a slight decrease in surface area and a slight enhancement of visible light absorption in the final sample, while the crystalline phase, morphology and separation efficiency of the collective carriers are severely affected by the EG. At the same time, the photocatalytic effect of CeVO$_4$-V$_2$O$_3$ composites was evaluated by the degradation rate of methylene blue (MB) under simulated visible light. The sample for 10 mL EG obtained the highest efficiency of 96.9%, while the one for 15 mL EG showed the lowest efficiency of 67.5% within 300 min. The trapping experiments and ESR experiment showed that the contribution of active species to the photocatalytic degradation of MB was ·OH > h$^+$ > ·O^{2-} in descending order, and a possible degradation mechanism was proposed.

Keywords: ethylene glycol; CeVO$_4$-V$_2$O$_3$; methylene blue oxidation; visible light

Citation: Wang, Y.; Jin, Y.; Jia, M.; Ruan, H.; Tao, X.; Liu, X.; Lu, G.; Zhang, X. Enhanced Visible-Light Photocatalytic Activities of CeVO$_4$-V$_2$O$_3$ Composite: Effect of Ethylene Glycol. *Catalysts* 2023, 13, 659. https://doi.org/10.3390/catal13040659

Academic Editor: Meng Li

Received: 27 February 2023
Revised: 20 March 2023
Accepted: 20 March 2023
Published: 27 March 2023

Copyright: © 2023 by the authors. Licensee MDPI, Basel, Switzerland. This article is an open access article distributed under the terms and conditions of the Creative Commons Attribution (CC BY) license (https://creativecommons.org/licenses/by/4.0/).

1. Introduction

Photocatalytic technology is a green and economical solar-driven pollutant removal technology that can be carried out at normal temperatures and pressure. Using semiconductors and related materials as photocatalysts, we can carry out relatively compound chemical conversion under relatively simple conditions, which has unparalleled advantages in pollutant removal technology by directly using reducing agents or oxidizing agents. Photocatalytic technology provides an environmentally friendly and efficient transformation path for pollutant removal, leading many researchers to develop in a more cutting-edge research direction, and provides an efficient solution for carbon neutrality and sustainable development [1–6]. Nevertheless, TiO$_2$ cannot be excited when the wavelength is higher than 420 nm, which greatly limits its practical application in the visible light region. Therefore, research is devoted to developing a series of visible light-responsive photocatalysts to effectively utilize solar energy [7–11].

Cerium vanadate (CeVO$_4$), as a rare earth vanadate, has good physical and chemical properties and is widely used in the fields of batteries, semiconductors and catalysis. After vanadate is combined with rare earth metals such as Ce, La and Pr, its electrochemical performance, thermal stability, specific surface area and magnetism are obviously enhanced [12]. In view of the excellent physical and chemical properties of CeVO$_4$, many researchers use the precipitation method, microwave radiation method, sonochemical method and hydrothermal method to synthesize CeVO$_4$ [13–15]. In the absence of any catalyst and template, Xie et al. prepared tetragonal CeVO$_4$ with the same precursor by hydrothermal method and ultrasonic methods, such as micro-rods, nanoparticles, nanorods and nanosheets [16]. Mosleh et al. reported a simple sonochemical method to synthesize $CeVO_4$ nanoparticles with the aid of ammonium metavanadate, using cerium (III) nitrate hexahydrate as the initial reagent and water as the solvent [17].

$CeVO_4$ has excellent catalytic performance, relatively stable chemical properties and a relatively simple preparation method, so it is a photocatalytic material with important research significance. Therefore, various micro/nanosized $CeVO_4$ samples have been prepared for use in the removal of organic compounds [18–21]. Phuruangrat et al. synthesized tetragonal zircon $CeVO_4$ photocatalytic particles by sol-gel method with tartaric acid as a complexing agent, which showed good activity in photocatalytic degradation of MB under ultraviolet light [22]. Liu et al. prepared egg yolk shell-like $CeVO_4$ microspheres composed of nanosheets by citric acid-assisted hydrothermal method, which realized the rare long cycle and high capacity of $CeVO_4$, and was considered as the next promising candidate negative electrode material for lithium-ion batteries [23]. However, considering the practical application, the photocatalytic activity of pure $CeVO_4$ still needs to be improved. The combination of $CeVO_4$ with other materials is considered to be an effective method to improve the separation efficiency of photo-generated carriers and improve photocatalytic performance [24]. Ma et al. prepared the flexible integrated composite of $CeVO_4$ and multi-walled carbon nanotubes (MWCNTs) by a simple hydrothermal technique. $CeVO_4$/MWCNTs composite can be applied to the detection of sulfadiazine residues in water samples of actual aquaculture [25]. Song et al. obtained the one-dimensional structure of AgNW@$CeVO_4$ composite photocatalyst by depositing $CeVO_4$ on the surface of a silver nanowire, which expanded the light absorption range of $CeVO_4$. In addition, compared with pure $CeVO_4$, AgNW@$CeVO_4$ composite photocatalyst showed excellent photocatalytic performance for the degradation of pollutants such as rhodamine B, MB and 4-nitrophenol under sunlight irradiation [26]. However, the above-mentioned combination methods were performed by adding additional materials, which undoubtedly increases the cost and complicates the process. Therefore, we try to study a simple and economical one-step preparation method; that is, cerium vanadate and vanadium anhydride can be obtained at the same time by controlling the reaction conditions, and they can form composites, thus improving the photocatalytic performance of $CeVO_4$.

In this paper, $CeVO_4$-V_2O_3 composites were prepared by a simple ethylene glycol (EG) assisted hydrothermal method. Because the solubility of ammonium vanadate is 4.8 at room temperature, and it is slightly soluble in water, adding organic solvent is a favorable dissolution method. Compared with ethanol, glycerol and other solvents, EG was chosen as the solvent. Moreover, the effects of EG addition on the crystal structure, morphology, separation efficiency of the recombination electron holes and photocatalytic activity have been studied in detail by XRD, N_2 adsorption–desorption, SEM, EDS, XPS, PL and UV-vis spectra. Trapping experiments and ESR are employed to study the active species in the photodegradation process. Based on the above results, the possible photodegradation mechanism of MB over $CeVO_4$-V_2O_3 composites is proposed.

2. Results and Discussion

2.1. Characterization and Analysis

The influence of EG on the crystal phase of the as-prepared samples was analyzed by X-ray powder diffraction (XRD). Figure 1 shows the XRD patterns of the composite samples. The four main peaks of these samples located at 2θ = 18.3, 24.2, 32.4 and 46.5 are assigned to the (101), (200), (112) and (312) planes of $CeVO_4$, respectively. All the diffraction peaks of the products obtained from different EG addition were assigned to the tetragonal phase of $CeVO_4$ (JCPDS No. 12-0757). The XRD patterns also showed two main diffraction peaks at 2θ = 28.6 and 36.9, corresponding to the (102) and (110) planes of the V_2O_3 phase (JCPDS No.01-071-0344).

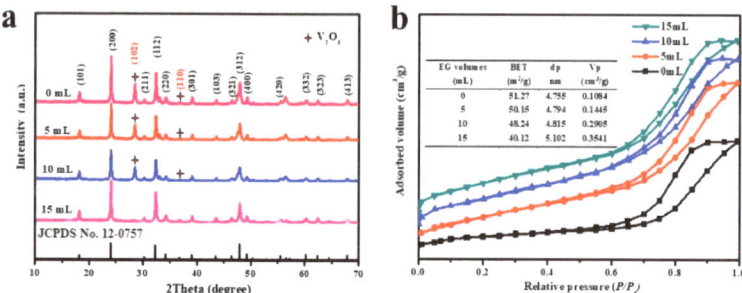

Figure 1. XRD pattern (**a**); N$_2$ adsorption–desorption isotherm curve of different samples (**b**).

It can also be seen from Figure 1a that the sample added with 15 mL EG only has the diffraction peak of tetragonal CeVO$_4$, and no impurity peak is detected, indicating that the CeVO$_4$-V$_2$O$_3$ composite has not been formed. In other samples, the diffraction peak corresponding to the V$_2$O$_3$ phase was detected, indicating the formation of the CeVO$_4$-V$_2$O$_3$ composites. Therefore, the addition of EG promotes the formation of the composites, but the excessive EG is not conducive to the formation of vanadic anhydride.

The specific surface area and porosity of the as-synthesized samples are determined by N$_2$ adsorption–desorption isotherms (Figure 1b). Obviously, the isotherms of all the samples were identified as type IV, with an H1 pattern hysteresis loop [27,28]. It could be deduced that as-prepared samples are members, and the specific surface area of obtained catalysts with additional amounts of 0, 5, 10 and 15 mL EG are 51.27, 50.15, 48.24 and 40.12 m$^2 \cdot$g^{-1}, respectively. These results show that the addition of EG leads to a slight decrease in specific surface area.

The morphology of as-prepared samples with different amounts of EG is investigated by SEM. As shown in Figure 2a, the product with no EG appears as a nanoparticle with an average diameter of about 230 nm. Adding 5 mL EG (as shown in Figure 2b), the product appears as a nano-tetrahedral bipyramid with an average diameter of about 550 nm. Upon increasing EG to 10 mL (as shown in Figure 2c), the product appears as microparticles with an average diameter of 1.0 μm, which is composed of several tetrahedrons and slices. Further increasing EG to 15 mL (as shown in Figure 2d), numerous leaf-like nano-slices are clustered together in groups to form a microsphere, caused by excessive EG probably. In addition, the average diameter increases to 1.6μm. The SEM results show that an increase in EG content leads to an increase in the average diameter of nanocomposites, which leads to a decrease in the specific surface area of nanocomposites. The chemical composition and element distribution of samples prepared in a solution containing 15 mL EG were studied by EDS. As shown in Figure 2e, the atomic ratio of O:V:Ce for the as-prepared sample is 66.46:17.82:15.72, which is larger than the theoretical stoichiometric ratio of 4:1:1 for O:V:Ce in CeVO$_4$.

X-ray photoelectron spectroscopy (XPS), as an important surface analysis technology, has the characteristics of simple preparation, no damage to the sample and can distinguish the chemical state information of elements, which has attracted more and more attention and use by researchers [29–31]. The chemical composition of the obtained catalysts is analyzed by XPS with 10 and 15 mL of EG addition. The survey scan spectra in Figure 3a is shown that both samples mainly contain Ce, V, C and O elements peaks (C 1s peak is assigned to the signal of the background hydrocarbon). As for the HR-XPS spectra of Ce 3d shown in Figure 3b, the Ce 3d$_{5/2}$ peaks locate at 881.1 and 884.9 eV and Ce 3d$_{3/2}$ peaks locate at 899.8 and 903.2 eV, indicating Ce in both samples mainly exists as Ce^{3+} [32]. The HR-XPS spectra of V 2p resolve into two spin-orbit doublets (Figure 3c), characteristic of V^{5+} and V^{3+}. The binding energies at 516.8 and 524.4 eV correspond to V 2p$_{3/2}$ and V 2p$_{1/2}$ of V^{5+}, and those at 515.4 and 522.6 eV can be attributed to V 2p$_{3/2}$ and V 2p$_{1/2}$ of V^{3+}. As for the HR-XPS spectra shown in Figure 3d, O 1s peaks are divided into the lattice oxygen of CeVO$_4$ at 529.4 eV and O^{2-} ions of V$_2$O$_3$ at 531.1 eV [33]. These results show that the final samples synthesized in solutions containing 10 and 15 mL of EG are CeVO$_4$-V$_2$O$_3$ composites, which are different from the XRD results. The possibility of such a result can be explained by the fact that V$_2$O$_3$ is distributed in amorphous form over the sample for 15 mL EG, which is not detectable by XRD.

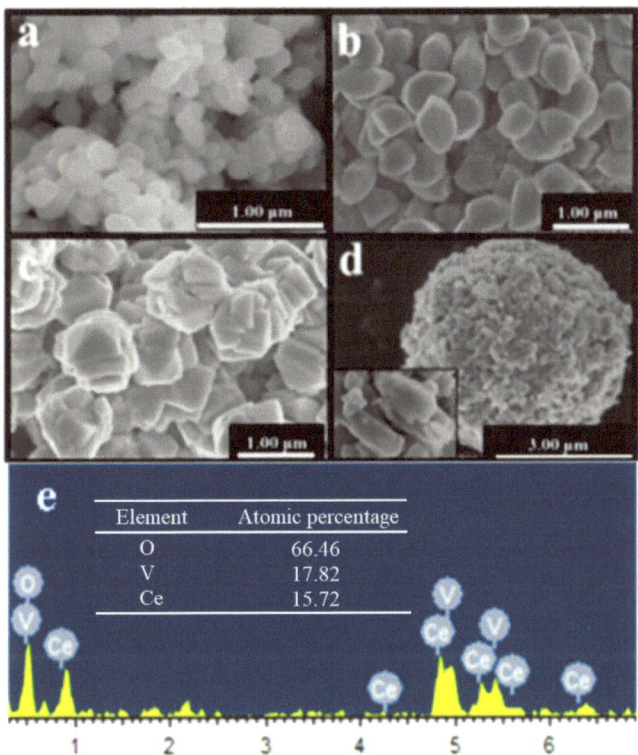

Figure 2. SEM of CeVO$_4$-V$_2$O$_3$ prepared with EG addition amounts of 0 (**a**), 5 (**b**), 10 (**c**) and 15 mL (**d**) and EDS of CeVO$_4$-V$_2$O$_3$ prepared with EG addition amounts of 15 mL (**e**).

2.2. Photocatalytic Performance of CeVO$_4$-V$_2$O$_3$ Composites

Photoluminescence spectra (PL) can be used to determine the recombination of electrons and holes induced by light indirectly. Higher fluorescence intensity means higher electron and hole recombination efficiency [34,35]. Figure 4a shows the PL spectra of the CeVO$_4$-V$_2$O$_3$ composites. The emission bands centered at 512, 521, 556, 559 and 638 nm correspond to 5D0→7Fi (i = 0, 1, 2, 3, 4) electronic transitions of Ce^{3+} ion, respectively [36]. This result indicates that the CeVO$_4$-V$_2$O$_3$ composites are excited by UV light (360 nm), transfer energy to Ce^{3+}, and then exhibit red emission through the f–f transition of Ce^{3+} [37]. In addition, the strength of the CeVO$_4$-V$_2$O$_3$ sample without EG is high, which shows that the photo-generated electrons and holes generated by the sample under the condition of light excitation are easier to recombine and reduce the photocatalytic efficiency. The PL emission intensity of the sample synthesized in the EG precursor solution shows a dramatic weakness compared to that of the sample without EG. The intensity of the emission peak is the lowest for an added EG of 10 mL, indicating that the photo-generated electron and hole separation efficiency at the surface of this catalyst is maximally enhanced, and therefore, many photo-generated electrons and holes are generated.

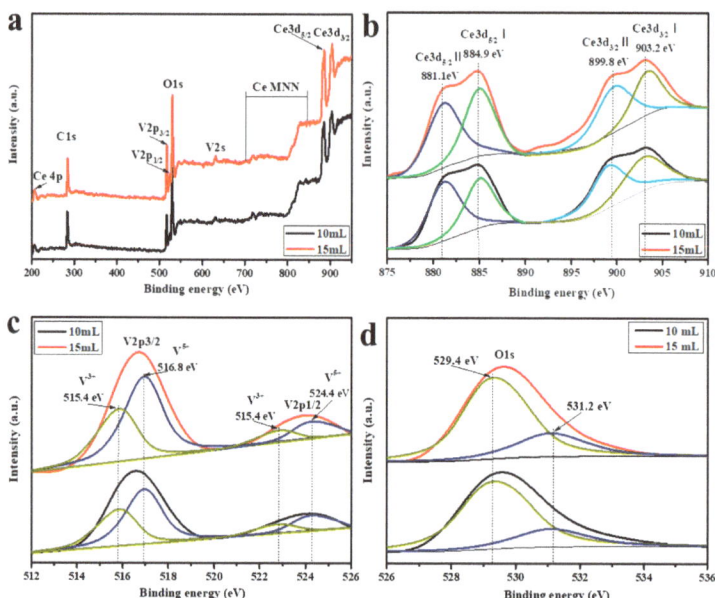

Figure 3. XPS results of the samples synthesized in the solutions containing 10 and 15 mL EG of survey (**a**), Ce 3d (**b**), V 2p (**c**) and O1s (**d**).

The optical properties of the as-synthesized samples were studied by UV-Vis absorption spectra (Figure 4b). The band gap of a semiconductor photocatalyst has a vital influence on its catalytic performance and determines the light absorption range in the process of photocatalysis. The band gap of the $CeVO_4$-V_2O_3 sample is calculated according to the following formula (Kubelka–Munk formula) [38,39]:

$$\alpha h\nu = A(h\nu - E_g)^2 \tag{1}$$

where α stands for absorption coefficient, ν corresponds to the optical frequency, h is Planck constant and E_g stands for the band gap of the sample. Figure 4b shows the curve of $(\alpha h\nu)^2$ versus $h\nu$, which is derived from the corresponding absorption spectra. Without EG, the four absorption peaks are located at 257, 353, 452 and 590 nm, respectively, and the energy gap is estimated to be 1.20 eV. When 5, 10 and 15 mL EG were added to the sample, the three absorption peaks were concentrated at 257, 452 and 568 nm, respectively, and the energy gap can be estimated as 1.03 eV. The results show that the addition of EG may enhance the absorption of visible light from the final products, indicating that they have an excellent UV-Vis response to visible light illumination. Comparatively speaking, the smaller the band gap of the composite, the less energy is needed for the electron transition reaction and the easier the photocatalytic reaction.

The catalytic activity of as-prepared $CeVO_4$-V_2O_3 photocatalysts was evaluated, and the effect of EG on the photocatalytic performance of the $CeVO_4$-V_2O_3 sample was also studied by simulating the photocatalytic conversion efficiency of MB under visible light (λ > 420 nm). Figure 4c shows the degradation efficiency of MB by photocatalyst in different irradiation times. Before photodegradation, the photocatalyst was dispersed in MB solution, and the adsorption–desorption equilibrium was completed in the dark for 60 min. It can be seen that the photocatalytic conversion efficiency of MB increases gradually with the increase of the amount of EG from 0 mL to 10 mL, but the photocatalytic conversion efficiency is the lowest when the amount of EG increases to 15 mL. It shows that the addition of EG is beneficial to the enhancement of the photocatalytic effect, but it is not suitable to add too much. As can be observed, the photodegradation rate of MB can reach 96.9% within 300 min under the irradiation of visible light. However, the photocatalytic effect of the as-prepared CeVO$_4$-V$_2$O$_3$ composites is not as good as that of Ag nanowire@CeVO$_4$ heterostructure photocatalyst. Maybe introducing Ag can strengthen O$_2$ adsorption on the CeVO$_4$ surface, which advances the photocatalytic activity of CeVO$_4$ [26].

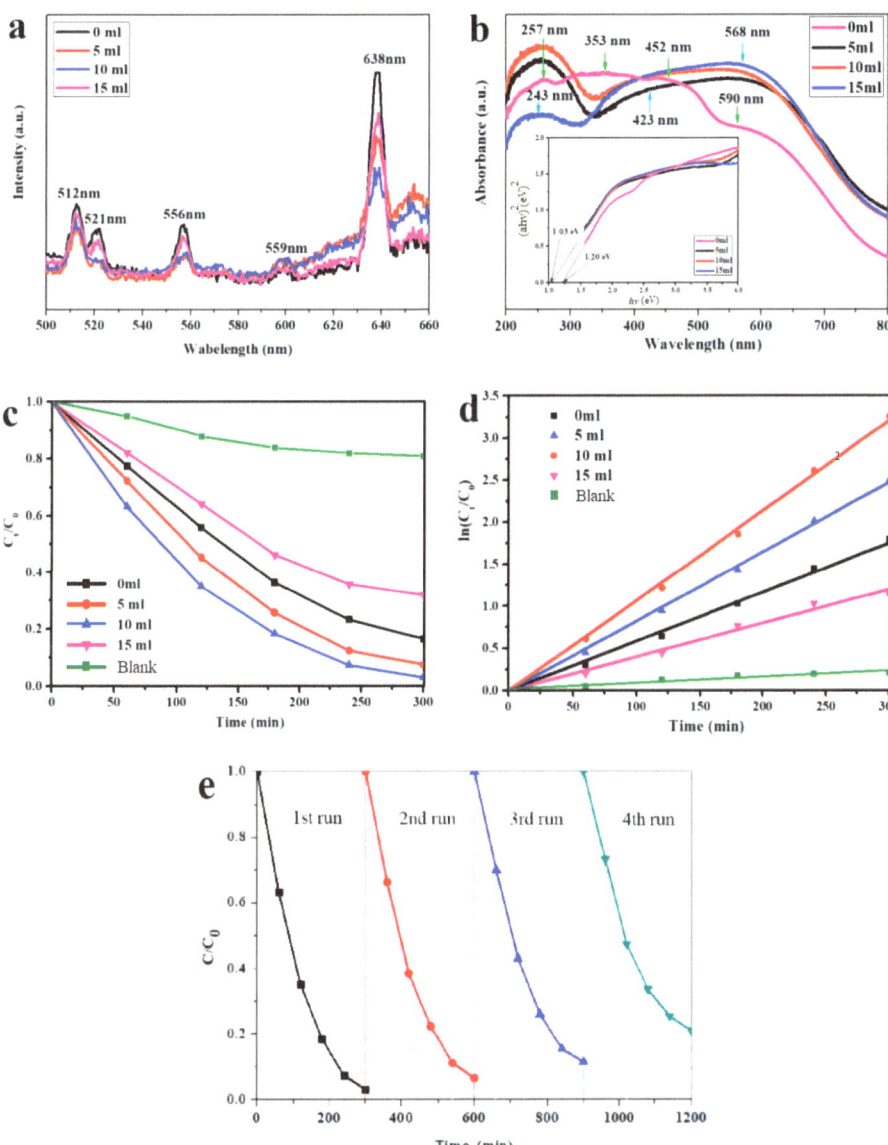

Figure 4. PL spectra (**a**), UV-vis absorption spectra (**b**), efficiency (**c**) and pseudo-first-order (**d**) of MB degradation using different CeVO$_4$-V$_2$O$_3$ samples under visible-light radiation. Cycling runs of MB photodegradation over CeVO$_4$-V$_2$O$_3$ samples with 10 mL EG (**e**).

In addition, based on the Langmuir–Hinshelwood (L-H) kinetic model, the photocatalysis degradation results conform to pseudo-first-order photocatalysis kinetics, and the corresponding reaction rate constants (k) of the different photocatalysts are calculated from the following equation [40,41]:

$$\ln(C_0/C) = kt \qquad (2)$$

where C_0 corresponds to the initial concentration of MB solution, C is the concentration of MB solution at t minutes and t stands for irradiation time. As can be clearly seen from Figure 4d, the $CeVO_4$-V_2O_3 photocatalyst prepared by adding 10 mL of EG has the highest k value (0.0107 min^{-1}), which shows that adding 10 mL EG is helpful in optimizing the photocatalytic activity of $CeVO_4$-V_2O_3 for MB. The result is similar to the rate constants of 6 wt% $CeVO_4/g$-C_3N_4 [42].

Cycling runs of photodegradation of MB were performed to determine the stability of the $CeVO_4$-V_2O_3 sample with 10 mL EG. As shown in Figure 4e, the removal rates of MB by $CeVO_4$-V_2O_3 photocatalyst after four cycles decrease from 96.9% to 87.8%, indicating excellent photocatalytic stability. The result is consistent with the cycle stability of T-$CeVO_4$ [34].

As can be seen from the characterization results, EG slightly reduces the surface area and slightly improves the absorption of visible light in the final sample while severely affecting the crystalline phase, morphology and electron–hole separation efficiency. When the additive content of EG is increased from 0 to 10 mL, the final $CeVO_4$-V_2O_3 composite is formed, which improves the electron–hole separation efficiency and thus enhances the photocatalytic activity. Further increasing the amount of EG will form amorphous V_2O_3, and the nanosheets will gather together, which will reduce the electron–hole separation efficiency, thus reducing the photocatalytic activity.

2.3. Photocatalytic Mechanism of $CeVO_4$-V_2O_3 Composites

In order to explore the reaction mechanism of $CeVO_4$-V_2O_3 in the photocatalytic process, the trapping test of active substances was carried out, as shown in Figure 5. TEOA, BQ and IPA are trapping agents for the hole (h^+), superoxide radical ($\cdot O_2^-$) and hydroxyl radical ($\cdot OH$), respectively. During the photocatalytic oxidation h^+, $\cdot OH$ and $\cdot O_2^-$ usually separately or together act as the main radicals for destroying and mineralizing pollutants [43]. Thence, we added TEOA, BQ and IPA scavengers during the removal reaction of MB to remove h^+, $\cdot O_2^-$ or $\cdot OH$ species generated in the $CeVO_4$-V_2O_3 system, respectively [44]. Figure 5 shows the degradation rate of MB is reduced from 96.9% to 70.4% by the addition of TEOA, illustrating that h^+ plays a significant role in photocatalytic oxidation. Then, after BQ was added, the degradation efficiency of MB decreased from 96.9% to 86.8%, demonstrating that fewer $\cdot O_2^-$ participated in the photocatalytic reaction, which indicated that $\cdot O_2^-$ did not play a major role in the photocatalytic process. However, with the addition of IPA, the degradation rate of MB decreased significantly from 96.9% to 35.4%, indicating that additional $\cdot OH$ is involved in the photocatalysis process, which illustrated that $\cdot OH$ was an important active substance produced in the degradation process of MB. Thus, the order of influence of active species on the photocatalytic degradation of MB by $CeVO_4$-V_2O_3 is $\cdot OH > h^+ > \cdot O_2^-$. MB is mainly degraded by $\cdot O_2^-$ and $\cdot OH$ active species in $CeVO_4$-V_2O_3 samples under the irradiation of visible light.

Electron paramagnetic resonance (EPR) has been extensively used for the identification of free radicals that are generated from advanced oxidation processes (AOPs) so as to establish the reaction mechanism. According to the energy band theory of semiconductors, when the semiconductor is irradiated by light with energy equal to or greater than the forbidden band, electrons (e^-) in the valence band are excited to transition to the conduction band, and corresponding holes (h^+) are generated in the valence band. Photo-induced holes have strong electron acquisition ability, which can capture the electrons of water or hydroxyl adsorbed on the surface of photocatalysis particles and produce hydroxyl radicals. In ESR, 5,5-dimethyl-1-pyrroline N-oxide (DMPO) is usually used as a trapping agent of ·OH, which can interact with photo-generated holes (h⁺) or ·OH on the surface of photocatalysis to form a stable DMPO-OH adduct [45].

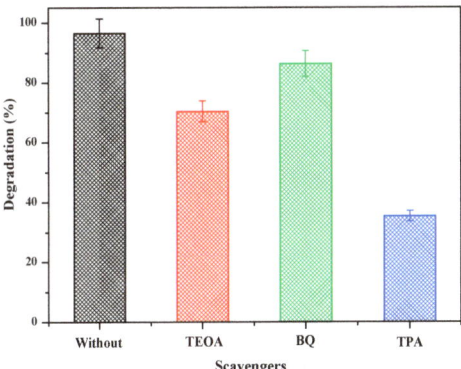

Figure 5. Trapping test of CeVO$_4$-V$_2$O$_3$ sample prepared with 10 mL EG.

To further identify the active species generated in the CeVO$_4$-V$_2$O$_3$ for 10 mL EG, the DMPO spin-trapping ESR technique was utilized to detect the production of ·OH and ·O$_2^-$. Figure 6a,b show the detection results of ·OH and ·O$_2^-$ generation, respectively. When it was not excited by visible light, ESR signals of DMPO-O$_2^-$ adducts and DMPO-OH adducts were not detected, indicating that no free radicals were produced in the CeVO$_4$-V$_2$O$_3$ sample. After being irradiated by visible light for 1 min, the characteristic peak of the DMPO-OH adduct can be clearly detected. Similarly, the typical signal of DMPO-O$_2^-$ adduct can also be detected, but the intensity is weaker than that of DMPO-OH, which is also consistent with the results of the active trapping experiment. The ESR signal intensity of spin adducts of DMPO-O$_2^-$ and DMPO-OH after 5 min of visible light irradiation is higher than that after 1 min of irradiation. Meanwhile, the signal intensity of DMPO-OH and DMPO-O$_2^-$ increase with light irradiation, which further confirms that ·OH and ·O$_2^-$ play a major role in promoting the degradation of MB. However, the intensities of DMPO-OH are obviously higher than that of DMPO-O$_2^-$, which indicates the CeVO$_4$-V$_2$O$_3$ for 10 mL EG produces additional ·OH active species. This result is consistent with the trapping test, which shows that ·OH was an important active substance produced in the degradation of MB.

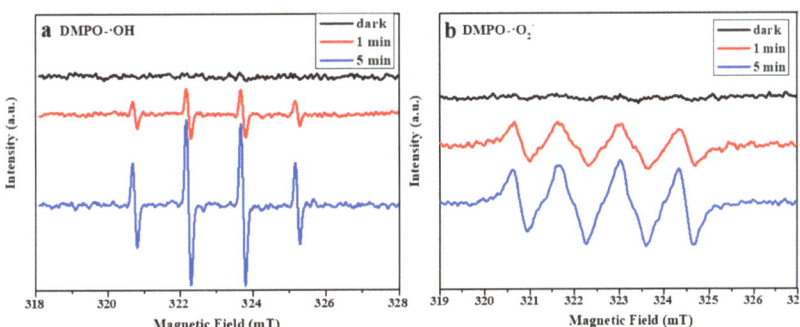

Figure 6. ESR spectra of the DMPO-·OH (**a**) and DMPO-·O$_2^-$ (**b**) for CeVO$_4$-V$_2$O$_3$ sample prepared with 10 mL EG.

Based on the above analysis, the tentative mechanism for the photocatalytic reaction of MB in CeVO$_4$-V$_2$O$_3$ samples prepared with 10 mL EG is proposed, as shown in Figure 7.

In the visible light irradiation, the photo-generated electrons in VB located in V_2O_3 and $CeVO_4$ migrate to CB in V_2O_3 and $CeVO_4$, respectively. After the coupling, the interface between V_2O_3 and $CeVO_4$ pushes the photo-generated electrons from the CB of $CeVO_4$ to the CB of V_2O_3. At the same time, the photo-generated holes are also transferred from the VB of V_2O_3 to the VB of $CeVO_4$. Therefore, faster electron transfer between $CeVO_4$ and V_2O_3 may lead to higher quantum efficiency and provide more photo-generated electrons for photocatalytic reactions. Then, the photo-generated electrons are captured by O_2 in water, resulting in superoxide radical $\cdot O^{2-}$, while the photo-generated holes are captured by the OH^- or H_2O, resulting in $\cdot OH$. The oxidizability of $\cdot OH$, $\cdot O^{2-}$ and h^+ is enough to effectively degrade MB into CO_2 and H_2O.

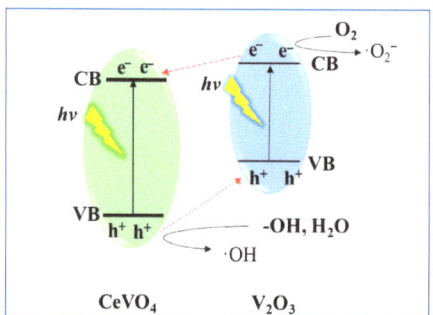

Figure 7. The photocatalytic mechanism of $CeVO_4$-V_2O_3 sample prepared with 10 mL EG.

3. Materials and Methods

3.1. Materials

All reagents used in the study, including ammonium metavanadate, ethylene glycol, cerium nitrate hexahydrate, ethylene glycol and anhydrous ethanol, are of analytical grade. They are purchased from Aladdin's company and used as received without further purification. All the solutions were prepared with deionized water obtained from a PURE ROUP 30 water purification system.

3.2. Synthesis of $CeVO_4$-V_2O_3 Composite

Typically, 2.17 g $Ce(NO_3)_3 \cdot 6H_2O$ and 0.59 g NH_4VO_3 were dispersed in 60 mL of deionized water at 90 °C under vigorous stirring for 1 h. Subsequently, 0–15 mL EG was added to the above suspension under stirring for an additional 1 h. Afterward, the obtained mixture was transferred into a Teflon-lined steel autoclave of 100 mL capacity, which was kept in an oven at 160 °C for 5 h. After naturally cooling in air, the obtained precipitates were alternately washed with deionized water and absolute ethanol three times and then dried at 100 °C for 12 h and calcined at 300 °C for 5 h (in N_2 atmosphere). A schematic representation of the synthesis of the $CeVO_4$-V_2O_3 composite is shown in Figure 8.

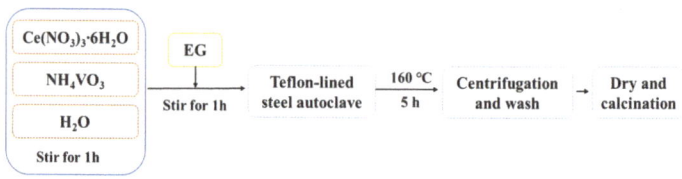

Figure 8. The schematic representation of synthesis of $CeVO_4$-V_2O_3 composite.

3.3. Characterization of $CeVO_4$-V_2O_3 Composite

The crystalline phase of the products was characterized by X-ray Diffractometer (Bruker Advance D-8, Billerica, MA, USA) with PIXel3D detector using CuKa radiation.

The diffraction patterns were matched with that of the recorded standards in JCPDS. The materials' surface morphology was studied with the aid of scanning electron microscopy-SEM (Hitachi S-4300 Field Emission SEM, Tokyo, Japan). The absorption spectra were determined by UV-Vis DRS (UV-2550), and the emission spectra were measured by photoluminescence spectra (PL, Spex Fluorolog-3, Metuchen, NJ, USA). The surface area was studied by N_2 adsorption–desorption (ASAP 2020). The elemental components and bonding energies of the synthesized samples were measured on X-ray photoelectron spectroscopy (XPS, VG Multilab 2000, London, UK).

3.4. Photocatalytic Performance Experiment

By studying the degradation of MB aqueous solution under visible light irradiation, the photocatalytic activities of $CeVO_4$-V_2O_3 composites with the addition of 0, 5, 10 and 15 were evaluated. A 500 W Xe discharge lamp with a 420 nm cut-off filter, equipped with a circulating water source and a 50 mg/L concentration of the dye, was prepared and used for the study. For each measurement, 50 mg of the catalyst was suspended in MB aqueous solution and stirred vigorously in the dark for 60 min to ensure the establishment of adsorption equilibrium between the catalyst surface and the dye. Then, the suspension was exposed to a 500 W Xe discharge lamp with a 420 nm cut-off filter and equipped with a circulating water source and was continuously stirred for 300 min. During this process, 5 mL of the suspension in the reaction system was taken out for testing at 60, 120, 180, 240 and 300 min, respectively. The resulting supernatants were examined using a UV-1100 spectrophotometer, and the absorbance of the MB solution was measured at its characteristic absorption wavelength of 664 nm. The degradation curve of the dye was studied by absorption spectrum. Use the formula in Equation (3) to calculate the degraded MB concentration.

$$\text{Degradation } (\%) = \frac{C_0 - C_i}{C_0} \times 100\% = \frac{A_0 - A_i}{A_0} \times 100 \, (\%) \tag{3}$$

where C_i is the concentration of MB at specific time; C_0 denotes the initial concentration of MB; A_i is the absorbance of MB at different times and A_0 denotes the blank absorbance of the original MB solution.

3.5. Photocatalytic Mechanism Tests

The trapping tests are similar to the photodegradation experiment. The difference is that 1 mmol of isopropanol (IPA), triethanolamine (TEOA) or benzoquinone (BQ) is added into the reaction solution before irradiation, and the degradation rate of MB after irradiation for 300 min is analyzed. Species (h^+, $\cdot O_2^-$ and $\cdot OH$) formed in the photodegradation process were studied with electron spin resonance (ESR, Bruker E500, Billerica, MA, USA) by adding 5,5-dimethyl-1-pyrroline N-oxide (DMPO, >99.0%) into ultrapure water and methanol, respectively.

4. Conclusions

The effects of EG on the crystalline phase, surface area, morphology, chemical composition, electron–hole separation efficiency and optical properties of the final product have been studied using XRD, N_2 adsorption–desorption, SEM, XPS, PL and DRS. The results show that a proper amount of EG can form $CeVO_4$-V_2O_3 composites. With the addition of EG, the specific surface area of the composites decreased slightly, the average particle size increased, and the visible light absorption increased. At the same time, the photocatalytic removal effect of $CeVO_4$-V_2O_3 composites on MB shows that adding moderate EG into the precursor solution (0–10 mL) can improve the electron–hole separation efficiency of the final products, thus increasing the photocatalytic activity (83.6% to 96.9%). However, excessive EG (15 mL) makes the nanosheets gather together to form aggregates, which reduces the separation efficiency of electron holes, thus reducing the photocatalytic activity (67.5%). According to the trapping experiments and ESR experiments, h^+ and $\cdot OH$ played a more significant part in the photocatalytic degradation of MB than $\cdot O_2^-$, and the order of influence of active species on the photodegradation of MB over $CeVO_4$-V_2O_3 sample prepared with 10 mL EG is $\cdot OH > h^+ > \cdot O_2^-$. Finally, the mechanism of photocatalytic degradation of MB by the $CeVO_4$-V_2O_3 composites was discussed. The coupling of V_2O_3 and $CeVO_4$ realizes the effective separation and faster transfer of photo-generated electrons and holes and leads to higher quantum efficiency. When they are captured by O_2 and OH^- in water, active species, $\cdot O_2^-$ and $\cdot OH$, are generated. The oxidizability of $\cdot OH$, $\cdot O_2^-$ and h^+ is enough to effectively degrade MB into CO_2 and H_2O.

Author Contributions: Data curation, Y.W.; formal analysis, Y.W., M.J. and X.T.; investigation, M.J. and H.R.; methodology, X.L.; resources, Y.W., G.L. and X.Z.; writing—original draft, Y.W.; writing—review and editing, Y.W., Y.J. and X.T. All authors have read and agreed to the published version of the manuscript.

Funding: This work was supported by the "School Enterprise Cooperation Project" for Domestic Visiting Engineers of Colleges and Universities (No. FG2022260), the project of Taizhou Science and technology planning (2101gy32) and the cultivating project of Taizhou vocational and technical college (2021PY04).

Informed Consent Statement: Written informed consent has been obtained from the patient(s) to publish this paper.

Data Availability Statement: Not applicable.

Conflicts of Interest: The authors declare that they have no known competing financial interest or personal relationship that could have appeared to influence the work reported in this paper.

References

1. Zhao, Z.; Ma, S.; Gao, B.; Bi, F.; Qiao, R.; Yang, Y.; Wu, M.; Zhang, X. A systematic review of intermediates and their characterization methods in VOCs degradation by different catalytic technologies. *Sep. Purif. Technol.* **2023**, *314*, 123510. [CrossRef]
2. Zhao, S.; Yang, Y.; Bi, F.; Chen, Y.; Wu, M.; Zhang, X.; Wang, G. Oxygen vacancies in the catalyst: Efficient degradation of gaseous pollutants. *Chem. Eng. J.* **2023**, *454*, 140376. [CrossRef]
3. Shen, Q.; Lu, Z.; Bi, F.; Fang, Y.; Song, L.; Yang, Y.; Wu, M.; Zhang, X. Effect of actual working conditions on catalyst structure and activity for oxidation of volatile organic compounds: A review. *Fuel* **2023**, *343*, 128012. [CrossRef]
4. Zhang, X.; Yue, K.; Rao, R.; Chen, J.; Liu, Q.; Yang, Y.; Bi, F.; Wang, Y.; Xu, J.; Liu, N. Synthesis of acidic MIL-125 from plastic waste: Significant contribution of N orbital for efficient photocatalytic degradation of chlorobenzene and toluene. *Appl. Catal. B* **2022**, *310*, 121300. [CrossRef]
5. Guo, W.; Sun, W.; Lv, L.; Kong, S.; Wang, Y. Microwave-assisted morphology evolution of Fe-based metal–organic frameworks and their derived Fe_2O_3 nanostructures for li-Ion storage. *ACS Nano.* **2017**, *11*, 4198–4205. [CrossRef]
6. Chen, X.; Zhang, H.; Ci, C.; Sun, W.; Wang, Y. Few-Layered Boronic Ester Based Covalent Organic Frameworks/Carbon Nanotube Composites for High-Performance K-Organic Batteries. *ACS Nano.* **2019**, *13*, 3600–3607. [CrossRef]
7. Yang, Y.; Zhao, S.; Bi, F.; Chen, J.; Li, Y.; Cui, L.; Xu, J.; Zhang, X. Oxygen-vacancy-induced O_2 activation and electron-hole migration enhance photothermal catalytic toluene oxidation. *Cell Rep. Phys. Sci.* **2022**, *3*, 101011. [CrossRef]
8. Yang, Y.; Zhao, S.; Bi, F.; Chen, J.; Wang, Y.; Cui, L.; Xu, J.; Zhang, X. Highly efficient photothermal catalysis of toluene over Co_3O_4/TiO_2 p-n heterojunction: The crucial roles of interface defects and band structure. *Appl. Catal. B* **2022**, *315*, 121550. [CrossRef]
9. Zou, X.; Yuan, C.; Dong, Y.; Ge, H.; Ke, J.; Cui, Y. Lanthanum orthovanadate/bismuth oxybromide heterojunction for enhanced photocatalytic air purification and mechanism exploration. *Chem. Eng. J.* **2020**, *379*, 122380. [CrossRef]
10. Chen, J.; Yang, Y.; Zhao, S.; Bi, F.; Song, L.; Liu, N.; Xu, J.; Wang, Y.; Zhang, X. Stable black phosphorus encapsulation in porous mesh-like UiO-66 promoted charge transfer for photocatalytic oxidation of toluene and o-Dichlorobenzene: Performance, degradation pathway, and mechanism. *ACS Catal.* **2022**, *12*, 8069–8081. [CrossRef]
11. Rao, R.; Ma, S.; Gao, B.; Bi, F.; Chen, Y.; Yang, Y.; Liu, N.; Wu, M.; Zhang, X. Recent advances of metal-organic frameworks-based and their derivatives materials in heterogeneous catalytic removal of VOCs. *J. Colloid Interf. Sci.* **2023**, *636*, 55–72. [CrossRef] [PubMed]
12. Fan, C.; Liu, Q.; Ma, T.; Shen, J.; Yang, Y.; Tang, H.; Wang, Y.; Yang, J. Fabrication of 3D $CeVO_4$/graphene aerogels with efficient visible-light photocatalytic activity. *Ceram. Int.* **2016**, *42*, 10487–10492. [CrossRef]
13. Gu, L.; Feng, B.; Xi, P.; Xu, J.; Chen, B.; Zhong, S.; Yang, W. Elucidating the role of Mo doping in enhancing the meta-Xylene ammoxidation performance over the $CeVO_4$ catalyst. *Chem. Eng. J.* **2023**, *459*, 141645. [CrossRef]

14. Sun, L.; Ye, X.; Cao, Z.; Zhang, C.; Yao, C.; Ni, C.; Li, X. Upconversion enhanced photocatalytic conversion of lignin biomass into valuable product over CeVO$_4$/palygorskite nanocomposite: Effect of Gd^{3+} incorporation. *Appl. Catal. A-Gen.* **2022**, *648*, 118923. [CrossRef]
15. Zonarsaghar, A.; Mousavi-Kamazani, M.; Zinatloo-Ajabshir, S. Sonochemical synthesis of CeVO$_4$ nanoparticles for electrochemical hydrogen storage. *Int. J. Hydrogen Energ.* **2022**, *47*, 5403–5417. [CrossRef]
16. Li, Y.; Zhang, Z.; Zhao, X.; Liu, Z.; Zhang, T.; Niu, X.; Zhu, Y. Effects of Nb-modified CeVO$_4$ to form surface Ce-O-Nb bonds on improving low-temperature NH$_3$-SCR deNOx activity and resistance to SO$_2$ & H$_2$O. *Fuel* **2023**, *331*, 125799.
17. Xie, B.; Lu, G.; Dai, Q.; Wang, Y.; Guo, Y.; Guo, Y. Synthesis of CeVO$_4$ Crystals with Different Sizes and Shapes. *J. Clust. Sci.* **2011**, *22*, 555–561. [CrossRef]
18. Mosleh, M. Mahinpour, Afsane. Sonochemical synthesis and characterization of cerium vanadate nanoparticles and investigation of its photocatalyst application. *J. Mater. Sci. Mater. Electron.* **2016**, *27*, 8930–8934. [CrossRef]
19. Phuruangrat, A.; Kuntalue, B.; Thongtem, S.; Thongtem, T. Effect of PEG on phase, morphology and photocatalytic activity of CeVO$_4$ nanostructures. *Mater. Lett.* **2016**, *174*, 138–141. [CrossRef]
20. Lu, G.; Zou, X.; Wang, F.; Wang, H.; Li, W. Facile fabrication of CeVO$_4$ microspheres with efficient visible-light photocatalytic activity. *Mater. Lett.* **2017**, *195*, 168–171. [CrossRef]
21. Othman, I.; Zain, J.; Haija, M.; Banat, F. Catalytic activation of peroxymonosulfate using CeVO$_4$ for phenol degradation: An insight into the reaction pathway. *Appl. Cata. B Envron.* **2020**, *266*, 118601. [CrossRef]
22. Phuruangrat, A.; Thongtem, S.; Thongtem, T. Synthesis, characterization, and UV light-driven photocatalytic properties of CeVO$_4$ nanoparticles synthesized by sol-gel method. *J. Aust. Ceram. Soc.* **2021**, *57*, 597–604. [CrossRef]
23. Liu, H.; Chang, J. CeVO$_4$ yolk-shell microspheres constructed by nanosheets with enhanced lithium storage performances. *J. Alloys Compd.* **2021**, *849*, 156682. [CrossRef]
24. Lu, G.; Lun, Z.; Liang, H.; Hui, W.; Zheng, L.; Wei, M. In situ fabrication of BiVO$_4$-CeVO$_4$ heterojunction for excellent visible light photocatalytic degradation of levofloxacin. *J. Alloys Compd.* **2019**, *772*, 122–131. [CrossRef]
25. Ma, J.; Zhang, C.; Hong, X.; Liu, J. Incorporating Cerium Vanadate into Multi-Walled Carbon Nanotubes for Fabrication of Sensitive Electrochemical Sensors toward Sulfamethazine Determination in Water Samples. *Chemosensors* **2023**, *11*, 64. [CrossRef]
26. Song, Y.; Wang, R.; Li, X.; Shao, B.; You, H.; He, C. Constructing a novel Ag nanowire@CeVO$_4$ heterostructure photocatalyst for promoting charge separation and sunlight driven photodegradation of organic pollutants. *Chin. Chem. Lett.* **2022**, *33*, 1283–1287. [CrossRef]
27. Zhang, X.; Zhao, Z.; Zhao, S.; Xiang, S.; Gao, W.; Wang, L.; Xu, J.; Wang, Y. The promoting effect of alkali metal and H$_2$O on Mn-MOF derivatives for toluene oxidation: A combined experimental and theoretical investigation. *J. Catal.* **2022**, *415*, 218–235. [CrossRef]
28. Bi, F.; Zhao, Z.; Yang, Y.; Liu, N.; Huang, Y.; Zhang, X. Chlorine-coordinated Pd single atom enhanced the chlorine-resistance for volatile organic compounds degradation: Mechanism study. *Environ. Sci. Technol.* **2022**, *56*, 17321–17330. [CrossRef]
29. Wang, H.; Zou, W.; Liu, C.; Sun, Y.; Xu, Y.; Sun, W.; Wang, Y. β-Ketoenamine-Linked Covalent Organic Framework with Co Intercalation: Improved Lithium-Storage Properties and Mechanism for High-Performance Lithium-Organic Batteries. *Batter. Supercaps* **2023**, *6*, e202200434. [CrossRef]
30. Bi, F.; Ma, S.; Gao, B.; Yang, Y.; Wang, L.; Fei, F.; Xu, J.; Huang, Y.; Wu, M.; Zhang, X. Non-oxide supported Pt-metal-group catalysts for efficiently CO and toluene co-oxidation: Difference in water resistance and degradation intermediates. *Fuel* **2023**, *264*, 118464. [CrossRef]
31. Zhang, X.; Hou, F.; Li, H.; Yang, Y.; Wang, Y.; Liu, N.; Yang, Y. A strawsheave-like metal organic framework Ce-BTC derivative containing high specific surface area for improving the catalytic activity of CO oxidation reaction. *Micropor. Mesopor. Mater.* **2018**, *259*, 211–219. [CrossRef]
32. Shah, A.; Liu, Y.; Jin, W.; Chen, W.; Mehmood, I.; Nguyen, V. Highly selective ethanol sensing properties of hydrothermally synthesized cerium orthovanadate (CeVO$_4$) nanorods. *Mater. Lett.* **2015**, *154*, 144–147. [CrossRef]
33. Lu, G.; Song, B.; Li, Z.; Liang, H.; Zou, X. Photocatalytic degradation of naphthalene on CeVO$_4$ Nanoparticles Under Visible Light. *Chem. Eng. J.* **2020**, *402*, 125645. [CrossRef]
34. Yin, B.; Du, W.; Zhang, Y.; Gao, X.; Ma, C.; Guo, M. Preparation and photocatalytic performance of biomimetic wood structure cerium vanadate. *J. For. Eng.* **2022**, *7*, 38–45.
35. Yadav, R.; Rai, S. Structural analysis and enhanced photoluminescence via host sensitization from a lanthanide doped BiVO$_4$ nano-phosphor. *J. Phys. Chem. Solids* **2017**, *110*, 211–217. [CrossRef]
36. Deng, H.; Yang, S.; Xiao, S.; Gong, H.; Wang, Q. Controlled synthesis and upconverted avalanche luminescence of cerium(III) and neodymium(III) orthovanadate nanocrystals with high uniformity of size and shape. *J. Am. Chem. Soc.* **2008**, *130*, 2032–2040. [CrossRef]
37. Zheng, L.; Han, S.; Liu, H.; Yu, P.; Fang, X. Hierarchical MoS$_2$ nanosheet@TiO$_2$ nanotube array composites with enhanced photocatalytic and photocurrent performances. *Small* **2016**, *12*, 1527–1536. [CrossRef]
38. Xie, Y.; Zhou, Y.; Gao, C.; Liu, L.; Zhang, Y.; Chen, Y.; Shao, Y. Construction of AgBr/BiOBr S-scheme heterojunction using ion exchange strategy for high-efficiency reduction of CO$_2$ to CO under visible light. *Sep. Purif. Technol.* **2022**, *303*, 122288. [CrossRef]
39. Niu, C.; Wang, S.; Li, X.; Chen, F.; Sun, J. Enhanced Photocatalytic Degradation of Tetracycline by AgI/BiVO$_4$ Heterojunction under Visible-Light Irradiation: Mineralization Efficiency and Mechanism. *ACS Appl. Mater. Inter.* **2016**, *8*, 32887–32900.

40. Li, C.; Zhang, P.; Lv, R.; Lu, J.; Wang, T.; Wang, S. Selective Deposition of Ag$_3$PO$_4$ on Monoclinic BiVO$_4$ (040) for Highly Efficient Photocatalysis. *Small* **2013**, *9*, 3951–3956. [CrossRef]
41. Boitumelo, M.; Opeyemi, A.; Sam, R.; Lydia, M.; Damian, C. Nanocomposite of CeVO$_4$/BiVO$_4$ Loaded on Reduced Graphene Oxide for the Photocatalytic Degradation of Methyl Orange. *J. Clust. Sci.* **2022**, *33*, 2707–2721.
42. Ren, J.; Wu, Y.; Zou, H.; Dai, Y.; Sha, D.; Chen, M.; Wang, J.; Pan, J.; Yan, X. Synthesis of a novel CeVO$_4$/graphitic C$_3$N$_4$ composite with enhanced visible-light photocatalytic property. *Mater. Lett.* **2016**, *183*, 219–222. [CrossRef]
43. Yang, L.; Li, X.; Zhang, X.; Huang, C. Supercritical solvothermal synthesis and formation mechanism of V$_2$O$_3$ microspheres with excellent catalytic activity on the thermal decomposition of ammonium perchlorate. *J. Alloys Compd.* **2019**, *806*, 1394–1402. [CrossRef]
44. Li, X.; Wang, Z.; Shi, H.; Dai, D.; Zuo, S.; Yao, C.; Ni, C. Full spectrum driven SCR removal of NO over hierarchical CeVO$_4$/attapulgite nanocomposite with high resistance to SO$_2$ and H$_2$O. *J. Hazard. Mater.* **2020**, *386*, 121977. [CrossRef] [PubMed]
45. Ai, J.; Yang, Y.; Xian, Y.; Li, J.; Li, L.; Jin, L. ESR study on the photoinduced production of hydroxyl radicals by nano-TiO$_2$. *J. East Chin. Norm. Univ. Nat. Sci. Ed.* **2004**, *3*, 76–81.

Disclaimer/Publisher's Note: The statements, opinions and data contained in all publications are solely those of the individual author(s) and contributor(s) and not of MDPI and/or the editor(s). MDPI and/or the editor(s) disclaim responsibility for any injury to people or property resulting from any ideas, methods, instructions or products referred to in the content.

Article

Sonocatalytic Degradation of Chrysoidine R Dye Using Ultrasonically Synthesized NiFe$_2$O$_4$ Catalyst

Yogesh M. Gote, Pankaj S. Sinhmar and Parag R. Gogate *

Department of Chemical Engineering, Institute of Chemical Technology, Matunga, Mumbai 400019, India
* Correspondence: pr.gogate@ictmumbai.edu.in

Abstract: The novel ultrasound-assisted co-precipitation method was successfully applied for the synthesis of the NiFe$_2$O$_4$ catalyst, which offered the advantages of lower particle size and better crystalline structure without affecting the phase planes. Furthermore, the efficacy of synthesized catalysts was evaluated using ultrasound-assisted catalytic degradation of Chrysoidine R dye. The study was designed to evaluate the effect of different parameters, such as pH, duty cycle, power output, and catalyst loading on Chrysoidine R dye degradation using a 5 wt% NiFe$_2$O$_4$ catalyst synthesized ultrasonically. At the optimized condition of 120 W ultrasonic power, 70% duty cycle, 3 pH, 0.5 g/L catalyst loading, and 160 min of reaction time, the best degradation of 45.01% was obtained. At similar conditions, the conventionally synthesized catalyst resulted in about 15% less degradation. Chrysoidine R dye degradation was observed to follow second-order kinetics. To accelerate the degradation, studies were performed using hydrogen peroxide at various loadings where it was elucidated that optimum use of 75 ppm loading showed the maximum degradation of 92.83%, signifying the important role of the co-oxidant in ultrasound-assisted catalytic degradation of Chrysoidine R dye. Overall, the present study clearly demonstrated the potential benefits of ultrasound in catalyst synthesis as well as in catalytic degradation.

Keywords: Chrysoidine R dye; ultrasound-assisted synthesis; NiFe$_2$O$_4$; sonocatalytic degradation

Citation: Gote, Y.M.; Sinhmar, P.S.; Gogate, P.R. Sonocatalytic Degradation of Chrysoidine R Dye Using Ultrasonically Synthesized NiFe$_2$O$_4$ Catalyst. *Catalysts* **2023**, *13*, 597. https://doi.org/10.3390/catal13030597

Academic Editor: Meng Li

Received: 18 February 2023
Revised: 11 March 2023
Accepted: 14 March 2023
Published: 16 March 2023

Copyright: © 2023 by the authors. Licensee MDPI, Basel, Switzerland. This article is an open access article distributed under the terms and conditions of the Creative Commons Attribution (CC BY) license (https://creativecommons.org/licenses/by/4.0/).

1. Introduction

Wastewater treatment is an essential element of sustainability as natural water resources are limited and purification of saline water is costly. The hazardous effluents discharged by various industrial sectors accelerate the pollution levels and result in contamination of freshwater resources, which has been a hot topic of discussion for over a century. Hazardous-colored wastewater produced by textile enterprises is one of the leading causes of water pollution as most of the color pigments are resistant to bio-degradation, temperature, light, and other chemicals [1]. The removal of these dye compounds is the key issue in effluent treatment for textile-processing companies. Since the traditional wastewater treatment methods often fail to provide sufficient pollutant degradation efficiency, produce secondary pollutants, and are time consuming processes that are not cost effective, new greener and innovative degradation methods have been researched actively lately. One such method is the sonocatalytic degradation approach, which has been reported to actively degrade different dye pollutants, such as methyl orange using MnO$_2$/CeO$_2$ [2], anazolene sodium using Dy-doped Cdse [3], methylene blue using CoFe$_2$O$_4$/SiO$_2$/CuMoF [4], polyenes using Se/Fe$_2$O$_4$ [5], brilliant green using CaFe$_2$O$_4$ [6], and Congo red using Ni-Co-MnO$_2$ [7] as catalysts. Ultrasound induces the formation, subsequent growth under the compression—rarefaction cycles, and intense collapse of micro-bubbles in an aqueous mixture leading to much higher temperature and pressure pulses, considered as local hot spots. During the cavity collapse, water molecules and oxidants undergo thermal dissociation, mainly generating the hydroxyl radicals and other oxidizing agents, depending on the type of oxidants used or the components present in the liquid mixture. Ultrasound alone, however, is not an effective approach due to lower degradation efficiency and higher processing times for the target dye molecule, and coupling with a catalyst and oxidant intensifies the degradation based on generation of higher quantum of active radicals, resulting in higher degradation of pollutants present in effluent.

Among the different semiconductor particles that are gathering attention, nickel ferrite is a promising catalyst in the field of catalytic degradation as it possesses a band gap of 1.53 eV [8], offering broad absorption in visible regions, which is beneficial to catalysis applications [9]. $NiFe_2O_4$ can also provide useful catalytic action during the sono-catalytic degradation and for the dissociation of hydrogen peroxide. Considering these aspects, the selection of $NiFe_2O_4$ as the catalyst complex in the current work is justified.

The synthesis of various metal-loaded semiconductors has been reported using various synthesis methods, such as solvo-thermal [10], molten salt synthesis [11], hydrothermal [12], and glycerol-assisted sol-gel [13]. The major demerits of these methods are large crystalline zone, high-temperature requirements, and lack of homogeneity. The simple and easy route of synthesis is chemical co-precipitation [14–16], which offers the benefits of mild operating conditions, but it is often observed that the particle size and catalytic activity are poor though it offers good homogeneity and better crystallinity than other methods. To overcome these demerits, we propose a modified ultrasound-assisted co-precipitation method for nickel ferrite synthesis. In the current study, the sonocatalytic degradation of Chrysoidine R dye has also been investigated using the synthesized nickel ferrite catalyst. The influence of various parameters, such as duty cycle, pH, power, and catalyst loading on the degradation of Chrysoidine R dye were evaluated. The catalytic activity of synthesized catalyst samples was examined under the optimized process conditions. The effect of use of an additional oxidant on sono-catalytic degradation was also investigated. Based on the literature assessment, it can be clearly stated that the degradation of Chrysoidine R dye using an ultrasound-assisted catalytic approach based on ultrasonically synthesized $NiFe_2O_4$ catalyst has not been evaluated, establishing the novelty of the current work.

2. Results and Discussions

2.1. Characterization of Nickel Ferrite Oxide Catalyst

To examine the potential advantages offered by the $NiFe_2O_4$ catalyst complex obtained using the ultrasound assisted synthesis method compared to that of the conventional method, the samples were characterized using particle size and XRD analysis.

2.1.1. Particle Size Analysis

Quantification of the particle size revealed the ultrasonically prepared catalyst complex resulted in a particle size of 15.58 µm, which was found to be approximately 4.31 times smaller than the conventionally synthesized catalyst (80.62 µm), as depicted in Figure 1a,b.

The obtained results clearly indicate that the proposed sonication method can significantly reduce the particle size, hence lowering post-micronizing costs. Use of the ultrasound in the catalyst synthesis generally aids in achieving lower particle size due to the turbulence and shear effects generated during the cavity collapse. Vardikar et al. [17] reported a similar finding of a reduced particle size of 293 nm for the ultrasonically produced KL-CTS-TiO_2 catalyst as opposed to a particle size of 439 nm for the conventional approach. Díez-García et al. [18] also reported that $NiTiO_3$ nanorods synthesized under sonication resulted in smaller size (1.8 µm in length and 0.6 µm in dia) nanorods as compared to the conventional approach of synthesis (2.2 µm in length and 0.7 µm in dia). Dalvi et al. [19] also observed a significant reduction in particle size of Fe-doped TiO_2 catalyst from 806.4 µm to 31.22 µm on shifting a conventional approach to the ultrasound-assisted synthesis approach.

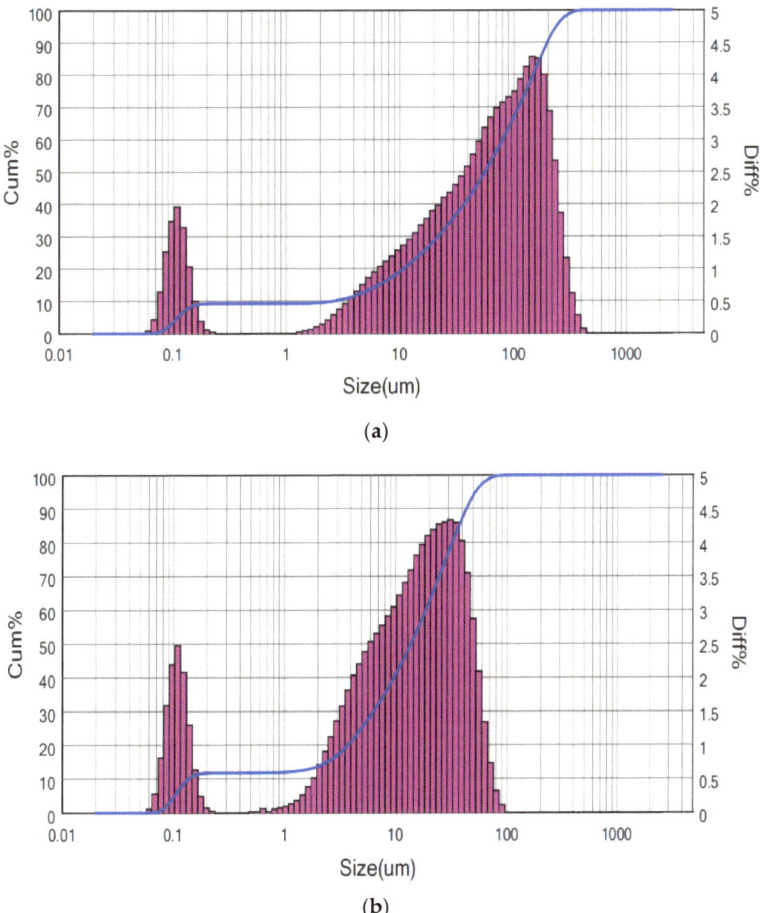

Figure 1. Particle size distribution graphs for (**a**) ultrasonically prepared and (**b**) conventionally prepared $NiFe_2O_4$ catalysts at 5 wt% Ni loading.

2.1.2. X-ray Diffraction Analysis

The phase purity and crystal structure of synthesized $NiFe_2O_4$ catalysts were established using the XRD diffraction analysis. The obtained XRD plots of conventional and ultrasound-assisted synthesized catalysts are depicted in Figure 2.

The prominent peaks were observed at 2θ = 18.51°, 30.98°, 36.01°, 37.17°, 43.68°, 54.21°, 57.69°, 63.35°, and 75.66° with planes as (111), (220), (311), (222), (400), (422), (511), (440), and (533), signifying the cubic spinel structure of the $NiFe_2O_4$ complex [20–22]. XRD spectrums confirmed that the novel ultrasound-assisted synthesis method does not alter the structure of the $NiFe_2O_4$ complex since both spectra have the same planes. When compared to the conventional synthesis approach, the peak intensities of the ultrasonically synthesized catalyst complex were found to be greater, indicating the creation of crystalline particles with higher crystallinity. Besides this, the absence of any intermediate additional peak revealed the formation of a highly pure catalyst complex. In general, it was observed that ultrasound facilitated the formation of an effective catalyst, especially with respect to crystalline content and lower particle size, which can offer better results in the catalytic reaction. With an objective to confirm this hypothesis, actual degradation studies for

Chrysoidine R dye were also performed, providing a detailed understanding into the effect of operating conditions.

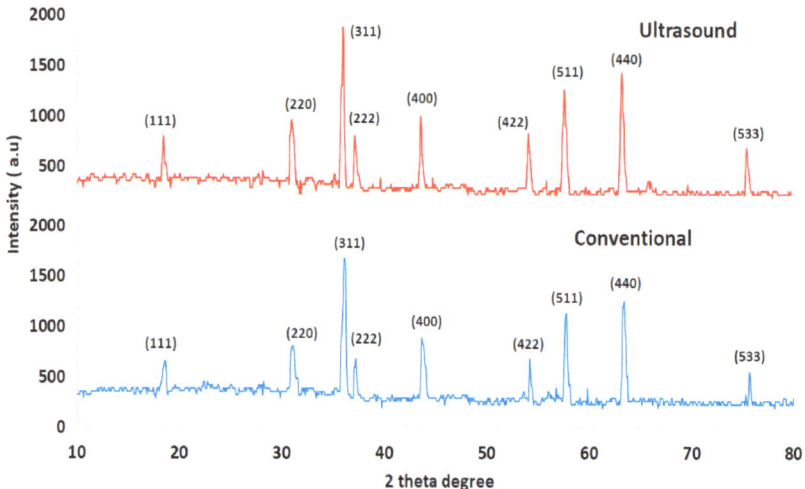

Figure 2. XRD spectrum of ultrasonically prepared and conventionally prepared $NiFe_2O_4$ catalysts at 5 wt% Ni loading.

2.2. Ultrasound-Assisted Chrysoidine R Dye Degradation Studies

2.2.1. Effect of pH on the Degradation of Chrysoidine R Dye

pH is a crucial factor to be considered while studying dye degradation as it is widely reported to influence the degradation rates. In the present study, the pH of the 20-ppm dye solution was varied between 3 to 9. The degradation studies were performed using 0.4 g/L catalyst loading (ultrasonically synthesized) under conditions of 40 °C, time duration of 180 min, ultrasound power of 100 W, and duty cycle of 70%. The obtained results are plotted in Figure 3a in terms of the extent of degradation and 3b in terms of kinetic studies.

It was observed that degradation increased when the solution was made acidic. At 9 pH, the maximum Chrysoidine R dye degradation observed was only 5.08%, which increased to 15.06% and 34.03% when the acidity of the dye solution increased to pH of 5 and 4, respectively. Furthermore, the degradation showed an increasing trend as the acidity of the dye solution further increased from pH of 4 to 3. At 3 pH, maximum dye degradation of 39.49% was noted, and pH 3 was chosen as the best operating condition for further experimental studies, keeping in mind the corrosion and life of the ultrasonic horn (still lower pH were not investigated in the work). The observed trend of higher degradation under acidic conditions can be attributed to the presence of a pollutant in the molecular state instead of the ionized state. In the molecular state, hydrophobicity shifts the molecules towards the liquid–gas interface where the concentration of •OH radicals is much higher to facilitate direct attack, resulting in higher degradation. In bulk liquid, the •OH radical concentration is less, and hence, lower extent of degradation is seen. In addition, the oxidation capacity of the hydroxyl radicals that are mainly involved in the degradation process is also higher under acidic conditions. It was also seen that Chrysoidine R dye degradation followed second-order reaction kinetics. The obtained kinetic data are tabulated in Table 1.

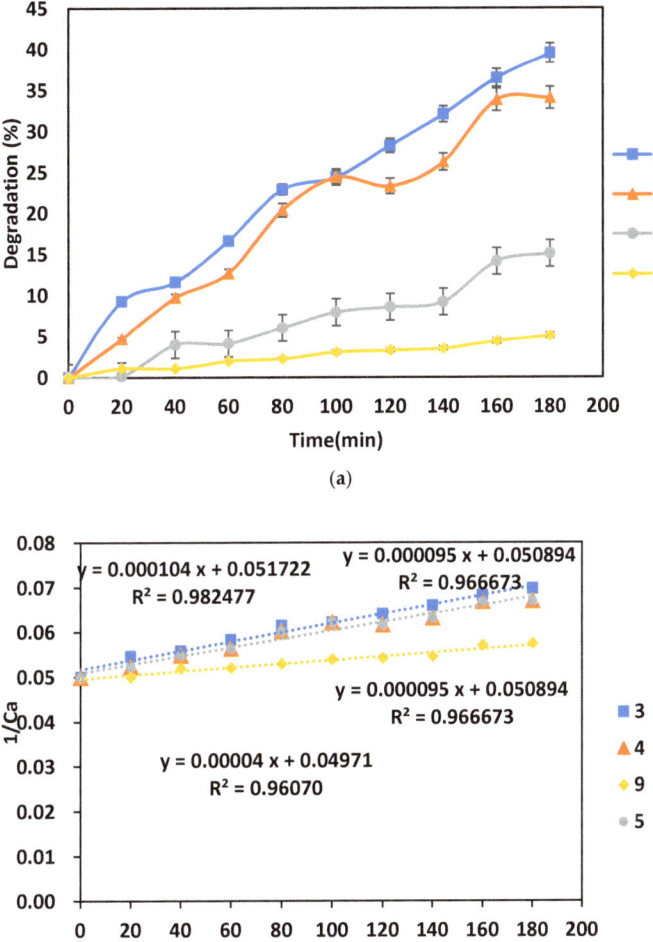

Figure 3. Effect of pH on the degradation of Chrysoidine R dye: (**a**) extent of degradation and (**b**) kinetic fitting.

Table 1. Extent of degradation and kinetic rate constant data at various pH.

pH	Degradation (%)	Second-Order Rate Constant ($K \times 10^{-5}$ mL^{-1}min^{-1})	R^2
3	39.49	10.4	0.9824
4	34.03	9.5	0.9667
5	15.06	9.5	0.9667
9	5.08	4.2	0.9606

It was again elucidated from Table 1 that the kinetic rate constant also showed similar trends to that of the extent of degradation variation with pH. Liu et al. [23] studied the rhodamine B dye degradation over the pH range of 2–4.5 and found at pH 3, the degradation was maximum with a value of 90%. Shiljashree et al. [24] studied Irgalite violet dye degradation using $NiFe_2O_4$ nanoparticles over the range of pH 2.0 to 4.0 and reported maximum degradation of 99.9% was seen at pH 3.0 with lower degradation as 78% for pH of 4. The effect of pH on Remazol Red RB and Direct Green B degradation under the condition of 5 mg/L as nickel ferrite nanoparticle loading, ozone concentration of 55 g/m^3, 1 L of dye solution with a concentration of dye as 150 mg/L, and temperature as 25 °C was analyzed by Mahmoodi [25] at pH values of 3, 5, 7, and 9. It was reported that a pH of 3 results in maximum degradation of 86% for RRRB and 90% for DGB. It was clearly elucidated that acidic conditions (best pH of 3) favored Chrysoidine R dye degradation more efficiently and hence was selected for further studies.

2.2.2. Ultrasonic Duty Cycle Effect on Degradation of Chrysoidine R Dye

The ultrasound duty cycle represents the on and off time of ultrasonic irradiation. Duty cycle effect on chrysoidine R dye degradation was investigated by varying the duty cycle over the range of 50–80%. The experiments were performed at a constant power output of 100 W, 0.4 g/L as the catalyst loading, 40 °C temperature, 180 min as treatment time, and pH of 3. Figure 4 represents the graphical plot of the obtained results in terms of extent of degradation and kinetic rate constants.

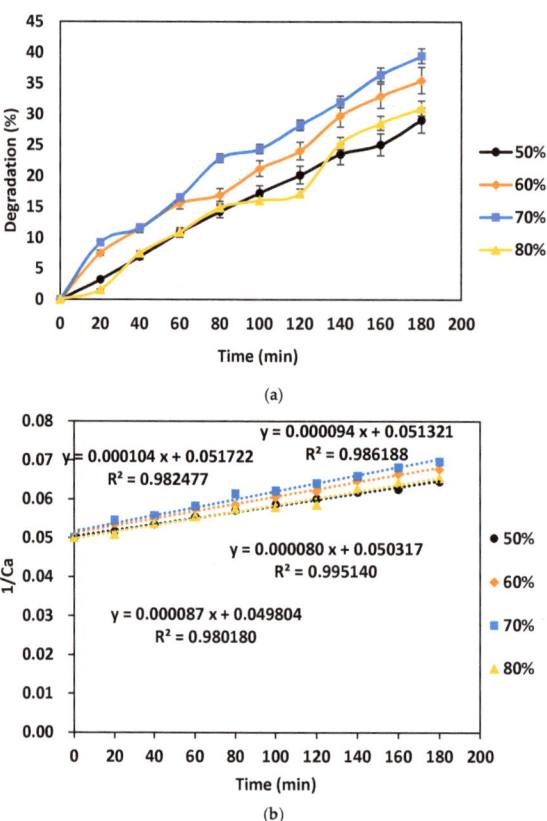

Figure 4. Effect of ultrasound duty cycle on the degradation of Chrysoidine R dye: (**a**) extent of degradation and (**b**) kinetic fitting.

It was seen that the degradation almost doubled from 29.05% to 39.48% when the duty cycle increased from 50% to 70%, indicating that the elevation of on time of ultrasonic irradiation can significantly improve the rate of degradation. This enhancement can be ascribed to the enhanced activity of cavitation resulting in an increased number of active hydroxyl radical generation. It was also seen that a subsequent increase of the duty cycle to 80% gave lower degradation of 35.56%, which can be attributed to the fact that beyond the optimum duty cycle, the number of cavity formations increases too much, leading to a possible coalescence of the cavities, yielding cushioned collapse of much lower intensity giving the lower extent of degradation [26]. The degradation of Chrysoidine R dye was also found to follow second-order kinetics, and the obtained data is presented in Figure 4b. The trend for the kinetic rate constant was also similar to the trend in the extent of degradation. Thakare et al. [27] also reported that COD reduction increased continuously until 60% of the duty cycle, giving the actual value as 21.59%, and a marginally lower value as 20.26% was observed at 80% duty cycle, indicating 60% as the optimum.

2.2.3. Effect of Catalyst Dosage

Selecting the optimum quantum of the catalyst is another important requirement in the sono-catalytic oxidation, and hence, the effect of dosage of 5 wt% Ni-loaded $NiFe_2O_4$ catalyst (Ni content was kept fixed during the synthesis) was studied using various values as 0.4, 0.5, and 0.6 g/L. The results shown in Figure 5a elucidate that using 0.4 g/L as the catalyst loading led to 39.49% dye degradation, which increased to about 43% when the catalyst loading was increased to 0.5 g/L.

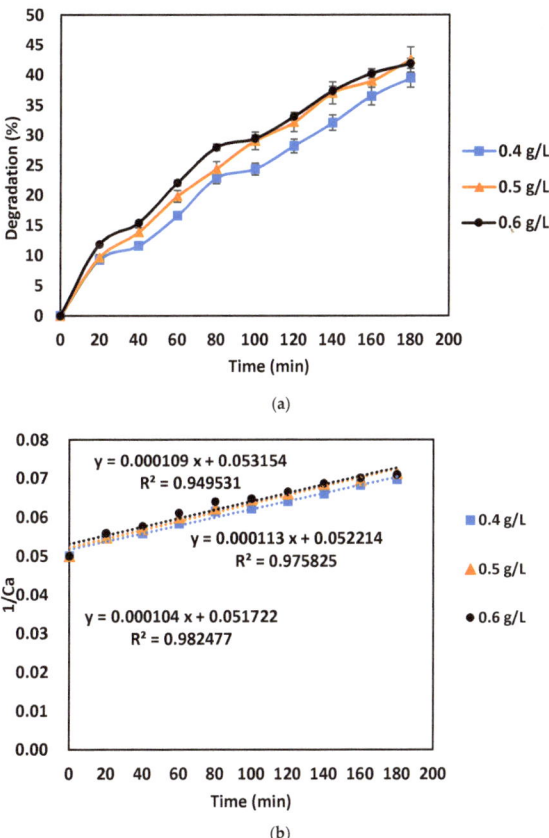

Figure 5. Effect of catalyst dosage on the degradation of Chrysoidine R dye: (**a**) extent of degradation and (**b**) kinetic fitting.

The trend was attributed to the fact that the added NiFe$_2$O$_4$ catalyst till an optimum loading provided additional heterogeneous nuclei for the cavitating bubble formation and hot spots, which subsequently generates more active hydroxyl radicals for dye degradation. However, a further increase in the loading to 0.6 g/L resulted in a marginal reduction in the degradation to 41.89%, as per the data shown in Figure 5a. The trend may be associated with the consideration that an excessive presence of catalyst can scatter the ultrasound, altering the efficient transmission, which reduces the cavitational intensity and hence the rate of sonocatalytic degradation [28]. In addition, the possible explanation would be that the excessive catalyst can interfere with the collapsing bubble, leading to a less powerful collapse [29]. The second-order kinetics was found to fit well to the dye degradation, as depicted in Figure 5b. It was observed that the second-order constant increased from 10.9×10^{-5} and 11.3×10^{-5} mL^{-1}min^{-1} for a change in catalyst dose from 0.4 g/L to 0.5 g/L, but it decreased to 10.4×10^{-5} mL^{-1}min^{-1} at a catalyst loading of 0.6 g/L. Bose et al. [30] also elucidated that increasing the MgFe$_2$O$_4$ catalyst dosage from 1 g/L to 1.25 g/L showed insignificant improvements in brilliant green dye degradation. Sobana et al. [31] reported similar trends for Direct Blue 53 dye degradation in the presence of an Ag-TiO$_2$ catalyst, whereas Abdellah et al. [32] reported a similar trend for the methylene blue degradation using the TiO$_2$ catalyst. The literature analysis revealed the existence of different optimum loading, confirming the importance of the current planned work for Chrysoidine R dye. Based on the results obtained in the present work, 0.5 g/L was considered as optimum dosage.

2.2.4. Effect of Power of Sonication

The effect of the power on the degradation of dye was studied over the range of 80 W to 120 W. The obtained Chrysoidine R dye degradation at different power outputs is shown in Figure 6.

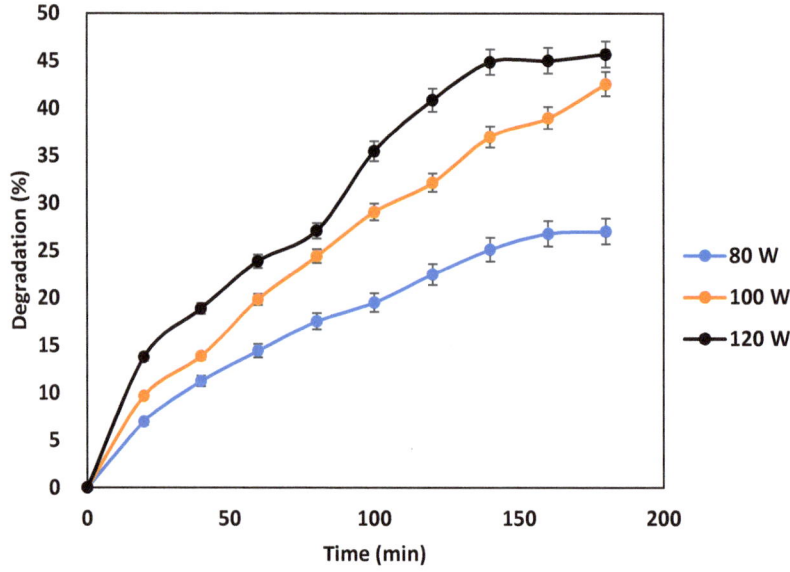

Figure 6. Effect of ultrasonic power on the degradation of Chrysoidine R dye.

It was observed that 80 W power resulted in the lowest degradation (27.03%). In comparison to 80 W power, higher degradation of Chrysoidine R dye as 42.56% and 45.68% was observed at 100 W and 120 W, respectively. The increasing trend with increasing power is due to the fact that the use of higher power dissipation results in a higher generation of cavities and active radical formation within the reaction mixture, thereby resulting in higher degradation. A comparable result was reported by Xu et al. [33] where a rise in degradation of rhodamine B from 91.89% to 97.77% was reported with an increase in power from 150 W to 300 W in 40 min of reaction time for 20 ppm rhodamine dye solution. In another study, Kodavatiganti et al. [34] studied the applicability of different power outputs for the degradation of acid violet 7 dye at a loading of 20 ppm. The authors found the optimum power of 100 W resulted in a maximum of 40.1% decolorization of the dye solution. The results obtained in the present work and comparison with the literature showed that the power has a significant effect on the dye degradation efficiency. Considering the operating limit set by the manufacturer and the maximum degradation efficiency obtained, 120 W power output was chosen as optimum for further studies.

2.2.5. Comparison of Degradation Using Conventionally and Ultrasonically Prepared $NiFe_2O_4$ Catalyst

Experiments were conducted to compare the efficiencies of the $NiFe_2O_4$ catalyst prepared using the ultrasound-assisted approach and conventional approach. Both experiments of sono-catalytic degradation were performed under previously optimized conditions of 120 W as the power, 0.5 g/L catalyst loading, 70% duty cycle, and 160 min as the treatment time. The ultrasonically prepared catalyst resulted in higher degradation of 45.68% compared to 29.45% obtained using the conventionally prepared catalyst, as per the data depicted in Figure 7.

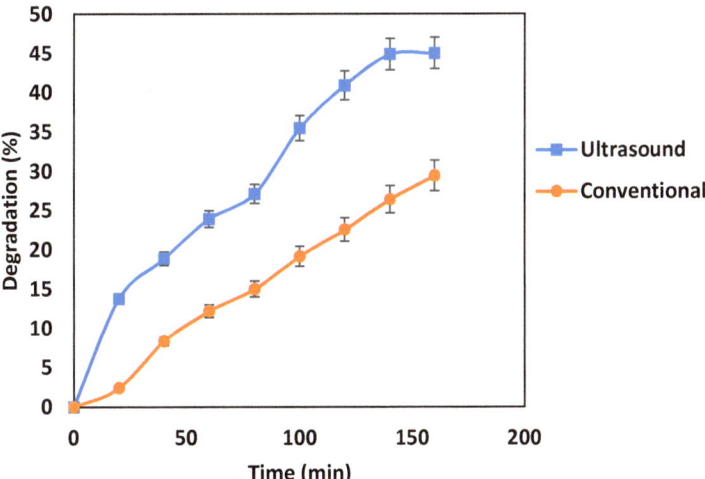

Figure 7. Comparison of ultrasonically synthesized and conventionally synthesized catalyst applied for the degradation of Chrysoidine R dye.

The elevated degradation obtained in the case of ultrasonically synthesized catalyst can be accredited to better crystallinity (as per XRD results) and lower particle size of 18.58 μm that results in increased surface area and availability of active sites for sonocatalytic degradation of dye molecules. Similar outcomes have been reported in the literature, highlighting the benefits of sonochemical synthesis. For example, the degradation of CV dye at an initial dye concentration of 50 ppm with 0.1 g rGO-ZnO-TiO2 nano-composite loading was studied using the conventional and sono-chemically synthesized catalyst, and it was reported that the catalyst obtained sono-chemically resulted in higher degradation as 87.06%, whereas the conventionally prepared catalyst yielded 72.10% degradation [35]. Sancheti et al. [36] reported similar results for brilliant green dye degradation where it was observed that the NiO-CeO$_2$ catalyst synthesized sono-chemically gave 82% of degradation, whereas the same catalyst obtained conventionally gave only 40% degradation. Satdeve et al. [37] evaluated the performance of an Ag-Zno nano-composite synthesized using ultrasonication and a conventional approach for photo-catalytic decolorization of methylene blue dye at 400 mg/L of catalyst loading. It was found that the sono-chemically synthesized catalyst had a higher decolorization efficiency of 96.2% compared to 89.76% for the conventional catalyst. Importantly, the degree of intensification is different in all these literature illustrations, clearly highlighting the importance of the current work for the chrysoidine R dye degradation.

2.2.6. Effect of Using H$_2$O$_2$ Combined with the SonoCatalytic Degradation

Hydrogen peroxide is an oxidant that is extensively used in advanced oxidation processes and has been reported to offer excellent oxidation potential when coupled with ultrasound, especially based on the enhanced hydroxyl radical production. To further enhance the degradation of chrysoidine R, hydrogen peroxide was also incorporated as the oxidant in a sonocatalytic study of dye degradation. The influence of different H$_2$O$_2$ loadings, such as 50, 75, and 100 ppm, has been examined at previously optimized conditions, and the obtained results are illustrated in Figure 8.

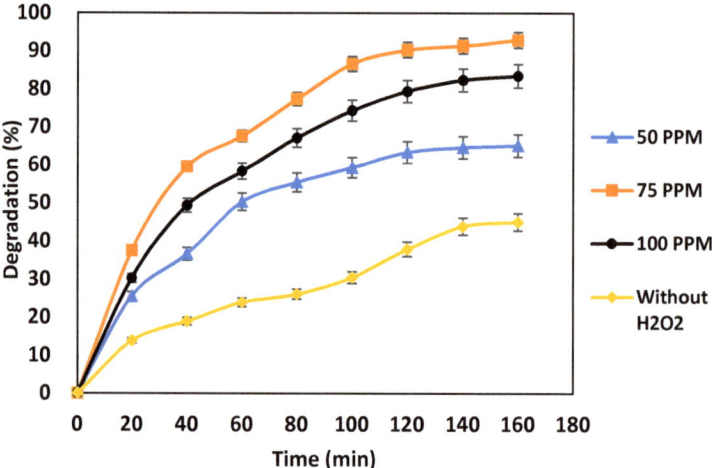

Figure 8. Effect of H$_2$O$_2$ addition on the degradation of Chrysoidine R dye.

Without hydrogen peroxide, the degradation observed was around 45%, which significantly rose to about 65.12% with the use of 50 ppm H$_2$O$_2$. A further increase in H$_2$O$_2$ loading to 75 ppm resulted in a further increase in the degradation to 92.83%. At 100 ppm loading, however, there was a reduction in the degradation of dye with an actual value of 83.41%, which is about 11% lower compared to 75 ppm loading. As 75 ppm yielded maximum degradation, it was chosen as the optimum. The declining degradation at the highest oxidant loading can be due to residual hydrogen peroxide, which generally acts as a hydroxyl radical scavenger and favors the generation of per-hydroxyl radical, which is known to possess less oxidation potential, resulting in a decrease in dye degradation [38,39].

$$H_2O_2 + \bullet OH \rightarrow H_2O + \bullet OOH$$

Prakash et al. [40] examined the AB 15 dye removal efficiency at various H_2O_2 concentrations ranging from 5 mM to 25 mM and reported that at an optimum of 10 mM loading, 99% dye removal can be achieved, whereas a further increase in H_2O_2 concentration to 15 mM, 20 mM, and, 25 mM resulted in lower values of degradation as 98%, 79%, and 75%, respectively. Despite similar trends, the exact optimum concentration value and level of intensification are different, elucidating the importance of the planned work. The presence of the $NiFe_2O_4$ catalyst enhanced the formation of active hydroxyl radials by the following reactions:

$$Ni^{3+} + H_2O_2 \rightarrow Ni^{2+} + H^+ + \bullet OOH$$

$$Fe^{3+} + H_2O_2 \rightarrow Fe^{2+} + H^+ + \bullet OOH$$

$$Ni^{2+} + H_2O_2 \rightarrow Ni^{3+} + \bullet OH + OH^-$$

$$Fe^{2+} + H_2O_2 \rightarrow Fe^{3+} + \bullet OH + OH^-$$

A possible degradation pathway of Chrysoidine R dye in the presence of an active hydroxyl radical has been proposed and illustrated in Figure 9.

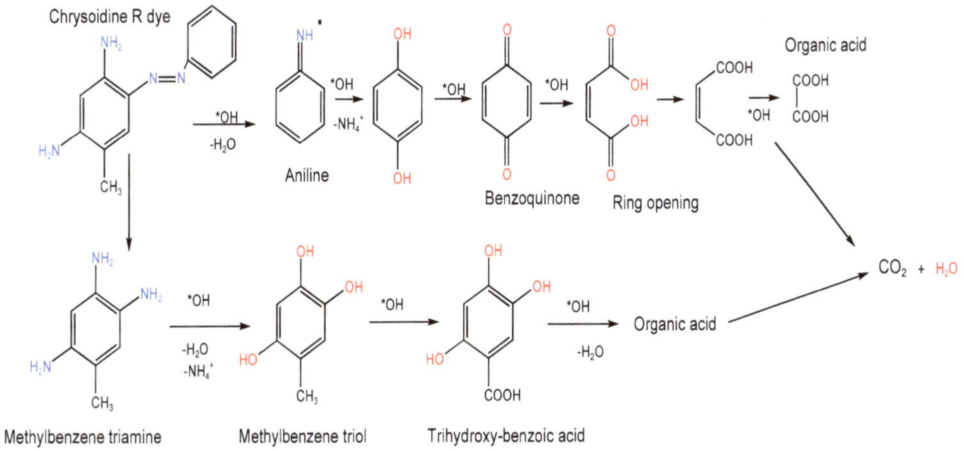

Figure 9. Proposed reaction mechanism of sono-catalytic degradation of Chrysoidine R dye (* or • denotes radical).

The breakdown of Chrysoidine R dye begins with the formation of two intermediates: aniline and methylbenzene triamine. Both intermediates follow different paths of degradation. Aniline molecule oxidation results in the formation of benzoquionone, which was broken down into organic acid followed by transformation into carbon dioxide and water. Simultaneously, the methylbenzenetriamine is transformed into methyl benzene triol by liberating water and ammonium cation. Hydroxyl radicals accelerated the conversion of methylbenzene triol to trihydroxybenoic acid, which is further converted to small organic acids, and eventually, the mineralization of small organic acids result in the formation of carbon dioxide and water.

It is important to note that hydroxyl radicals are the main oxidizing species generally observed in the case of ultrasound-induced degradation of contaminants. The presence of hydrogen peroxide at optimum loading further enhances the formation of hydroxyl radicals, which has been confirmed in many of the earlier studies [41,42] based on the use of radical scavengers, such as 2,2,6,6-tetramethylpiperidine-1-oxyl (TEMPO) or hydroquinone (HQ), though not explicitly confirmed in the current work.

3. Materials & Methods

3.1. Materials

Analytical grade nickel acetate (tetrahydrate), ferric chloride, and sodium hydroxide, which were procured from Molychem Pvt. Ltd., Mumbai, India were used to synthesize the catalyst complex. Chrysoidine R [Basic orange 1 (BO)] dye was purchased from Huntsman Corporation, India. An oxidant, hydrogen peroxide (commercial H_2O_2 solution with strength as 30% w/v), was procured from Thomas Baker Chemicals Pvt. Ltd. in Mumbai, India. For the preparation of different solutions, distilled water was used, which was freshly prepared using the Milipore distillation unit available in our laboratory.

3.2. Experimental Methodology for Catalyst Synthesis

3.2.1. Synthesis of Nickel Ferrite Oxide

The nickel ferrite oxide catalyst complex was synthesized by dissolving the known amounts of nickel acetate tetrahydrate and ferric chloride in 70 mL of water and then mixing the solution with stirring for 30 min. After that, 2 M NaOH was gradually added drop by drop to the homogeneous mixture at a constant stirring rate of 600 rpm until the desired pH of 12 was attained. The resulting pH-balanced solution was then agitated for 2 h at a constant temperature of 40 °C to result in a dark brown color precipitate. The solution was then subjected to sonication (fixed frequency 20 kHz, Dakshin, Mumbai, India) at room temperature. Furthermore, the mixture was washed with distilled water and then filtered with Whatman filter paper. The wet cake was then dried at 80 °C for 6 h and then calcined at 800 °C for 4 h in a muffle furnace. For comparison, the catalyst was also synthesized using the conventional co-precipitation method (without ultrasound) under the same conditions specified for ultrasound-assisted catalyst synthesis, as illustrated in Figure 10. X-ray diffraction was also used as a characterization method to confirm the formation of crystalline $NiFe_2O_4$ catalyst complex as well as the phase purity and crystal structure.

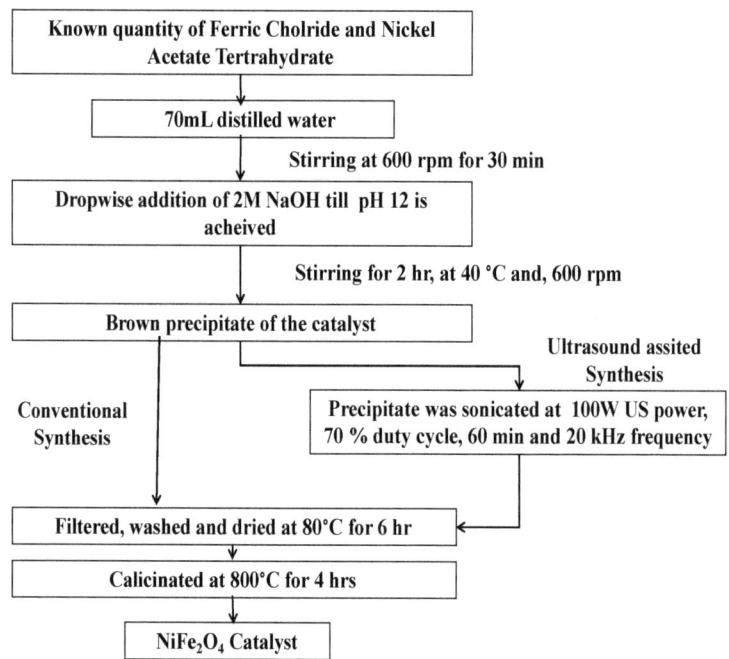

Figure 10. Flowchart of conventional and ultrasound-assisted synthesis approaches.

3.2.2. Experimental Methodology for Degradation Study

A fixed frequency ultrasonic horn was purchased from M/s Dakshin, Mumbai, India with rated power dissipation as 150 W and a tip diameter of 1 cm. A quartz beaker of 250 mL capacity with a diameter of 68 mm and height of 95 mm was used throughout the experimental study. Figure 11 shows the setup of the sono-catalytic reactor. A well-mixed dye solution (200 mL as fixed quantity in all the experiments) having known concentration kept constant at 20 ppm was used for degradation study. The temperature was maintained at 40 °C using a water bath. The ultrasonic horn tip was dipped at height of 0.5 cm below the liquid level. The process variables investigated were the pH of the solution, duty cycle, catalyst dosage, output power of the ultrasonic horn, irradiation time, and oxidant loading.

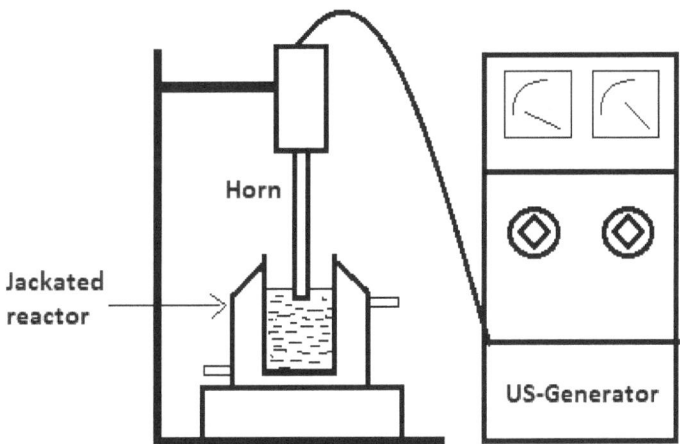

Figure 11. Schematic representation of ultrasonic horn used for catalyst synthesis and dye degradation.

3.3. UV-Visible Spectroscopy Analysis for Dye Concentration

A double-beam UV spectrophotometer unit (Model UV 1900, Shimadzu, Tokyo, Japan) was used to analyze the samples collected at different time intervals. The wavelength of 452.5 nm was found to have the highest absorbance of Chrysoidine R dye in the visible region. The standard calibration curve was created for dye concentrations ranging from 5 to 25 ppm. The least square method was used for curve fitting, yielding a linear equation of $y = 0.0402x$ with $R^2 = 0.9981$. The percentage degradation was calculated using Equation (1), as shown below:

$$Percentage\ degradation = \left[\frac{(C_i - C_f)}{C_i}\right] \times 100 \qquad (1)$$

where, C_i and C_f are the initial and final Chrysoidine R dye concentrations, respectively.

3.4. Characterization of Nickel Ferrite Oxide Catalyst

3.4.1. Particle Size Analysis

To quantify the influence of the ultrasonic synthesis method as compared to the conventional catalyst synthesis approach, the particle size was analyzed using Bettersizer 2600 E (wet) analyzer.

3.4.2. X-ray Diffraction Analysis

The diffraction patterns of the different catalyst samples were logged using a powder X-ray diffractometer (Philips PW 1800, Phillips, Amsterdam, The Netherlands) equipped with Ni-filtered Cu-Kα radiation (λ = 1.5418 Å) at ambient temperature. The XRD patterns of the sample were recorded at angles between 10° and 80° with a scanning rate of 2°/min.

4. Conclusions

The conducted research aimed at the development of an advanced nickel ferrite synthesis method that is more efficient than the current conventional process and at establishing its catalytic performance for the degradation of Chrysoidine R dye. The modification of the conventional method using ultrasound successfully demonstrated that the ultrasound-assisted synthesis method has great potential to improve structural properties and catalytic degradation activity. The study also optimized the ultrasound-assisted catalytic degradation of dye solution using the synthesized catalyst. The results obtained indicated degradation efficiency can be greatly affected by power output and other process conditions. Maximum sono-degradation of 92.83% was obtained under the optimized conditions as 75 ppm H_2O_2 loading, pH 3, 70% US duty cycle, 120 W US power output, 160 min of reaction time, and 0.5 g/L ultrasonically synthesized catalyst loading. Thus, it can be concluded that the developed approach of ultrasound-assisted catalytic degradation in the presence of hydrogen peroxide oxidant and sono-chemically synthesized catalyst is an effective sustainable approach for chrysoidine R dye degradation.

Author Contributions: Conceptualization, P.R.G.; methodology, Y.M.G. and P.R.G.; investigation, Y.M.G.; data curation, Y.M.G. and P.S.S.; writing—original draft preparation, Y.M.G.; writing—review and editing, P.S.S. and P.R.G.; supervision, P.R.G.; All authors have read and agreed to the published version of the manuscript.

Funding: This research received no external funding.

Data Availability Statement: Data will be made available on request.

Conflicts of Interest: The authors declare no conflict of interest.

References

1. Nirmal, P.; Paulraj, R.; Ramasamy, P.; Vijayan, N. One Step Synthesis of Tin Oxide Nanomaterials and Their Sintering Effect in Dye Degrdation. *Optik* **2017**, *135*, 434–445. [CrossRef]
2. Zhang, N.; Zhang, G.; Chong, S.; Zhao, H.; Huang, T.; Zhu, J. Ultrasonic Impregnation of MnO_2/CeO_2 and Its Application in Catalytic Sono-Degradation of Methyl Orange. *J. Environ. Manag.* **2018**, *205*, 134–141. [CrossRef] [PubMed]
3. Khataee, A.; Mohamadi, F.T.; Rad, T.S.; Vahid, B. Heterogeneous Sonocatalytic Degradation of Anazolene Sodium by Synthesized Dysprosium Doped CdSe Nanostructures. *Ultrason. Sonochem.* **2018**, *40*, 361–372. [CrossRef]
4. Saemian, T.; Hossaini Sadr, M.; Tavakkoli Yaraki, M.; Gharagozlou, M.; Soltani, B. Synthesis and Characterization of $CoFe_2O_4/SiO_2/Cu$-MOF for Degradation of Methylene Blue through Catalytic Sono-Fenton-like Reaction. *Inorg. Chem. Commun.* **2022**, *138*, 109305. [CrossRef]
5. Chen, X.; Mao, J.; Liu, C.; Chen, C.; Cao, H.; Yu, L. An Unexpected Generation of Magnetically Separable Se/Fe_3O_4 for Catalytic Degradation of Polyene Contaminants with Molecular Oxygen. *Chinese Chem. Lett.* **2020**, *31*, 3205–3208. [CrossRef]
6. Mukhopadhyay, A.; Tripathy, B.K.; Debnath, A.; Kumar, M. Enhanced Persulfate Activated Sono-Catalytic Degradation of Brilliant Green Dye by Magnetic $CaFe_2O_4$ Nanoparticles: Degradation Pathway Study, Assessment of Bio-Toxicity and Cost Analysis. *Surf. Interfaces* **2021**, *26*, 101412. [CrossRef]
7. JothiRamalingam, R.; Periyasami, G.; Ouladsmane, M.; ALOthman, Z.A.; Arunachalam, P.; Altalhi, T.; Radhika, T.; Alanazi, A.G. Ultra-Sonication Assisted Metal Chalcogenide Modified Mesoporous Nickel-Cobalt Doped Manganese Oxide Nanocomposite Fabrication for Sono-Catalytic Dye Degradation and Mechanism Insights. *J. Alloys Compd.* **2021**, *875*, 160072. [CrossRef]
8. Lazarova, T.; Georgieva, M.; Tzankov, D.; Voykova, D.; Aleksandrov, L.; Cherkezova-Zheleva, Z.; Kovacheva, D. Influence of the Type of Fuel Used for the Solution Combustion Synthesis on the Structure, Morphology and Magnetic Properties of Nanosized $NiFe_2O_4$. *J. Alloys Compd.* **2017**, *700*, 272–283. [CrossRef]
9. Kodama, R.H.; Berkowitz, A.E.; McNiff, E.J.; Foner, S. Surface Spin Disorder in Ferrite Nanoparticles. *Mater. Sci. Forum* **1997**, *235–238*, 643–650. [CrossRef]
10. Wang, J.; Ren, F.; Jia, B.; Liu, X. Solvothermal Synthesis and Characterization of $NiFe_2O_4$ Nanospheres with Adjustable Sizes. *Solid State Commun.* **2010**, *150*, 1141–1144. [CrossRef]
11. Liu, J.; Yang, H.; Xue, X. Structure, Morphology, and Magnetic Properties of $NiFe_2O_4$ Powder Prepared by Molten Salt Method. *Powder Technol.* **2019**, *355*, 708–715. [CrossRef]
12. Zhang, Y.; Zhang, W.; Yu, C.; Liu, Z.; Yu, X.; Meng, F. Synthesis, Structure and Supercapacitive Behavior of Spinel $NiFe_2O_4$ and $NiO@NiFe_2O_4$ Nanoparticles. *Ceram. Int.* **2021**, *47*, 10063–10071. [CrossRef]
13. Taha, T.A.; Azab, A.A.; Sebak, M.A. Glycerol-Assisted Sol-Gel Synthesis, Optical, and Magnetic Properties of $NiFe_2O_4$ Nanoparticles. *J. Mol. Struct.* **2019**, *1181*, 14–18. [CrossRef]

14. Maaz, K.; Karim, S.; Mumtaz, A.; Hasanain, S.K.; Liu, J.; Duan, J.L. Synthesis and Magnetic Characterization of Nickel Ferrite Nanoparticles Prepared by Co-Precipitation Route. *J. Magn. Magn. Mater.* **2009**, *321*, 1838–1842. [CrossRef]
15. Feng, S.; Yang, W.; Wang, Z. Synthesis of Porous $NiFe_2O_4$ Microparticles and Its Catalytic Properties for Methane Combustion. *Mater. Sci. Eng. B Solid-State Mater. Adv. Technol.* **2011**, *176*, 1509–1512. [CrossRef]
16. Gherca, D.; Pui, A.; Cornei, N.; Cojocariu, A.; Nica, V.; Caltun, O. Synthesis, Characterization and Magnetic Properties of MFe_2O_4 (M=Co, Mg, Mn, Ni) Nanoparticles Using Ricin Oil as Capping Agent. *J. Magn. Magn. Mater.* **2012**, *324*, 3906–3911. [CrossRef]
17. Vardikar, H.S.; Bhanvase, B.A.; Rathod, A.P.; Sonawane, S.H. Sonochemical Synthesis, Characterization and Sorption Study of Kaolin-Chitosan-TiO_2 Ternary Nanocomposite: Advantage over Conventional Method. *Mater. Chem. Phys.* **2018**, *217*, 457–467. [CrossRef]
18. Díez-García, M.I.; Manzi-Orezzoli, V.; Jankulovska, M.; Anandan, S.; Bonete, P.; Gómez, R.; Lana-Villarreal, T. Effects of Ultrasound Irradiation on the Synthesis of Metal Oxide Nanostructures. *Phys. Procedia* **2015**, *63*, 85–90. [CrossRef]
19. Dalvi, P.; Dey, A.; Gogate, P.R. Ultrasound-Assisted Synthesis of a N-TiO_2/Fe_3O_4@ZnO Complex and Its Catalytic Application for Desulfurization. *Sustainability* **2022**, *14*, 16201. [CrossRef]
20. Shetty, K.; Renuka, L.; Nagaswarupa, H.P.; Nagabhushana, H.; Anantharaju, K.S.; Rangappa, D.; Prashantha, S.C.; Ashwini, K. A Comparative Study on $CuFe_2O_4$, $ZnFe_2O_4$ and $NiFe_2O_4$: Morphology, Impedance and Photocatalytic Studies. *Mater. Today Proc.* **2017**, *4*, 11806–11815. [CrossRef]
21. Wang, Y.; Wang, H.; Yang, Y.; Xin, B. Magnetic $NiFe_2O_4$ 3D Nanosphere Photocatalyst: Glycerol-Assisted Microwave Solvothermal Synthesis and Photocatalytic Activity under Microwave Electrodeless Discharge Lamp. *Ceram. Int.* **2021**, *47*, 14594–14602. [CrossRef]
22. Hariani, P.L.; Said, M.; Rachmat, A.; Riyanti, F.; Pratiwi, H.C.; Rizki, W.T. Preparation of $NiFe_2O_4$ Nanoparticles by Solution Combustion Method as Photocatalyst of Congo Red. *Bull. Chem. React. Eng. Catal.* **2021**, *16*, 481–490. [CrossRef]
23. Liu, S.Q.; Feng, L.R.; Xu, N.; Chen, Z.G.; Wang, X.M. Magnetic Nickel Ferrite as a Heterogeneous Photo-Fenton Catalyst for the Degradation of Rhodamine B in the Presence of Oxalic Acid. *Chem. Eng. J.* **2012**, *203*, 432–439. [CrossRef]
24. Vijay, S.; Balakrishnan, R.M.; Rene, E.R.; Priyanka, U. Photocatalytic Degradation of Irgalite Violet Dye Using Nickel Ferrite Nanoparticles. *J. Water Supply Res. Technol.-AQUA* **2019**, *68*, 666–674. [CrossRef]
25. Mahmoodi, N.M. Photocatalytic Degradation of Textile Dyes Using Ozonation and Magnetic Nickel Ferrite Nanoparticle. *Prog. Color. Color. Coatings* **2016**, *9*, 163–171. [CrossRef]
26. Gogate, P.R. Cavitation: An Auxiliary Technique in Wastewater Treatment Schemes. *Adv. Environ. Res.* **2002**, *6*, 335–358. [CrossRef]
27. Thakare, Y.D. Reduction of COD of Textile Industry Waste Water by Using Acoustic Cavitation Coupled with Advanced Oxidation Processes. *Int. J. Res. Appl. Sci. Eng. Technol.* **2018**, *6*, 686–697. [CrossRef]
28. Pang, Y.L.; Abdullah, A.Z. Fe^{3+} Doped TiO_2 Nanotubes for Combined Adsorption-Sonocatalytic Degradation of Real Textile Wastewater. *Appl. Catal. B Environ.* **2013**, *129*, 473–481. [CrossRef]
29. Eren, Z.; Ince, N.H. Sonolytic and Sonocatalytic Degradation of Azo Dyes by Low and High Frequency Ultrasound. *J. Hazard. Mater.* **2010**, *177*, 1019–1024. [CrossRef]
30. Bose, S.; Kumar Tripathy, B.; Debnath, A.; Kumar, M. Boosted Sono-Oxidative Catalytic Degradation of Brilliant Green Dye by Magnetic $MgFe_2O_4$ Catalyst: Degradation Mechanism, Assessment of Bio-Toxicity and Cost Analysis. *Ultrason. Sonochem.* **2021**, *75*, 105592. [CrossRef]
31. Sobana, N.; Subash, B.; Swaminathan, M. Effect of Operational Parameters on Photodegradation of Direct Blue 53 by Silver Loaded-Titania under Ultraviolet and Solar Illumination. *Mater. Sci. Semicond. Process.* **2015**, *36*, 149–155. [CrossRef]
32. Abdellah, M.H.; Nosier, S.A.; El-Shazly, A.H.; Mubarak, A.A. Photocatalytic Decolorization of Methylene Blue Using TiO_2/UV System Enhanced by Air Sparging. *Alexandria Eng. J.* **2018**, *57*, 3727–3735. [CrossRef]
33. Xu, D.; Ma, H. Degradation of Rhodamine B in Water by Ultrasound-Assisted TiO_2 Photocatalysis. *J. Clean. Prod.* **2021**, *313*, 127758. [CrossRef]
34. Kodavatiganti, S.; Bhat, A.P.; Gogate, P.R. Intensified Degradation of Acid Violet 7 Dye Using Ultrasound Combined with Hydrogen Peroxide, Fenton, and Persulfate. *Sep. Purif. Technol.* **2021**, *279*, 119673. [CrossRef]
35. Potle, V.D.; Shirsath, S.R.; Bhanvase, B.A.; Saharan, V.K. Sonochemical Preparation of Ternary RGO-ZnO-TiO_2 Nanocomposite Photocatalyst for Efficient Degradation of Crystal Violet Dye. *Optik* **2020**, *208*, 164555. [CrossRef]
36. Sancheti, S.V.; Saini, C.; Ambati, R.; Gogate, P.R. Synthesis of Ultrasound Assisted Nanostuctured Photocatalyst (NiO Supported over CeO_2) and Its Application for Photocatalytic as Well as Sonocatalytic Dye Degradation. *Catal. Today* **2018**, *300*, 50–57. [CrossRef]
37. Satdeve, N.S.; Ugwekar, R.P.; Bhanvase, B.A. Ultrasound Assisted Preparation and Characterization of Ag Supported on ZnO Nanoparticles for Visible Light Degradation of Methylene Blue Dye. *J. Mol. Liq.* **2019**, *291*, 111313. [CrossRef]
38. Aravindhan, R.; Fathima, N.N.; Rao, J.R.; Nair, B.U. Wet Oxidation of Acid Brown Dye by Hydrogen Peroxide Using Heterogeneous Catalyst Mn-Salen-Y Zeolite: A Potential Catalyst. *J. Hazard. Mater.* **2006**, *138*, 152–159. [CrossRef]
39. Hassan, H.; Hameed, B.H. Oxidative Decolorization of Acid Red 1 Solutions by Fe-Zeolite Y Type Catalyst. *Desalination* **2011**, *276*, 45–52. [CrossRef]
40. Prakash, L.V.; Gopinath, A.; Gandhimathi, R.; Velmathi, S.; Ramesh, S.T.; Nidheesh, P.V. Ultrasound Aided Heterogeneous Fenton Degradation of Acid Blue 15 over Green Synthesized Magnetite Nanoparticles. *Sep. Purif. Technol.* **2021**, *266*, 118230. [CrossRef]

41. Deng, X.; Qian, R.; Zhou, H.; Yu, L. Organotellurium-catalyzed oxidative deoximation reactions using visible-light as the precise driving energy. *Chin. Chem. Lett.* **2021**, *32*, 1029–1032. [CrossRef]
42. Li, W.; Wang, F.; Shi, Y.; Yu, L. Polyaniline-supported tungsten-catalyzed oxidative deoximation reaction with high catalyst turnover number. *Chin. Chem. Lett.* **2023**, *34*, 107505. [CrossRef]

Disclaimer/Publisher's Note: The statements, opinions and data contained in all publications are solely those of the individual author(s) and contributor(s) and not of MDPI and/or the editor(s). MDPI and/or the editor(s) disclaim responsibility for any injury to people or property resulting from any ideas, methods, instructions or products referred to in the content.

Article

A Comparison of Different Reagents Applicable for Destroying Halogenated Anionic Textile Dye Mordant Blue 9 in Polluted Aqueous Streams

Barbora Kamenická and Tomáš Weidlich *

Chemical Technology Group, Institute of Environmental and Chemical Engineering, Faculty of Chemical Technology, University of Pardubice, Studentska 573, 532 10 Pardubice, Czech Republic
* Correspondence: tomas.weidlich@upce.cz; Tel.: +420-46-603-8049

Citation: Kamenická, B.; Weidlich, T. A Comparison of Different Reagents Applicable for Destroying Halogenated Anionic Textile Dye Mordant Blue 9 in Polluted Aqueous Streams. *Catalysts* 2023, 13, 460. https://doi.org/10.3390/catal13030460

Academic Editor: Meng Li

Received: 6 January 2023
Revised: 27 January 2023
Accepted: 31 January 2023
Published: 22 February 2023

Copyright: © 2023 by the authors. Licensee MDPI, Basel, Switzerland. This article is an open access article distributed under the terms and conditions of the Creative Commons Attribution (CC BY) license (https://creativecommons.org/licenses/by/4.0/).

Abstract: This article aimed to compare the degradation efficiencies of different reactants applicable for the oxidative or reductive degradation of a chlorinated anionic azo dye, Mordant Blue 9 (MB9). In this article, the broadly applied Fenton oxidation process was optimized for the oxidative treatment of MB9, and the obtained results were compared with other innovative chemical reduction methods. In the reductive degradation of MB9, we compared the efficiencies of different reductive agents such as Fe^0 (ZVI), Al^0, the Raney Al-Ni alloy, $NaBH_4$, $NaBH_4/Na_2S_2O_5$, and other combinations of these reductants. The reductive methods aimed to reduce the azo bond together with the bound chlorine in the structure of MB9. The dechlorination of MB9 produces non-chlorinated aminophenols, which are more easily biodegradable in wastewater treatment plants (WWTPs) compared to their corresponding chlorinated aromatic compounds. The efficiencies of both the oxidative and reductive degradation processes were monitored by visible spectroscopy and determined based on the chemical oxygen demand (COD). The hydrodechlorination of MB9 to non-chlorinated products was expressed using the measurement of adsorbable organically bound halogens (AOXs) and controlled by LC–MS analyses. Optimally, 28 mol of H_2SO_4, 120 mol of H_2O_2, and 4 mol of $FeSO_4$ should be applied per one mol of dissolved MB9 dye for a practically complete oxidative degradation after 20 h of action. On the other hand, the application of the Al-Ni alloy/NaOH (100 mol of Al in the Al-Ni alloy + 100 mol of NaOH per one mol of MB9) proceeded smoothly and seven-times faster than the Fenton reaction, consumed similar quantities of reagents, and produced dechlorinated aminophenols. The cost of the Al-Ni alloy/NaOH-based method could be decreased significantly by applying a pretreatment with Al^0/NaOH and a subsequent hydrodechlorination using smaller Al-Ni alloy doses. The homogeneous reduction accompanied by HDC using in situ produced $Na_2S_2O_4$ (by the action of $NaBH_4/Na_2S_2O_5$) was an effective, rapid, and simple treatment method. This reductive system consumed quantities of reagents that are almost twice as low (66 mol of $NaBH_4$ + 66 mol of $Na_2S_2O_5$ + 18 mol of H_2SO_4 per one mol of MB9) in comparison with the other oxidative/reductive systems and allowed the effective and fast degradation of MB9 accompanied by the effective removal of AOX. A comparison of the oxidative and reductive methods for chlorinated acid azo dye MB9 degradation showed that an innovative combination of reduction methods offers a smooth, simple, and efficient degradation and hydrodehalogenation of chlorinated textile MB9 dye.

Keywords: acid azo dye; Mordant Blue 9; chlorophenol; Fenton oxidation; degradation; hydrodehalogenation; hydrodechlorination; $NaBH_4$; $Na_2S_2O_5$; Raney Al-Ni alloy

1. Introduction

Azo dyes are the most important synthetic colorants and have been used for a wide spectrum of textile dyeing [1,2]. Acid dyes soluble in aqueous solutions are usually applied for dyeing natural fibers [3,4]. Mordant dyes derived from acid dyes have been widely used for the coloration of textile fibers such as wool, silk, polyester, cotton, and some modified cellulose fibers [5–8]. One example of the mentioned mordant azo dyes is the acid dye Mordant Blue 9 (MB9, Figure 1). However, the effective treatment of dye baths and wastewater from dye production and utilization before they are released to the environment has received significant attention because of their potential toxic effects based on the fact that Mordant dyes are difficult to degrade [9].

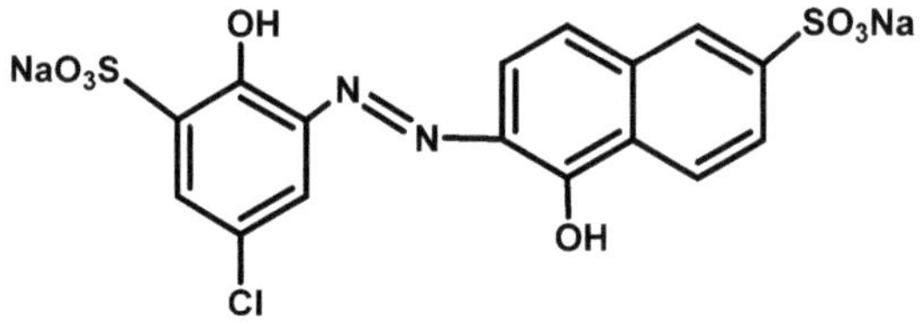

Figure 1. The chemical structure of anionic azo dye Mordant Blue 9.

On average, the annual production of dyes exceeds 7×10^5 tons. However, in the textile industry, typically about 10% of the used dyes are discharged, with the effluent creating both environmental and economic issues due to the limited biodegradability of azo dyes, especially halogenated ones [10]. The occurrence of especially halogenated non-biodegradable azo dyes in wastewater not only deteriorates water aesthetically due to its coloration but also increases its chemical oxygen demand (COD) and adsorbable organic halogen content (AOX) [11].

Various removal methods have been developed to remove the dyes produced during the dyeing of textile fabrics from wastewater. However, the effective destruction of non-biodegradable azo dyes in effluents does not only comprise decolorization [12–17]. The practical removal of spent dyes in wastewater streams should be accompanied by the formation of easily biodegradable products or by the complete mineralization of the original dye to CO_2, H_2O, nitrate or N_2, sulphate, etc. [16,17].

Fenton-based advanced oxidation processes (AOPs) are among the most effective processes because of their relative ease of operation, low-cost facilities, and high removal efficiency [18].

Fenton oxidation degrades organic pollutants by applying hydroxyl radicals (HO·), which are generated from a hydrogen peroxide (H_2O_2) + ferrous sulfate reaction in an acidic medium [19]. However, the common application of Fenton processes is limited by several limitations, such as acidic pH conditions, iron-based sludge generation, and the large consumption of hydrogen peroxide [20].

The large consumption of H_2O_2 in Fenton oxidation is caused by the high reactivity (low stability) of the produced OH· radicals together with the easier degradability of the organic oxidation products compared to the starting pollutant [18]. In the case of chemical oxidation, a complete mineralization is typically necessary for the removal of toxic and non-biodegradable halogenated organic by-products such as AOX [13].

Some improvements in the utilization of the oxidant used in chemical oxidation processes provides the substitution of hydrogen peroxide with persulfate ($S_2O_8^{2-}$) [18].

The formation of the sulfate radical ·SO_4^- from alkali metal persulfate is commonly catalyzed by the action of transition metal ions such as Fe^{2+}, Co^{2+}, etc. The sulfate radical works as a strong oxidant possessing a higher redox potential and an even higher span of its half-life stability compared with hydroxyl radicals [21–27].

However, the higher price of persulfate (800–1160 USD/t of $Na_2S_2O_8$ [28]) compared to hydrogen peroxide (350–450 USD/t of H_2O_2 [29]) and the secondary contamination of treated water with inorganic sulphates are the limitations of persulfate-based AOPs.

Compared with chemical oxidation, the chemical reduction method enables more selective reductive cleavage of azo or other simply reducible bonds (NO_2, etc.) [30]. Applying appropriate reductants, it facilitates the effective hydrodehalogenation of carbon-halogen bonds that occur in many synthetic dyes (the reductive removal of AOX-producing inorganic halides) [31].

Several papers [32–35] have described the reductive degradation of azo dyes using zero-valent iron (Fe^0 or ZVI) or nano ZVI. However, ZVI is not able to effectively reduce carbon–halogen bonds in halogenated aromatic compounds [33].

As we recently published, the Raney Al-Ni alloy is an excellent hydrodechlorination (HDC) agent [30,31]. Hydrodehalogenation using Al-based alloys is based on dissolving aluminium in the sodium hydroxide solution. Hydrogen is thus produced and adsorbed on the catalytically acting nickel surface which subsequently converts chlorinated organic compounds into non-halogenated products, which are easily biodegradable [36].

The other reduction method for effective azo dye degradation mentioned in the literature is based on sodium borohydride ($NaBH_4$) reduction or on sodium dithionite ($Na_2S_2O_4$) produced in situ by the co-action of $NaBH_4$ and $Na_2S_2O_5$ [37].

With the emphasis on MB9, only a few articles were focused on the removal of this mentioned dye [13,18,38–41]. The previous article of our research group [13] reported the separation of MB9 using ionic liquids. As we presented in this article, the isolation method of MB9 using ionic liquids provides the effective removal of the tested dye [13]. The separation of MB9 by adsorption was consequently studied in several publications [38–40]. As shown by Singh et al., the microbial degradation of MB9 dye enables the effective discoloration of aqueous solutions contaminated with this dye. The level of MB9 HDC is low [41].

Only one article [18] focused on the oxidative degradation of MB9 using the Fenton-activated persulfate system. Pervez et al. reported the effective degradation of MB9 using the above-mentioned method expressed with discoloration and TOC removal efficiency. However, the authors did not control the possible dehalogenation of MB9 [18].

The aim of this article is to compare the effectiveness of chlorinated anionic azo dye Mordant Blue 9 (Figure 1) degradation in an aqueous solution using oxidation or reduction techniques. We compare the broadly applied Fenton reaction with innovative reduction methods.

2. Results and Discussion

2.1. Chemical Oxidation of MB9 Using H_2O_2

All the experiments were performed using a model aqueous solution of MB9 simulated industrial (real) wastewater from the production of acid azo dyes [42]. The applied final concentration of MB9 in reaction mixtures was 0.25 mmol/L (125 mg/L).

The initial experiments were conducted using an excess of hydrogen peroxide with overnight vigorous stirring (20 h) for the possible mineralization of MB9 dye according to Equation (1):

$$C_{16}H_9ClN_2O_8S_2Na_2 + 35\ H_2O_2 \rightarrow 16\ CO_2 + 38\ H_2O + 1\ HCl + N_2 + 2\ NaHSO_4 \quad (1)$$

In experiments of MB9 oxidation using H_2O_2, we demonstrated the degradation of MB9 using absorbance and discoloration efficiency measurements. The COD_{Cr} was applied to determine the removal of organic compounds after oxidation quenching by the addition of 0.1 g MnO_2 which causes the decomposition of residual H_2O_2 (see the Materials and Methods section). Due to the study of chlorinated compound MB9 degradation, AOX was then determined, which expressed the content of chlorinated compounds in samples before and after degradation.

The fragmentation products of MB9 oxidation were not utilized due to complex oxidation pathways. As shown in Equation (1), theoretically at least 35 mol of H_2O_2 is necessary for the complete mineralization of 1 mol of MB9.

Figure 2 shows that using 20 mol of H_2O_2 per mol of MB9, the MB9 dye molecule was destroyed with 92% efficiency after 20 h of action according to the colour removal (measured as absorbance at absorption maximum in the visible spectrum). According to the COD determination, however, less than 10% of oxidizable organic compounds were mineralized under these reaction conditions (using 1 mmol of H_2O_2 per 0.05 mmol of MB9). The gradual increase in hydrogen peroxide dosage only caused a slow increase in COD removal. Using a 40-fold molar excess of H_2O_2 above the MB9 dye, only around 10% of COD (and around 93% of MB9) was removed, even after 20 h of action. This observation is in apparent contradiction with the theoretical consumption of H_2O_2 necessary for the complete mineralization of MB9 to inorganic products (Equation (1)). Even when 200 mol of H_2O_2 per mol of MB9 was used, the COD value decreased to only 40% of the initial quantity (10 mmol of H_2O_2 used per 0.05 mmol of MB9, as shown in Figure 2).

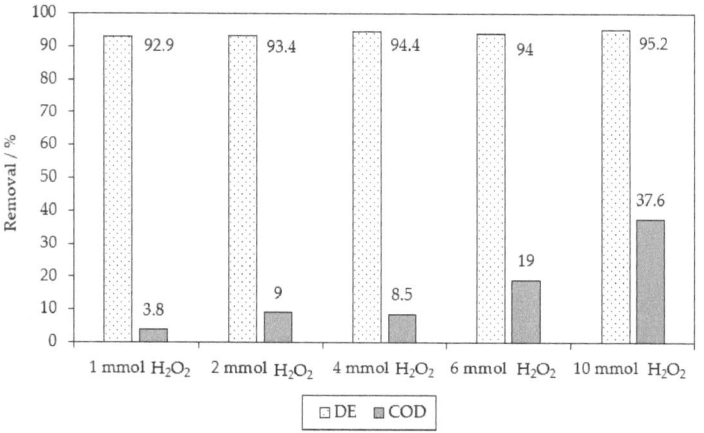

Figure 2. Results of the oxidative degradation of MB9 using an excess of H_2O_2. (Appropriate quantity of H_2O_2 was added to 0.05 mmol of MB9 dissolved in 200 mL of H_2O, with a reaction time of 20 h).

In the subsequent reaction experiments, the ferrous sulphate was added to the acidified mixture of MB9 (2 mL of 16% H_2SO_4 was added to 200 mL of aqueous MB9 solution containing 0.05 mmol of MB9, and the pH was adjusted in pH interval 2–2.5) and stirred with an appropriate quantity of H_2O_2 with the aim of utilizing the oxidation ability of OH· radical produced from the Fenton reaction (Equation (2)):

$$Fe^{2+} + H_2O_2 \rightarrow OH\cdot + Fe^{3+} + OH^- \qquad (2)$$

The applied low pH value was used with the aim of minimizing the possibility of iron (hydr)oxide precipitation during the Fenton oxidation process. As Kremer [43] reported, the optimum pH can also be lower than 3 for an efficient Fenton reaction. The studied dependence of discoloration efficiency/COD removal on the pH value is illustrated by Figure S1 in the Supplementary Materials. It was demonstrated that that there are no significant changes in discoloration efficiencies/COD removal in the range of pH values 2–4. Based on these reasons, the pH value was adjusted to pH = 2–2.5. Our experiments were therefore focused on the adjustment of the optimum dosage of H_2SO_4 for ensuring the acidic environment and efficient oxidation of MB9.

The addition of 2 mL of 16% H_2SO_4 per 200 mL of aqueous 0.25 mM MB9 solution (with a pH value of 2–2.5 and an efficient Fenton oxidation, as shown in Figure 3).

As shown in Figure 3, the Fenton oxidation initialized by the addition of $FeSO_4$ to the acidified MB9 solution significantly increased the COD removal process (compared with the action of sole H_2O_2). In the case of a 40-fold molar excess of H_2O_2 towards MB9, the ferrous sulphate addition enabled the 60% removal of oxidizable matter. By using 6 mmol of H_2O_2 and 0.06 mmol of $FeSO_4$ per 0.05 mmol of MB9 (molar ratio $[Fe^{2+}]/[H_2O_2]/[MB9]$ = 1.2:120:1), the removal of MB9 was practically quantitative and subsequently around 82% of oxidizable matter was removed by Fenton oxidation (according to the COD removal efficiency). Similar $[Fe^{2+}]/[H_2O_2]$ ratios were reported as most effective by Gulkaya et al. for spent dye bath oxidative treatment [19]. Without FeSO$_4$, the COD removal was low (with a COD removal efficiency below 20%).

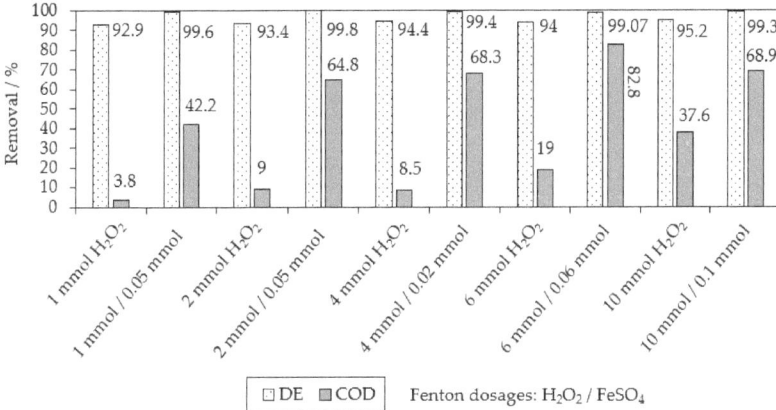

Figure 3. A comparison of the effect of adding FeSO$_4$ to the mixture of H$_2$O$_2$ and MB9 on COD and MB9 removal efficiency. (H$_2$O$_2$ and subsequently FeSO$_4$ were added to 0.05 mmol of MB9 dissolved in 200 mL of H$_2$O).

For a subsequent biological wastewater treatment plant (WWTP), AOX removal during the chemical oxidation of recalcitrant chemicals is most important, especially due to the biocidal effect of many chlorinated compounds such as halogenated phenols [44,45].

The MB9 removal efficiency caused by Fenton oxidation was compared with AOX removal efficiency using an excess of the Fenton reagent in an optimized molar ratio 1 mol of FeSO$_4$ per 100 mol of H$_2$O$_2$. As is illustrated in Figure 4, both AOXs and decolorization efficiencies were quite similar (88% of MB9 removal and 86% of AOX removal) and high enough using the molar ratio [Fe^{2+}]/[H$_2$O$_2$]/[MB9] = 0.2:20:1, even after 4 h of action.

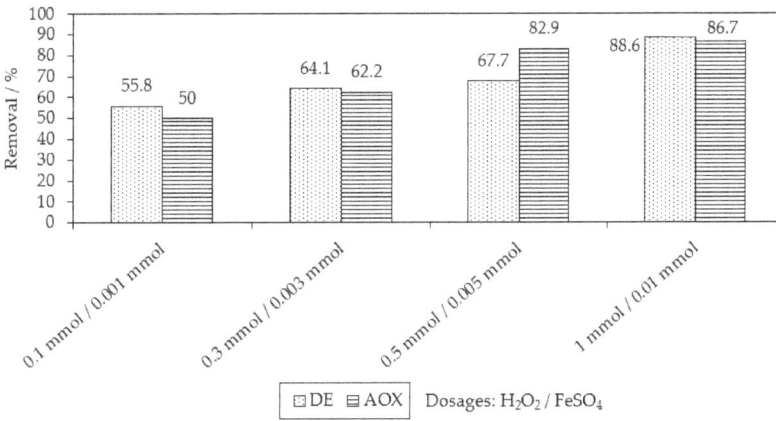

Figure 4. A comparison of MB9 oxidation using the Fenton reaction after 4 h of action. (H$_2$O$_2$ and subsequently FeSO$_4$ were added to 0.05 mmol of MB9 dissolved in 200 mL of H$_2$O).

We observed that the prolonged reaction time of Fenton oxidation is an important parameter for the maximization of AOX and COD removal (Figure 5 below). Due to this reason reaction time, 20 h was applied in the subsequent experiments.

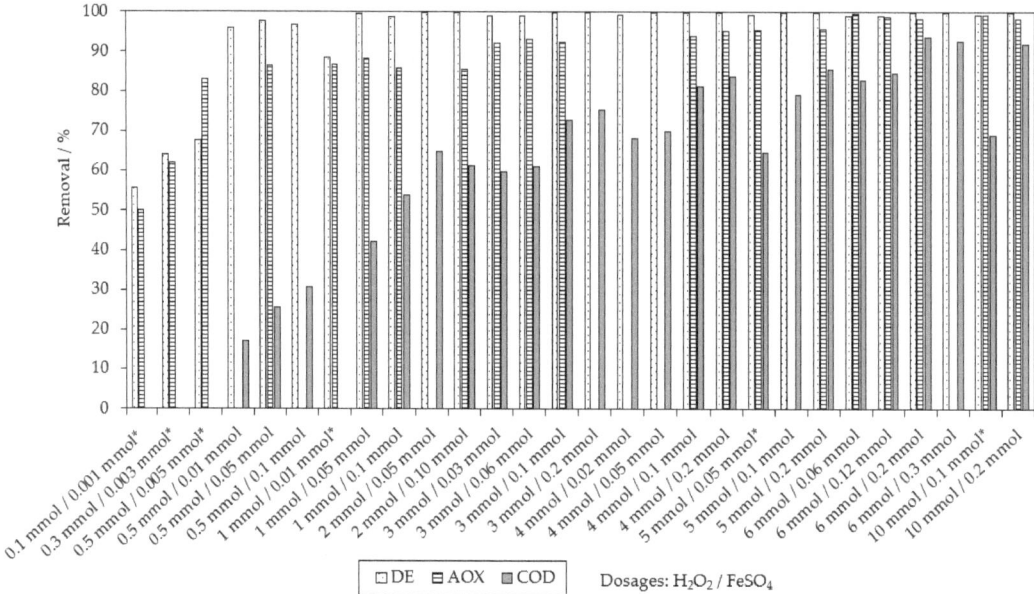

Figure 5. Summary of removal efficiencies obtained using different molar ratios between [Fe^{2+}]/[H_2O_2]/[MB9]. (* H_2O_2 and subsequently $FeSO_4$ were added to acidified 0.05 mmol of MB9 in 200 mL of H_2O, with a reaction time of 20 h or 4 h).

The effect of different molar ratios [Fe^{2+}]/[H_2O_2]/[MB9] = 1–4:10–120:1 is illustrated in Figures 5 and 6. Comparing the relationship between the molar ratios of $FeSO_4$ and H_2O_2, [Fe^{2+}]/[H_2O_2] = 1:20–30 seems to be more effective. It is evident that for the fairly complete removal of MB9 and AOX, at least 120 mol of H_2O_2, together with 4 mol of $FeSO_4$ per mol of MB9, should be used (Figures 6 and 7). This means that the actual minimal excess of H_2O_2 is more than 3–3.4 times higher compared to the theoretical consumption of H_2O_2 for the complete mineralization of MB9 (see Equation (1)).

Despite the high discoloration efficiency of MB9 using Fenton reaction after 4 h (see Figure 7), the removal of COD and AOX did not amount to satisfactory results, as shown in Figure 6. Therefore, the prolonged reaction time (20 h) in combination with high dosages of H_2O_2/Fe^{2+} was proven to be necessary for the effective degradation of MB9 (including AOX and COD removal).

Under described conditions, however, the [Fe^{2+}]/[H_2O_2] ratio seems to be far from optimal [46], likely accompanied by a high consumption of H_2O_2. The real excess of H_2O_2 was more than 3–3.5-times higher compared to the theoretical consumption (see Equation (1)).

Due to the reasons mentioned above, optimization experiments were performed with the aim of rationalizing the dosage of components of the Fenton reagent.

Compared with the persulfate-based method [18], the required excess of H_2O_2 for the practically complete removal of AOX and COD seems to be high. However, the published persulfate method only determined residual TOC and MB9 removal, not a concentration of hardly biodegradable and toxic AOXs. For 95% discoloration, a similar 10-fold molar excess of H_2O_2 and an equimolar quantity of $FeSO_4$ toward MB9 should be used (Figure 5, second column from the left).

Finally, it is evident that the subsequent neutralization of the reaction mixture after Fenton oxidation increases the additional removal of MB9 (almost a 20% increase in the COD removal efficiency) caused by the coagulation of iron (hydr)oxides, as shown in the Supplementary Materials, Table S1. By neutralizing the acidic reaction mixture, in contrast, the produced solid by-product (iron (hydr)oxides containing slurry) is potentially contaminated with adsorbed oxidation (chlorinated) products. This slurry is categorized as a dangerous waste and must be disposed in a safe manner [47].

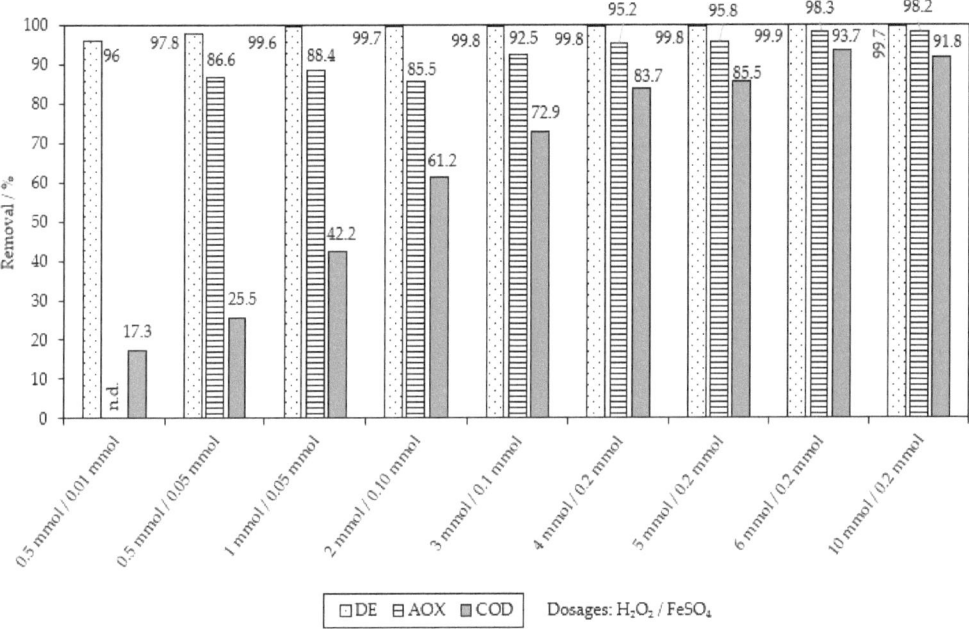

Figure 6. The effect of different FeSO$_4$ and/or H$_2$O$_2$ dosages on the removal efficiencies. (H$_2$O$_2$ and subsequently FeSO$_4$ were added to acidified 0.05 mmol of MB9 dissolved in 200 mL of H$_2$O, with a reaction time of 20 h).

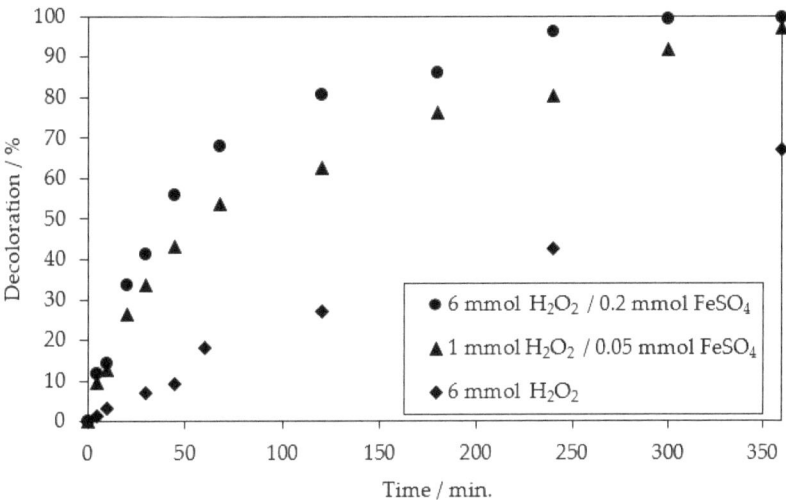

Figure 7. A comparison of the dependence of MB9 oxidative degradation on time under different dosages of H$_2$O$_2$ and FeSO$_4$. (Appropriate quantities of H$_2$O$_2$ and FeSO$_4$ were added to 0.05 mmol of MB9 dissolved in 200 mL of H$_2$O).

2.2. The Chemical Reduction of MB9

Compared with chemical oxidation, the chemical reduction of pollutants was described as more selective [30,31]. A higher selectivity is joined with a lower consumption of reagents.

In the first set of experiments, iron turnings (ZVI) were used for the reductive degradation of MB9.

The obtained results are illustrated in Figure 8. ZVI is a strong reducing agent, and moreover ZVI is easily produced and inexpensive [32]. It has already been rendered to be effective in the reductive degradation of selected chlorinated organic derivatives [48]. In some studies [33,34], methods of dye reduction using micro or nano iron have been described.

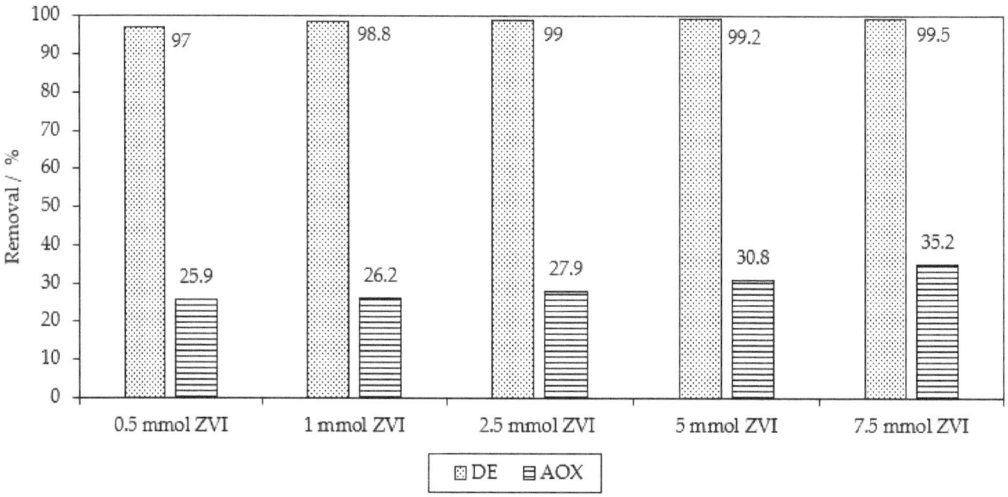

Figure 8. The reductive degradation of MB9 using ZVI. (ZVI was added to aqueous solution containing 0.05 mmol of MB9 in 100 mL of H_2O acidified with 40 mmol of H_2SO_4, with a reaction time of 3 h).

Even using a 10-fold molar excess of ZVI, 97% of MB9 was removed. The AOX removal efficiency, however, was low (25.9%). This corresponds with the known fact that ZVI is not a very effective dechlorination agent in the case of chlorinated aromatic compounds. In the case of a 150-fold molar excess of ZVI, only a 35.2% AOX removal efficiency was observed (together with a 99.5% MB9 removal efficiency).

Apart from ZVI, metallic aluminum (granules, powder, or foil) was tested for the reductive degradation of MB9 dye. Aluminum was reported to be more effective in wastewater treatment compared with ZVI [35]. Figure 9 illustrates differences in Al^0-based reduction efficiencies using diluted sulfuric acid. In addition to Al^0, the Raney Al-Ni alloy, known to be very effective reductant, was used [30,31]. The obtained degradation efficiencies were slightly worse using Al^0 compared with Fe^0.

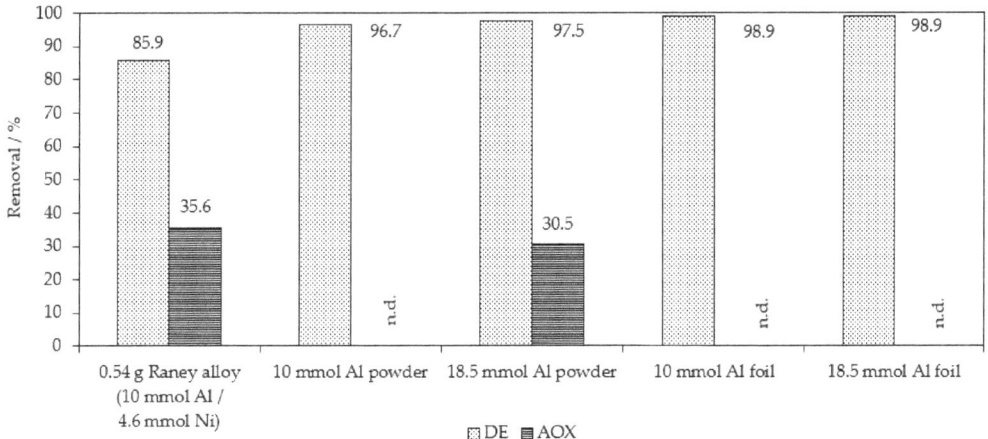

Figure 9. The reductive degradation of MB9 using different types of Al^0 in an acidic environment. (Al^0 was added to the solution of 0.05 mmol of MB9 and 40 mmol of H_2SO_4 in 100 mL of H_2O, with a reaction time of 20 h) (n.d. = not determined).

It is well known that Al^0 is a powerful reductant, especially in diluted alkali metal hydroxide solutions [35]. For this reason, alkaline conditions were used for subsequent experiments. The obtained results are presented in Figure 10.

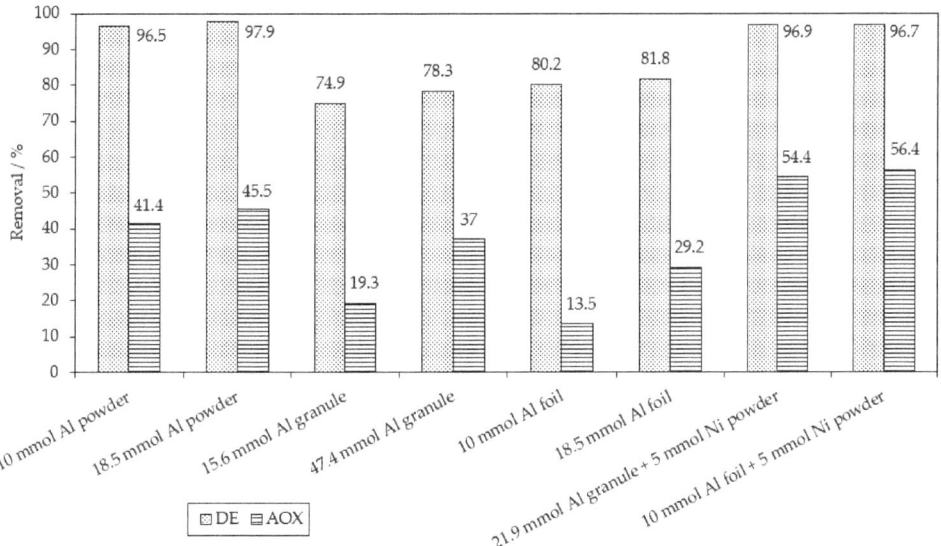

Figure 10. The reductive degradation of MB9 using different types of Al^0 and their mixtures with nickel in an aqueous 50 mM NaOH solution. (Al^0 and Ni^0 were added to the solution of 0.05 mmol of MB9 and 5 mmol of NaOH in 100 mL of H_2O, with a reaction time of 3 h).

When using Al powder in particular, both MB9 and AOX removals were higher compared with ZVI/H_2SO_4 or Al^0/H_2SO_4 actions. In addition, the used Al granules were similarly active as Al foil, although both above-mentioned forms of Al were less active compared with Al powder. In the case of the used Al granules, a low specific surface seems to be the reason for the low reduction activity. In addition, the kinetic comparison indicates that the reaction rate of Al^0/NaOH was higher (Figure 11).

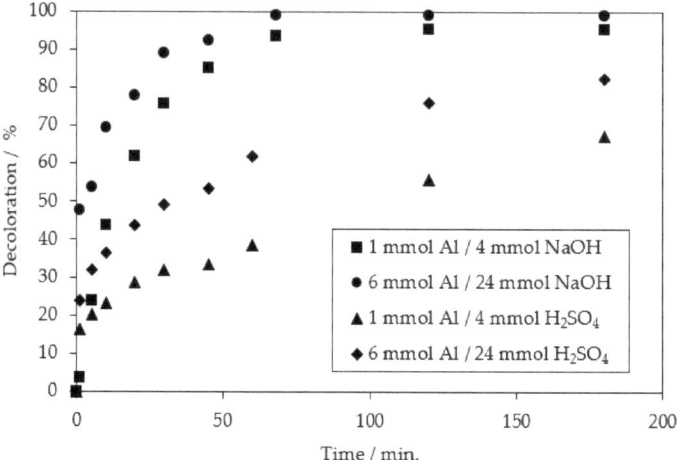

Figure 11. A comparison of rates of MB9 reductive degradation under different dosages of Al and NaOH or H_2SO_4. (Appropriate quantities of Al and NaOH/H_2SO_4 were added to 0.05 mmol of MB9 dissolved in 200 mL of H_2O).

Due to our excellent experience with the hydrodehalogenation activity of Al-Ni alloys [30,31], we tried to use a mixture of Ni powder together with Al^0 (Figure 10) and Raney Al-Ni alloys for MB9 reductive treatment (Figure 12). After the addition of Ni powder, both the MB9 and AOX removal efficiencies increased in comparison with the action of Al^0 alone. Using the Raney Al-Ni alloy, however, the HDC of MB9 reduction products was much higher compared with the action of a mixture of Al^0 and Ni^0 powders.

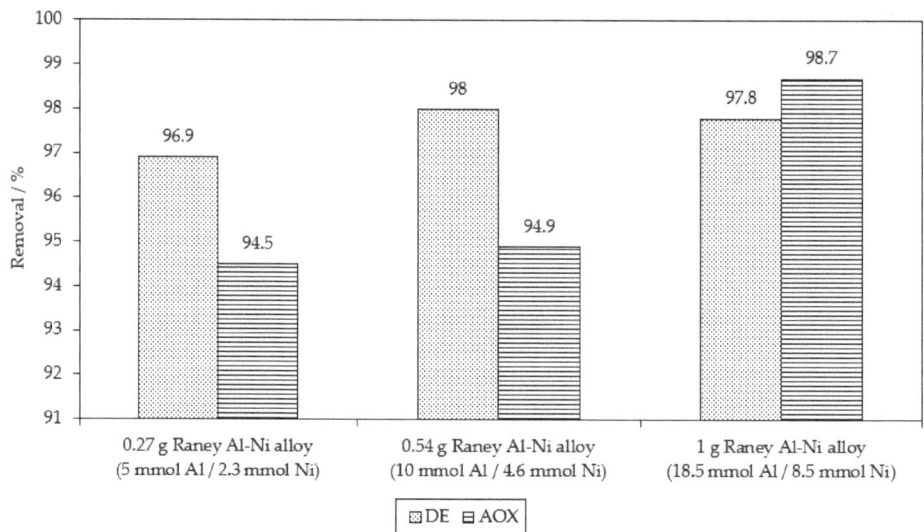

Figure 12. A comparison of AOX and MB9 removal caused by Raney Al-Ni alloy action. (Al-Ni alloy added to aqueous solutions of 0.05 mmol of MB9 and 5 mmol of NaOH in 100 mL of H_2O, with a reaction time of 3 h).

In addition, by using the Raney Al-Ni alloy instead of $Al^0 + Ni^0$ powders, the removal of AOX (hydrodechlorination, HDC) was very effective (Figure 12) [30,31]. HDC was completed after 3 h of Al-Ni alloy/NaOH action. In comparison with the Fenton reaction, HDC using Al-Ni alloy/NaOH processed almost seven times faster in the degradation of MB9. The proposed reaction pathway for the reductive treatment of MB9 using $Al^0 + Ni^0$ or Al/Ni alloys is depicted in Scheme 1.

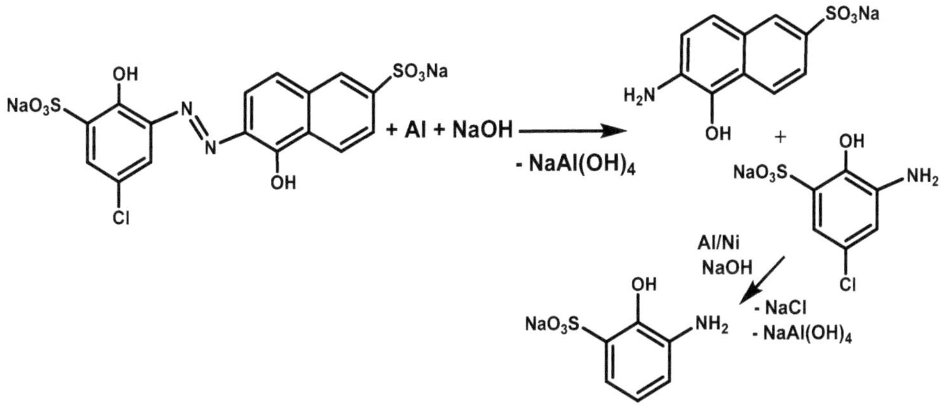

Scheme 1. Proposed reaction pathway for the reduction of MB9 accompanied by hydrodechlorination.

A combination of the rapid azo bond reduction of the MB9 structure using cheaper Al^0 powder (pretreatment) and subsequently HDC using the Al-Ni alloy (see Scheme 1) serves as a simple cost-effective application in the degradation of non-biodegradable chlorinated dye in aqueous streams. Non-chlorinated products based on the reduction/HDC of this promising reductive method were proved using the LC-MS method. As indicated by LC-MS analysis (see Supplementary Material, Figure S2), the HDC of MB9 using Al-Ni (Figure 12, first column from the left, 3 h of action) proceeded smoothly according to the HDC reaction depicted in Scheme 1.

The starting Raney Al-Ni alloy and its reaction products were studied by X-ray powder diffraction in our earlier work [30]. The changes in surface morphology were also documented by SEM [30]. As we published earlier, the starting Raney Al-Ni alloy contained 41% Ni_2Al_3, 57% $NiAl_3$, and 2% aluminum. Briefly, after the beginning of the reaction (after 30 min of HDC of chloroaromates (using 0.27 g of Al-Ni/1 mmol Ar-Cl + 50 mmol of NaOH), the original Raney Al-Ni alloy changed slightly, the pure aluminum disappeared, and the $NiAl_3$ alloy quantity decreased (34% $NiAl_3$ and 66% Ni_2Al_3). After 60 min of the reaction, the $NiAl_3$ diffraction lines disappeared completely, and the isolated solid part only contained Ni_2Al_3. At the end of the HDC reaction, apart from the rest of Ni_2Al_3 (18%), the isolated solid mainly contained $Ni_{0.92}Al_{0.08}$ alloy (61%) and pure metallic nickel (10%), together with a low quantity of bunsenite (2% NiO) [30].

In addition, after the HDC, the produced nickel sludge was simply removable by sedimentation and/or filtration. The deactivated Ni catalyst can be regenerated and recycled with several benefits [49].

The dissolved aluminum in the alkaline reaction mixture after HDC can be utilized for the subsequent sorption of HDC products via the coagulation and flocculation of $Al(OH)_3$ produced by alkaline aqueous phase neutralization after the HDC procedure [30,31,49–51].

The above-mentioned coagulation also increased the COD removal efficiency by up to 15% (see the Supplementary Materials, Table S1).

The HDC of MB9 using Al-Ni therefore effectively removed AOX, and, subsequently, the neutralization of the reaction mixture, accompanied by coagulation, facilitated additional COD removal. In contrast with the Fenton reaction, this slurry after coagulation contains more biodegradable (non-chlorinated) HDC products and it can be used as an additive for cement production.

In the above-mentioned Al-based MB9 removal methods, the sorption of MB9 on the surface of Al-based reductive agents could play an important role. The sorption experiments were performed to compare the sorption effect on MB9 removal using different sources of Al^0.

A comparison of sorption efficiencies is illustrated in Figure 13 (the contact time was 3 h). The highest removal efficiency of MB9 was obtained with Al^0 powder (MB9 65.8%) or Al^0 foil (59.5%). A similar removal efficiency was achieved by the Raney Al-Ni alloy (60%), as shown in Figure 13. Al^0 granules were not very efficient for MB9 sorption, possibly due to the low specific surface area of Al granules.

In addition, the calculated sorption parameter—adsorption capacity q (mg/kg)—demonstrated that the sorption capacities of Al powder and Al foil for MB9 dye were 65.8 mg/kg and 59.5 mg/kg, respectively. In contrast, the adsorption capacity of the Raney Al-Ni alloy was only 30 mg/kg.

Despite this fact, the results indicate that MB9 is selectively adsorbed, especially on powders of aluminum or Raney Al-Ni alloys or Al foil. From the above-mentioned results, it can be deduced that the main drawback of the used Al foil in the reduction process is its very rapid dissolution (consumption) under the used reaction conditions (over a few minutes, as was observed), which prevents the effective MB9 reductive removal.

A comparison of decolorization rates based on the application of Al^0 in alkaline and acidic reaction conditions is illustrated in Figure 14. It is evident that the removal of MB9 dye is the most rapid when the Al^0-based reduction process is used in aqueous NaOH. The reaction rates of the Fenton oxidation and Al^0-based reduction of aqueous H_2SO_4 were quite similar.

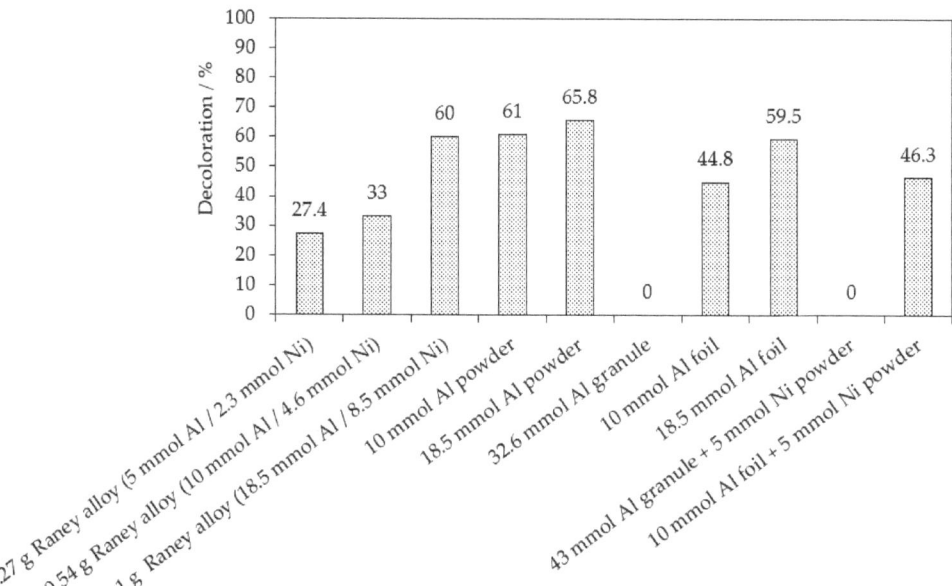

Figure 13. A comparison of sorption abilities of tested Al-based reduction agents. (Mentioned quantities of metal(s) were added to the 0.05 mmol of MB9 dissolved in 100 mL of H_2O, with a sorption time of 3 h).

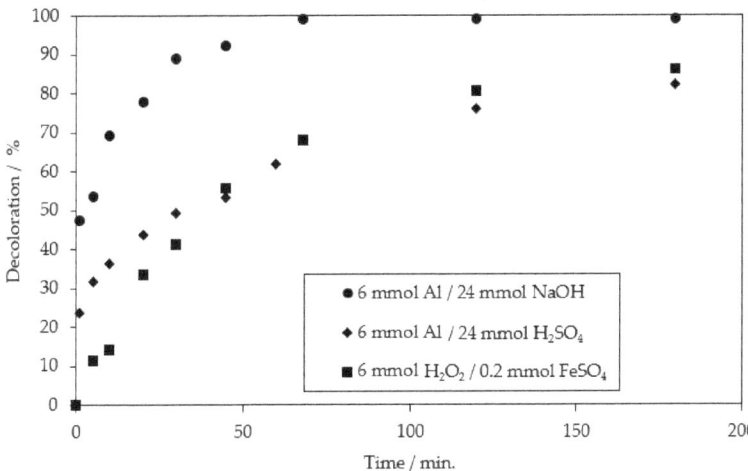

Figure 14. A comparison of discoloration of MB9 using of 6 mmol of oxidant/reductant. (An appropriate quantity of reagents was added to 0.05 mmol of MB9 dissolved in 200 mL of H_2O).

Another reduction method for effective azo dye degradation mentioned in the literature is based on sodium borohydride ($NaBH_4$) or sodium dithionite ($Na_2S_2O_4$, see reactions 3 and 4) produced in situ by the reaction of $NaBH_4$ with $Na_2S_2O_5$ [37] (Scheme 2). It was verified that sodium borohydride caused efficient MB9 reduction when applied alone; however, the AOX removal was sluggish (Figure 15).

$$S_2O_5^{2-} + H_2O \rightarrow 2\ HSO_3^- \tag{3}$$

$$BH_4^- + 8\ HSO_3^- + H^+ \rightarrow 4\ S_2O_4^{2-} + B(OH)_3 + 5\ H_2O \tag{4}$$

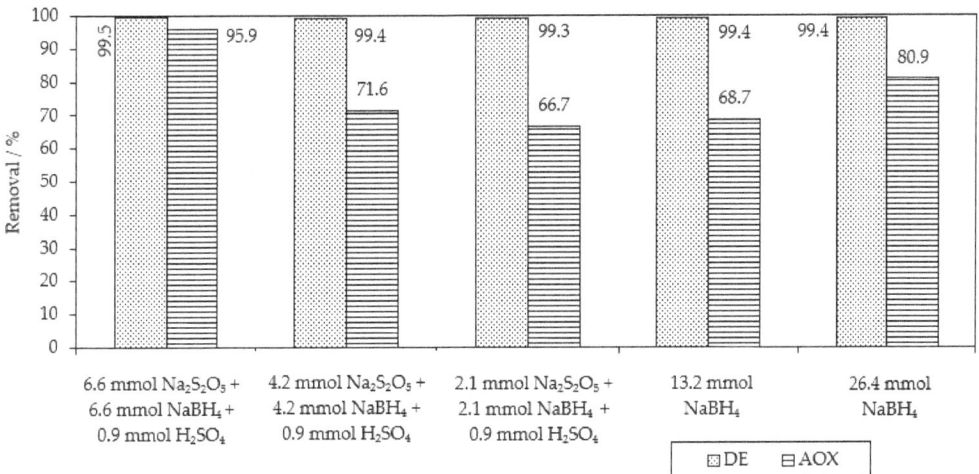

Figure 15. Reductive removal of MB9 and AOX using reductant(s) based on $NaBH_4$ action. (The above-mentioned reduction agents were added to the aq. solution of 0.1 mmol of MB9 in 200 mL of H_2O, with a reaction time of 20 h).

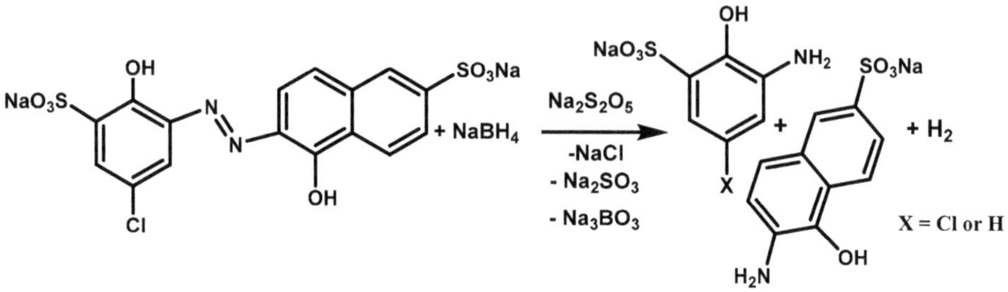

Scheme 2. The proposed reaction scheme of MB9 reductive degradation using $NaBH_4/Na_2S_2O_5$.

When applying a combination of cheap $Na_2S_2O_5$ accompanied by the subsequent addition of $NaBH_4$, both MB9 and AOX were removed very effectively (Figure 15). The proposed reaction scheme of MB9 reductive degradation using $NaBH_4/Na_2S_2O_5$ is illustrated in Scheme 2. As indicated by LC-MS analysis (see Supplementary Material, Figure S3), a smooth reduction of MB9 is depicted in Scheme 2.

Table 1 shows a comparison of the above-described results of the oxidation and reduction of MB9 (calculated per one mol of tested dye for the demonstration of practical application). The reaction rates of the tested reduction processes are compared in Figure 16. It is evident that the homogeneous reduction using $NaBH_4/Na_2S_2O_5$ is the most rapid process weighted by decolorization. A comparison of reagents consumption and degradation efficiencies in performed optimized experiments is depicted in Figure 17. The pretreatment of MB9 solutions using $Al^0/NaOH$ and subsequently HDC with Al-Ni/NaOH also facilitates the effective and smooth removal of MB9 dye.

In summary, the Fenton reaction is cheaper than other tested reductive systems (see Supplementary Materials, Table S2) due to the low cost of hydrogen peroxide. The reaction rate, however, is slow, even when using an excess of H_2O_2 that is more than 3.5 times higher than that using the theoretical consumption. On the other hand, for the practical rapid removal of AOX necessary for scale-up, reductive systems such as $NaBH_4/Na_2S_2O_5$ or a combination of Al^0/Al-Ni alloys are more utilizable.

The above-mentioned combination of $NaBH_4/Na_2S_2O_5$ or Al^0/Al-Ni alloys provides an increase in the removal of MB9 from model wastewater. These methods also offer a decrease in price compared with less efficient/more expensive reductions using $NaBH_4$ or Al-Ni alloy alone (see the Supplementary Materials, Table S2).

Moreover, HDC using Al-Ni alloys proceeds smoothly in a few hours in contrast with HDC using the Fenton reaction (reaction time for effective MB9 removal 20 h, as shown in Table 1). The lower reaction time of the proposed reductive process saves energy and further decreases economic costs. In addition, the application of the Al-Ni alloy enables the recycling of produced Ni-slurry [49–51]. This recycling of used Al-Ni alloys also potentially decreases the price of this effective HDC method.

Table 1. A comparison of MB9 removal efficiencies caused by oxidation or reduction using different reactants at room temperature (calculated per mol of MB9).

Applied Degradation Method	Quantity of Used Reagents per mol of MB9	Efficiency of MB9 Discoloration	Other Removal Efficiencies (%)	Ref.
Persulfate-based AOP at pH = 4.88	16 mol of $Na_2S_2O_8$ + 14 mol of $FeSO_4$	97% (after 30 min)	60% of TOC (after 30 min)	[18]
Fenton oxidation at pH = 4.88	16 mol of H_2O_2 + 14 mol of $FeSO_4$	48% of MB9 (after 30 min)	38% of TOC (after 30 min)	[18]

Table 1. *Cont.*

Applied Degradation Method	Quantity of Used Reagents per mol of MB9	Efficiency of MB9 Discoloration	Other Removal Efficiencies (%)	Ref.
Fenton oxidation at pH = 2–2.5	28 mol of H_2SO_4 120 mol of H_2O_2 + 4 mol of $FeSO_4$	86.1% of MB9 (after 3 h)	37.6% of COD (after 3 h)	This work
Fenton oxidation at pH = 2–2.5	28 mol of H_2SO_4 120 mol of H_2O_2 + 4 mol of $FeSO_4$	>99% of MB9 (after 20 h)	98.3% of AOX and 93.7% of COD (after 20 h)	This work
Chemical reduction ZVI	150 mol of Fe^0 800 mol of H_2SO_4	99.5% of MB9 (after 3 h)	35.2% of AOX (after 3 h)	This work
Chemical reduction Al^0 powder	370 mol of Al^0 800 mol of H_2SO_4	97.5% of MB9 (after 20 h)	30.5% of AOX (after 20 h)	This work
Chemical reduction Al^0 powder	200 mol of Al^0 100 mol of NaOH	96.5% of MB9 (after 3 h)	41.4% of AOX (after 3 h)	This work
Chemical reduction Al^0 powder + Ni^0 powder	200 mol of Al^0 50 mol of Ni^0 100 mol of NaOH	96.7% of MB9	56.4% of AOX (after 3 h)	This work
Chemical reduction Raney Al-Ni	100 mol of Al^0 used in Al-Ni 100 mol of NaOH	96.9% of MB9 (after 3 h)	94.5% of AOX (after 3 h)	This work
Chemical reduction Al^0 powder + Raney Al-Ni	120 mol of Al^0 480 mol of NaOH 44 mol of Al^0 (in Al-Ni) 480 mol of NaOH	99% of MB9 (after 3 + 20 h)	91.7% of AOX (after 3 + 20 h)	This work
Chemical reduction $NaBH_4$	132 mol of $NaBH_4$	99.4% of MB9 (after 20 h)	68.7% of AOX (after 20 h)	This work
Chemical reduction $NaBH_4/Na_2S_2O_5$	66 mol of $NaBH_4$ + 66 mol of $Na_2S_2O_5$ + 18 mol of H_2SO_4	99.5% of MB9 (after 30 min)	95.9% of AOX (after 20 h)	This work

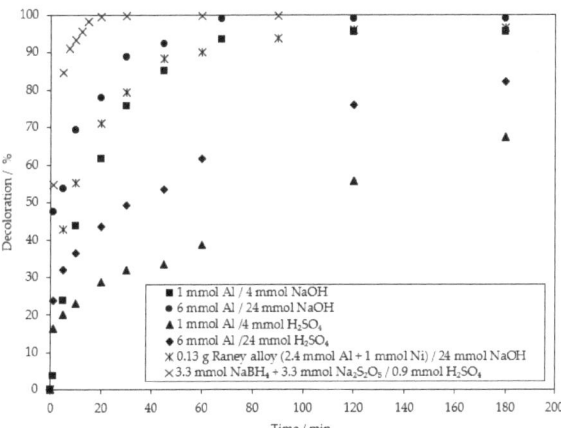

Figure 16. A comparison of the reaction rates of MB9 reductive degradation under different dosages of reductants. (An appropriate quantity of reagents was added to 0.05 mmol of MB9 dissolved in 200 mL of H_2O).

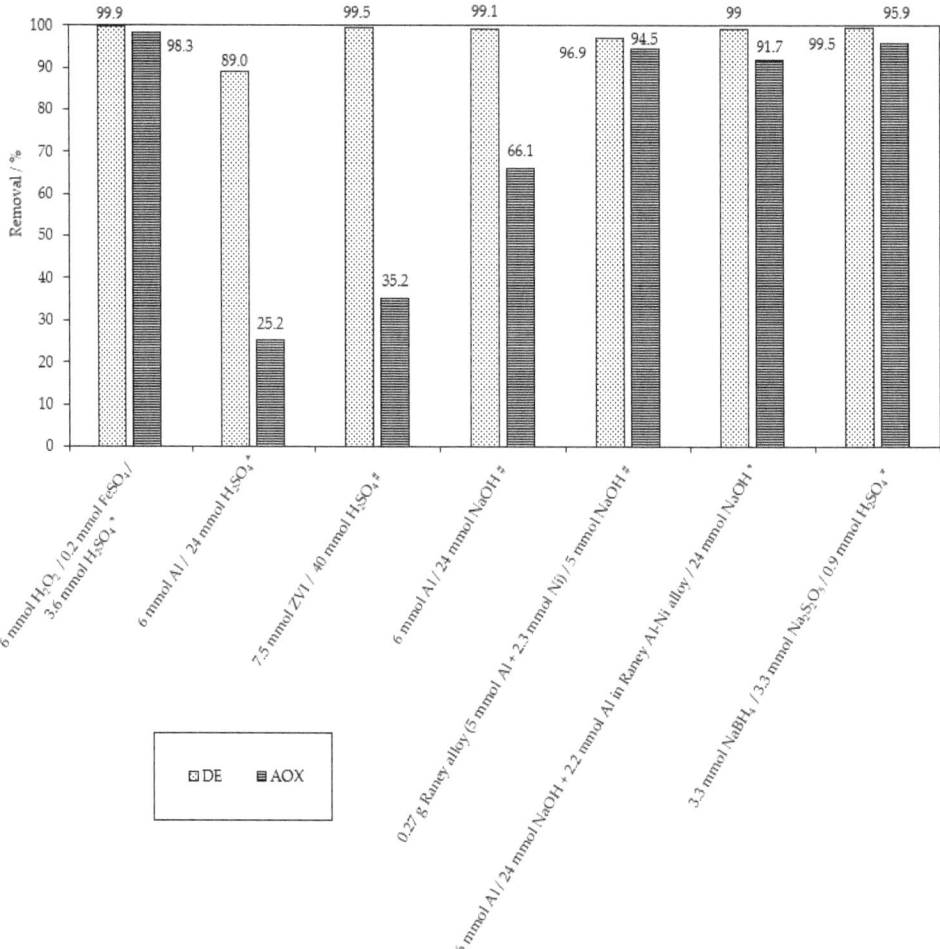

Figure 17. A comparison of reagent consumption and effectiveness of MB9 degradation. (An appropriate quantity of reagents was added to 0.05 mmol of MB9 dissolved in 200 mL of H_2O, with a reaction time of # 3 h or * 20–21 h).

3. Materials and Methods

3.1. Reagents and Materials

The acidic azo dye Mordant Blue 9 (MB9) was purchased from commercial sources in a defined purity 50% of MB9 (Sigma Aldrich Co., Prague, Czech Republic). Sigma-Aldrich does not provide the exact composition of the other ingredients. It can be assumed, however, that other ingredients are inorganic salts from dye manufacturing [6].

H_2O_2 (30%) and $FeSO_4 \cdot 7H_2O$ (p.a. quality) were obtained from Lach-Ner Co., Neratovice, Czech Republic. The Raney Al-Ni alloy, $NaBH_4$, and $Na_2S_2O_5$ were supplied from Sigma Aldrich Co., Prague, Czech Republic. The powdered Al^0 (Labo MS Co., Prague, Czech Republic), granulated Al (Lachema Co., Brno, Czech Republic), Al foil, powdered Ni (Alfa Aesar, Heysham, Lancashire, UK), and zero-valent Fe^0 (ZVI, local supplier Synthesia Co., Pardubice, Czech Republic) were purchased in purity higher than 95%. Additional chemicals (H_2SO_4, NaOH and MnO_2) in p.a. quality were obtained from Lach-Ner Co.,

Neratovice, Czech Republic. Finally, all the aqueous solutions were prepared from demineralized water.

3.2. Oxidative and Reductive Degradation of MB9 Procedures

The comparative degradation experiments were performed in a 250 mL round-bottomed flasks equipped with magnetic stirring on Starfish equipment (Radleys Discovery Technologies, Saffron Walden, UK), installed on a magnetic stirrer Heidolph Heistandard for parallel reactions. The reaction flasks were closed by a tube filled with granulated charcoal. After the appropriate time period of stirring, the reaction mixtures were filtered and analyzed.

In the case of the oxidative degradation of MB9 using hydrogen peroxide, all the samples were mixed before filtration with 0.1 g of MnO_2 used as the catalyst for the rapid decomposition of unreacted H_2O_2. The final measurement of residual H_2O_2 in the reaction mixtures was not required and the interferences of COD determination were negligible (unreacted H_2O_2 significantly increased the COD value, as verified). For the Fenton reaction, the pH values of reaction mixtures were modified using the addition of 10 mL of 16% H_2SO_4 per liter of 0.5 mM aqueous MB9 solution (2 mL of 16% H_2SO_4 per 200 mL of H_2O containing 0.05 mmol of MB9). The pH values were measured and/or controlled by a contact pH meter Orio Star 10 (Thermo Fisher Scientific, Pardubice, Czech Republic).

In the case of the ZVI application, due to the limitation of the magnetic stirring influence, the reaction mixtures were intensively stirred. Thus, the sufficient contact of ZVI with the aqueous solutions of MB9 was ensured.

3.3. Chemical Analysis

A Hach DR2800 (Austria) VIS spectrophotometer was employed for the absorbance measurements using 1 cm glass cuvettes. The concentrations of MB9 were determined by measuring at a wavelength 516 nm. Analyses of the COD were carried out in accordance with the ISO EN 9562 standard and aqueous solutions were determined using the Hach Lange cuvette test using the Hach DR2800 (Austria) VIS spectrometer. The AOXs were analyzed according to the European ISO 9562 standard using the Multi X 2500 analyzer (Analytic Jena GmbH., Jena, Germany).

The discoloration of MB9 solutions was calculated according to Equation (5):

$$DE = \left(1 - \frac{A}{A_0}\right) \times 100 \quad (5)$$

where DE is the discoloration efficiency of MB9 aqueous solution (%), A is the measured absorbance after the degradation process, and A_0 is the initial absorbance of MB9 solution.

The removal efficiency (%) of COD or AOX from the model aqueous solutions of MB9 was evaluated by Equation (6):

$$RE = \left(1 - \frac{c}{c_0}\right) \times 100 \quad (6)$$

where RE is removal efficiency of AOX or COD (%), c is the concentration of AOX or COD in the solution after the degradation process (mg/L), and c_0 is the initial concentration of AOX or COD in the MB9 solution before the degradation process (mg/L).

The reduction products after the reductive degradation of MB9 were detected using the LC-MS technique with additional UV detection. The water samples were properly diluted and 2 μL of the diluted sample was injected in the Hypersil Gold C4 column (100 × 3 mm). The LC separation of the detected compounds was performed in a LC-MSD TRAP XCT Plus system (Agilent Technologies, Santa Clara, CA, USA) and was carried out with a gradient program using (A) 5 mM CH_2COONH_2, (B) 95% CH_3COONH_4, and (C) acetonitrile at a flow rate of 0.5 mL/min and at temperature of 40 °C. The detected compounds were completely separated under the established gradient program in 40 min. The UV detection was carried out at a wavelength 254 nm. The LC system was connected to an Agilent 6490 Triple Q MS with an electrospray interface (ESI) and was operated in both positive and negative ionization mode, using the above-described gradient program in both cases. The MS settings were as follows: a drying gas temperature of 350 °C; a drying gas of 10 L/min; ion source gas of 50 psi; and a mass range of 50–1200 Da. Nitrogen was used as the nebulizer gas, curtain gas, and collision gas.

4. Conclusions

This paper aims to compare the efficiency of common Fenton oxidation together with innovative chemical reduction of chlorinated anionic azo dye Mordant Blue 9 to simply biodegradable nonchlorinated products.

The classic Fenton oxidation is broadly used and is a frequently evaluated treatment method using cheap and low toxic reagents, such as hydrogen peroxide and ferrous sulphate. These reagents cause the negligible secondary contamination of treated and subsequently neutralized water free of insoluble impurities such as iron (hydr)oxides.

We demonstrated that an appropriate excess of hydrogen peroxide (120 mol), H_2SO_4 (28 mol), and $FeSO_4$ (4 mol) is able to destroy chlorinated acid azo dye Mordant Blue 9 to non-chlorinated by-products with the subsequent removal of oxidizable organic compounds after 20 h of action. By neutralizing the acidic reaction mixture, the produced solid by-product, including iron (hydr)oxides containing slurry, is potentially contaminated with adsorbed oxidation products. This slurry is categorized as dangerous waste and must be disposed in a safe manner. In addition, the large consumption of reagents could disfavor the Fenton oxidation for potential large-scale application.

The tested reductive wastewater treatment methods are less common compared with chemical oxidation. Using appropriate reduction agents, the chemical reduction serves as a simple and economically attractive method for the destruction of a hardly biodegradable compound, such as chlorinated azo dye MB9, to non-chlorinated by-products suitable for subsequent biodegradation in common wastewater treatment plants.

Both ZVI/H_2SO_4 and Al^0/H_2SO_4 are effective azo reduction bonds, but poor HDC agents. The same inconvenience was observed in the case of $Al^0/NaOH$ action. In contrast, the heterogeneous reductant Raney Al-Ni alloy in NaOH and the homogeneous reductant $NaBH_4/Na_2S_2O_5$ alloy are very effective for both azo bond reduction and subsequent HDC.

There is still uncertainty around (i) the minimal quantity and the price of effective reductants applicable for both the azo bond reduction and effective HDC of produced chlorinated aminophenol- and aminonaphtol-based sulphonic acids and (ii) the accompanying treatment of produced by-products descended from used reductants/HDC agents.

From the consumption of the reagents point of view, the homogeneous reduction accompanied by HDC using $Na_2S_2O_4$ produced in situ (by the action of $NaBH_4/Na_2S_2O_5$) is the most effective, rapid, and simple destruction method.

An innovative heterogeneous reduction combining pretreatment with $Al^0/NaOH$ with the subsequent addition of effective HDC agents (Raney Al-Ni alloy/NaOH) is more exacting. In the case of the used Al-Ni alloy, the recycling of produced Ni-slurry was demonstrated earlier with the aim of obtaining an effective hydrodehalogenation agent.

Supplementary Materials: The following supporting information can be downloaded at: https://www.mdpi.com/article/10.3390/catal13030460/s1, Figure S1: The dependence of discoloration efficiency/COD removal on the pH value for the Fenton reaction using 6 mmol of H_2O_2 + 0.2 mmol of $FeSO_4$ per 0.05 mmol of MB9 in 200 mL of H_2O (reaction time 20 h). Figure S2: LC chromatogram of LC-MS analysis of the reaction mixture obtained after the reductive degradation of MB9 using Al-Ni/NaOH. Figure S3a: LC chromatogram of LC-MS analysis of reaction mixture obtained after the reductive degradation of MB9 using $NaBH_4/Na_2S_2O_5$. Figure S3b: Qualitative LC-MS analysis of the reaction mixture obtained after the reductive degradation of MB9 using $NaBH_4/Na_2S_2O_5$. Figure S3c: Mass spectrums (EI, 70 eV) of analyzed components. Table S1: The effect of subsequently coagulation on Fenton reaction and reduction/HDC using the Al-Ni alloy. Table S2: A comparison of economic viability of MB9 removal using different oxidative/reductive agents [52–62].

Author Contributions: Conceptualization, T.W.; methodology, T.W.; software, B.K.; validation, B.K.; formal analysis, B.K.; investigation, B.K. and T.W.; resources, T.W.; data curation, B.K.; writing—original draft preparation, T.W.; writing—review and editing, T.W.; visualization, B.K.; supervision, T.W. All authors have read and agreed to the published version of the manuscript.

Funding: This research received no external funding.

Data Availability Statement: Not applicable.

Acknowledgments: The authors acknowledge the financial support for excellent technological teams from the Faculty of Chemical Technology, University of Pardubice.

Conflicts of Interest: The authors declare no conflict of interest.

References

1. Varjani, S.; Rakholiya, P.; Shindhal, T.; Shah, A.V.; Ngo, H.H. Trends in dye industry effluent treatment and recovery of value-added products. *J. Water Process Eng.* **2021**, *39*, 101734. [CrossRef]
2. Sen, S.K.; Raut, S.; Bandyopadhyay, P.; Raut, S. Fungal decolouration and degradation of azo dyes: A review. *Fungal Biol. Rev.* **2016**, *30*, 112–133. [CrossRef]
3. Selvaraj, V.; Karthika, T.S.; Mansiya, C.; Alagar, M. An over review on recently developed techniques, mechanisms and intermediate involved in the advanced azo dye degradation for industrial applications. *J. Mol. Struct.* **2021**, *1224*, 129195. [CrossRef]
4. Gong, R.; Li, M.; Yang, C.; Sun, Y.; Chen, J. Removal of cationic dyes from aqueous solution by adsorption on peanut hull. *J. Hazard. Mater.* **2005**, *121*, 247–250. [CrossRef]
5. Cai, Y.; David, S.; Pailthorpe, M. Dyeing of jute and jute/cotton blend fabrics with 2:1 pre-metallised dyes. *Dye. Pigment.* **2000**, *45*, 161–168. [CrossRef]
6. Ding, Y.; Freeman, H.S. Mordant dye application on cotton: Optimisation and combination with natural dyes. *Color. Technol.* **2017**, *133*, 369–375. [CrossRef]
7. Sheikh, M.; Karim, R.; Farouqui, A.N.; Yahya, R.; Hassan, A. Effect of Acid Modification on Dyeing Properties of Rajshahi Silk Fabric with Different Dye Classes. *Fibers Polym.* **2011**, *12*, 642–647. [CrossRef]
8. Sheikh, R.K.; Faruqui, A.N. A Comparative Study on Dyeing Behavior of Modified Rajshahi Silk Fabric with Reactive Orange 14, Direct Yellow 29 and Mordant Blue 9 Dyes. *Man-Made Text. India* **2010**, *53*, 10.
9. Edwards, L.C.; Freeman, H.S. Synthetic dyes based on environmental considerations. Part 3: Aquatic toxicity of iron-complexed azo dyes. *Color. Technol.* **2005**, *121*, 265–270. [CrossRef]
10. Gao, H.W.; Lin, J.; Li, W.Y.; Hu, Z.J.; Zhang, Y.L. Formation of shaped barium sulfate-dye hybrids: Waste dye utilization for eco-friendly treatment of wastewater. *Environ. Sci. Pollut. Res.* **2010**, *17*, 78–83. [CrossRef]
11. Angelakis, A.; Snyder, S. Wastewater treatment and reuse: Past, present, and future. *Water* **2015**, *7*, 4887–4895. [CrossRef]
12. Ran, J.H.; Shushina, I.; Priazhnikova, V.; Telegin, F. Inhibition of Mordant Dyes Destruction in Fenton Reaction. *Adv. Mater. Res.* **2013**, *821*, 493–496. [CrossRef]
13. Šimek, M.; Mikulášek, P.; Kalenda, P.; Weidlich, T. Possibilities for removal of chlorinated dye Mordant Blue 9 from model wastewater. *Chem. Pap.* **2016**, *70*, 470–476. [CrossRef]
14. Mishra, G.; Tripathy, M. A critical review of the treatments for decolourization of textile effluent. *Colourage* **1993**, *40*, 35–38.
15. Xavier, S.; Gandhimathi, R.; Nidheesh, P.V.; Ramesh, S.T. Comparison of homogeneous and heterogeneous Fenton processes for the removal of reactive dye Magenta MB from aqueous solution. *Desalin. Water Treat.* **2015**, *53*, 109–118. [CrossRef]
16. Nidheesh, P.V.; Gandhimathi, R.; Ramesh, S.T. Degradation of dyes from aqueous solution by Fenton processes: A review. *Environ. Sci. Pollut. Res.* **2013**, *20*, 2099–2132. [CrossRef]
17. Nidheesh, P.V.; Gandhimathi, R. Trends in electro-Fenton process for water and wastewater treatment: An overview. *Desalination* **2012**, *299*, 1–15. [CrossRef]
18. Pervez, M.N.; Telegin, F.Y.; Cai, Y.; Xia, D.; Zarra, T.; Naddeo, V. Efficient Degradation of Mordant Blue 9 Using the Fenton-Activated Persulfate System. *Water* **2019**, *11*, 2532. [CrossRef]
19. Gulkaya, I.; Surucu, G.A.; Dilek, F.B. Importance of H_2O_2/Fe^{2+} ratio in Fenton's treatment of a carpet dyeing wastewater. *J. Hazard. Mater.* **2006**, *136*, 763–769. [CrossRef]
20. Olmez-Hanci, T.; Arslan-Alaton, I. Comparison of sulfate and hydroxyl radical based advanced oxidation of phenol. *Chem. Eng. J.* **2013**, *224*, 10–16. [CrossRef]
21. Rastogi, A.; Al-Abed, S.R.; Dionysiou, D.D. Sulfate radical-based ferrous–peroxymonosulfate oxidative system for PCBs degradation in aqueous and sediment systems. *Appl. Catal. B Environ.* **2009**, *85*, 171–179. [CrossRef]
22. Rao, Y.; Qu, L.; Yang, H.; Chu, W. Degradation of carbamazepine by Fe (II)-activated persulfate process. *J. Hazard. Mater.* **2014**, *268*, 23–32. [CrossRef]
23. Wang, S.; Wu, J.; Lu, X.; Xu, W.; Gong, Q.; Ding, J.; Dan, B.; Xie, P. Removal of acetaminophen in the Fe2+/persulfate system: Kinetic model and degradation pathways. *Chem. Eng. J.* **2019**, *358*, 1091–1100. [CrossRef]

24. Shang, W.; Dong, Z.; Li, M.; Song, X.; Zhang, M.; Jiang, C.; Feiyun, S. Degradation of diatrizoate in water by Fe (II)-activated persulfate oxidation. *Chem. Eng. J.* **2019**, *361*, 1333–1344. [CrossRef]
25. Zhu, J.-P.; Lin, Y.-L.; Zhang, T.-Y.; Cao, T.-C.; Xu, B.; Pan, Y.; Zhang, X.-T.; Gao, N.-Y. Modelling of iohexol degradation in a Fe (II)-activated persulfate system. *Chem. Eng. J.* **2019**, *367*, 86–93. [CrossRef]
26. Kusic, H.; Peternel, I.; Ukic, S.; Koprivanac, N.; Bolanca, T.; Papic, S.; Bozic, A.L. Modeling of iron activated persulfate oxidation treating reactive azo dye in water matrix. *Chem. Eng. J.* **2011**, *172*, 109–121. [CrossRef]
27. Rodriguez, S.; Vasquez, L.; Costa, D.; Romero, A.; Santos, A. Oxidation of Orange G by persulfate activated by Fe (II), Fe (III) and zero valent iron (ZVI). *Chemosphere* **2014**, *101*, 86–92. [CrossRef]
28. Made in China, Persulphate Price. Available online: https://www.made-in-china.com/products-search/hot-china-products/Persulphate_Price.html#:~:text=Sodium%20Persulfate%20Na2s2o8%20Factory%20Supply%20Plant%207775-27-1,PriceSodium%20Persulphate%20FOB%20Price%3A%20US%20%24%20900-1160%2F%20T (accessed on 11 November 2022).
29. Procurement Resource, Hydrogen Peroxide Price Trends. Available online: https://www.procurementresource.com/resource-center/hydrogen-peroxide-price-trends (accessed on 11 November 2022).
30. Weidlich, T.; Kamenická, B.; Melánová, K.; Čičmancová, V.; Komersová, A.; Čermák, J. Hydrodechlorination of different chloroaromatic compounds at room temperature and ambient pressure—Differences in reactivity of Cu-and Ni-Based Al alloys in an alkaline aqueous solution. *Catalysts* **2021**, *10*, 994. [CrossRef]
31. Bendová, H.; Kamenická, B.; Weidlich, T.; Beneš, L.; Vlček, M.; Lacina, P.; Švec, P. Application of Raney Al-Ni alloy for simple hydrodehalogenation of Diclofenac and other halogenated biocidal contaminants in alkaline aqueous solution under ambient conditions. *Materials* **2022**, *15*, 3939. [CrossRef]
32. Lin, Y.T.; Weng, C.H.; Chen, F.Y. Effective removal of AB24 dye by nano/micro-size zero-valent iron. *Sep. Purif. Technol.* **2008**, *64*, 26–30. [CrossRef]
33. Raman, C.D.; Kanmani, S. Textile dye degradation using nano zero valent iron: A review. *J. Environ. Manag.* **2016**, *177*, 341–355. [CrossRef]
34. Fan, J.; Guo, Y.; Wang, J.; Fan, M. Rapid decolorization of azo dye methyl orange in aqueous solution by nanoscale zerovalent iron particles. *J. Hazard. Mater.* **2009**, *166*, 904–910. [CrossRef]
35. Nidheesh, P.V.; Khatri, J.; Singh, T.A.; Gandhimathi, R.; Ramesh, S.T. Review of zero-valent aluminium based water and wastewater treatment methods. *Chemosphere* **2018**, *200*, 621–631. [CrossRef]
36. Yang, B.; Zhang, F.; Deng, S.; Yu, G.; Zhang, H.; Xiao, J.; Shen, J. A facile method for the highly efficient hydrodechlorination of 2-chlorophenol using Al-Ni alloy in the presence of fluorine ion. *J. Chem. Eng.* **2012**, *209*, 79–85. [CrossRef]
37. Reife, A.; Freeman, H.S. *Environmental Chemistry of Dyes and Pigments*, 1st ed.; John Wiley & Sons:: Hoboken, NJ, USA, 1996; p. 352, ISBN 978-0-471-58927-3.
38. Martins, A.; Nelson, N. Adsorption of a textile dye on commercial activated carbon: A simple experiment to explore the role of surface chemistry and ionic strength. *J. Chem. Educ.* **2015**, *92*, 143–147. [CrossRef]
39. Tomic, N.M.; Dohcevic-Mitrovic, Z.D.; Paunović, N.M.; Mijin, D.Z.; Radić, N.D.; Grbic, B.V.; Bajuk-Bogdanovic, D.V. Nanocrystalline $CeO_2 - \delta$ as effective adsorbent of azo dyes. *Langmuir* **2014**, *30*, 11582–11590. [CrossRef]
40. Kumar, M.; Venugopal, A.K.P.; Pakshirajan, K. Novel biologically synthesized metal nanopowder from wastewater for dye removal application. *Environ. Sci. Pollut. Res.* **2022**, *29*, 38478–38492. [CrossRef]
41. Singh, S.; Pakshirajan, K.; Daverey, A. Screening and optimization of media constituents for decolourization of Mordant Blue-9 dye by Phanerochaete chrysosporium. *Clean Technol. Environ. Policy* **2010**, *12*, 313–323. [CrossRef]
42. Yassen, D.A.; Scholz, M. Textile dye wastewater characteristics and constituents of synthetic effluents: A critical review. *Int. J. Environ. Sci. Technol.* **2019**, *16*, 1193–1226. [CrossRef]
43. Kremer, M.L. The Fenton reaction. Dependence of the rate on pH. *J. Phys. Chem. A* **2003**, *107*, 1734–1741. [CrossRef]
44. Igbinosa, E.O.; Odjadjare, E.E.; Chigor, V.N.; Igbinosa, I.H.; Emoghene, A.O.; Ekhaise, F.O.; Igiehon, N.O.; Idemudia, O.G. Toxicological profile of chlorophenols and their derivatives in the environment: The public health perspective. *Sci. World J.* **2013**, *11*, 460215. [CrossRef]
45. Weidlich, T.; Prokes, L.; Pospisilova, D. Debromination of 2,4,6-tribromophenol coupled with biodegradation. *Cent. Eur. J. Chem.* **2013**, *11*, 979. [CrossRef]
46. Bousalah, D.; Zazoua, H.; Boudjemaa, A.; Benmounah, A.; Bachari, K. Degradation of Indigotine food dye by Fenton and photo-Fenton processes. *Int. J. Environ. Anal. Chem.* **2020**, *102*, 4609. [CrossRef]
47. Kang, S.F.; Liao, C.H.; Chen, M.C. Pre-oxidation and coagulation of textile wastewater by the Fenton process. *Chemosphere* **2002**, *46*, 923–928. [CrossRef]
48. Zhang, W.X. Nanoscale iron particles for environmental remediation: An overview. *J. Nanoparticle Res.* **2003**, *5*, 323–332. [CrossRef]
49. Weidlich, T.; Kamenicka, B. Recycling of Spent Hydrodehalogenation Catalysts–Problems Dealing with Separation of Aluminium. *Inz. Miner.* **2019**, *21*, 177–182. [CrossRef]
50. Weidlich, T.; Krejcova, A.; Prokes, L. Hydrodebromination of 2,4,6-tribromophenol in aqueous solution using Devarda's alloy. *Monatsh. Chem.* **2013**, *144*, 155–162. [CrossRef]
51. Bendova, H.; Weidlich, T. Application of diffusion dialysis in hydrometallurgical separation of nickel from spent Raney Ni catalyst. *Sep. Sci. Technol.* **2018**, *53*, 1218–1222. [CrossRef]

52. Weidlich, T.; Oprsal, J.; Krejcova, A.; Jasurek, B. Effect of glucose on lowering Al-Ni alloy consumption in dehalogenation of halogenoanilines. *Monatsh. Chem.* **2015**, *146*, 613–620. [CrossRef]
53. Made in China. H_2SO_4 Price. Available online: https://kunyabiological.en.made-in-china.com/product/kZMtfiCTMFWw/China-Sulfuric-Acid-with-Competitive-Price-and-CAS-No-7664-93-9.html (accessed on 23 January 2023).
54. Sigma Aldrich. H_2O_2 Price. Available online: https://www.sigmaaldrich.com/CZ/en/substance/hydrogenperoxide34017722841 (accessed on 23 January 2023).
55. Sigma Aldrich. $FeSO_4$ Price. Available online: https://www.sigmaaldrich.com/CZ/en/substance/ironiisulfateheptahydrate278017782630 (accessed on 23 January 2023).
56. AliExpress. Fe^0 Price. Available online: https://www.aliexpress.com/item/1005003952039242.html?spm=a2g0o.productlist.main.1.666b1ebeXVNzaz&algo_pvid=567a960c-734e-414b-979d-f5ea646666da&algo_exp_id=567a960c-734e-414b-979d-f5ea646666da-0&pdp_ext_f=%7B%22sku_id%22%3A%2212000030178526057%22%7D&pdp_npi=2%40dis%21CZK%21367.31%21367.31%21%21%21%21%21%40214528be16744809367621410d06b6%2112000030178526057%21sea&curPageLogUid=hPMiO1R1R4b7 (accessed on 23 January 2023).
57. Amazon. Al^0 Price. Available online: https://www.amazon.com/Aluminum-Powder-Pound-MS-MetalShipper/dp/B0849WHSTY/ref=sr_1_1?crid=32BXCVP7ZHZH1&keywords=aluminum+powder&qid=1674482314&sprefix=%2Caps%2C1251&sr=8-1 (accessed on 23 January 2023).
58. Made in China. NaOH Price. Available online: https://highmountainchem.en.made-in-china.com/product/HElUSyYxsrhD/China-Sodium-Hydroxide-Caustic-Soda-Flakes-Alkali-99-Naoh-CAS-1310-73-2.html (accessed on 23 January 2023).
59. AliExpress. Ni^0 Price. Available online: https://www.aliexpress.com/item/1005004187537341.html?spm=a2g0o.productlist.main.1.6f4ePBR5PBR5j9&algo_pvid=bab4e3ef-19f2-475e-b186-02000b20af84&algo_exp_id=bab4e3ef-19f2-475e-b186-02000b20af84-0&pdp_ext_f=%7B%22sku_id%22%3A%2212000028321382122%22%7D&pdp_npi=2%40dis%21CZK%21457.29%21251.53%21%21%21%21%21%4021452885167448112722585567d0761%2112000028321382122%21sea&curPageLogUid=sc5TkfWdj7Qm (accessed on 23 January 2023).
60. Fisher Scientific. Al-Ni Alloy Price. Available online: https://www.fishersci.fi/shop/products/aluminum-nickel-raney-type-alloy-powder-al-ni-50-50-thermo-scientific/11946951 (accessed on 23 January 2023).
61. SRL CHEM. $NaBH_4$ Price. Available online: https://srlchem.com/products/product_details/productId/1132 (accessed on 23 January 2023).
62. India Global Mart. $Na_2S_2O_5$ Price. Available online: https://dir.indiamart.com/impcat/sodium-metabisulfite.html (accessed on 23 January 2023).

Disclaimer/Publisher's Note: The statements, opinions and data contained in all publications are solely those of the individual author(s) and contributor(s) and not of MDPI and/or the editor(s). MDPI and/or the editor(s) disclaim responsibility for any injury to people or property resulting from any ideas, methods, instructions or products referred to in the content.

Article

Synthesis of Novel Zn₃V₂O₈/Ag Nanocomposite for Efficient Photocatalytic Hydrogen Production

Fahad A. Alharthi *, Alanood Sulaiman Ababtain, Hend Khalid Aldubeikl, Hamdah S. Alanazi and Imran Hasan *

Department of Chemistry, College of Science, King Saud University, Riyadh 11451, Saudi Arabia
* Correspondence: fharthi@ksu.edu.sa (F.A.A.); iabdulateef@ksu.edu.sa (I.H.)

Abstract: In this study, we fabricated $Zn_3V_2O_8$ and a Ag-modified $Zn_3V_2O_8$ composite ($Zn_3V_2O_8$/Ag) by utilizing effective and benign approaches. Further characterization techniques such as powder X-ray diffraction (XRD) and scanning electron microscopy (SEM) were explored to examine the phase and structural properties, respectively, of the synthesized $Zn_3V_2O_8$/Ag and $Zn_3V_2O_8$/Ag composite materials. The oxidation states and elemental composition of the synthesized $Zn_3V_2O_8$/Ag and $Zn_3V_2O_8$/Ag were characterized by adopting X-ray photoelectron spectroscopy (XPS) and energy-dispersive X-ray spectroscopy (EDX). The optical band gaps of the synthesized $Zn_3V_2O_8$/Ag and $Zn_3V_2O_8$/Ag were examined by employing ultraviolet–visible (UV-vis) diffuse reflection spectroscopy. HRTEM images clearly show that ZnV@Ag NC has a hexagonal plate-like morphology. Subsequently, $Zn_3V_2O_8$ and $Zn_3V_2O_8$/Ag were used as photocatalysts for photocatalytic hydrogen (H_2) production. It was observed that after Ag doping, the energy band gap of ZnV was reduced from 2.33 eV to 2.19 eV. EDX mapping images also show the presence of Ag, O, Zn, and V elements and confirm the formation of ZnV@Ag NC with good phase purity. Observations clearly showed the presence of excellent photocatalytic properties of the synthesized photocatalyst. The $Zn_3V_2O_8$/Ag photocatalyst exhibited H_2 generation of 37.52 $\mu mol g^{-1} h^{-1}$, which is higher compared to pristine $Zn_3V_2O_8$. The $Zn_3V_2O_8$/Ag photocatalyst also demonstrated excellent reusability, including decent stability. The reusability experiments suggested that ZnV@Ag NC has excellent cyclic stability for up to six cycles.

Keywords: $Zn_3V_2O_8$/Ag nanocomposite; photocatalytic H_2 production; water splitting; surface plasmon resonance; sacrificial reagents

Citation: Alharthi, F.A.; Ababtain, A.S.; Aldubeikl, H.K.; Alanazi, H.S.; Hasan, I. Synthesis of Novel Zn₃V₂O₈/Ag Nanocomposite for Efficient Photocatalytic Hydrogen Production. *Catalysts* **2023**, *13*, 455. https://doi.org/10.3390/catal13030455

Academic Editor: Meng Li

Received: 29 December 2022
Revised: 21 January 2023
Accepted: 8 February 2023
Published: 21 February 2023

Copyright: © 2023 by the authors. Licensee MDPI, Basel, Switzerland. This article is an open access article distributed under the terms and conditions of the Creative Commons Attribution (CC BY) license (https://creativecommons.org/licenses/by/4.0/).

1. Introduction

In the present scenario, the energy crisis and environmental pollution are major concerns [1–6]. Energy demand has increased drastically all over the world [4]. Energy also plays a crucial role in economic growth, progress, and development [2]. Although conventional fuels such as coal, natural gas, and oil are still regarded as the primary sources of energy, their depletion and unexpected environmental risks have prompted people to search for alternate energy sources [1]. In this context, various alternative energy technologies, such as solar cells, energy storage, and hydrogen (H_2) production, have been developed [7,8]. Various approaches and methods (electrolysis, thermolysis, and photocatalysis) have been developed for H_2 production [9]. The increasing global energy demand has directed the attention of researchers and scientists toward H_2 as a clean and green energy source for the next generations. Photocatalytic water splitting has become a hot area of research for converting cheap solar energy into hydrogen energy [10]. Photocatalytic H_2 production is an important and widely used method. In this process, the charge of a catalytic material under irradiation is transported to a counter electrode and thus initiates a redox reaction for hydrogen production [9]. PEC water splitting was first carried out by Fujishima and Honda et al. [11] at 0.8 V under ultraviolet light using a titanium dioxide

(TiO$_2$) photoanode, opening up a new area of study. TiO$_2$ has a wide band gap of 3.2 eV, which makes it unsuitable for practical PEC applications [12]. Thus, it is highly necessary to design and develop low-band-gap materials for photocatalytic H$_2$ production applications. According to the reported literature, a suitable photocatalyst should have a narrow band gap for better charge transfer during the photocatalytic process [13,14]. It has also been noted in the available literature on photocatalytic H$_2$ evolution that surface properties such as morphological characteristics largely influence the H$_2$ evolution rate and photocatalytic H$_2$ evolution by providing a better path for electron transport [14]. Thus, it is noted that the mechanical stability, surface structural properties, and band gap of the photocatalyst can have a direct influence on the H$_2$ evolution rate [15]. Nanostructured photocatalysts can easily separate electron–hole pairs during a photocatalytic reaction and may generate a suitable and easy path for electron transport, which can significantly enhance the photocatalytic H$_2$ evolution [9]. Recently, various novel photocatalyst materials with unique and nanostructured surface morphologies have been widely used for photocatalytic H$_2$ evolution reactions [10]. In this context, TiO$_2$, zinc oxide, molybdenum disulfide, cadmium sulfide, and graphite-like carbon nitride materials have been widely used for H$_2$ production [6,9,10]. Although various nanostructured materials have been reported for the H$_2$ evolution process, there is still room for the design and fabrication of novel photocatalysts for the H$_2$ evolution process.

In the past few years, metal vanadates have received extensive attention because of their exceptional charge transport/separation and optical properties [13–16]. Additionally, metal vanadates have band gaps of 2.2 to 2.7 eV, which enables them to effectively absorb solar energy in the visible range [17–19]. Moreover, metal vanadates are relatively robust against photocorrosion [20]. In addition, the valence band location is close to the thermodynamic potential of oxygen evolution, making them good catalysts for the visible-range photo-oxidation of water [21]. Zinc vanadate (Zn$_3$V$_2$O$_8$) has excellent optoelectronic features and has been extensively used in supercapacitors, batteries, H$_2$ storage, catalysis, photocatalysis, and magnetic devices [22–25]. However, there are still some challenges that limit their photocatalytic efficiency, such as the minimum energy band gap needed to break water into hydrogen and oxygen, as most semiconductors have wider band gaps. Secondly, the electron–hole recombination rate hinders the light absorption of the semiconductor material, and hence, low photocatalytic efficiency is achieved [19,23,24]. To overcome these barriers, several procedures and efforts have been reported, and out of these, merging the properties of two different materials is of great significance for a variety of applications [26]. Thus, it is worth designing and fabricating hybrid composite materials containing Zn$_3$V$_2$O$_8$. Silver nanoparticles (Ag NPs) possess excellent catalytic and conducting properties [27]. Ag NPs have been widely used for the preparation of hybrid composite materials [28].

There are very few reports available on metal-vanadate-based photocatalytic H$_2$ production. Very low hydrogen production of 11.5 µmolg^{-1}h^{-1} was reported for a bismuth vanadate-modified reduced graphene oxide composite (BiVO$_4$/rGO) [18]. Another work also demonstrated the use of BiVO$_4$/rGO as a photocatalyst for improved H$_2$O$_2$ evolution [14]. Thus, it is revealed that metal vanadate has excellent properties and can be explored in the photocatalytic H$_2$ evolution process.

In this study, our group successfully fabricated a Ag-NP-decorated Zn$_3$V$_2$O$_8$ composite material. Furthermore, photocatalytic H$_2$ production studies were carried out using the Zn$_3$V$_2$O$_8$/Ag NP composite. The investigations demonstrated reasonable improvements in the photocatalytic activity of the Zn$_3$V$_2$O$_8$/Ag NP composite compared to pristine Zn$_3$V$_2$O$_8$. According to our literature survey, we report for the first time the photocatalytic activity of a Zn$_3$V$_2$O$_8$/Ag NP composite for H$_2$ production applications. In the present study, we propose a simple synthetic strategy for the preparation of a Zn$_3$V$_2$O$_8$/Ag NP composite and explored it as a cost-effective, environmentally friendly, highly stable, and efficient photocatalyst for H$_2$ production applications.

2. Results and Discussion

2.1. Materials Characterization

The XRD patterns of ZnV and ZnV@Ag NC were obtained in the 2θ range of 10–100°. The obtained XRD data of ZnV and ZnV@Ag NC are displayed in Figure 1. The XD pattern of ZnV shows various diffraction peaks at 2θ values of 10–100°. The obtained diffraction peaks can be assigned to the well-defined (001), (010), (002), (110), (012), (111), (020), (021), (112), (013), (022), (023), (122), (220), and (021) diffraction planes of ZnV. The XRD pattern is in agreement with the reported JCPDS number of 50-0570. The strong diffraction peaks indicate that ZnV has decent crystallinity. The XRD pattern of ZnV@Ag NC demonstrates the presence of (002), (110), (012), (020), (021), (013), (122), and (024) diffraction planes. No diffraction plane or peak was observed for Ag NPs, which is due to the low amount of Ag NPs in the synthesized ZnV@Ag NC sample. Further information about the crystallite size can be obtained by using the Debye–Scherer formula given by Equation (1) [29]:

$$D_c = \frac{K\lambda}{\beta \times \cos\theta} \quad (1)$$

where λ is the wavelength of the X-ray source, K is a shape factor and is usually ~0.9, θ is the corresponding angle, and β is the breadth of the observed diffraction line at its half-intensity maximum. Using Equation (1), the crystallite sizes of ZnV and ZnV@Ag were found to be 27.50 nm and 24.60 nm. The overall XRD results confirmed that ZnV and ZnV@Ag NC are formed with decent phase purity. The crystallite sizes of the synthesized ZnV and ZnV@Ag NC are presented in Table 1. The observations showed that the insertion of Ag reduced the crystallite size of ZnV@Ag NC.

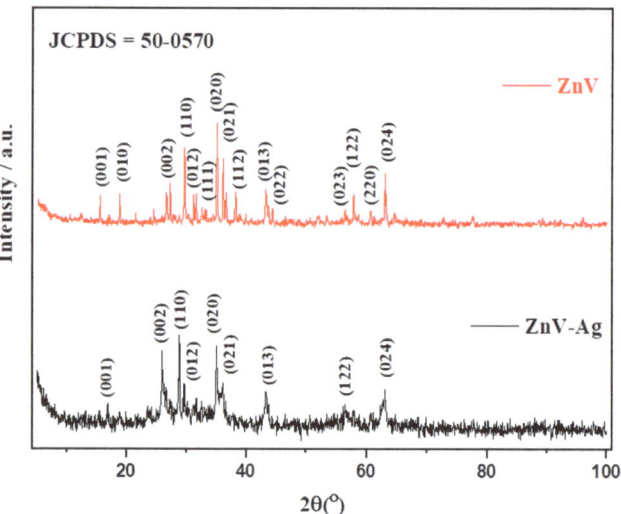

Figure 1. XRD data of ZnV and ZnV-Ag NC.

The FTIR spectra of the prepared ZnV and ZnV@Ag NC were collected to further authenticate the formation of ZnV and ZnV@Ag NC. Figure 2 shows the FTIR spectra of ZnV and ZnV-Ag NC. The FTIR spectrum of ZnV shows absorption bands at 417 and 498 cm^{-1}, which are related to the vibration and stretching modes of Zn-O bonds [30]. The presence of vibrational peaks at 838, 648, and 793 cm^{-1} is attributed to the presence of tetrahedral VO$_4$ vibrational modes and asymmetric vibration modes of V-O-Zn and V-O-V in the ZnV sample [30,31]. The FTIR spectrum of ZnV@Ag shows an absorption band at 415 cm^{-1}, which can be ascribed to the Zn-O bond [30]. Similarly, the presence of vibrational peaks at 823, 637, and 782 cm^{-1} are attributed to the presence of tetrahedral VO$_4$ vibrational modes and asymmetric vibration modes of V-O-Zn and V-O-V in the ZnV sample [31]. The observed results indicate that the presence of Ag NPs in the ZnV@Ag sample slightly shifted the absorption band positions. The obtained results are similar to those in a previous report [31].

Table 1. XRD parameters such as crystallite size, lattice parameters, unit volume, and band-gap energy for ZnV and ZnV@Ag NC.

Material	Crystallite Size (nm)	Lattice Parameters	Unit Volume	Energy Band Gap
ZnV	27.50	a = 6.18 Å, b = 11.76 Å, c = 8.38 Å	609.03 Å^3	2.33 eV
ZnV@Ag	24.60	a = 5.88 Å, b = 11.38 Å, c = 8.18 Å	547.97 Å^3	2.19 eV

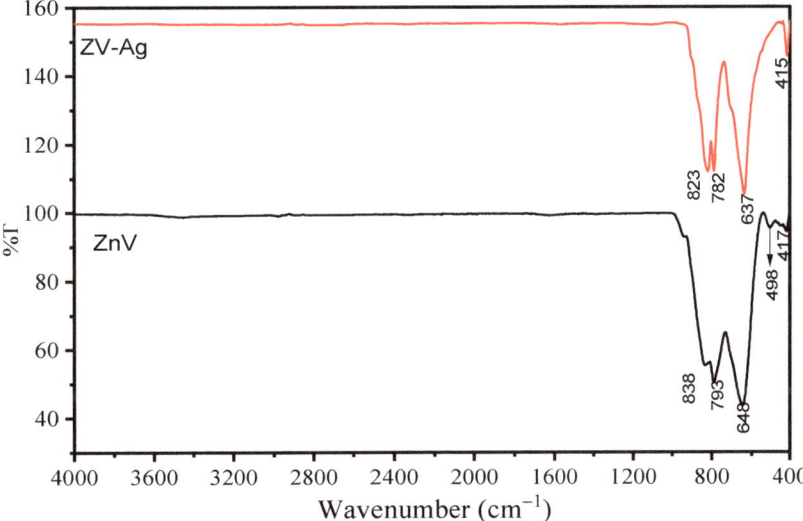

Figure 2. FTIR of ZnV (black line) and ZnV@Ag NC (red line).

The topological and surface structural features of ZnV and ZnV@Ag NC were studied by obtaining their SEM images. The obtained SEM results for ZnV and ZnV@Ag NC are provided in Figure 3. From the SEM investigations, it was observed that a plate-like surface with a uniform particle distribution was formed. Some particles were agglomerated and interconnected (Figure 3a). The average particle size of ZnV was less than 100 nm (Figure 3a). In further investigations, SEM pictures of ZnV@Ag NC were also recorded under the microscope. Figure 3b shows that Ag nanoparticles are embedded in the surface of the ZnV photocatalyst. The Ag particles have a thin needle-like surface morphology and are very well connected to the ZnV surface, as shown in Figure 3b.

Further, the presence of Ag on the ZnV surface was confirmed by the EDX study. The recorded EDX spectrum of ZnV@Ag NC is presented in Figure 3c. The EDX results for ZnV@Ag NC revealed the presence of Ag, O, Zn, and V elements. This further verified that Ag NPs had been successfully grown on the ZnV surface. EDX mapping images of ZnV@Ag NC were also obtained and are displayed in Figure 4a–e. The EDX mapping images also show the presence of Ag, O, Zn, and V elements and confirm the formation of ZnV@Ag NC with good phase purity.

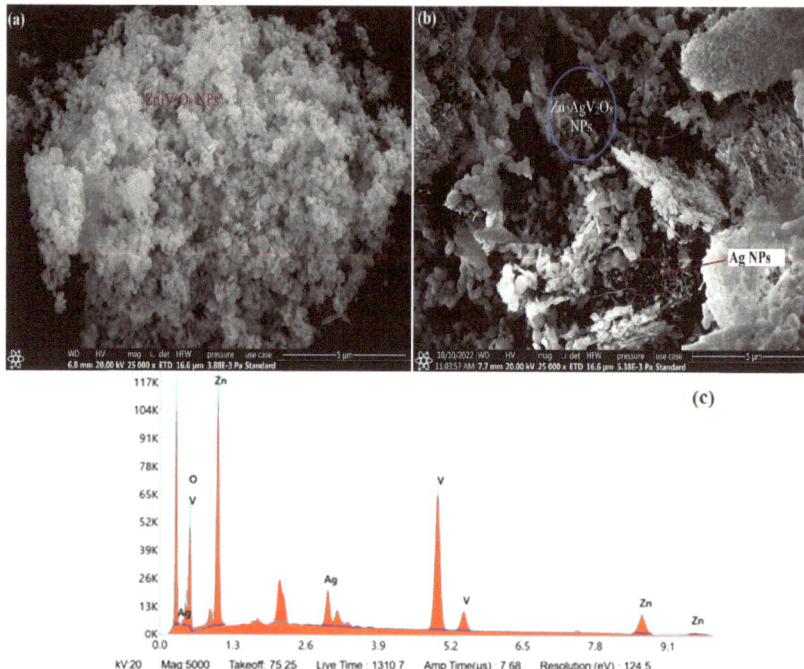

Figure 3. SEM images of ZnV (**a**) and ZnV@Ag NC (**b**) and EDX spectrum of ZnV@Ag NC (**c**).

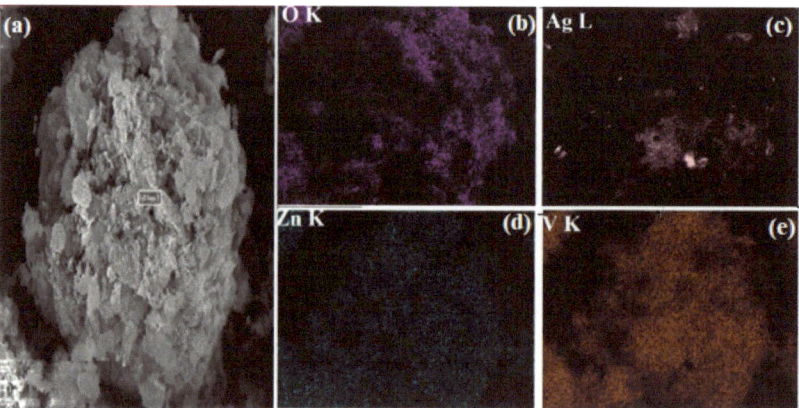

Figure 4. Selected-area SEM image (**a**) and mapping images of O (**b**), Ag (**c**), Zn (**d**), and V (**e**) elements present in ZnV@Ag NC.

Furthermore, TEM analysis was also conducted to further identify the structural properties of the prepared ZnV@Ag NC. The obtained TEM results for ZnV@Ag NC are presented in Figure 5. Figure 5a shows that Ag NPs are present on the ZnV surface with particle sizes less than 100 nm. The high-resolution TEM image is displayed in Figure 5b. The HRTEM image clearly shows that ZnV@Ag NC has a hexagonal plate-like morphology.

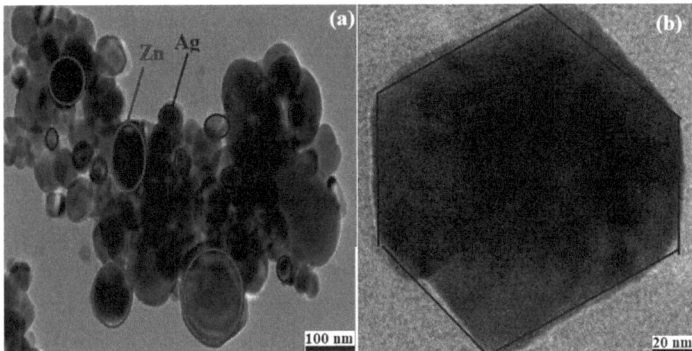

Figure 5. (a) TEM image of ZnV@Ag NC. (b) High-resolution TEM image of ZnV@Ag NC.

We can conclude that Ag NPs were successfully grown on the hexagonal plate-like surface of ZnV. The UV-vis reflectance spectra of ZnV and ZnV@Ag NC are presented in Figure 6a. The UV spectra of ZnV exhibited two peaks at 288 nm and 393 nm. The 288 nm peak is attributed to the transfer of charge from oxygen to the central metal V atom, while 393 nm is due to the transfer of charge from oxygen to the zinc atom [32,33]. The UV spectrum of ZnV@Ag exhibited the same type of pattern with an extra peak at 480 nm associated with Ag coordinated with an oxygen atom [34]. The reflectance curve for ZnV@Ag NC was found to have higher reflectance in the visible region as compared to pristine ZnV NPs due to the surface plasmon resonance (SPR) effect of doped Ag NPs. The SPR effect generally arises in metal nanoparticles due to collective oscillations of free electronic charge under the influence of electromagnetic radiation [35]. The SPR effect is generally affected by the particle size, size distribution, and shape of the particle in a particular medium. The doping of Ag in ZnV results in the origination of the SPR effect in the material, which partially contributed to the enhancement of photocatalytic activity by reducing the electron–hole pair recombination rate [36]. The band gaps of ZnV and ZnV@Ag NC were determined by extrapolating the linear portion of the $(F(R)h\nu)^2$ curve versus the photon energy (hv) using the Kubelka–Munk equation, which is given by Equation (2) [37]:

$$(F(R)h\nu) = A(h\nu - E_g)^n \qquad (2)$$

where R is diffused reflectance, h is the Planck constant, ν is the frequency of radiation, A is a constant, E_g is the energy band gap, and n is an integer representing the magnitude of direct and indirect energy band gaps. The synthesized ZnV and ZnV@Ag NC show band gaps of 2.33 and 2.19 eV, respectively. The observed results show that the presence of Ag reduced the band gap of ZnV, which makes it a more suitable candidate for photocatalytic applications.

Furthermore, we also used the XPS technique to identify the existence of the valence state of Ag in the synthesized ZnV@Ag sample. The XPS survey scan of ZnV@Ag NC is shown in Figure 7a. The survey spectrum of ZnV@Ag shows the presence of Ag3d, V2p, Zn2p, and O1s, which suggested that ZnV@Ag had been formed. The presence of binding energies at ~1023 eV and 1046 eV is attributed to $Zn2p_{3/2}$ and $Zn2p_{1/2}$ of Zn^{2+}, respectively (Figure 7b). According to the high-resolution V2p spectrum, the appearance of two peaks at binding energy values of 522.5 and 515.7 eV can be ascribed to $V2p_{5/2}$ and $V2p_{3/2}$ of V^{5+} in ZnV, respectively (Figure 7c). The presence of two peaks at binding energies of 365.4 and 372.4 eV can be assigned to $Ag3d_{5/2}$ and $Ag3d_{3/2}$, respectively (Figure 7d). The obtained high-resolution XPS spectrum of Ag3d suggested the presence of metallic Ag (0). Figure 7e shows the O1s spectrum of ZnV@Ag, and the presence of a peak at the binding energy value of 528.34 eV can be assigned to the presence of lattice oxygen in the ZnV@Ag sample. The XPS results for ZnV@Ag are well matched with the published literature [31].

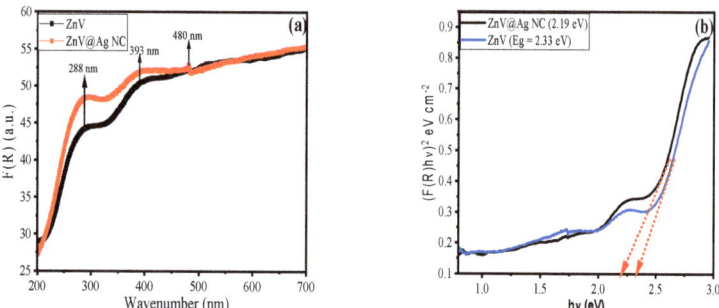

Figure 6. UV-vis reflectance spectra (**a**) and band-gap plot (**b**) of ZnV and ZnV@Ag NC.

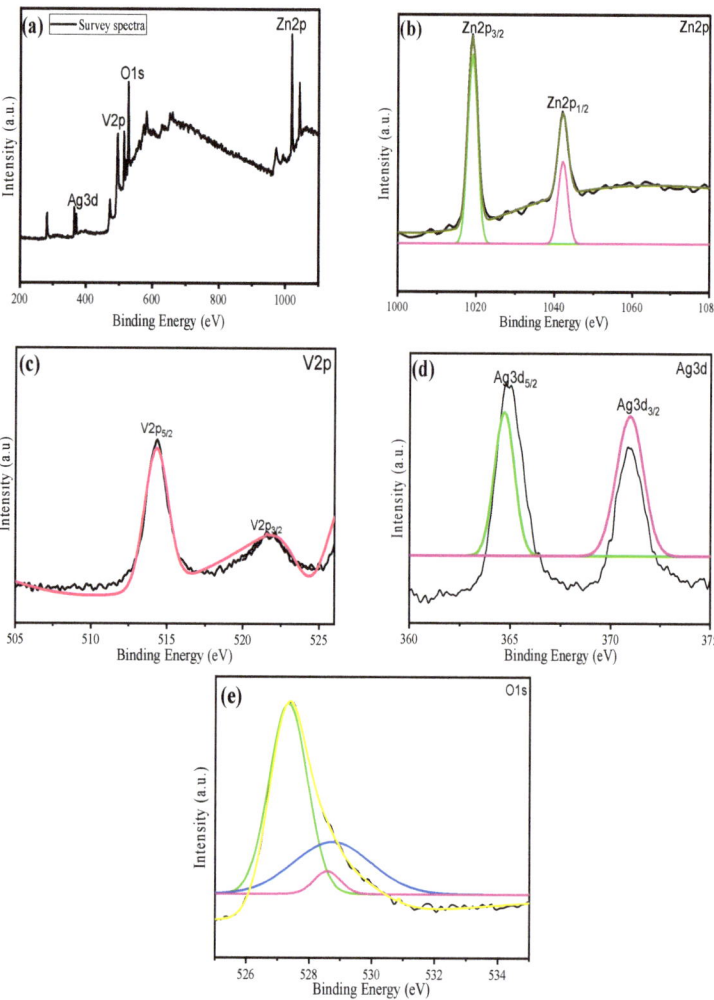

Figure 7. XPS survey spectrum of ZnV@Ag NC: (**a**) Zn2P (**b**), V2p (**c**), Ag3d (**d**), and O1s (**e**) of ZnV@Ag.

2.2. Photocatalytic H_2 Production Activities

The photocatalytic activities of the prepared ZnV and ZnV@Ag NC photocatalysts were studied in the presence of lactic acid. The obtained results are presented in Figure 8a. The ZnV photocatalyst exhibits good photocatalytic activity toward H_2 generation, and a reasonably good amount (16.44 $\mu mol g^{-1} h^{-1}$) was obtained. Further studies showed interesting photocatalytic properties of ZnV@Ag NC for H_2 generation. The improved H_2 generation amount of 37.52 $\mu mol g^{-1} h^{-1}$ was obtained for ZnV@Ag NC. This suggests that ZnV@Ag NC has excellent photocatalytic properties, which may be ascribed to the narrow band gap and synergistic interactions between Ag NPs and ZnV. In the above studies, the pH of the solution was 3. Since pH may significantly alter the performance of the photocatalyst, we further checked the effect of various pH values on photocatalytic H_2 generation. The optimal photocatalytic performance of ZnV and ZnV@Ag NC was found at pH 3. The lowest H_2 generation was observed at pH 7 (Figure 8b). The optimized pH of 3 was utilized for further experiments. The type of solvent may play a vital role in activating the photocatalyst. Thus, the photocatalytic activities of the prepared ZnV and ZnV@Ag NC photocatalysts were examined in the presence of various solvents: water, methanol, lactic acid, Na_2S/Na_2SO_3, and triethanolamine. The observations indicate that ZnV and ZnV@Ag NC have poor photocatalytic activities in water. This activity was slightly enhanced when water was replaced with triethanolamine. Further enhancements were observed for methanol- and Na_2S/Na_2SO_{30}-based systems. However, the highest photocatalytic activities of ZnV and ZnV@Ag NC were observed in lactic acid compared to the other solvents (Figure 8c). The amount of photocatalyst affects its performance; thus, we optimized the amount of the photocatalyst for improved H_2 generation.

Various amounts (10, 20, 30, 40, 50, 60, 70, 80, 90, and 100 mg) of ZnV and ZnV@Ag NC photocatalysts were used to optimize the photocatalytic performance for effective H_2 generation in lactic acid. The obtained data are summarized in Figure 8d. The obtained results revealed that 60 mg of the photocatalyst has the most effective and efficient photocatalytic performance (Figure 8d). The amount of lactic acid was also optimized, and the obtained results are depicted in Figure 8e. The highest amount of H_2 was produced in the presence of 25 mL of lactic acid, as shown in Figure 8e. The above investigations demonstrated excellent photocatalytic activities of ZnV@Ag NC, which suggests its potential applications at a large scale. It is believed that a photocatalyst should have excellent stability or reusability for H_2 production. In this context, we examined the reusability of ZnV@Ag NC in the presence of 25 mL of lactic acid. The obtained results are displayed in Figure 8f. It can be clearly observed that ZnV@Ag NC has excellent cyclic stability for up to 6 cycles.

The obtained H_2 production activities of ZnV and ZnV@Ag NC are summarized in Table 2.

2.3. Mechanism of H_2 Production

Scheme 1 presents the plausible mechanism for the water-splitting and lactic acid oxidation/reforming reaction over ZnV@Ag NC under visible-light irradiation. When the material is placed under the light, the formation of electron–hole pairs occurs (Equation (3)). The photogenerated electrons can move from the conduction band (CB) of ZnV to Ag metal via the SPR effect and thereby get trapped, thus hindering electron–hole recombination. The photogenerated holes in the valence band (VB) interact with water to form H^+ and •OH radicals (Equation (4)). The H^+ ions are reduced to H_2 gas on the metal surface (Equation (5)), while the •OH radicals react with lactic acid to further produce H^+ ions and other oxidized products. The oxidized products further degrade to form CO_2 and H_2O (Equation (6)), and H^+ ions are reduced to produce H_2 gas on the catalyst surface. Thus, the sacrificial reagent lactic acid performs a dual role as a H+ ion producer and a suppressor of the electron–hole pair recombination rate by reacting with hydroxyl radicals and stabilizing the photogenerated holes, hence increasing the hydrogen production efficiency [38,39].

$$ZnV@Ag \xrightarrow{h\nu} ZnV@Ag^*(h_{VB}^+ + e_{CB}^-) \quad (3)$$

$$ZnV@Ag^*(h_{VB}^+) + H_2O \rightarrow H^+ + {}^\bullet OH \quad (4)$$

$$ZnV@Ag^*(e_{CB}^-) + H^+ \rightarrow 1/2 H_2 \quad (5)$$

$${}^\bullet OH + \text{Lactic acid} \rightarrow H_2O + CO_2 \quad (6)$$

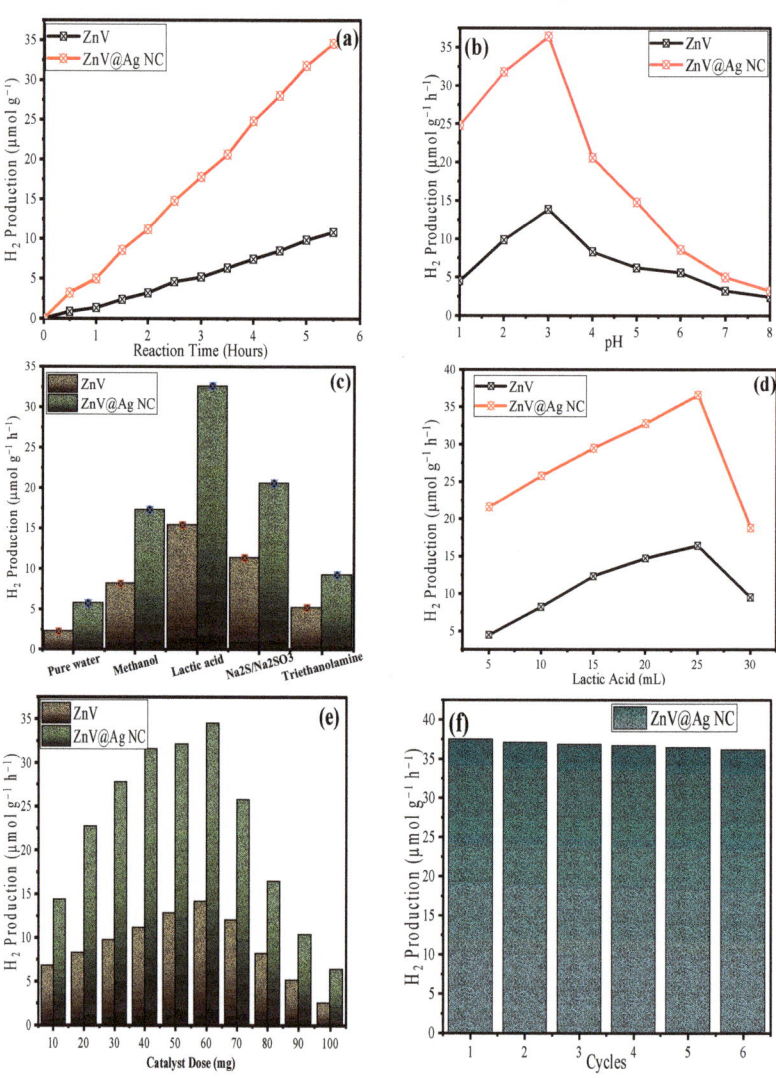

Figure 8. Photocatalytic H_2 generation activities of ZnV and ZnV@Ag NC at different times (**a**), pH values (**b**), and solvents (**c**). Photocatalytic H_2 generation activities of ZnV and ZnV@Ag NC with different doses of catalyst (**d**) and lactic acid (**e**). Stability study of ZnV@Ag NC for H_2 production (**f**).

Table 2. H_2 production activities of **ZnV** and ZnV@Ag NC.

Photocatalyst Materials	H_2 Production Efficiency	Light Source
ZnV@Ag NC	37.52 μmolg^{-1}h^{-1}	Xenon lamp (λ = 420 nm)
ZnV	16.44 μmolg^{-1}h^{-1}	Xenon lamp (λ = 420 nm)

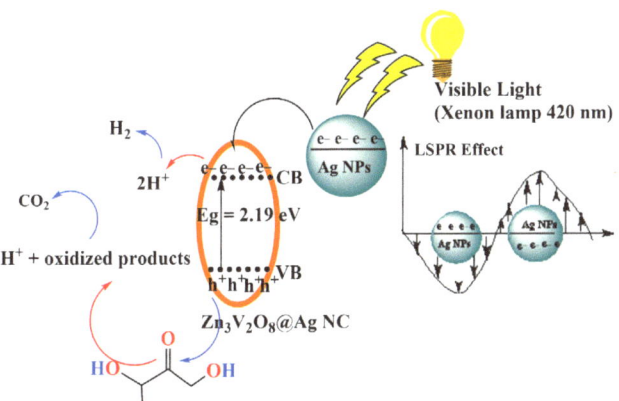

Scheme 1. Illustration of H_2 gas production by ZnV@Ag NC using lactic acid as sacrificial reagent.

3. Experimental Section

3.1. Chemicals and Reagents

Zinc nitrate (Zn (NO$_3$)$_2$·4H2O, 99%), vanadium pentoxide (V$_2$O$_5$, 99.99%), and silver nitrate (AgNO$_3$, >98%) were purchased from Merck, Germany. Hydrogen peroxide (30%, H$_2$O$_2$), sodium hydroxide (NaOH pellets, 98%), and ethanol (C$_2$H$_5$OH, 99%) were supplied by Loba Chemie, Mumbai, India. All of the chemicals were used as received without further purification.

3.2. Synthesis of $Zn_3V_2O_8$/Ag NPs

We adopted a benign approach for the preparation of Zn$_3$V$_2$O$_8$/Ag NPs and Zn$_3$V$_2$O$_8$ [28]. In a 100 mL capacity beaker, 0.75 g of V$_2$O$_5$ dissolved in 20 mL of DI water was added, followed by the addition of 30% H$_2$O$_2$ drop by drop. The color of the solution changed from bright yellow to orange. Amounts of 0.25 g of Zn (NO$_3$)$_2$·6H$_2$O and 0.12g of AgNO$_3$ in 30 mL of DI water were added to the above solution and mixed under magnetic stirring for 60 min. After 60 min, a 0.5 M solution of NaOH (20 mL) was added to the above mixture to maintain a pH of 8–9. The mixture was transferred to a 100 mL Teflon-lined autoclave and heated at 185 °C for 24 h. After the completion of the reaction, the autoclave was allowed to naturally cool at room temperature, and the precipitate was collected by centrifugation. The material was washed several times with DI water and ethanol to remove any unreacted reactant species, dried at 80 °C for 5 h in a hot air oven, and finally calcined at 600 °C for 3 h at a heating rate of 5 °C/min. Similarly, pristine ZnV was also prepared by the same method but without using AgNO$_3$. In further studies, Zn$_3$V$_2$O$_8$ and the Zn$_3$V$_2$O$_8$/Ag nanocomposite are labeled as ZnV and ZnV@Ag NC.

3.3. H_2 Production Assembly

Hydrogen production experiments were carried out in a photocatalytic reactor using a 350 W Xenon lamp with a cutoff wavelength of 420 nm and an intensity of 180 mW cm^{-2} as a visible-light source. The synthesized ZnV and ZnV@Ag NC materials were immersed in a solution of lactic acid/water as a sacrificial agent and placed in a photocatalytic reactor equipped with a double-walled quartz reaction vessel connected to a closed gas circuit with a water jack flushing cold water to maintain the temperature of the reaction at 10 °C. The produced H_2 gas was analyzed using a multichannel analyzer (Emerson) equipped with a thermal conductivity detector. Scheme 2 presents the schematic diagram of photocatalytic hydrogen production.

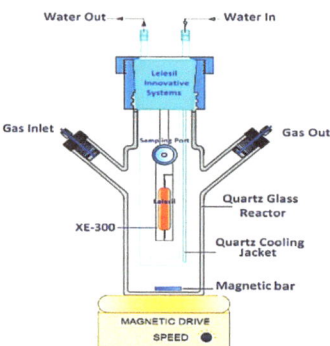

Scheme 2. Schematic diagram of a photocatalytic hydrogen production reactor.

3.4. Apparatus

In the present study, a Rigaku (Rigaku Ultima IV, Austin, TX, USA) powder X-ray diffractometer (XRD) was utilized for recording XRD patterns. A Zeiss (Zeiss Gemini SEM, Carl Zeiss, Jena, Germany) microscope was used to capture the field-emission scanning electron microscopy (FESEM) images of the prepared materials. A Thermo-Scientific instrument (K-alpha, MA USA) was used to obtain the X-ray photoelectron spectroscopy (XPS) spectra of the synthesized materials. The ultraviolet–visible (UV-vis) spectra were recorded on an Agilent Cary Instrument (Cary 60, Santa Clara, CA, USA). Energy-dispersive X-ray spectroscopic (EDX) studies were performed on an Oxford EDX instrument connected to SEM. Transmission electron microscopy (TEM) images of the obtained samples were collected on a Tecnai G2, F30 instrument (Netherlands). Photocatalytic H_2 production studies were performed on a Joel gas chromatograph. Fourier transform infrared (FTIR) spectra of the samples were obtained on a BRUKER spectrometer (Alpha II, Billerica, MA, USA).

4. Conclusions

In the present study, a facile synthesis was employed for the preparation of ZnV and a ZnV@Ag nanocomposite. The physiochemical and optical properties of the prepared ZnV and ZnV@Ag NC were evaluated by utilizing various advanced techniques. The optical band gap of ZnV was reduced from 2.33 eV to 2.19 eV with the introduction of Ag NPs to the ZnV matrix/surface. This indicates that ZnV@Ag NC, with a narrow band gap of 2.19 eV, will absorb light, and improved photocatalytic H_2 production can be observed. The excellent H_2 amount of 37.52 μmolg^{-1}h^{-1} was obtained using ZnV@Ag NC as a photocatalyst compared to ZnV. ZnV@Ag NC also showed excellent cyclic stability, which can be attributed to the presence of electroactive sites and a better ion/electron transport path in the prepared ZnV@Ag NC with improved conductivity. This study proposes a new eco-friendly and cost-effective photocatalyst for photocatalytic H_2 production applications.

Author Contributions: Conceptualization, F.A.A. and I.H.; Methodology, F.A.A. and I.H.; Software, H.S.A. and I.H.; Validation, H.S.A. and I.H.; Formal analysis, A.S.A. and H.K.A.; Investigation, A.S.A. and H.K.A.; Resources, A.S.A. and H.K.A.; Data curation, I.H.; Writing—original draft, A.S.A.; Writing—review & editing, I.H.; Visualization, H.S.A.; Supervision, F.A.A. and H.S.A.; Project administration, F.A.A.; Funding acquisition, F.A.A. All authors have read and agreed to the published version of the manuscript.

Funding: This work was funded by deputyship of Research and Innovation, Ministry of Education in Saudi Arabia, project number IFK-SURG-2-1326.

Data Availability Statement: Data is contained within the article.

Acknowledgments: The authors extend their appreciation to the deputyship of Research and Innovation, Ministry of Education in Saudi Arabia, for funding this research. All the authors have consented to the acknowledgment.

Conflicts of Interest: The authors declare that there is no conflict of interest related to this research.

References

1. Ishaq, H.; Dincer, I.; Crawford, C. A review on hydrogen production and utilization: Challenges and opportunities. *Int. J. Hydrogen Energy* **2022**, *47*, 26238–26264. [CrossRef]
2. Ahmad, K.; Mobin, S.M. Recent Progress and Challenges in $A_3Sb_2X_9$-Based Perovskite Solar Cells. *ACS Omega* **2020**, *5*, 28404–28412. [CrossRef]
3. Alam, M.W.; Al Qahtani, H.S.; Souayeh, B.; Ahmed, W.; Albalawi, H.; Farhan, M.; Abuzir, A.; Naeem, S. Novel Copper-Zinc-Manganese Ternary Metal Oxide Nanocomposite as Heterogeneous Catalyst for Glucose Sensor and Antibacterial Activity. *Antioxidants* **2022**, *11*, 1064. [CrossRef]
4. Ahmad, K.; Shinde, M.A.; Song, G.; Kim, H. Design and fabrication of $MoSe_2/WO_3$ thin films for the construction of electrochromic devices on indium tin oxide based glass and flexible substrates. *Ceram. Int.* **2021**, *47*, 34297–34306. [CrossRef]
5. Alam, M.W.; Azam, H.; Khalid, N.R.; Naeem, S.; Hussain, M.K.; BaQais, A.; Farhan, M.; Souayeh, B.; Zaidi, N.; Khan, K. Enhanced Photocatalytic Performance of Ag_3PO_4/Mn-ZnO Nanocomposite for the Degradation of Tetracycline Hydrochloride. *Crystals* **2022**, *12*, 1156. [CrossRef]
6. Suresh Philip, C.; Nivetha, A.; Sakthivel, C.; Veena, C.G.; Prabha, I. Novel fabrication of cellulose sprinkled crystalline nanocomposites using economical fibrous sources: High performance, compatible catalytic and electrochemical properties. *Microporous Mesoporous Mater.* **2021**, *318*, 111021. [CrossRef]
7. Ahmad, K.; Shinde, M.A.; Kim, H. Molybdenum disulfide/reduced graphene oxide: Progress in synthesis and electro-catalytic properties for electrochemical sensing and dye sensitized solar cells. *Microchem. J.* **2021**, *169*, 106583. [CrossRef]
8. Low, W.H.; Khiew, P.S.; Lim, S.S.; Siong, C.W.; Chia, C.H.; Ezeigwe, E.R. Facile synthesis of graphene-$Zn_3V_2O_8$ nanocomposite as a high performance electrode material for symmetric supercapacitor. *J. Alloys Compd.* **2019**, *784*, 847–858. [CrossRef]
9. Liu, W.; Wang, X.; Yu, H.; Yu, J. Direct Photoinduced Synthesis of Amorphous $CoMoS_x$ Cocatalyst and Its Improved Photocatalytic H_2-Evolution Activity of CdS. *ACS Sustain. Chem. Eng.* **2018**, *6*, 12436–12445. [CrossRef]
10. Sharma, R.; Almáši, M.; Nehra, S.P.; Rao, V.S.; Panchal, P.; Paul, D.R.; Jain, I.P.; Sharma, A. Photocatalytic hydrogen production using graphitic carbon nitride (GCN): A precise review. *Renew. Sustain. Energy Rev.* **2022**, *168*, 112776. [CrossRef]
11. Fujishima, A.; Honda, K. Electrochemical Photolysis of Water at a Semiconductor Electrode. *Nature* **1972**, *238*, 37–38. [CrossRef] [PubMed]
12. Shanmugaratnam, S.; Velauthapillai, D.; Ravirajan, P.; Christy, A.A.; Shivatharsiny, Y. CoS_2/TiO_2 Nanocomposites for Hydrogen Production under UV Irradiation. *Materials* **2019**, *12*, 3882. [CrossRef] [PubMed]
13. Sekar, K.; Kassam, A.; Bai, Y.; Coulson, B.; Li, W.; Douthwaite, R.E.; Sasaki, K.; Lee, A.F. Hierarchical bismuth vanadate/reduced graphene oxide composite photocatalyst for hydrogen evolution and bisphenol A degradation. *Appl. Mater. Today* **2021**, *22*, 100963. [CrossRef]
14. Dhabarde, N.; Carrillo-Ceja, O.; Tian, S.; Xiong, G.; Raja, K.; Subramanian, V.R. Bismuth Vanadate Encapsulated with Reduced Graphene Oxide: A Nanocomposite for Optimized Photocatalytic Hydrogen Peroxide Generation. *J. Phys. Chem. C* **2021**, *125*, 23669–23679. [CrossRef]
15. Marberger, A.; Ferri, D.; Elsener, M.; Sagar, A.; Artner, C.; Schermanz, K.; Kröcher, O. Relationship between structures and activities of supported metal vanadates for the selective catalytic reduction of NO by NH_3. *Appl. Catal. B Environ.* **2017**, *218*, 731–742. [CrossRef]
16. Lashari, N.R.; Zhao, M.; Zheng, Q.; Gong, H.; Duan, W.; Xu, T.; Wang, F.; Song, X. Excellent cycling stability and capability of novel mixed-metal vanadate coated with V_2O_5 materials in an aqueous solution. *Electrochim. Acta* **2019**, *314*, 115–123. [CrossRef]
17. Vignesh, K.; Hariharan, R.; Rajarajan, M.; Suganthi, A. Visible light assisted photocatalytic activity of TiO_2–metal vanadate (M = Sr, Ag and Cd) nanocomposites. *Mater. Sci. Semicond. Process.* **2013**, *16*, 1521–1530. [CrossRef]
18. Muthurasu, A.; Tiwari, A.P.; Chhetri, K.; Dahal, B.; Kim, H.Y. Construction of iron doped cobalt-vanadate-cobalt oxide with metal-organic framework oriented nanoflakes for portable rechargeable zinc-air batteries powered total water splitting. *Nano Energy* **2021**, *88*, 106238. [CrossRef]
19. Zhang, W.; Zhang, Y.; Yuan, H.; Li, J.; Ding, L.; Chu, S.; Wang, L.; Zhai, W.; Jiao, Z. Carbon hollow matrix anchored by isolated transition metal atoms serving as a single atom cocatalyst to facilitate the water oxidation kinetics of bismuth vanadate. *J. Colloid Interface Sci.* **2022**, *616*, 631–640. [CrossRef]
20. Yao, X.; Zhao, X.; Hu, J.; Xie, H.; Wang, D.; Cao, X.; Zhang, Z.; Huang, Y.; Chen, Z.; Sritharan, T. The Self-Passivation Mechanism in Degradation of $BiVO_4$ Photoanode. *IScience* **2019**, *19*, 976–985. [CrossRef]
21. Su, J.; Bai, Z.; Huang, B.; Quan, X.; Chen, G. Unique three dimensional architecture using a metal-free semiconductor cross-linked bismuth vanadate for efficient photoelectrochemical water oxidation. *Nano Energy* **2016**, *24*, 148–157. [CrossRef]

22. Rajaji, U.; Govindasamy, M.; Sha, R.; Alshgari, R.A.; Juang, R.-S.; Liu, T.-Y. Surface engineering of 3D spinel $Zn_3V_2O_8$ wrapped on sulfur doped graphitic nitride composites: Investigation on the dual role of electrocatalyst for simultaneous detection of antibiotic drugs in biological fluids. *Compos. Part B Eng.* **2022**, *242*, 110017. [CrossRef]
23. Yin, Z.; Qin, J.; Wang, W.; Cao, M. Rationally designed hollow precursor-derived $Zn_3V_2O_8$ nanocages as a high-performance anode material for lithium-ion batteries. *Nano Energy* **2017**, *31*, 367–376. [CrossRef]
24. Rajkumar, S.; Elanthamilan, E.; Princy Merlin, J. Facile synthesis of $Zn_3V_2O_8$ nanostructured material and its enhanced supercapacitive performance. *J. Alloys Compd.* **2021**, *861*, 157539. [CrossRef]
25. Gan, L.; Deng, D.; Zhang, Y.; Li, G.; Wang, X.; Jiang, L.; Wang, C. $Zn_3V_2O_8$ hexagon nanosheets: A high-performance anode material for lithium-ion batteries. *J. Mater. Chem. A* **2013**, *2*, 2461–2466. [CrossRef]
26. Mirsadeghi, S.; Ghoreishian, S.M.; Zandavar, H.; Behjatmanesh-Ardakani, R.; Naghian, E.; Ghoreishian, M.; Mehrani, A.; Abdolhoseinpoor, N.; Ganjali, M.R.; Huh, Y.S.; et al. In-depth insight into the photocatalytic and electrocatalytic mechanisms of $Mg_3V_2O_8@Zn_3V_2O_8@ZnO$ ternary heterostructure toward linezolid: Experimental and DFT studies. *J. Environ. Chem. Eng.* **2023**, *11*, 109106. [CrossRef]
27. Khan, F.U.; Chen, Y.; Khan, Z.U.H.; Khan, A.U.; Ahmad, A.; Tahir, K.; Wang, L.; Khan, M.R.; Wan, P. Antioxidant and catalytic applications of silver nanoparticles using Dimocarpus longan seed extract as a reducing and stabilizing agent. *J. Photochem. Photobiol. B Biol.* **2016**, *164*, 344–351. [CrossRef]
28. Liu, P.; Yi, J.; Bao, R.; Fang, D. A flower-like $Zn_3V_2O_8$/Ag composite with enhanced visible light driven photocatalytic activity: The triple-functional roles of Ag nanoparticles. *New J. Chem.* **2019**, *43*, 7482–7490. [CrossRef]
29. Scherrer, P. Estimation of the Size and Internal Structure of Colloidal Particles by Means of Rontgen Rays. *Nachr. Von Der Ges. Der Wiss. Zu Göttingen* **1918**, *26*, 98–100.
30. Luo, J.; Ning, X.; Zhan, L.; Zhou, X. Facile construction of a fascinating Z-scheme AgI/$Zn_3V_2O_8$ photocatalyst for the photocatalytic degradation of tetracycline under visible light irradiation. *Sep. Purif. Technol.* **2021**, *255*, 117691. [CrossRef]
31. Jiang, Y.H.; Liu, P.P.; Tian, S.J.; Liu, Y.; Peng, Z.Y.; Li, F.; Ni, L.; Liu, Z.C. Sustainable visible-light-driven Z-scheme porous $Zn_3(VO_4)_2$/g-C_3N_4 heterostructure toward highly photoredox pollutant and mechanism insight. *J. Taiwan Inst. Chem. Eng.* **2017**, *78*, 517–529. [CrossRef]
32. Pei, L.Z.; Lin, N.; Wei, T.; Liu, H.D.; Yu, H.Y. Zinc Vanadate Nanorods and Their Visible Light Photocatalytic Activity. *J. Alloys Compd.* **2015**, *631*, 90–98. [CrossRef]
33. Ahmed, R.; Mukhtar, S.; Gao, W.; Zafar Ilyas, S. Ab-Initio Calculations of Structural, Electronic, and Optical Properties of $Zn_3(VO_4)_2$. *Chin. Phys. B* **2018**, *27*, 033101. [CrossRef]
34. Du, M.; Xiong, S.; Wu, T.; Zhao, D.; Zhang, Q.; Fan, Z.; Zeng, Y.; Ji, F.; He, Q.; Xu, X. Preparation of a Microspherical Silver-Reduced Graphene Oxide-Bismuth Vanadate Composite and Evaluation of Its Photocatalytic Activity. *Materials* **2016**, *9*, 160. [CrossRef]
35. Ijaz, M. Plasmonic Hot Electrons: Potential Candidates for Improved Photocatalytic Hydrogen Production. *Int J. Hydrogen Energy* **2022**. [CrossRef]
36. Arif Sher Shah, M.S.; Zhang, K.; Park, A.R.; Kim, K.S.; Park, N.G.; Park, J.H.; Yoo, P.J. Single-Step Solvothermal Synthesis of Mesoporous Ag–TiO_2–Reduced Graphene Oxide Ternary Composites with Enhanced Photocatalytic Activity. *Nanoscale* **2013**, *5*, 5093–5101. [CrossRef] [PubMed]
37. Kubelka, P. Ein Beitrag Zur Optik Der Farbanstriche (Contribution to the Optic of Paint). *Z. Fur Tech. Phys.* **1931**, *12*, 593–601.
38. Kondarides, D.I.; Daskalaki, V.M.; Patsoura, A.; Verykios, X.E. Hydrogen Production by Photo-Induced Reforming of Biomass Components and Derivatives at Ambient Conditions. *Catal. Lett.* **2008**, *122*, 26–32. [CrossRef]
39. Wang, Y.; Liu, T.; Tian, W.; Zhang, Y.; Shan, P.; Chen, Y.; Wei, W.; Yuan, H.; Cui, H. Mechanism for Hydrogen Evolution from Water Splitting Based on a MoS_2/WSe_2 Heterojunction Photocatalyst: A First-Principle Study. *RSC Adv.* **2020**, *10*, 41127–41136. [CrossRef]

Disclaimer/Publisher's Note: The statements, opinions and data contained in all publications are solely those of the individual author(s) and contributor(s) and not of MDPI and/or the editor(s). MDPI and/or the editor(s) disclaim responsibility for any injury to people or property resulting from any ideas, methods, instructions or products referred to in the content.

Article
Hydrothermal Synthesis of Bimetallic (Zn, Co) Co-Doped Tungstate Nanocomposite with Direct Z-Scheme for Enhanced Photodegradation of Xylenol Orange

Fahad A. Alharthi *, Wedyan Saud Al-Nafaei, Alanoud Abdullah Alshayiqi, Hamdah S. Alanazi and Imran Hasan *

Department of Chemistry, College of Science, King Saud University, Riyadh 11451, Saudi Arabia
* Correspondence: fharthi@ksu.edu.sa (F.A.A.); iabdulateef@ksu.edu.sa (I.H.)

Abstract: In the present study, pristine $ZnWO_4$, $CoWO_4$, and mixed metal $Zn_{0.5}Co_{0.5}WO_4$ were synthesized through the hydrothermal process using a Teflon-lined autoclave at 180 °C. The synthesized nanomaterials were characterized by various spectroscopic techniques, such as TEM, FTIR, UV–vis, XRD, and SEM-EDX-mapping to confirm the formation of nanocomposite material. The synthesized materials were explored as photocatalysts for the degradation of xylenol orange (XO) under a visible light source and a comparative study was explored to check the efficiency of the bimetallic co-doped nanocomposite to the pristine metal tungstate NPs. XRD analysis proved that reinforcement of Co^{2+} in $ZnWO_4$ lattice results in a reduction in interplanar distance from 0.203 nm to 0.185 nm, which is reflected in its crystallite size, which reduced from 32 nm to 24 nm. Contraction in crystallite size reflects on the optical properties as the energy bandgap of $ZnWO_4$ reduced from 3.49 eV to 3.33 eV in $Zn_{0.5}Co_{0.5}WO_4$, which is due to the formation of a Z-scheme for charge transfer and enhancement in photocatalytic efficiency. The experimental results suggested that $ZnWO_4$, $CoWO_4$, and $Zn_{0.5}Co_{0.5}WO_4$ NPs achieved a photocatalytic efficiency of 97.89%, 98.10%, and 98.77% towards XO in 120 min of visible solar light irradiation. The kinetics of photodegradation was best explained by pseudo-first-order kinetics and the values of apparent rate const (k_{app}) also supported the enhanced photocatalytic efficiency of mixed metal $Zn_{0.5}Co_{0.5}WO_4$ NPs towards XO degradation.

Keywords: nanocomposites; bimetallic Z-scheme; photoreduction; hydrothermal reaction; organic pollutants

1. Introduction

The development of societies and industries has led to the deterioration of freshwater qualities through the release of untreated effluents containing hazardous chemical contaminants, such as dyes and pharmaceutical products, etc. [1,2]. The toxic organic compounds, especially dyes and pharmaceuticals, contaminated wastewater in the environment, which can cause a variety of health issues to humans [3,4]. Among various organic pollutants, xylenol orange (XO; $C_{31}H_{32}N_2O_{13}S$) belongs to the acidic dye family commonly used in the textile, paper, and printing industries, but beyond the permissible limit, it causes nervous system breakdown, liver–kidney damage, blood disorders, and skin irritation [5,6]. So, it has become a matter of serious concern to develop new technology-based methods and resolutions which can treat these hazardous chemicals in the water and regulate their limits in the aquatic environment. Among various chemical, physical, and biological wastewater treatment methods, photocatalysis enables the adsorbed organic molecules to be degraded at a low cost into H_2O and CO_2 using solar irradiation, which is abundant, renewable, and environmentally friendly [7,8]. Moreover, advanced oxidation processes (AOPs) were created to promote the removal or degradation of contaminant species through redox processes, focused on the remediation of contaminants in wastewater; they are a group of techniques that use reactive oxidant species (ROS) to mineralize a wide range of resistant organics [9,10].

Metal tungstate nanoparticles (NPs) have received a lot of interest among the different types of semiconductor photocatalysts [11]. For this reason, the study and synthesis of single photocatalysts that successfully use visible light to stimulate photocatalysis have recently become a promising idea. Until now, various semiconductors with sufficient energy band gaps to absorb visible light, such as Fe_2O_3, Ag_2O, Bi_2WO_6, $g-C_3N_4$, $MnWO_4$, $CdWO_4$, CdS, $NiMoO_4$, Cu_2O, In_2S_3, and Si_3N_4, have been effectively synthesized and utilized for photocatalysis, because of its high average refractive index, high chemical stability, excellent light absorbing property, and high catalytic activity [12–14]. In the tungstate family, zinc tungstate ($ZnWO_4$) has been a widely known semiconductor photocatalyst owing to its high physical and chemical stability and difficulty to dissolve [15]. However, due to homogeneity, it is only UV light active and possesses a high rate of electron–hole pair recombination, which poses an obstacle in the practical applications of $ZnWO_4$ [16]. Various methods have been applied previously to increase the photocatalytic efficiency of $ZnWO_4$, such as semiconductor coupling [17], noble metal loading [18], and metal deposition [19]. The metal deposition of $ZnWO_4$ with other metals having a narrow band gap could effectively inhibit the electron–hole pair recombination by expanding the light absorption range, and thus a robust enhancement in photocatalytic efficiency will appear [20,21]. So, the Co deposition method was applied in this study to synthesize the Co-doped nanocomposite $Zn_{0.5}Co_{0.5}WO_4$ with improved energy bandgap, optical properties (UV–vis light active), and photocatalytic efficiency. Since the ionic radii of both Co^{2+} and Zn^{2+} are approximately the same, it can be easily deposited in the solid matrix of $ZnWO_4$ and result in the reduction of the energy bandgap from 3.87 eV ($ZnWO_4$) to 1.95 eV ($Zn_{0.5}Co_{0.5}WO_4$). $ZnWO_4$ was identified as a large band gap semiconductor (band gap = 3.87 eV) with the appropriate conduction band and valance band locations based on its electronic characteristics, while, on the other hand, cobalt tungstate ($CoWO_4$) has been recognized as having excellent phase composition, strong chemical stability, and a low energy bandgap, which makes it adaptable to degrade the dyestuff in the visible light range [22].

In the present work, a hydrothermal method was used for the synthesis of pristine metal tungstate nanomaterials ($ZnWO_4$, $CoWO_4$ NPs) and mixed metal tungstate nanocomposite ($Zn_{0.5}Co_{0.5}WO_4$ NPs) using a Teflon line autoclave. The synthesized NPs were explored for the photocatalytic degradation of XO dye and a comparative study was designed to explore the photocatalytic mechanism and efficiency of the photocatalyst.

2. Results and Discussion

2.1. Material Characterization

Figure 1 represents the FTIR of $ZnWO_4$, $CoWO_4$, and $Zn_{0.5}Co_{0.5}WO_4$, in which the characteristic peaks of $ZnWO_4$ are 3448 and 1644 cm^{-1} (stretching and bending vibrations of –OH group) due to surface adsorbed water, 473 to 881 cm^{-1} (W_2O_8 bonds (WO_6) groups shared at edges and WO_2 groups), and 473 cm^{-1} (stretching vibrational of the Zn–O) [23]. Three peaks at 541, 685, and 716 cm^{-1} correspond to symmetric and asymmetric stretching vibrations of the bridging atom of the WO_2 groups of distorted octahedral (WO_6) clusters [24]. Broad absorption bands at 816 cm^{-1} and 881 cm^{-1} are caused by symmetrical vibrations of bridge oxygen atoms of Zn–O–W groups [24]. Similar types of stretching and vibrational bands are also shown by $CoWO_4$ with different peak values and 469 cm^{-1} belongs to the Co–O bond. As compared to $ZnWO_4$, the FTIR of mixed metal $Zn_{0.5}Co_{0.5}WO_4$ showed the difference in peak shapes and intensities due to the partial replacement of Zn^{2+} ions by Co^{2+} ions [25].

The formation of $ZnWO_4$, $CoWO_4$, and $Zn_{0.5}Co_{0.5}WO_4$ was confirmed by the XRD analysis. Figure 2 shows the XRD pattern of the $ZnWO_4$, $CoWO_4$, and $Zn_{0.5}Co_{0.5}WO_4$ prepared by the hydrothermal method at 180°C for 24 h. The XRD spectra of $ZnWO_4$ shows characteristic peaks at 2θ value of 15.58°, 19.04°, 23.94°, 24.66°, 30.59°, 31.38°, 36.54°, 38.47°, 41.41°, 44.40°, 46.55°, 50.39 51.84°, 53.76°, 54.11°, 61.85°, 64.90°, and 68.34°, which belong to miller indices (010), (100), (011), (110), (111), (020), (021), (200), (121), (112), (211), (220) (130), (221), (202), (113), (132), and (041), respectively (JCPDs card no. 96-210-1675). Then, the XRD spectra of CoWO$_4$ shows characteristic peaks at 2θ value of 19.02°, 23.88°, 27.72°, 31.50°, 36.44°, 38.69°, 48.52°, 52.17°, 61.89°, and 65.14°, which belong to miller indices (100), (011), (110), (020), (002), (200), (022), (031), (310), and (040), respectively (JCPDs card no. 96-591-0317). Finally, the XRD pattern of Zn$_{0.5}$Co$_{0.5}$WO$_4$ NPs shows peaks ascribed to the ZnWO$_4$ at 15.54° (010), 36.41° (021), 41.35° (211), 51.95° (220), 53.94° (130), and 68.56° (041), and peaks ascribed to the CoWO$_4$ at 48.81° (022), 61.81° (310), and 64.96° (040), respectively, which suggests that Co has been successfully doped in the solid matrix of ZnWO$_4$. Moreover, the peak intensities of Zn$_{0.5}$Co$_{0.5}$WO$_4$ NPs are greater than pristine ZnWO$_4$ suggesting the increase in crystallinity of the complex upon being mixed with Co^{2+}, which has been proved by data given in Table 1. There are several extra peaks observed in the XRD spectra of Zn$_{0.5}$Co$_{0.5}$WO$_4$ which do not belong to either ZnWO$_4$ or CoWO$_4$. These peaks are due to the formation of Na$_2$W$_2$O$_7$ and Na$_2$W$_4$O$_{13}$ phases with the Zn$_{0.5}$Co$_{0.5}$WO$_4$ phase [26,27].

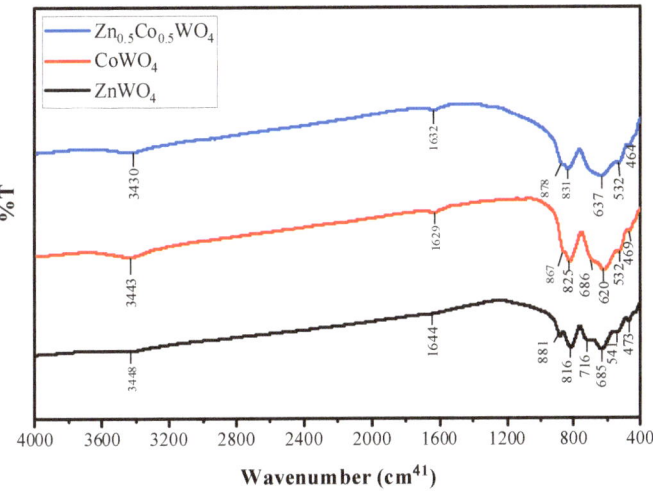

Figure 1. FTIR spectra of ZnWO$_4$ (black line), CoWO$_4$ (purple line), and Zn$_{0.5}$Co$_{0.5}$WO$_4$ (light blue line).

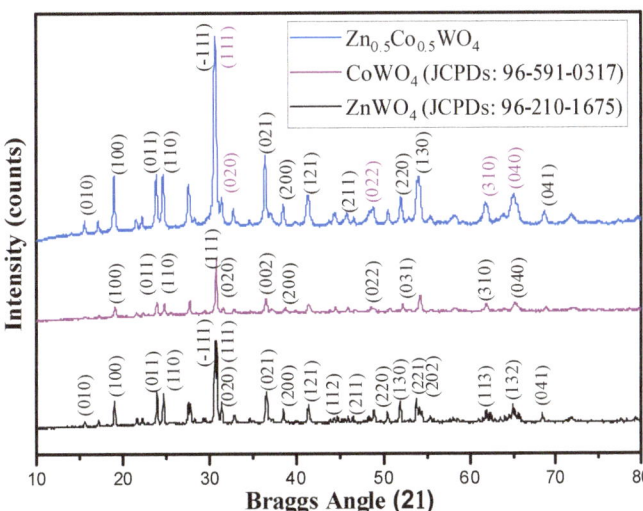

Figure 2. X-ray diffraction pattern of ZnWO$_4$, CoWO$_4$, and Zn$_{0.5}$Co$_{0.5}$WO$_4$ calcined at 600 °C.

Table 1. XRD parameter of $ZnWO_4$, $CoWO_4$, and $Zn_{0.5}Co_{0.5}WO_4$.

Crystallinity (%)	Dislocation Density (δ) Lines ($10^{14} \times m^2$)	Crystallite Size (nm) at 2θ value	Interlayer Spacing (nm) at 2θ	FWHM (β_{hkl})	2θ	Component
45.24	9.78	31.98	0.203	0.45	30.72	$ZnWO_4$
48.46	9.65	32.19	0.195	0.81	30.05	$CoWO_4$
42.18	17.02	24.24	0.198	0.76	30.33	$Zn_{0.5}Co_{0.5}WO_4$

Further information about crystallite size and dislocation density; Scherrer's equation was taken into consideration [28].

$$D = \frac{0.9\lambda}{\beta \cos\theta} \quad (1)$$

$$\text{Dislocation Density}(\delta) = \frac{1}{D^2} \quad (2)$$

$$\text{Interlayer Spacing}(d_{111}) = \frac{n\lambda}{2\sin\theta} \quad (3)$$

$$\%\text{Crystallinity} = \frac{\text{Area under the crystalline peaks}}{\text{Total area}} \times 100 \quad (4)$$

where D is the crystallite size, λ is the characteristic wavelength of the X-ray, β represents the angular width in radian at an intensity equal to half of its maximum (HWMI) of the peak, and θ is the diffraction angle. Using Equation (1), the average particle size of $ZnWO_4$, $CoWO_4$, and $Zn_{0.5}Co_{0.5}WO_4$ was 31.98, 32.19, and 24.24 nm suggesting a contraction in particle size upon Co inclusion in the $ZnWO_4$ solid lattice.

The surface morphology of the $ZnWO_4$, $CoWO_4$, and $Zn_{0.5}Co_{0.5}WO_4$ were studied using SEM analysis given in (Figure 3). $ZnWO_4$ reveals an aggregation of irregularly shaped crystals with plenty of nanorods (Figure 3a). $CoWO_4$ exhibit uniform nanoparticles of various size (Figure 3b), while $Zn_{0.5}Co_{0.5}WO_4$ revealed an assembly of fine nanoparticles with a porous surface (Figure 3c). The elemental composition of the synthesized $Zn_{0.5}Co_{0.5}WO_4$ NPs was observed by EDX analysis and the results showed the presence of the element (weight%) as O (28.52%), Zn (4.55%), Co (4.22%), and W (62.81%), which confirms the successful deposition of Co^{2+} in the $ZnWO_4$ lattice (Figure 3d). The high atomic percentage of W in the EDX spectra is due to the presence of impurity in the form of $Na_2W_2O_7$ and $Na_2W_4O_{13}$ phases with the $Zn_{0.5}Co_{0.5}WO_4$, which resulted in an extra amount of W in the EDX spectra [27].

Furthermore, to figure out the size of the $ZnWO_4$, $CoWO_4$, and $Zn_{0.5}Co_{0.5}WO_4$, TEM analysis was performed (Figure 4a–c). $ZnWO_4$ (Figure 4a) is composed of particles with a rod-like shape; $CoWO_4$ (Figure 4b) presented a morphologically spherical shaped particle. The synthesized material $Zn_{0.5}Co_{0.5}WO_4$ (Figure 4c) exhibited particles mixed of nanorods with some spherical particles with size in a range from 5 to 60 nm, and the Gaussian distribution of the average size was found to be 31 nm given in Figure 4d. Figure 4e represents the TEM image of $Zn_{0.5}Co_{0.5}WO_4$, which suggested the shape of particles to be distorted monoclinic due to the doping of Co into the $ZnWO_4$ lattice. Selected area electron diffraction (SAED) analysis (Figure 4f) was further considered to simulate the TEM data with XRD to assess the purity and shape of the nanoparticles. It contains patterns originating from diffraction points in the TEM screen belonging to randomly oriented nanoparticles. The occurrence of diffraction patterns in Figure 4f suggested that the material is crystalline and pure. The hkl values belonging to yellow fringes simulate the XRD hkl planes of $Zn_{0.5}Co_{0.5}WO_4$ suggesting the shape of the nanoparticles to be monoclinic.

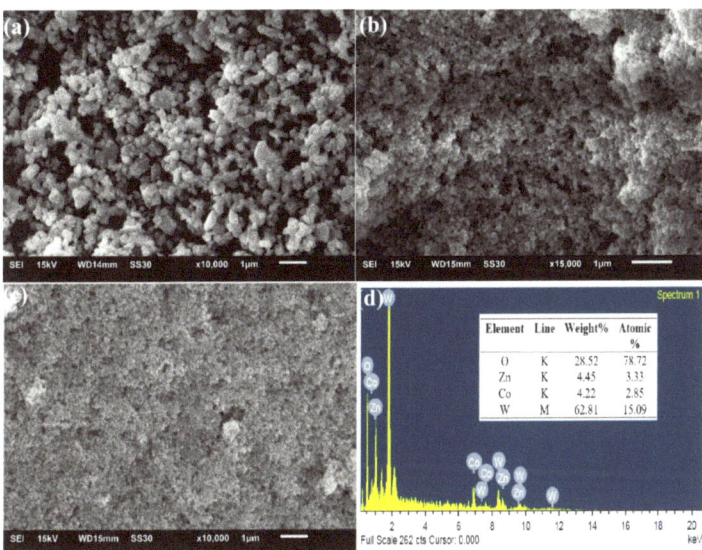

Figure 3. SEM micrographs of (**a**) ZnWO$_4$ NPs; (**b**) CoWO$_4$ NPs; (**c**) Zn$_{0.5}$Co$_{0.5}$WO$_4$ at 1 µm scale; (**d**) EDX spectra of Zn$_{0.5}$Co$_{0.5}$WO$_4$ in 0–20 keV range.

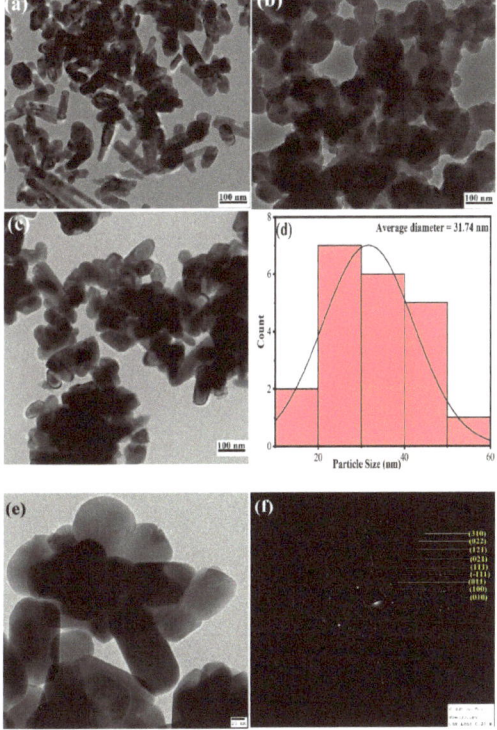

Figure 4. TEM images of (**a**) ZnWO$_4$; (**b**) CoWO$_4$; (**c**) Zn$_{0.5}$Co$_{0.5}$WO$_4$ at 100 nm magnification range; (**d**) particle size histogram; (**e**) TEM image of Zn$_{0.5}$Co$_{0.5}$WO$_4$ at 20 nm magnification range; (**f**) selected area electron diffraction (SAED) image of Zn$_{0.5}$Co$_{0.5}$WO$_4$.

The optical properties of the synthesized NPs and their energy bandgap were estimated by using UV–vis spectroscopy in the range of 200–700 nm and the results are given in Figure 5. The absorption spectra of all the synthesized NPs $ZnWO_4$, $CoWO_4$, and $Zn_{0.5}Co_{0.5}WO_4$ exhibited two peaks, one around 260–312 nm and another broad peak around 530–650 nm, suggesting the UV–vis activeness of the NPs. The inset in Figure 5 shows Tauc's plot, which is used to calculate band gap energy (Eg) of the prepared samples by using the Equation (5) [29];

$$(\alpha h\nu) = A(h\nu - E_g)^n \tag{5}$$

where α is the absorption coefficient, A is constant, h is plank constant, ν is the frequency of radiations, and n is a constant of transition variations, i.e., n = 1/2 for direct transitions and n = 2 for the indirect transitions. Tauc's plot indicated that the synthesized $ZnWO_4$, $CoWO_4$, and $Zn_{0.5}Co_{0.5}WO_4$ were found to have an energy bandgap value of 3.49 eV, 3.42 eV, and 3.33 eV, respectively. Results showed that the deposition of Co^{2+} in the $ZnWO_4$ solid lattice results in the increase in absorption edge by decreasing the energy bandgap, which inhibits the rate of electron–hole pair recombination, and thus improves the photocatalytic efficiency of the synthesized material [30]. The doping of Co^{2+} into $ZnWO_4$ leads to the induction of deformations in the pure monoclinic lattice of $ZnWO_4$, which creates mobile oxygen vacancies in the cluster and makes an improvement in the catalytic [31,32].

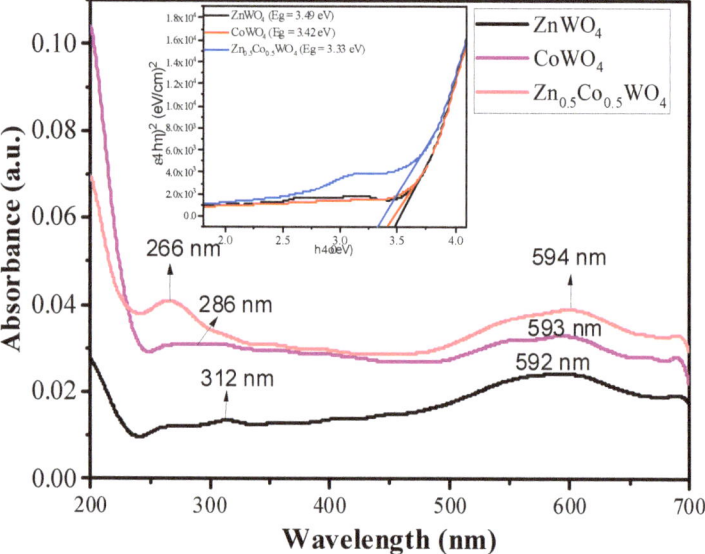

Figure 5. UV–vis plot of $ZnWO_4$ NPs, $CoWO_4$, and $Zn_{0.5}Co_{0.5}WO_4$ in the wavelength range of 200–700 nm (inset is the Tauc's plot for calculating the band gap energy (E_g) of the material).

Figure 6 consists of the photoluminescence emission spectra of $ZnWO_4$, $CoWO_4$, and $Zn_{0.5}Co_{0.5}WO_4$ recorded at a 594 nm excitation wavelength. It can be seen that the PL spectra spans in the range of 545 to 575 nm with a prominent emission peak at 558 nm. The information obtained from the PL spectra suggested that the PL intensity of synthesized $Zn_{0.5}Co_{0.5}WO_4$ was found to be lower than that of $ZnWO_4$ and $CoWO_4$. The inclusion of Co in the wolframite monoclinic structure of $ZnWO_4$ results in emissions due to the radiative transition between W and O in the WO_6^{6-} molecular complex, which effectively slows down the electron–hole [33–35].

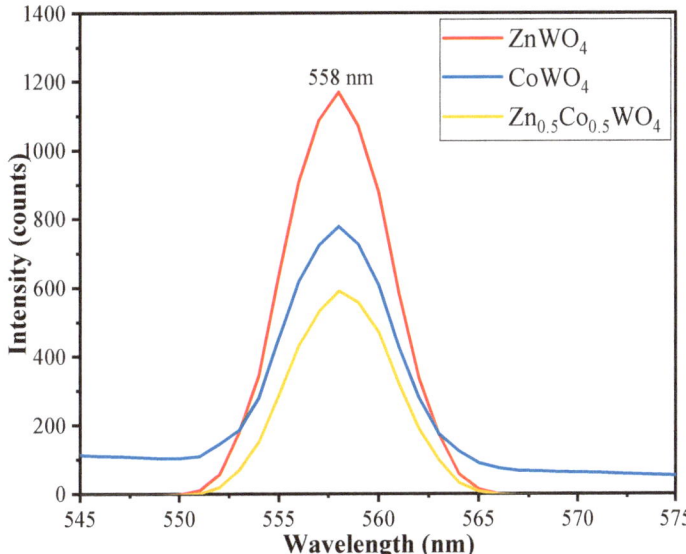

Figure 6. Photoluminescence (PL) spectra of $ZnWO_4$, $CoWO_4$ and $Zn_{0.5}Co_{0.5}WO_4$ were recorded at a 594 nm excitation wavelength dispersed in methanol.

2.2. Photocatalytic Experiments and Optimization of Reaction Parameters

2.2.1. Effect of Variable XO Concentration

The concentration of the dye solution is an important parameter for elucidating the photocatalytic efficiency of the material. Photocatalytic experiments were conducted by taking 10 mg of NPs as a catalyst with variable (20–100 ppm) XO concentration for 120 min of visible light irradiation. The results obtained are given in Figure 7a–d, which suggest that $Zn_{0.5}Co_{0.5}WO_4$ possesses better photocatalytic efficiency (97.17%) as compared to $CoWO_4$ (91.81%), followed by $ZnWO_4$ (89.51%) for 60 ppm of XO, but as the concentration of XO increases beyond 60 ppm, the percentage degradation decreases (Figure 7d). The causes could include: (1) XO may cover more $ZnWO_4$, $CoWO_4$, and $Zn_{0.5}Co_{0.5}WO_4$ active sites, which inhibits the production of oxidants ($^{\bullet}OH$ or $^{\bullet}O_2^-$ radicals) and lowers the efficiency of degradation: and (2) higher XO concentration absorbs a greater number of photons, and as a result, there are not enough photons to activate the surfaces of $ZnWO_4$, $CoWO_4$, and $Zn_{0.5}Co_{0.5}WO_4$ nanoparticles, which slows down XO's degradation at higher concentrations [36]. So, 60 ppm XO concentration was chosen as the optimum concentration for further photocatalytic experiments.

2.2.2. Effect of Variable Catalyst Dose

The physicochemical characteristics of nanomaterials, such as surface area, phase structure, particle size, and interface charge, make them able to complete adsorption action [27]. To efficiently degrade XO through catalysis using the catalyst, the photocatalytic experiment was performed by varying catalyst doses, such as 5, 10, 15, and 20 mg with 10 mL of 10 ppm XO. The results given in Figure 8a–d suggested that the photocatalytic efficiency increased from 5 mg catalyst dose to 10 mg and starts declining with further increase. All the synthesized NPs exhibit maximum photocatalytic efficiency at 10 mg catalyst dose as 97.89% for $ZnWO_4$, 98.10% for $CoWO_4$, and 98.77% for $Zn_{0.5}Co_{0.5}WO_4$. More active sites are produced on the catalyst's surface at lower concentrations, which promotes the creation of $^{\bullet}OH$ or $^{\bullet}O_2^-$ radicals. After that, increasing the catalytic dose causes the particles to agglomerate and sediment, which increases the turbidity and opacity of the slurry. Hence, the generation of •OH or •O$_2^-$ radicals automatically decrease [24]. Therefore, the catalyst dose of 10 mg was chosen to be the optimum dose for the degradation.

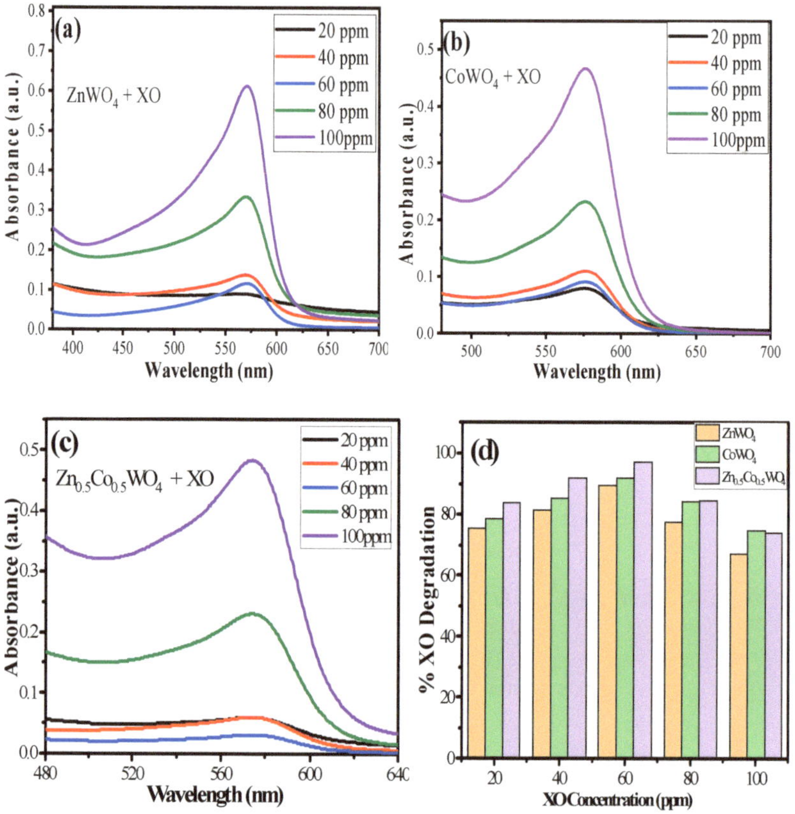

Figure 7. Effect of initial concentration of XO dye on photocatalytic degradation by (**a**) ZnWO$_4$; (**b**) CoWO$_4$; (**c**) Zn$_{0.5}$Co$_{0.5}$WO$_4$; (**d**) XO degradation (%) vs. XO concentration bar graph.

2.2.3. Effect of pH

Experiments were performed to evaluate the impact of pH on the photocatalytic efficiencies of ZnWO$_4$, CoWO$_4$, and Zn$_{0.5}$Co$_{0.5}$WO$_4$. A total of 10 mg of the catalyst was dispersed in 20 mL of 60 ppm XO solution with a variable pH range from 1–7, and irradiated for 120 min under the visible light source. The obtained results after the experiments are given in Figure 9, which suggests that the photocatalytic efficiency of all the catalysts increased until pH 3, giving a maximum value of 93.35% for ZnWO$_4$, 96.21% for CoWO$_4$, and 97.97% for Zn$_{0.5}$Co$_{0.5}$WO$_4$. Further, an increase in pH value beyond 3 results in a decrease in photocatalytic efficiency with a continuous fall until pH 7. So, pH 3 was taken as the optimum pH value for the photocatalytic experiments. The trend can be explained by the fact that as XO is an anionic dye, it can easily bind with the positive surface of the catalyst at low pH through electrostatic interactions, and thereby, in the presence of light, it gets degraded by •OH or •O$_2^-$ radicals [7]. As the pH values increase, the surface becomes negative, which will exert a repulsive force on the anionic XO molecule, and hence, the photocatalytic efficiency decreases.

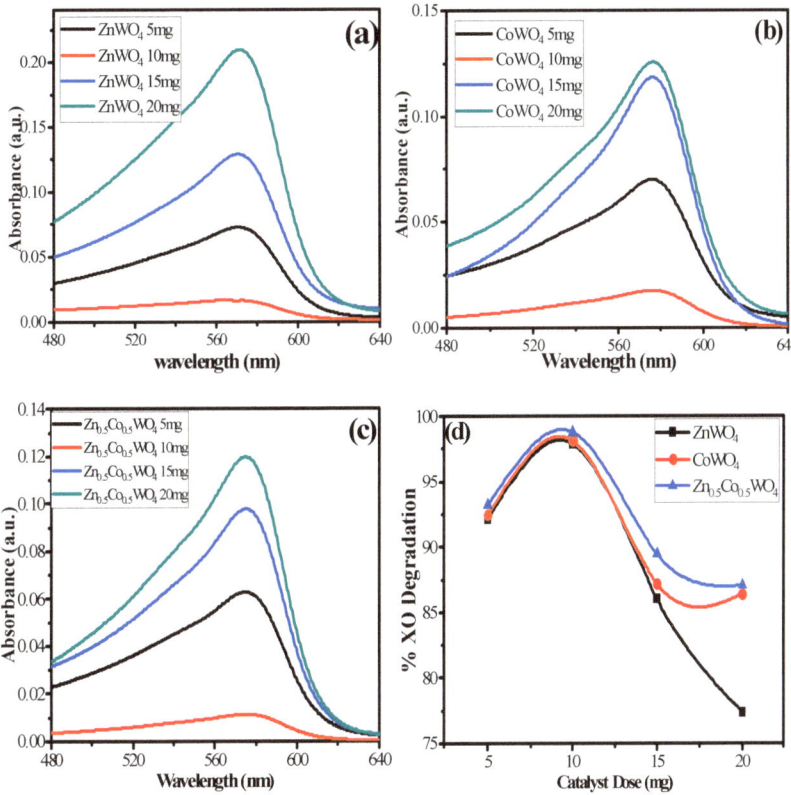

Figure 8. Photocatalytic degradation of XO with a different catalytic dose of (**a**) ZnWO$_4$; (**b**) CoWO$_4$; and (**c**) Zn$_{0.5}$Co$_{0.5}$WO$_4$; and (**d**) %degradation vs. catalyst dose graph.

2.3. Kinetics of Photodegradation

A variety of experiments were carried out by taking 10 mL of 20 ppm XO with 10 mg of catalyst at varying irradiation time from 0 to 120 min under visible solar radiation, and the data obtained was adjusted to the Langmuir–Hinshelwood (L–H) pseudo-first-order kinetic model for the quantitative analysis of XO degradation [37,38].

$$-\ln\left(\frac{C_e}{C_o}\right) = k_{app} \times t \qquad (6)$$

where C_o represents the initial concentration and C_t represents the final concentration of XO at time t. The results are given in Figure 10a–d, in which it was observed that with an increase in irradiation time, the photocatalytic efficiency increases. This may be due to an increase in irradiation time, the surface of the catalyst becomes more activated by absorbing the solar radiation and generates a greater number of ROS •OH or •O$_2^-$ radicals which interact with dye molecules and degrade them [39]. The slope of the pseudo-first-order reaction (Figure 10d) can be utilized to determine the value of k_{app} from the plot of $-\ln(C_e/C_0)$ vs. time (min), which shows a linear response. The obtained values of k_{app} given in Table 2 as 0.018 min^{-1} for ZnWO$_4$, 0.021 min^{-1} for CoWO$_4$, and 0.025 min^{-1} for Zn$_{0.5}$Co$_{0.5}$WO$_4$ proved the high efficiency of Zn$_{0.5}$Co$_{0.5}$WO$_4$ as compared to pristine ZnWO$_4$ and CoWO$_4$. The obtained value of $R^2 = 0.99$ for all the nanoparticles suggested that the photocatalytic data can be realistically simulated by the L–H model. The outcomes demonstrate that the deposition of Co^{2+} in ZnWO$_4$ of both catalysts considerably improves the degrading ability compared to each catalyst individually. Finally, $Zn_{0.5}Co_{0.5}WO_4$ was the novel composite having distinctive characteristics when compared to other composites.

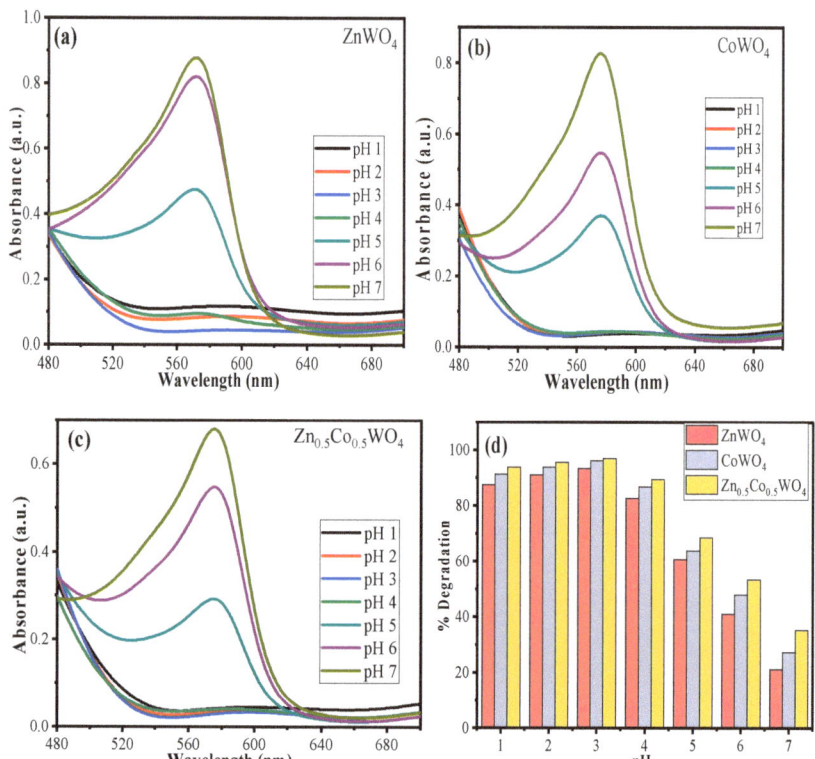

Figure 9. Effect of solution pH on photocatalytic degradation by (**a**) $ZnWO_4$; (**b**) $CoWO_4$; (**c**) $Zn_{0.5}Co_{0.5}WO_4$; (**d**) XO degradation (%) vs. pH bar graph.

Table 2. The kinetic parameters for the degradation of XO dye (20ppm) using 10 mg $ZnWO_4$, $CoWO_4$, and $Zn_{0.5}Co_{0.5}WO_4$ with variable irradiation time (5–120 min).

Material	k_1 (min^{-1})	R^2	$t_{1/2}$ (min)
$ZnWO_4$	0.018	0.99	38.50
$CoWO_4$	0.021	0.99	33.00
$Zn_{0.5}Co_{0.5}WO_4$	0.025	0.99	27.72

2.4. Scavenging Study and Mechanism of Photodegradation

It was suggested that a variety of reactive species, including the hydroxyl radical (•OH), superoxide (•O_2^-), valence band hole (h^+), and electron (e^-) play important roles in the photodegradation of organic pollutants, and their effectiveness depends upon the band structure and chemical nature of the photocatalysts [40]. So, scavenger studies were performed using a 1.3 mM solution of Benzoic acid (BA), Potassium dichromate ($K_2Cr_2O_7$), (AA), ethylenediaminetetraacetic acid (EDTA), and benzoquinone (BQ) by taking 10 mL of 20ppm XO with 10 mg at pH 3 for an irradiation time of 120 min. We know that BA is a scavenger for •OH, EDTA for h^+, $K_2Cr_2O_7$ for e^-, and BQ for •O_2^- [41,42]. Figure 11 displays the results, which show that without the use of a scavenger, the degrading efficiency of $Zn_{0.5}Co_{0.5}WO_4$ towards XO achieved 98.27%. After adding BA and EDTA, there appears no appreciable change in photocatalytic efficiency, suggesting no primary role of •OH and h$^+$ on the degradation of XO at the optimized conditions. On the contrary, when BQ and K$_2$Cr$_2$O$_7$ are added, an appreciable decrease in photocatalytic efficiency happens, demonstrating that photogenerated electrons (e$^-$) and $^\bullet$O$_2^-$ radicals are predominant species in this photodegradation reaction scheme. The results showed that the degradation of XO was primarily controlled by e$^-$ and $^\bullet$O$_2^-$ radicals (majorly).

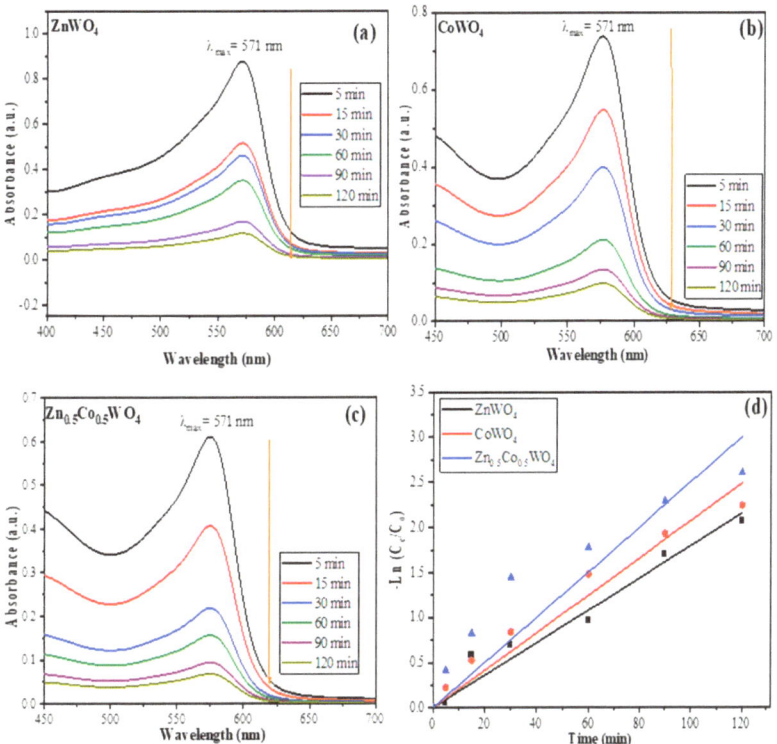

Figure 10. UV–vis absorption spectra of degraded XO dye (60 ppm) using 10 mg of (**a**) ZnWO$_4$; (**b**) CoWO$_4$; (**c**) Zn$_{0.5}$Co$_{0.5}$WO$_4$; and (**d**) L–H model pseudo-first-order kinetic graph with variable irradiation time (5–120 min).

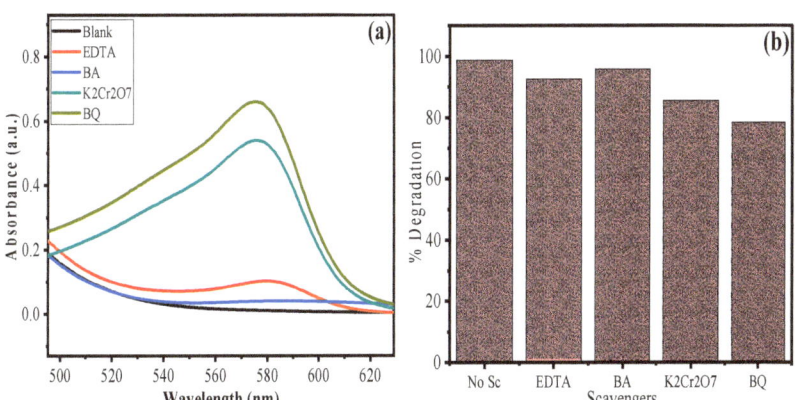

Figure 11. (**a**) UV–vis spectra of XO degradation in the presence of various scavengers; (**b**) % XO degradation vs. bar graph associated with each scavenger.

The plausible mechanism of degradation of XO by $^{\bullet}O_2{}^{-}$ radicals is given in Figure 12 [43,44].

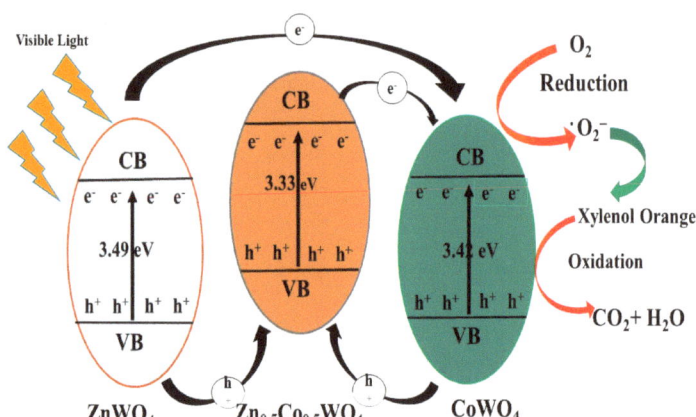

Figure 12. A plausible mechanism of degradation of XO by synthesized $Zn_{0.5}Co_{0.5}WO_4$.

2.5. Comparison with the Literature

The outcomes of this study were compared with the information available in the literature, and a comparison in Table 3 was made to investigate the upgraded scientific information added by the present study. It was found that there is no research where tungstate base nanomaterial is used for the photocatalytic degradation of xylenol orange.

Table 3. Comparison of the literature information with the present study.

Catalysts	Irradiation Time (min)	Light Source	Dye Used	% Degradation	References
$BiOBr/ZnWO_4$	170	UV-A light	RhB	99.40	[45]
La: $ZnWO_4$	90	UV light	MB	97.00	[46]
$CoWO_4/ZnWO_4$ p-n heterojunction	40	Xe lamp	RhB	93%	[47]
$ZnWO_4/WO_3$ heterojunction	120	Visible light	MB	83.60	[48]
$ZnWO_4$-(ZnO)	120	UV-illumination	MO	99.00	[49]
CDs-$ZnWO_4$	150	UV-illumination	NGB	93.00	[50]
$Zn_{0.5}Co_{0.5}WO_4$	120	Visible Solar Light	XO	98.77	Present Study

2.6. Reusability Test

The reusability test was performed for the synthesized $Zn_{0.5}Co_{0.5}WO_4$ up to five cycles to assess the stability of the material towards XO degradation, and the results are given in Figure 13a. It can be seen that there is no appreciable change in photocatalytic efficiency up to the first two cycles with greater than 98% of XO degradation. Furthermore, from cycle 3 to cycle 5, there appears some appreciable decrease in photocatalytic efficiency attaining 85% of XO degradation in cycle 5. The overall results suggest that the synthesized $Zn_{0.5}Co_{0.5}WO_4$ have good stability toward the XO degradation under optimized reaction conditions. To support the stability of the material, XRD analysis of the synthesized material before and after the photocatalytic reaction was taken into consideration, which is given in Figure 13b. The XRD spectra of $Zn_{0.5}Co_{0.5}WO_4$ after five consecutive cycles of reuse represented almost similar hkl planes, but with reduced intensity as compared to pristine $Zn_{0.5}Co_{0.5}WO_4$ before the photocatalytic reaction. Only one change was observed, that after the photocatalytic reaction the (-111) plane diminished, which may be due to the absorption of water.

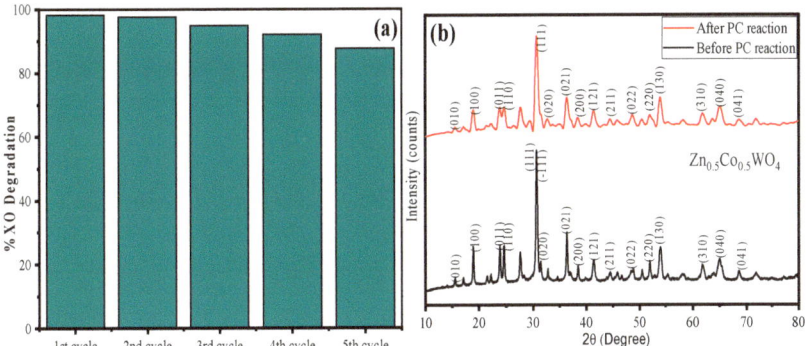

Figure 13. (a) Reusability test for synthesized $Zn_{0.5}Co_{0.5}WO_4$ NPs for XO degradation (b) XRD spectra of $Zn_{0.5}Co_{0.5}WO_4$ before and after the photocatalytic (PC) reaction.

3. Materials and Methods

3.1. Chemicals and Reagents

Sodium tungstate dihydrate ($Na_2WO_4 \cdot 2H_2O$, ACS grade, 99%) was purchased from Merck, Rahway, NJ, USA. Cobalt (II) nitrate hexahydrate (Co $(NO_3)_2 \cdot 6H_2O$ ACS grade 98%), Zinc nitrate hexahydrate (Zn $(NO_3)_2 \cdot 6H_2O$, reagent grade 98%), were purchased from Sigma Aldrich, St. Louis, Missouri USA. Xylenol orange ($C_{31}H_{28}N_2Na_4O_{13}S$, GR) and Ammonia solution (25%) were acquired from Otto Chemie, Mumbai, India. All the solutions were prepared in deionized (DI) water and used throughout all the photocatalytic experiments. Figure 14 represents the molecular structure of xylenol orange.

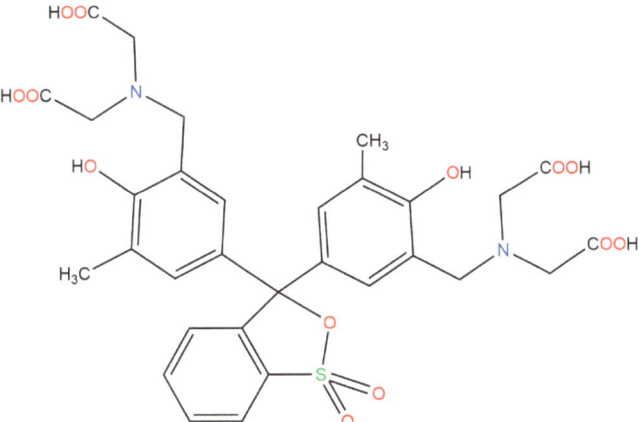

Figure 14. Structure of xylenol orange (XO).

3.2. Synthesis of $ZnWO_4$, $CoWO_4$, and $Zn_{0.5}Co_{0.5}WO_4$ Nanoparticles

The synthesis of pristine metal tungstate nanomaterials ($ZnWO_4$, $CoWO_4$ NPs) and mixed metal tungstate nanocomposite ($Zn_{0.5}Co_{0.5}WO_4$) was done according to the hydrothermal method reported elsewhere [51]. Three Erlenmeyer flasks of 50 mL capacity were equipped with 25 mL of 3mmol of sodium tungstate dihydrate ($Na_2WO_4 \cdot 2H_2O$), Zinc nitrate hexahydrate ($Zn(NO_3)_2 \cdot 6H_2O$; Flask 1), Cobalt(II)nitrate hexahydrate (Co $(NO_3)_2 \cdot 6H_2O$; Flask 2), and 1:1 ratio, i.e., 1.5 mmol each, of $Zn(NO_3)_2 \cdot 6H_2O$ and Co $(NO_3)_2 \cdot 6H_2O$ (Flask 3). All the flasks were placed on a magnetic stirring plate for 30 min to completely dissolve the salts and attain homogeneity. Then, the mixtures (F1, F2, and F3) were transferred into a Teflon-lined steel autoclave and heated in a convection oven at

180 °C for 24 h. The resulting precipitates of $ZnWO_4$, $CoWO_4$, and $Zn_{0.5}Co_{0.5}WO_4$ were centrifuged at 8000 rpm for 5 min and purified several times with DI water and finally with ethanol (2 times). After that, the materials were dried at 100 °C for 5 h and calcined at 600 °C for 4 h, then stored in a desiccator using a polystyrene Petri dish for further characterization and photocatalytic experiments.

3.3. Characterization Techniques

The $ZnWO_4$, $CoWO_4$, and $Zn_{0.5}Co_{0.5}WO_4$ were characterized using various techniques. The change in crystallite size and interplanar distance on Co deposition in the crystal lattice of $ZnWO_4$ was analyzed by an X-ray powder diffractometer (XRD, Rigaku Ultima IV, Tokyo, Japan) at Cu Kα wavelength of 1.5418 Å as a source of radiation. The change of WO_6^{6-} octahedron vibrations, Zn–O, and formation of Co–O bonding was assessed through Fourier transform infrared (FTIR) spectroscopy using a Perkin Elmer Spectrum 2 ATR spectrometer (Perkin Elmer, Waltham, MA, USA). The optical properties, energy bandgap of the synthesized NPs, and remaining concentration of XO in the effluent after the photocatalytic experiment was calculated by using a Shimadzu UV–1900 double-beam spectrophotometer (Kyoto, Japan). The surface morphology of the synthesized NPs was observed by scanning electron microscopy (SEM; JEOL GSM 6510LV, Tokyo, Japan) associated with an X-ray energy dispersive detector (EDX) to investigate the elemental composition and mapping. The morphological change in shape and size of the nanocrystalline solid upon Co deposition in $ZnWO_4$ was investigated using a transmission electron microscope (TEM; JEM 2100, Tokyo, Japan).

3.4. Photocatalysis Experiment

The photocatalytic efficiency of $ZnWO_4$, $CoWO_4$, and $Zn_{0.5}Co_{0.5}WO_4$ was analyzed towards the degradation of xylenol orange under visible light irradiation in a photocatalytic reactor. An aliquot of 10 mL of 50 ppm XO was taken in a 20 mL 1.8 cm glass tube with a magnetic bar of 10 mg of synthesized NPs in a photocatalytic reactor using Xe lamp (420 W cm^{-1}) for 5–120 min of irradiation. The aliquots are taken out of the photoreactor at a regular time interval, centrifuged to separate the nanocatalyst, and then placed in a UV–vis spectrophotometer at λ_{max} = 580 nm to check the photocatalytic efficiency of the synthesized NPs, which is given by Equation (7);

$$\text{XO Degradation}(\%) = \left(\frac{C_0 - C_e}{C_0}\right) \times 100 \qquad (7)$$

where C_0 and C_e represents the initial concentration of an organic pollutant at time t = 0 and at any time t, respectively. The experiments were repeated by changing the parameters, for instance, organic pollutant concentration (ppm), irradiation time (min), catalytic load (mg), and pH of the media to optimize the reaction condition and predict the mechanism of the degradation.

4. Conclusions

The present study reveals the hydrothermal synthesis of pristine $ZnWO_4$, $CoWO_4$, and mixed metal $Zn_{0.5}Co_{0.5}WO_4$ nanoparticles (NPs) in a Teflon-lined autoclave at 180 °C for 24 h. The structural, morphological, elemental, and optical analyses supported the formation of $Zn_{0.5}Co_{0.5}WO_4$. The XRD analysis revealed a contraction in crystallite size from 32 nm to 24 nm with a decrease in crystallinity from 45% to 42% in $ZnWO_4$ to $Zn_{0.5}Co_{0.5}WO_4$ suggesting the successful deposition of Co^{2+} in the $ZnWO_4$ lattice. The TEM analysis suggested the nanorod-like shape of the $Zn_{0.5}Co_{0.5}WO_4$ NPs with an average particle size of 31.74 nm. The optical studies revealed the UV–vis light activeness of materials with the energy band gap of $ZnWO_4$ (3.49 eV), $CoWO_4$ (3.42 eV), and mixed metal $Zn_{0.5}Co_{0.5}WO_4$ (3.33 eV). The material was explored as a catalyst for the photodegradation of XO dye under visible solar radiation and the results suggested the order of photocatalytic efficiency as $Zn_{0.5}Co_{0.5}WO_4$ (98.77%) > $CoWO_4$ (98.10%) > $ZnWO_4$ (97.89%) utilizing 10 mg of catalyst for 60 ppm XO for 120 min of solar irradiation. The photocatalytic data were well explained by L–H pseudo–first–order kinetics with $R^2 = 0.99$, and apparent rate constant (k_{app}) as 0.018 min^{-1} for ZnWO$_4$, 0.021 min^{-1} for CoWO$_4$, and 0.025 min^{-1} for Zn$_{0.5}$Co$_{0.5}$WO$_4$. The scavenger experiments disclosed that superoxide anion radicals (•O$_2^-$) and photogenerated electrons (e$^-$) played important roles in the degradation of XO. The outcomes of this study revealed that the eco-friendly and economical hydrothermal synthesized material (Zn0.5Co0.5WO4) can be used more efficiently for the treatment of XO-contaminated water in industries without the generation of any secondary pollution.

Author Contributions: Conceptualization, F.A.A.; Methodology, F.A.A., H.S.A. and I.H.; Software, I.H.; Validation I.H.; Investigation, W.S.A.-N. and A.A.A.; Resources, W.S.A.-N. and A.A.A.; Data curation, I.H.; Writing—original draft, W.S.A.-N.; Writing—review & editing, H.S.A. and I.H.; Visualization, I.H.; Supervision, F.A.A. and H.S.A.; Project administration, F.A.A.; Funding acquisition, F.A.A. All authors have read and agreed to the published version of the manuscript.

Funding: The authors extend their appreciation to the Deputyship of Research and Innovation, Ministry of Education in Saudi Arabia for funding this research work through project number (IFKSURG-2-255).

Conflicts of Interest: The authors declare no conflict of interest.

References

1. You, J.; Wang, L.; Zhao, Y.; Bao, W. A Review of Amino-Functionalized Magnetic Nanoparticles for Water Treatment: Features and Prospects. *J. Clean. Prod.* **2021**, *281*, 124668. [CrossRef]
2. Surendra, B.S.; Shashi Shekhar, T.R.; Veerabhadraswamy, M.; Nagaswarupa, H.P.; Prashantha, S.C.; Geethanjali, G.C.; Likitha, C. Probe Sonication Synthesis of ZnFe$_2$O$_4$ NPs for the Photocatalytic Degradation of Dyes and Effect of Treated Wastewater on Growth of Plants. *Chem. Phys. Lett.* **2020**, *745*, 137286. [CrossRef]
3. Akbari, A.; Sabouri, Z.; Hosseini, H.A.; Hashemzadeh, A.; Khatami, M.; Darroudi, M. Effect of Nickel Oxide Nanoparticles as a Photocatalyst in Dyes Degradation and Evaluation of Effective Parameters in Their Removal from Aqueous Environments. *Inorg. Chem. Commun.* **2020**, *115*, 107867. [CrossRef]
4. Sarkar, S.; Ponce, N.T.; Banerjee, A.; Bandopadhyay, R.; Rajendran, S.; Lichtfouse, E. Green Polymeric Nanomaterials for the Photocatalytic Degradation of Dyes: A Review. *Environ. Chem. Lett.* **2020**, *18*, 1569–1580. [CrossRef] [PubMed]
5. Xu, P.P.; Zhang, L.; Jia, X.; Wang, X.; Cao, Y.; Zhang, Y. Visible-Light-Enhanced Photocatalytic Activities for Degradation of Organics by Chromium Acetylacetone Supported on UiO-66-NH2. *ChemistrySelect* **2020**, *5*, 14877–14883. [CrossRef]
6. Garg, N.; Bera, S.; Rastogi, L.; Ballal, A.; Balaramakrishna, M.V. Synthesis and Characterization of L-Asparagine Stabilised Gold Nanoparticles: Catalyst for Degradation of Organic Dyes. *Spectrochim. Acta Part A Mol. Biomol. Spectrosc.* **2020**, *232*, 118126. [CrossRef]
7. Kumar, O.P.; Shahzad, K.; Nazir, M.A.; Farooq, N.; Malik, M.; Ahmad Shah, S.S.; ur Rehman, A. Photo-Fenton Activated C3N4x/AgOy@Co1-XBi0.1-YO7 Dual s-Scheme Heterojunction towards Degradation of Organic Pollutants. *Opt. Mater.* **2022**, *126*, 112199. [CrossRef]
8. Jamshaid, M.; Nazir, M.A.; Najam, T.; Shah, S.S.A.; Khan, H.M.; ur Rehman, A. Facile Synthesis of Yb^{3+}-Zn^{2+} Substituted M Type Hexaferrites: Structural, Electric and Photocatalytic Properties under Visible Light for Methylene Blue Removal. *Chem. Phys. Lett.* **2022**, *805*, 139939. [CrossRef]
9. Shahzad, K.; Hussain, S.; Nazir, M.A.; Jamshaid, M.; ur Rehman, A.; Alkorbi, A.S.; Alsaiari, R.; Alhemiary, N.A. Versatile Ag$_2$O and ZnO Nanomaterials Fabricated via Annealed Ag-PMOS and ZnO-PMOS: An Efficient Photocatalysis Tool for Azo Dyes. *J. Mol. Liq.* **2022**, *356*, 119036. [CrossRef]
10. Xu, L.; Wang, X.; Xu, M.L.; Liu, B.; Wang, X.F.; Wang, S.H.; Sun, T. Preparation of Zinc Tungstate Nanomaterial and Its Sonocatalytic Degradation of Meloxicam as a Novel Sonocatalyst in Aqueous Solution. *Ultrason. Sonochem.* **2020**, *61*, 104815. [CrossRef]
11. Taneja, P.; Sharma, S.; Umar, A.; Mehta, S.K.; Ibhadon, A.O.; Kansal, S.K. Visible-Light Driven Photocatalytic Degradation of Brilliant Green Dye Based on Cobalt Tungstate (CoWO$_4$) Nanoparticles. *Mater. Chem. Phys.* **2018**, *211*, 335–342. [CrossRef]
12. Karthikeyan, C.; Arunachalam, P.; Ramachandran, K.; Al-Mayouf, A.M.; Karuppuchamy, S. Recent Advances in Semiconductor Metal Oxides with Enhanced Methods for Solar Photocatalytic Applications. *J. Alloys Compd.* **2020**, *828*, 154281. [CrossRef]
13. Chawla, H.; Chandra, A.; Ingole, P.P.; Garg, S. Recent Advancements in Enhancement of Photocatalytic Activity Using Bismuth-Based Metal Oxides Bi2MO6 (M = W, Mo, Cr) for Environmental Remediation and Clean Energy Production. *J. Ind. Eng. Chem.* **2021**, *95*, 1–15. [CrossRef]
14. Ikram, M.; Rashid, M.; Haider, A.; Naz, S.; Haider, J.; Raza, A.; Ansar, M.T.; Uddin, M.K.; Ali, N.M.; Ahmed, S.S.; et al. A Review of Photocatalytic Characterization, and Environmental Cleaning, of Metal Oxide Nanostructured Materials. *Sustain. Mater. Technol.* **2021**, *30*, e00343. [CrossRef]

15. Lin, J.; Lin, J.; Zhu, Y. Controlled Synthesis of the ZnWO$_4$ Nanostructure and Effects on the Photocatalytic Performance. *Inorg. Chem.* **2007**, *46*, 8372–8378. [CrossRef] [PubMed]
16. Bai, X.; Wang, L.; Zhu, Y. Visible Photocatalytic Activity Enhancement of ZnWO4 by Graphene Hybridization. *ACS Catal.* **2012**, *2*, 2769–2778. [CrossRef]
17. Ke, J.; Niu, C.; Zhang, J.; Zeng, G. Significantly Enhanced Visible Light Photocatalytic Activity and Surface Plasmon Resonance Mechanism of Ag/AgCl/ZnWO$_4$ Composite. *J. Mol. Catal. A Chem.* **2014**, *395*, 276–282. [CrossRef]
18. Dong, T.; Li, Z.; Ding, Z.; Wu, L.; Wang, X.; Fu, X. Characterizations and Properties of Eu3+-Doped ZnWO4 Prepared via a Facile Self-Propagating Combustion Method. *Mater. Res. Bull.* **2008**, *43*, 1694–1701. [CrossRef]
19. Zhang, X.; Wang, B.; Wang, X.; Xiao, X.; Dai, Z.; Wu, W.; Zheng, J.; Ren, F.; Jiang, C. Preparation of M@BiFeO3 Nanocomposites (M = Ag, Au) Bowl Arrays with Enhanced Visible Light Photocatalytic Activity. *J. Am. Ceram. Soc.* **2015**, *98*, 2255–2263. [CrossRef]
20. Wang, F.; Li, W.; Gu, S.; Li, H.; Liu, X.; Wang, M. Fabrication of FeWO4@ZnWO4/ZnO Heterojunction Photocatalyst: Synergistic Effect of ZnWO$_4$/ZnO and FeWO$_4$@ZnWO$_4$/ZnO Heterojunction Structure on the Enhancement of Visible-Light Photocatalytic Activity. *ACS Sustain. Chem. Eng.* **2016**, *4*, 6288–6298. [CrossRef]
21. Li, P.; Zhao, X.; Jia, C.J.; Sun, H.; Sun, L.; Cheng, X.; Liu, L.; Fan, W. ZnWO4/BiOI Heterostructures with Highly Efficient Visible Light Photocatalytic Activity: The Case of Interface Lattice and Energy Level Match. *J. Mater. Chem. A Mater.* **2013**, *1*, 3421–3429. [CrossRef]
22. Mgidlana, S.; Nyokong, T. Asymmetrical Zinc(II) Phthalocyanines Cobalt Tungstate Nanomaterial Conjugates for Photodegradation of Methylene Blue. *J. Photochem. Photobiol. A Chem.* **2021**, *418*, 113421. [CrossRef]
23. Siriwong, P.; Thongtem, T.; Phuruangrat, A.; Thongtem, S. Hydrothermal Synthesis, Characterization, and Optical Properties of Wolframite ZnWO4 Nanorods. *Crystengcomm* **2011**, *13*, 1564–1569. [CrossRef]
24. Pavithra, N.S.; Nagaraju, G.; Patil, S.B. Ionic Liquid-Assisted Hydrothermal Synthesis of ZnWO$_4$ Nanoparticles Used for Photocatalytic Applications. *Ionics* **2021**, *27*, 3533–3541. [CrossRef]
25. Han, S.; Xiao, K.; Liu, L.; Huang, H. Zn$_{1-x}$Co$_x$WO$_4$ ($0 \leq x \leq 1$) Full Range Solid Solution: Structure, Optical Properties, and Magnetism. *Mater. Res. Bull.* **2016**, *74*, 436–440. [CrossRef]
26. Rahmani, M.; Sedaghat, T. Nitrogen-Doped ZnWO4 Nanophotocatalyst: Synthesis, Characterization and Photodegradation of Methylene Blue under Visible Light. *Res. Chem. Intermed.* **2019**, *45*, 5111–5124. [CrossRef]
27. Rahmani, M.; Sedaghat, T. A Facile Sol–Gel Process for Synthesis of ZnWO$_4$ Nanopartices with Enhanced Band Gap and Study of Its Photocatalytic Activity for Degradation of Methylene Blue. *J. Inorg. Organomet. Polym. Mater.* **2019**, *29*, 220–228. [CrossRef]
28. Scherrer, P. Estimation of the Size and Internal Structure of Colloidal Particles by Means of Rontgen Rays. *Nachr. Ges. Wiss. Göttingen* **1918**, *26*, 98–100.
29. Tauc, J. *Optical Properties of Solids*; Abelès, F., Ed.; Elsevier: Amsterdam, The Netherlands, 1970.
30. Sadeghfar, F.; Zalipour, Z.; Taghizadeh, M.; Taghizadeh, A.; Ghaedi, M. Photodegradation Processes. *Interface Sci. Technol.* **2021**, *32*, 55–124. [CrossRef]
31. Li, X.; Yu, J.; Low, J.; Fang, Y.; Xiao, J.; Chen, X. Engineering Heterogeneous Semiconductors for Solar Water Splitting. *J. Mater. Chem. A Mater.* **2015**, *3*, 2485–2534. [CrossRef]
32. Fujito, H.; Kunioku, H.; Kato, D.; Suzuki, H.; Higashi, M.; Kageyama, H.; Abe, R. Layered Perovskite Oxychloride Bi4NbO8Cl: A Stable Visible Light Responsive Photocatalyst for Water Splitting. *J. Am. Chem. Soc.* **2016**, *138*, 2082–2085. [CrossRef]
33. Li, Y.; Hua, S.; Zhou, Y.; Dang, Y.; Cui, R.; Fu, Y. Activating ZnWO$_4$ Nanorods for Efficient Electroanalysis of Bisphenol A via the Strategy of In Doping Induced Band Gap Change. *J. Electroanal. Chem.* **2020**, *856*, 113613. [CrossRef]
34. Yu, X.; Williams, C.T. Recent Advances in the Applications of Mesoporous Silica in Heterogeneous Catalysis. *Catal. Sci. Technol.* **2022**, *12*, 5765–5794. [CrossRef]
35. Malik, J.; Kumar, S.; Srivastava, P.; Bag, M.; Mandal, T.K. Cation Disorder and Octahedral Distortion Control of Internal Electric Field, Band Bending and Carrier Lifetime in Aurivillius Perovskite Solid Solutions for Enhanced Photocatalytic Activity. *Mater. Adv.* **2021**, *2*, 4832–4842. [CrossRef]
36. Kirankumar, V.S.; Sumathi, S. Copper and Cerium Co-Doped Cobalt Ferrite Nanoparticles: Structural, Morphological, Optical, Magnetic, and Photocatalytic Properties. *Environ. Sci. Pollut. Res.* **2019**, *26*, 19189–19206. [CrossRef] [PubMed]
37. Pan, Y.M.; Zhang, W.; Hu, Z.F.; Feng, Z.Y.; Ma, L.; Xiong, D.P.; Hu, P.J.; Wang, Y.H.; Wu, H.Y.; Luo, L. Synthesis of Ti^{4+}-Doped ZnWO$_4$ Phosphors for Enhancing Photocatalytic Activity. *J. Lumin.* **2019**, *206*, 267–272. [CrossRef]
38. Otrokov, M.M.; Klimovskikh, I.I.; Calleja, F.; Shikin, A.M.; Vilkov, O.; Rybkin, A.G.; Estyunin, D.; Muff, S.; Dil, J.H.; Vázquez De Parga, A.L.; et al. Evidence of Large Spin-Orbit Coupling Effects in Quasi-Free-Standing Graphene on Pb/Ir(1 1 1). *2d Mater* **2018**, *5*, 035029. [CrossRef]
39. Magdalane, C.M.; Kaviyarasu, K.; Vijaya, J.J.; Siddhardha, B.; Jeyaraj, B.; Kennedy, J.; Maaza, M. Evaluation on the Heterostructured CeO$_2$/Y$_2$O$_3$ Binary Metal Oxide Nanocomposites for UV/Vis Light Induced Photocatalytic Degradation of Rhodamine–B Dye for Textile Engineering Application. *J. Alloys Compd.* **2017**, *727*, 1324–1337. [CrossRef]
40. Zheng, X.; Yuan, J.; Shen, J.; Liang, J.; Che, J.; Tang, B.; He, G.; Chen, H. A Carnation-like RGO/Bi$_2$O$_2$CO$_3$/BiOCl Composite: Efficient Photocatalyst for the Degradation of Ciprofloxacin. *J. Mater. Sci. Mater. Electron.* **2019**, *30*, 5986–5994. [CrossRef]
41. Balu, S.; Velmurugan, S.; Palanisamy, S.; Chen, S.W.; Velusamy, V.; Yang, T.C.K.; El-Shafey, E.S.I. Synthesis of α-Fe$_2$O$_3$ Decorated g-C$_3$N$_4$/ZnO Ternary Z-Scheme Photocatalyst for Degradation of Tartrazine Dye in Aqueous Media. *J. Taiwan Inst. Chem. Eng.* **2019**, *99*, 258–267. [CrossRef]

42. Naresh, G.; Malik, J.; Meena, V.; Mandal, T.K. PH-Mediated Collective and Selective Solar Photocatalysis by a Series of Layered Aurivillius Perovskites. *ACS Omega* **2018**, *3*, 11104–11116. [CrossRef] [PubMed]
43. Saher, R.; Hanif, M.A.; Mansha, A.; Javed, H.M.A.; Zahid, M.; Nadeem, N.; Mustafa, G.; Shaheen, A.; Riaz, O. Sunlight-Driven Photocatalytic Degradation of Rhodamine B Dye by Ag/FeWO$_4$/g-C$_3$N$_4$ Composites. *Int. J. Environ. Sci. Technol.* **2021**, *18*, 927–938. [CrossRef]
44. Zhang, P.; Liang, H.; Liu, H.; Bai, J.; Li, C. A Novel Z-Scheme BiOI/BiOCl Nanofibers Photocatalyst Prepared by One-Pot Solvothermal with Efficient Visible-Light-Driven Photocatalytic Activity. *Mater. Chem. Phys.* **2021**, *272*, 125031. [CrossRef]
45. Santana, R.W.R.; Lima, A.E.B.; de Souza, L.K.C.; Santos, E.C.S.; Santos, C.C.; de Menezes, A.S.; Sharma, S.K.; Cavalcante, L.S.; Maia da Costa, M.E.H.; Sales, T.O.; et al. BiOBr/ZnWO$_4$ Heterostructures: An Important Key Player for Enhanced Photocatalytic Degradation of Rhodamine B Dye and Antibiotic Ciprofloxacin. *J. Phys. Chem. Solids* **2023**, *173*, 111093. [CrossRef]
46. Geetha, G.V.; Sivakumar, R.; Slimani, Y.; Sanjeeviraja, C.; Kannapiran, E. Rare Earth (RE: La and Ce) Elements Doped ZnWO$_4$ Nanoparticles for Enhanced Photocatalytic Removal of Methylene Blue Dye from Aquatic Environment. *Phys. B Condens. Matter* **2022**, *639*, 414028. [CrossRef]
47. Liu, X.; Shu, J.; Wang, H.; Jiang, Z.; Xu, L.; Liu, C. One-Pot Preparation of a Novel CoWO4/ZnWO4 p-n Heterojunction Photocatalyst for Enhanced Photocatalytic Activity under Visible Light Irradiation. *J. Phys. Chem. Solids* **2023**, *172*, 111061. [CrossRef]
48. Kumar, G.M.; Lee, D.J.; Jeon, H.C.; Ilanchezhiyan, P.; Deuk Young, K.; Tae Won, K. One Dimensional ZnWO$_4$ Nanorods Coupled with WO$_3$ Nanoplates Heterojunction Composite for Efficient Photocatalytic and Photoelectrochemical Activity. *Ceram. Int.* **2022**, *48*, 4332–4340. [CrossRef]
49. Jaramillo-Páez, C.; Navío, J.A.; Puga, F.; Hidalgo, M.C. Sol-Gel Synthesis of ZnWO$_4$-(ZnO) Composite Materials. Characterization and Photocatalytic Properties. *J. Photochem. Photobiol. A Chem.* **2021**, *404*, 112962. [CrossRef]
50. Jayamani, G.; Shanthi, M. An Efficient Nanocomposite CdS-ZnWO$_4$ for the Degradation of Naphthol Green B Dye under UV-A Light Illumination. *Nano-Struct. Nano-Objects* **2020**, *22*, 100452. [CrossRef]
51. Geetha, G.V.; Keerthana, S.P.; Madhuri, K.; Sivakumar, R. Effect of Solvent Volume on the Properties of ZnWO$_4$ Nanoparticles and Their Photocatalytic Activity for the Degradation of Cationic Dye. *Inorg. Chem. Commun.* **2021**, *132*, 108810. [CrossRef]

Disclaimer/Publisher's Note: The statements, opinions and data contained in all publications are solely those of the individual author(s) and contributor(s) and not of MDPI and/or the editor(s). MDPI and/or the editor(s) disclaim responsibility for any injury to people or property resulting from any ideas, methods, instructions or products referred to in the content.

Article

G-C$_3$N$_4$ Dots Decorated with Hetaerolite: Visible-Light Photocatalyst for Degradation of Organic Contaminants

Zahra Lahootifar [1], Aziz Habibi-Yangjeh [1,*], Shima Rahim Pouran [2] and Alireza Khataee [3,4]

1 Department of Chemistry, Faculty of Science, University of Mohaghegh Ardabili, Ardabil 56199-11367, Iran
2 Social Determinants of Health Research Center, Department of Environmental and Occupational Health, Ardabil University of Medical Sciences, Ardabil 85991-56189, Iran
3 Research Laboratory of Advanced Water and Wastewater Treatment Processes, Department of Applied Chemistry, Faculty of Chemistry, University of Tabriz, Tabriz 16471-51666, Iran
4 Department of Environmental Engineering, Faculty of Engineering, Gebze Technical University, Gebze 41400, Turkey
* Correspondence: ahabibi@uma.ac.ir

Citation: Lahootifar, Z.; Habibi-Yangjeh, A.; Rahim Pouran, S.; Khataee, A. G-C$_3$N$_4$ Dots Decorated with Hetaerolite: Visible-Light Photocatalyst for Degradation of Organic Contaminants. *Catalysts* 2023, *13*, 346. https://doi.org/10.3390/catal13020346

Academic Editor: Meng Li

Received: 5 January 2023
Revised: 27 January 2023
Accepted: 27 January 2023
Published: 3 February 2023

Copyright: © 2023 by the authors. Licensee MDPI, Basel, Switzerland. This article is an open access article distributed under the terms and conditions of the Creative Commons Attribution (CC BY) license (https://creativecommons.org/licenses/by/4.0/).

Abstract: In this paper, a facile hydrothermal approach was used to integrate graphitic carbon nitride dots (CNDs) with hetaerolite (ZnMn$_2$O$_4$) at different weight percentages. The morphology, microstructure, texture, electronic, phase composition, and electrochemical properties were identified by field emission scanning electron microscopy (FESEM), X-ray photoelectron spectroscopy (XPS), transmission electron microscopy (TEM), high-resolution TEM (HRTEM), energy dispersive X-ray spectroscopy (EDX), X-ray diffraction (XRD), Fourier transform-infrared (FT-IR), ultraviolet-visible diffuse reflectance (UV-vis DR), photoluminescence (PL), electrochemical impedance spectroscopy (EIS), Brunauer–Emmett–Teller (BET), Barrett–Joyner–Halenda (BJH), and photocurrent density. The results of XRD, FT-IR, EDX, and XPS analyses confirmed the synthesis of CNDs/ZnMn$_2$O$_4$ (20%) nanocomposite. As per PL, EIS, and photocurrent outcomes, the binary CNDs/ZnMn$_2$O$_4$ nanocomposite revealed superior features for interfacial transferring of charge carriers. The developed p–n heterojunction at the interface of CNDs and ZnMn$_2$O$_4$ nanoparticles partaken a significant role in the impressive charge segregation and migration. The binary nanocomposites were employed for the photodegradation of several dye pollutants, including rhodamine B (RhB), fuchsin, malachite green (MG), and methylene blue (MB) at visible wavelengths. Amongst the fabricated photocatalysts, the CNDs/ZnMn$_2$O$_4$ (20%) nanocomposite gave rise to about 98% RhB degradation efficiency within 45 min with the rate constant of 747×10^{-4} min^{-1}, which was 66.5-, 3.44-, and 2.72-fold superior to the activities of CN, CNDs, and ZnMn$_2$O$_4$ photocatalysts, respectively. The impressive photodegradation performance of this nanocomposite was not only associated with the capacity for impressive visible-light absorption and boosted separation and transport of charge carriers, but also with its large surface area.

Keywords: graphitic carbon nitride dots; ZnMn$_2$O$_4$; p-n heterojunction photocatalyst; organic pollutants

1. Introduction

Over the past few decades, the shortage of energy and water pollution issues raised by organic compounds have become global concerns of the scientific community [1]. Among the primary water pollutants are organic dyes, which have been a considerable part of water pollution due to their high consumption rate in various industries and resistance to biodegradation. To tackle the detrimental effects of these organic molecules on human health and the environment [2], especially on marine life, numerous advanced materials [3] and strategies have been developed and implemented over time. In particular, the focus has been paid to bio-compatible materials and practical strategies, in addition to being effective and efficient from energy and cost perspectives [2,3]. Advanced oxidation processes (AOPs) are on top of the wastewater treatment techniques that can fulfill the effective degradation of various organic pollutants [4]. The visible-light-driven heterogeneous photocatalysis is of particular interest among AOPs, since it uses solar energy to drive oxidation reactions for the degradation of organic molecules and eco-clean energy generation [5]. In this regard, the utilized photocatalyst undertakes the central role because its optical, structural, and surface properties chiefly determine the efficiency of a photocatalytic system [6].

On account of the appropriate energy band edges and band gap, tunable electronic structure, high stability, and biocompatibility, graphitic carbon nitride (CN) quickly became the subject of research for various photocatalytic reactions, including CO_2 reduction, synthesis of organic compounds, hydrogen production, removal of pollutants, and nitrogen fixation [7–9]. Nonetheless, the solar-light utilization ability of pristine CN is poor and generates an inefficient amount of charge carriers. Moreover, the rapid recombination of electron/hole pairs, insufficient surface-active centers, and low conductivity confine its practical applications [10]. Hence, numerous attempts have been made to improve the activity of CN through some structural/morphological modifications or heterojunction/s development by integration with other semiconductor/s [11]. Recently, dots and quantum dots of CN have received much interest due to their high disperse ability, quantum confinement effect, and improved heterojunction construction, in addition to the intrinsic properties of the bulk CN [12,13]. For example, Ma et al. realized that anchoring CN quantum dots on 2D g-C_3N_4 improved photocatalytic activity [13]. Moreover, the latest research activities on the improvement of carbon nitride photocatalytic activity have been reviewed [14,15].

Spinel zinc manganese oxide ($ZnMn_2O_4$, E_g = 1.70 eV) is a p-type semiconductor active in the visible region. Among spinels, $ZnMn_2O_4$ has attracted much attention due to its outstanding technological importance as catalyst, solid electrolyte, etc., [16]. Reduction in CO_2 and destruction of pollutants are photocatalytic applications of this spinel [16].

Among water pollutants, rhodamine B (RhB), with carcinogenic and neurotoxic properties, has a high potential to cause diseases in humans [17]. Fushin can affect the central nervous system and cause drowsiness and dizziness in humans [18]. Malachite Green (MG) is a carcinogenic and mutagenic agent in living organisms [19]. Moreover, methylene blue (MB) has a severe impact on human health and the environment, because it is carcinogenic and non-biodegradable in nature. Among the dangers of MB, its effect on the respiratory system, blindness, digestive and mental disorders, shock, gastritis, jaundice, tissue necrosis, and an increase in heart rate has been highlighted [20].

The appealing properties of CN dots (CNDs) and complementary characteristics of $ZnMn_2O_4$ brought the idea of designing a binary heterostructure based on these semiconductors. For this purpose, a hydrothermal method was used to deposit various weight percentages of $ZnMn_2O_4$ nanoparticles over CNDs. The photocatalytic activities were also determined for the degradation of several organic dye contaminants, including RhB, fuchsin, MG, and MB molecules. Finally, the nanocomposite with the best optical and photocatalytic activity was introduced, and the corresponding degradation mechanism was suggested.

2. Results and Discussions

The fabricated CN, CNDs, $ZnMn_2O_4$, and CNDs/$ZnMn_2O_4$ (20%) materials were characterized by XRD to expose the corresponding crystalline phases (Figure 1a). In the pattern of CN, two sharp peaks appeared at 2θ values of 13.1° and 27.4° (JCPDS No. 01-087-1526), corresponding to stacking between the surfaces of the CN layers [21]. The pattern of CNDs shows only one peak at 27.4°, indicating that CNDs has the same intrinsic crystal structure as CN. The peak at 13.1° disappeared and the intensity of the peak at 27.4° diminished compared to CN. This decrease is due to the small size of CNDs particles [22]. On the other hand, the peaks appeared at 2θ = 18.62°, 29.47°, 31.42°, 33.22°, 36.47°, 39.27°, 44.77°, 52.22°, 54.42°, 56.87°, 58.98°, 60.92°, and 65.25°, respectively, correspond to the reflections from (101), (112), (200), (103), (211), (004), (220), (105), (312), (303), (321), (224), and (400) planes of $ZnMn_2O_4$ (JCPDS file No. 01-077-0470) [23]. The XRD pattern of

CNDs/ZnMn$_2$O$_4$ (20%) nanocomposite included all the diffraction peaks relevant to CNDs and the tetragonal phase of ZnMn$_2$O$_4$, indicating the accuracy of the synthesis procedure. According to the correct position of the XRD peaks, it is concluded that the synthesis of the nanocomposite has been correctly performed. The surface functional groups of the CNDs/ZnMn$_2$O$_4$ (20%) sample were recorded by FT-IR spectroscopy and elucidated with respect to CN and CNDs spectra (Figure 1b). The CN and CNDs samples gave rise to similar FT-IR spectra. In both of them, the wide-ranging peaks at 2900–3450 cm^{-1} assign to the O–H and N–H stretching modes [24]. The peaks that appeared at wavenumbers of 1200–1650 cm^{-1} are relevant to the C–N and C=N groups [22,25]. Moreover, the bands related to the atmospheric CO$_2$ and C–O stretching modes emerged at 2385 cm^{-1} and 1060 cm^{-1}, respectively [26]. The peak related to the stretching vibration of heptazine units was also identified at 800 cm^{-1} [26]. In the case of CNDs/ZnMn$_2$O$_4$ (20%) nanocomposite, not only the peaks attributing to CNDs appeared in the spectrum, but also the peaks assigning to the M–O bonds of the corresponding metal oxides (M = Zn and Mn) showed up at 640 cm^{-1} and 545 cm^{-1}, respectively [27,28]. Hence, all the functional groups are visible in the figure, which confirmed the synthesis of the binary photocatalyst.

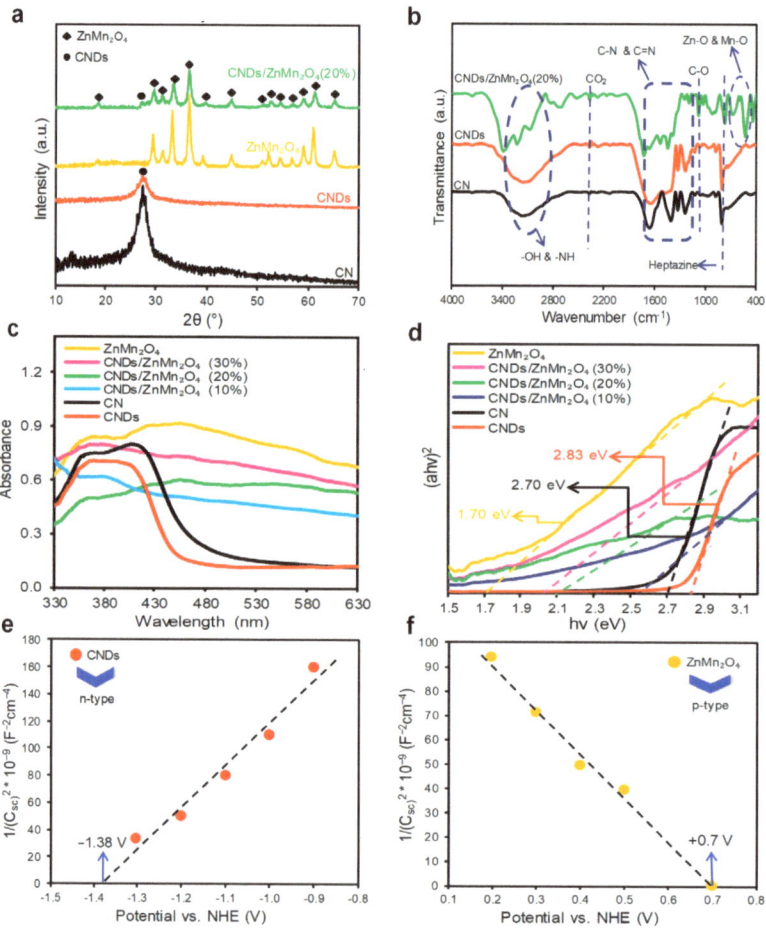

Figure 1. (**a**) XRD, (**b**) FT-IR, (**c**) Optical absorbance, (**d**) Tauc plots, and (**e**,**f**) Mott-Schottky plots of the specified samples.

The electronic properties of CN, CNDs, $ZnMn_2O_4$, and CNDs/$ZnMn_2O_4$ composites were studied through their UV–vis DR spectra (Figure 1c). As per the data, although the CN and CNDs photocatalysts showed absorptions at visible wavelengths, the visible-light absorption capability of CNDs/$ZnMn_2O_4$ nanocomposites is highly promoted following the incorporation of the small band gap $ZnMn_2O_4$ with medium band gap CNDs. The E_g values were afterward obtained using Tauc's plots [29], which were 2.70, 2.83, 1.70, 2.54, 2.10, and 2.02 eV for CN, CNDs, $ZnMn_2O_4$, CNDs/$ZnMn_2O_4$ (10%), CNDs/$ZnMn_2O_4$ (20%), and CNDs/$ZnMn_2O_4$ (30%) photocatalysts, respectively (Figure 1d). Moreover, the flat-band potentials and the type of semiconductors were determined via applying the Mott-Schottky (M-S) analysis [30] as illustrated in Figure 1e,f. As noticed, the CNDs and $ZnMn_2O_4$ exhibited n-type and p-type characteristics with a positive and negative slopes, respectively. The calculated E_{CB} and E_{VB} values for CNDs were −1.48 eV and +1.35 eV, and for $ZnMn_2O_4$ were −0.90 eV and +0.80 eV, respectively.

The morphological features of the CNDs/$ZnMn_2O_4$ (20%) nanocomposite were studied through FESEM, TEM, and HRTEM images. The FESEM and TEM images of the sample present aggregated particles composed of nearly spherical particles (Figure 2a,b). In the HRTEM image of CNDs/$ZnMn_2O_4$ (20%) nanocomposite presented in Figure 2c, the lattice fringes of CNDs and $ZnMn_2O_4$ were identified with inter-planar distances of 0.340 and 0.490 nm, respectively [22,31]. This figure confirms the combination of CNDs and $ZnMn_2O_4$ components to construct a binary p-n heterojunction photocatalyst. The elements of the binary nanocomposite were detected by EDX analyses (Figure 2d). The C, N, O, Mn, and Zn elements were the only elements identified in the CNDs/$ZnMn_2O_4$ (20%) nanocomposite. The EDX mapping images of the nanocomposite were also collected, which displayed elements with almost homogeneous dispersion in the nanocomposite structure (Figure 2e).

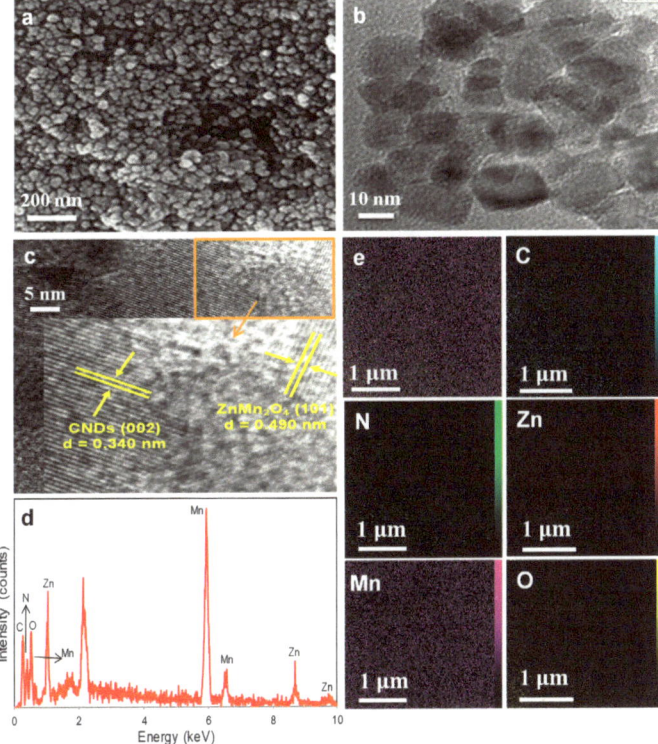

Figure 2. (a) SEM, (b) TEM, and (c) HRTEM images. (d) EDX spectrum and (e) EDX mapping of CNDs/$ZnMn_2O_4$ (20%) nanocomposite.

Figure 3 shows the oxidation states of the elements in the CNDs/ZnMn$_2$O$_4$ (20%) nanocomposite as per XPS analysis. The survey spectrum consisted of the peaks related to C 1s, N 1s, Zn 2p, Mn 2p, and O 1s. The narrow scan of carbon was completed in the 283–292 eV region and produced two dominant peaks, which were deconvoluted into three peak components at binding energies (B.E) of 284.7, 285.5, and 288.5 eV, which were, respectively, associated with the sp^2 graphitic carbon (C–C), sp^2 aromatic C attached to –NH$_2$, and the carbon corresponded to heptazine/triazine C-N-C coordination (Figure 3b) [24]. The N 1s spectrum generated in the corresponding narrow scan was deconvoluted into four peak components positioned at 398.5, 399.1, and 400.5 eV (Figure 3c), which correspond to the nitrogen as sp^2-hybridized (C=N-C), tertiary form (N–(C)$_3$) and hydrogen-bonded (C–N–H), respectively [24]. The Zn 2p high-resolution spectrum produced two peaks at B.E. of 1043.6 eV and 1020.5 eV, respectively, matched with Zn 2p$_{1/2}$ and Zn 2p$_{3/2}$ and indicated the existence of Zn element in +2 oxidation state (Figure 3d) [32]. On the other hand, the Mn 2p narrow scan generated two strong peaks, one associated with Mn 2p$_{3/2}$ centered at binding energy of 641.8 eV and the other one ascribed to Mn 2p$_{1/2}$ at 653.4 eV (Figure 3e) [32]. The O 1s spectrum deconvolution resulted in three peaks at 530.2, 531.8, and 533 eV, respectively, related to ZnMn$_2$O$_4$ lattice oxygen, surface hydroxyl content, and residual adsorbed H$_2$O, respectively (Figure 3f) [31,33].

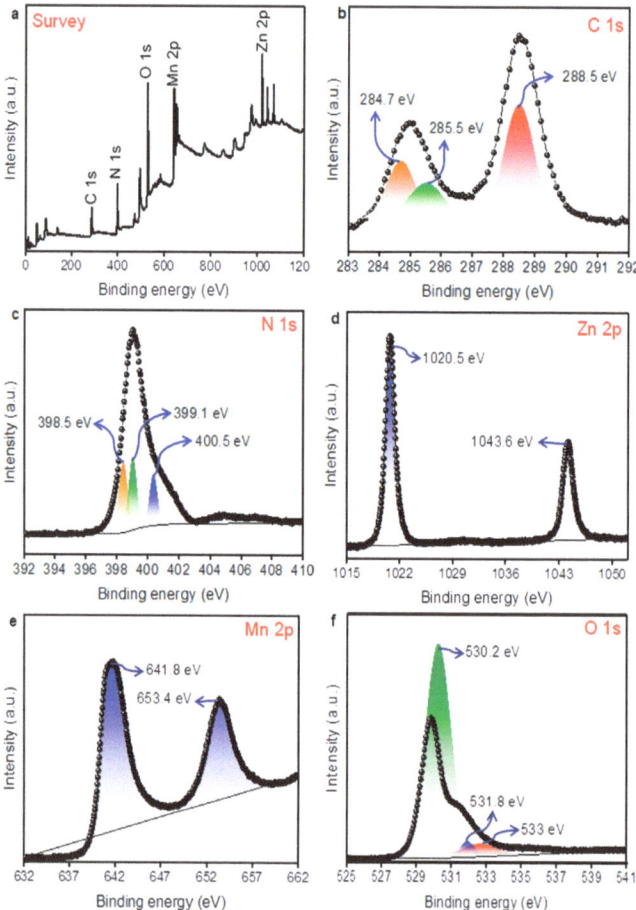

Figure 3. XPS spectra of CNDs/ZnMn$_2$O$_4$ (20%) photocatalyst: (**a**) Survey, (**b**) C1s, (**c**) N1s, (**d**) Zn 2p, (**e**) Mn 2p, and (**f**) O1s spectra.

The photocatalytic degradation efficiency of the samples was examined for photo-oxidation of RhB under visible light. Figure 4a depicts the changes in the RhB concentration (C_t/C_0) within diverse photocatalytic systems per irradiation period. Prior to the experiments, the adsorption capacity of the samples was investigated through the stirring of the dye solution with the photocatalyst under dark conditions for 60 min. As per results, a negligible amount (8%) of RhB was removed through only light illumination, i.e., the absence of photocatalyst, indicating the inefficacy of visible light in breaking the chemical bonds of RhB molecules. The system including the pristine CN resulted in low RhB removal efficiency of about 20%, which can be strongly associated with the poor harvesting of light in visible region and rapid recombination of charges. However, the CNDs photocatalyst showed much higher activity with removal of 73% at the same time. By contrast, almost complete removal of RhB has been achieved over the CNDs/ZnMn$_2$O$_4$ (20%) nanocomposite within 45 min. The order of photoactivity was as CNDs/ZnMn$_2$O$_4$ (20%) > CNDs/ZnMn$_2$O$_4$ (30%) > CNDs/ZnMn$_2$O$_4$ (10%) > CNDs/ZnMn$_2$O$_4$ (40%) > CNDs/ZnMn$_2$O$_4$ (5%) > ZnMn$_2$O$_4$ > CNDs > CN. The plots of ln (C_0/C) vs. time for degradation of RhB over the studied samples revealed that the photocatalytic reactions obeyed the pseudo-first-order kinetic model (see Figure 4b). The rate constants for RhB removal using different photocatalysts were calculated to be 11.2×10^{-4} min^{-1} (CN), 217×10^{-4} min^{-1} (CNDs), 275×10^{-4} min^{-1} (ZnMn$_2$O$_4$), 284×10^{-4} min^{-1} (CNDs/ZnMn$_2$O$_4$ (5%)), 329×10^{-4} min^{-1} (CNDs/ZnMn$_2$O$_4$ (40%)), 333×10^{-4} min^{-1} (CNDs/ZnMn$_2$O$_4$ (10%)), 429×10^{-4} min^{-1} (CNDs/ZnMn$_2$O$_4$ (30%)), and 747×10^{-4} min^{-1} (CNDs/ZnMn$_2$O$_4$ (20%)) (Figure 4c). As per results, the nanocomposites showed higher photodegradation rates than pristine CN, wherein the activity of CNDs/ZnMn$_2$O$_4$ (20%) nanocomposite was, respectively, 66.5, 3.44, and 2.72-fold superior to the activities of CN, CNDs, and ZnMn$_2$O$_4$ photocatalysts, respectively. The boosted photocatalytic activities of the binary samples implied that the combination of CNDs with ZnMn$_2$O$_4$ was the right action, which should be taken place with optimum weight percentages.

The reactive species produced in a photocatalytic system are the central elements involved in the photodegradation of organic molecules. To estimate the type of the reactive species, a group of experiments using scavengers, including AOX, BQ, and 2-PrOH was performed to, respectively, probe h$^+$, •O$_2^-$, and •OH species. Figure 4d depicts the RhB photo-oxidation rate constants using the CNDs/ZnMn$_2$O$_4$ (20%) nanocomposite, without or with the scavengers. The rate constant of 747×10^{-4} min^{-1}, which was obtained in the absence of the scavengers, was significantly dropped to 154×10^{-4}, 53.4×10^{-4}, and 1.03×10^{-4} min^{-1} upon 2-PrOH, AOX, and BQ supplementation into the system, respectively. As can be inferred from the result, all three probed species contributed to the RhB photodegradation process and their role are as •O$_2^-$ > h$^+$ > •OH.

The photocatalytic activities of CN, CNDs, and CNDs/ZnMn$_2$O$_4$ (20%) samples were further examined for degradation of some other dye pollutants, including MG, fuchsin and MB and the outcomes were presented as compared with the RhB results (Figure 4e). As per the results, the highest removal efficiencies of all the studied dyes were achieved over the CNDs/ZnMn$_2$O$_4$ (20%) system. The photoactivity of the binary sample for degradation of RhB, MG, fuchsin, and MB was 66.5, 15.8, 13.2, and 31.2-fold greater than CN, and 3.44, 1.5, 18.1, and 11.5 times premier than CNDs, respectively. Accordingly, the CNDs/ZnMn$_2$O$_4$ (20%) sample showed an outstanding photocatalytic activity for the treatment of various recalcitrant dye wastewater. The binary photocatalyst underwent four consecutive runs under the same operational conditions for RhB removal in order to assess its stability and potential reusability (Figure 4f). As noticed, the CNDs/ZnMn$_2$O$_4$ (20%) nanocomposite demonstrated excellent efficiency for RhB photodegradation under visible light in all the conducted experiments, and the loss of performance was insignificant at the end of the cycle, confirming its significant stability.

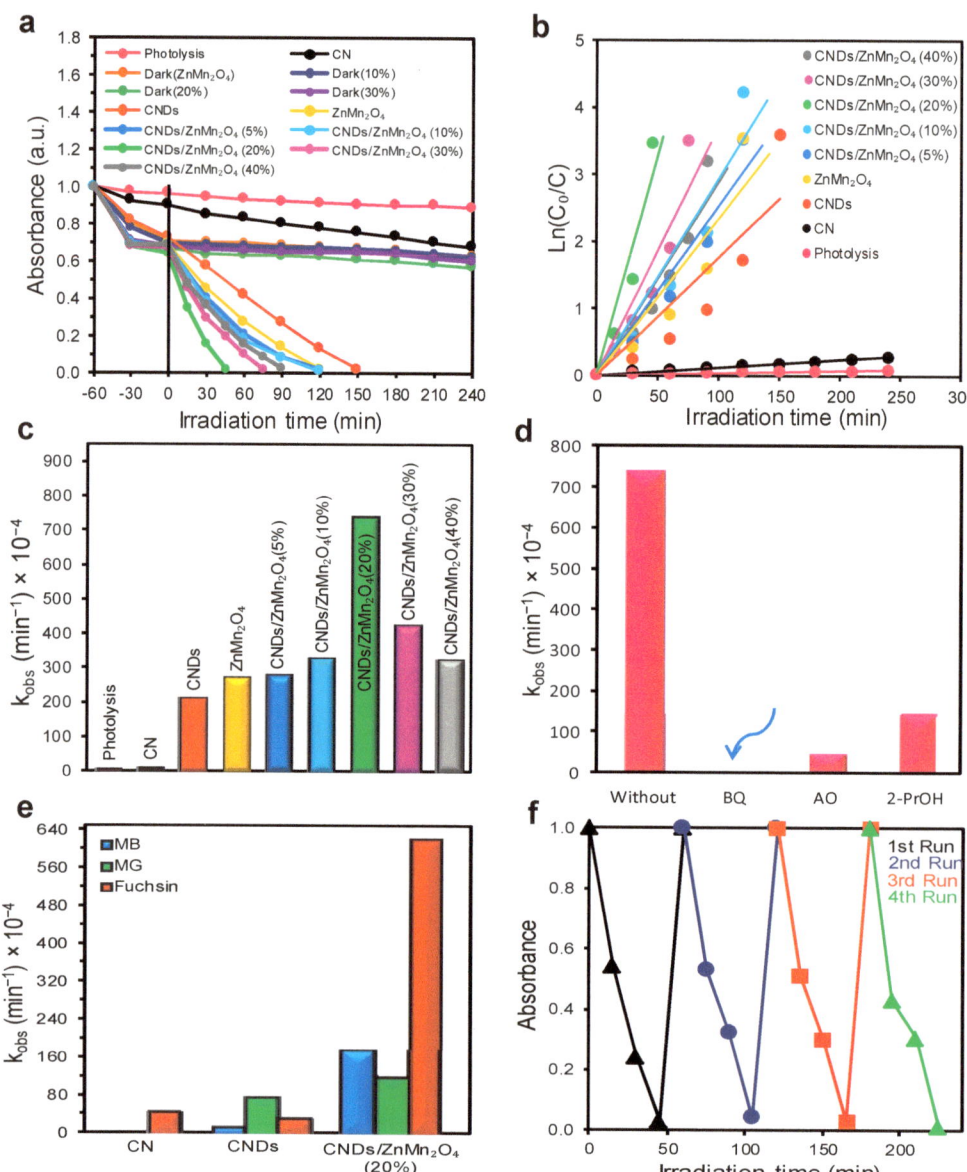

Figure 4. (**a**) RhB photodegradation, (**b**) First-order-kinetic plots for the degradation rates of RhB, (**c**) degradation rate constants of RhB, (**d**) Effect of trapping agents on the degradation of RhB, (**e**) Degradation constants of MB, MG, and fuchsin, and (**f**) the results of recycling experiments of CNDs/ZnMn$_2$O$_4$ (20%) system for degradation of RhB.

The ability for adequate segregation and transfer of the photoinduced charge carriers is among the crucial factors connected with immense photocatalytic activities [34]. On this basis, electrochemical impedance spectroscopy, photoluminescence, and photocurrent spectroscopy analyses were conducted to figure out the characteristics of the CNDs/ZnMn$_2$O$_4$ (20%) nanocomposite with respect to the ZnMn$_2$O$_4$, CNDs, and CN materials. Figure 5a shows the EIS responses of the samples, wherein the least impedance was detected for the

CNDs/ZnMn$_2$O$_4$ (20%) sample compared to those of ZnMn$_2$O$_4$, CNDs, and CN photocatalysts. This indicated the notable reduction in the resistance towards charge transfer in the CNDs/ZnMn$_2$O$_4$ (20%) nanocomposite. The PL results were also in line with the EIS response of the binary nanocomposite, wherein the weakest PL belonged to the CNDs/ZnMn$_2$O$_4$ (20%) nanocomposite (Figure 5b). Since the photoluminescence intensity shows the amount of photo energy released from returning the photo-excited material to the ground state, the higher released energy means more recombination of photoinduced charges. As per the results, the CNDs/ZnMn$_2$O$_4$ (20%) sample had a minimal recombination rate or the best performance for transferring the charge carriers compared to the pristine photocatalysts. The best optical performance of the binary sample was further confirmed via photocurrent analyses, as presented in Figure 5c. The stronger photocurrent intensity of the CNDs/ZnMn$_2$O$_4$ (20%) photocatalyst in comparison to the CN, CNDs, and ZnMn$_2$O$_4$ samples indicated its exceptional performance in extending the lifespan of the photoinduced charges. Therefore, the EIS, PL, and photocurrent results confirmed the high potential of the CNDs/ZnMn$_2$O$_4$ (20%) nanocomposite for photocatalytic applications.

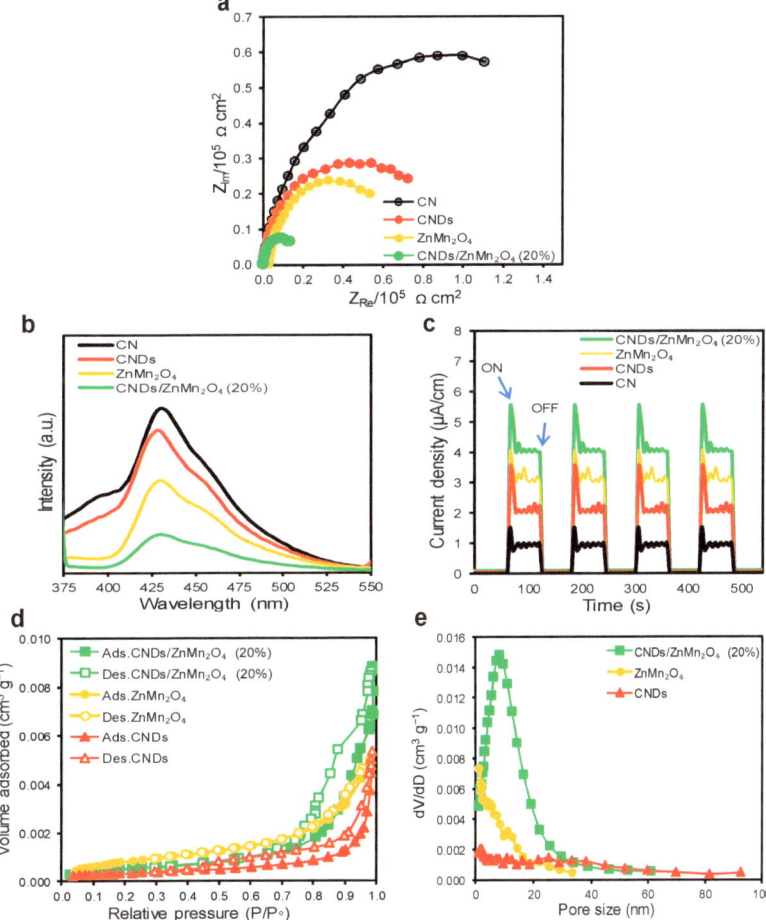

Figure 5. (**a**) EIS, (**b**) PL, and (**c**) Photocurrent of CN, CNDs, ZnMn$_2$O$_4$, and CNDs/ZnMn$_2$O$_4$ (20%) photocatalysts. (**d**) N$_2$ sorption data and (**e**) pore-size distributions for CNDs, ZnMn$_2$O$_4$, and CNQDs/ZnMn$_2$O$_4$ (20%) photocatalysts.

The surface properties of a photocatalyst are also influential factors as photocatalytic reactions take place on the surface [35]. Moreover, the charge transfer is also affected by surface characteristics. Therefore, the active sites and pore features of the studied photocatalysts were identified via N_2 adsorption–desorption isotherms (Figure 5d,e). From the BET isotherms, the N_2 sorption behavior of CNDs, $ZnMn_2O_4$, and CNDs/$ZnMn_2O_4$ (20%) photocatalysts followed type II with H_3 hysteresis loops. As per the outcomes (Table 1), the integration of CNDs (12.03 $m^2\ g^{-1}$) and $ZnMn_2O_4$ (33.8 $m^2\ g^{-1}$) contributed to an expanded active surface of the binary sample (44.6 $m^2\ g^{-1}$). The enlarged surface area could play a remarkable role in its promoted photocatalytic activity.

Table 1. Textural properties of the CNDs, $ZnMn_2O_4$, and CNDs/$ZnMn_2O_4$ (20%) photocatalysts.

Sample	Surface Area ($m^2\ g^{-1}$)	Mean Pore Diameter (nm)	Total Pore Volume ($cm^3\ g^{-1}$)
CNDs	12.03	27.73	0.08
$ZnMn_2O_4$	33.8	9.09	0.08
CNDs/$ZnMn_2O_4$ (20%)	44.6	27.5	0.31

Based on the outcomes, a mechanism was suggested to illustrate the route through which the pollutant molecules undergo a visible-light photodegradation reaction over the binary CNDs/$ZnMn_2O_4$ nanocomposites. As a result of the constructed p–n heterojunction between n-type CNDs and p-type $ZnMn_2O_4$, the electrons more from the Fermi level of CNDs to that of $ZnMn_2O_4$ until the Fermi levels reach an equilibrium state, resulting in, respectively, positive and negative charges on CNDs and $ZnMn_2O_4$ in the junction regions [36,37]. Upon the excitation of CNDs and $ZnMn_2O_4$ semiconductors by visible light, the electrons on the VBs of both components move to the corresponding CBs, while the holes remain on the VBs (Figure 6). After that, due to the influence of the inner electric field, the electrons on the $CB_{ZnMn2O4}$ rapidly transfer to the CB_{CNQDs}, where they can be trapped by oxygen molecules to produce $^\bullet O_2^-$ and $^\bullet OH$ species. Additionally, the contaminants can get oxidized directly by the holes remaining at the VB of $ZnMn_2O_4$. The rapid migration of the photo generated charges can be facilitated via the formed p–n heterojunction, leading to extending the lifetime of the charge carriers and diminishing recombination of them, which is highly beneficial for impressive photocatalytic performance [38,39].

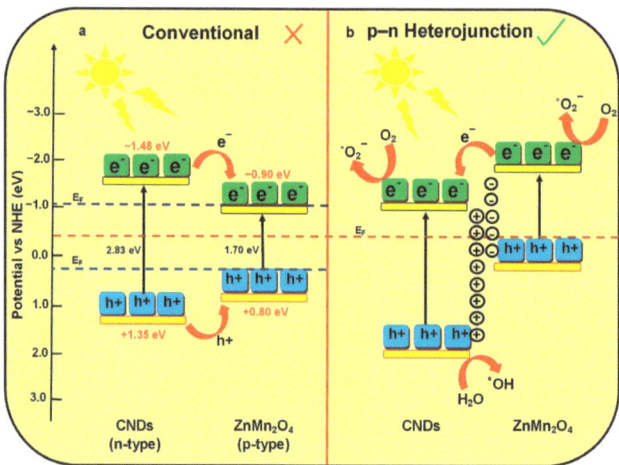

Figure 6. Photocatalytic mechanism for improved activity of CNDs/$ZnMn_2O_4$ nanocomposites: (**a**) Conventional and (**b**) p-n Heterojunction.

3. Experimental Part

3.1. Materials Section

The chemicals utilized in this research are melamine ($C_3H_6N_6$, 99%, Central Drug House), ethanol (96%, Merck), $Zn(NO_3)_2 \cdot 6H_2O$ (96%, Loba Chemie), $Mn(CH_3COO)_2 \cdot 4H_2O$ (99%, Merck), NaOH (98%, Merck), H_2O_2 (35%, Neutron).

3.2. Photocatalyst Synthesis

The CN was synthesized via the previously reported method [40]. For the synthesis of the CNDs sample, the mixture of CN with 20 mL of ethanol was stirred for half an hour. Then the solution of 25 mL of water and 25 mL of ethanol was added dropwise to the previous mixture. After stirring for half an hour, it was heated at 180 °C for 4 h in an autoclave.

The CNDs/$ZnMn_2O_4$ systems were fabricated using a hydrothermal procedure (Scheme 1). For the synthesis of CNDs/$ZnMn_2O_4$ (20%) nanocomposite, as the optimal photocatalyst with the best performance, 0.4 g of CNDs was mixed with 100 mL of H_2O under sonication for 10 min. Meanwhile, an aqueous solution of zinc nitrate ($Zn(NO_3)_2 \cdot 6H_2O$, Loba Chemie, 0.124 g in 10 mL H_2O) was mixed with the above-obtained suspension, while stirring for 60 min at 25 °C. On the other hand, an aqueous solution of manganese (II) acetate (Mn ($CH_3COO)_2 \cdot 4H_2O$, Merck, 0.205 g in 10 mL H_2O) was added to the resultant suspension and stirred for 60 min. At the next step, 20 mL NaOH (Merck, 0.4 M) and 1 mL H_2O_2 (Neutron, 14.9 M) solutions were combined with the suspension under stirring for more 15 min. In the end, the suspension was heated in a stainless-steel autoclave for 18 h at 160 °C. After cooling and washing, the resultant composite was dried in an oven.

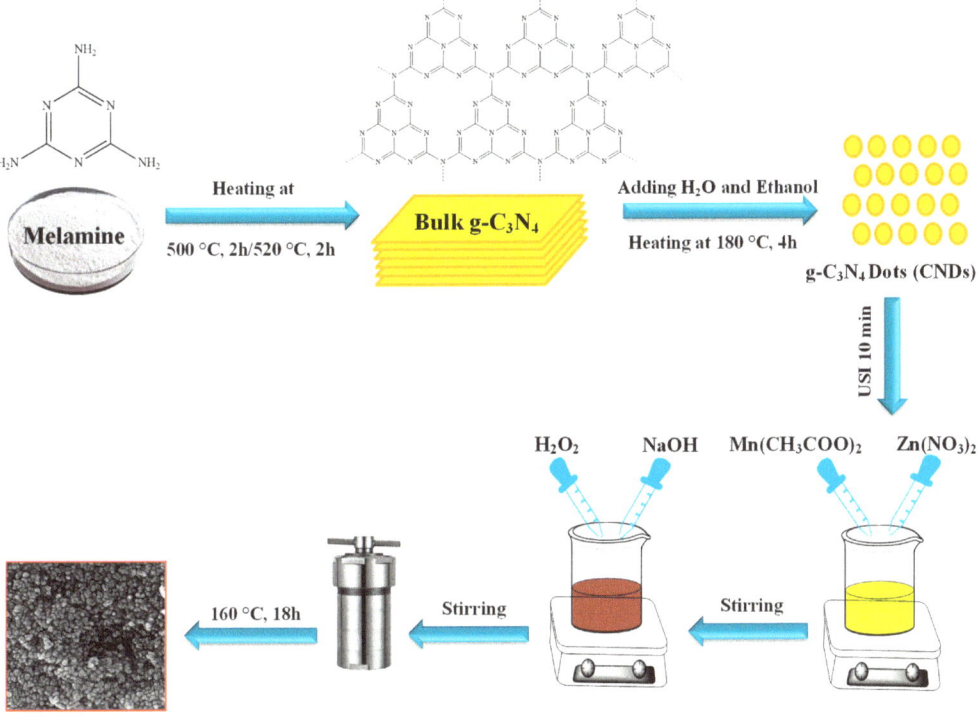

Scheme 1. Schematic illustration for the synthesis of CNDs/$ZnMn_2O_4$ photocatalysts.

3.3. Characterization Techniques

The EDX analyses and FESEM images were obtained by a Tescan Mira3 instrument. The XRD analyses were carried out on a Philips Xpert X-ray instrument with a Cu Kα source. A Scinco 4100 spectrophotometer was employed to record the UV-vis DR spectra. The FT-IR spectra were collected on a PerkinElmer Spectrum RX I apparatus. A PerkinElmer LS 55 fluorescence spectrophotometer was employed to record the PL spectra. The concentration of the dye pollutants was determined by UV-vis spectrophotometer (Cecile 9000). The TEM and HRTEM images were photographed using a HighTech HT7700 (Tokyo, Japan) instrument. The XPS analyses were conducted by a Specs-Flex XPS (Berlin, Germany). A BELSORP mini II (York, PA, USA) instrument was applied to fulfill the N_2 adsorption-desorption measurements. A μAutolabIII/FRA2 EIS (Utrecht, The Netherlands) was employed to accomplish the electrochemical analyses. The results of photocurrent data were also obtained with μAutolabIII/FRA2.

3.4. Photodegradation Tests

The photodegradation experiments were conducted in a photoreactor including a glass reactor furnished with a water-circulating device, a magnetic bar, and an air-purging system. An installed LED lamp (50 W, λ = 450–650 nm) on the top of the reactor supplied the visible light. For the experiments, the photocatalyst powder (0.1 g) was supplemented into 250 mL of the aqueous suspensions of the pollutant (RhB: 1.0×10^{-5} M, MB: 1.0×10^{-5} M, MG: 1.0×10^{-5} M, or fuchsin: 1.0×10^{-5} M), and vigorously stirred for 60 min in the absence of light to accomplish the adsorption–desorption equilibrium. Next, the lamp was switched on for a predetermined time and almost 4 mL of the suspension was sampled at fixed time pauses, and photocatalyst particles were separated using a centrifuge (at 4000 rpm). The concentrations of RhB, fuchsin, MG, and MB were monitored by a UV–vis spectrophotometer, respectively, at 553, 540, 610 and 664 nm. To show the stability of the nanocomposite, we performed the recycling tests for four times consecutively. After each photocatalysis experiment, the photocatalyst was removed and utilized in the next test. Furthermore, to identify the reactive species generated during the degradation of RhB, ammonium oxalate (AOX, h^+ quencher, 0.035 g), benzoquinone (BQ, $^\bullet O_2^-$ quencher, 0.027 g), and 2-propanol (2-PrOH, $^\bullet OH$ quencher, 20 μL) were added to the system to capture, respectively, h^+, $^\bullet O_2^-$, and $^\bullet OH$ species.

4. Conclusions

In brief, CNDs/$ZnMn_2O_4$ nanocomposites were synthesized via a facile hydrothermal route. The successful integration of $ZnMn_2O_4$ nanoparticles over the CNDs was confirmed by various analyses. The photodegradation experiments revealed that the prepared binary nanocomposites had higher efficiencies for degradations of the selected pollutants under visible light compared to the efficiencies of $ZnMn_2O_4$, CNDs, and bulk CN photocatalysts. Amongst the binary photocatalysts, the nanocomposite with 20 wt.% of $ZnMn_2O_4$ had admirable activity in the removal of RhB with degradation constant of 747×10^{-4} min^{-1}, mainly resulting from the enhanced specific surface area, significant absorption of visible light, and premiere segregation and transfer of the photoinduced charge carriers through developed p-n heterojunction. Moreover, the activity of the binary sample for degradation of RhB, MG, fuchsin, and MB was 66.5, 15.8, 13.2, and 31.2-fold greater than CN, and 3.44, 1.5, 18.1, and 11.5 times premier than CNDs, respectively. As per radical trapping experiments, the $^\bullet O_2^-$, h^+, and $^\bullet OH$ radicals were recognized as the core reactive species involved in the photocatalytic processes. The outcomes of this work disclosed that the combination of $ZnMn_2O_4$ and CNDs components through p–n heterojunction not only provoked better absorption efficiencies at visible-light region but also improved the lifespan of the photoinduced charges and their movement on the surface of the photocatalyst for promoted degradation of various organic contaminants.

Author Contributions: Z.L.: Validation; Formal analysis; Investigation; Resources; Writing—Original Draft. A.H.-Y.: Conceptualization; Methodology; Writing—Review & Editing; Supervision; Project administration. S.R.P.: Writing—Review & Editing. A.K.: Formal analysis; Review & Editing. All authors have read and agreed to the published version of the manuscript.

Funding: This research was supported by University of Mohaghegh Ardabili.

Data Availability Statement: The data will be available on request.

Acknowledgments: We are grateful to the University of Mohaghegh Ardabili for supporting this work.

Conflicts of Interest: Authors stated that there is no conflict of interest.

References

1. Ramalingam, G.; Perumal, N.; Priya, A.; Rajendran, S. A review of graphene-based semiconductors for photocatalytic degradation of pollutants in wastewater. *Chemosphere* **2022**, *300*, 134391. [CrossRef]
2. Solangi, N.H.; Karri, R.R.; Mazari, S.A.; Mubarak, N.M.; Jatoi, A.S.; Malafaia, G.; Azad, A.K. MXene as emerging material for photocatalytic degradation of environmental pollutants. *Coord. Chem. Rev.* **2023**, *477*, 214965. [CrossRef]
3. He, J.; Han, L.; Wang, F.; Ma, C.; Cai, Y.; Ma, W.; Xu, E.G.; Xing, B.; Yang, Z. Photocatalytic strategy to mitigate microplastic pollution in aquatic environments: Promising catalysts, efficiencies, mechanisms, and ecological risks. *Crit. Rev. Environ. Sci. Technol.* **2022**, *53*, 504–526. [CrossRef]
4. Brillas, E. A critical review on ibuprofen removal from synthetic waters, natural waters, and real wastewaters by advanced oxidation processes. *Chemosphere* **2021**, *286*, 131849. [CrossRef] [PubMed]
5. Ganiyu, S.O.; Sable, S.; El-Din, M.G. Advanced oxidation processes for the degradation of dissolved organics in produced water: A review of process performance, degradation kinetics and pathway. *Chem. Eng. J.* **2021**, *429*, 132492. [CrossRef]
6. Ch-Th, T.; Manisekaran, R.; Santoyo-Salazar, J.; Schoefs, B.; Velumani, S.; Castaneda, H.; Jantrania, A. Graphene oxide decorated TiO_2 and $BiVO_4$ nanocatalysts for enhanced visible-light-driven photocatalytic bacterial inactivation. *J. Photochem. Photobiol. A Chem.* **2021**, *418*, 113374. [CrossRef]
7. Lin, J.; Tian, W.; Zhang, H.; Duan, X.; Sun, H.; Wang, H.; Fang, Y.; Huang, Y.; Wang, S. Carbon nitride-based Z-scheme heterojunctions for solar-driven advanced oxidation processes. *J. Hazard. Mater.* **2022**, *434*, 128866. [CrossRef] [PubMed]
8. Bai, L.; Huang, H.; Yu, S.; Zhang, D.; Huang, H.; Zhang, Y. Role of transition metal oxides in $g-C_3N_4$-based heterojunctions for photocatalysis and supercapacitors. *J. Energy Chem.* **2021**, *64*, 214–235. [CrossRef]
9. Huang, H.; Jiang, L.; Yang, J.; Zhou, S.; Yuan, X.; Liang, J.; Wang, H.; Bu, Y.; Li, H. Synthesis and modification of ultrathin $g-C_3N_4$ for photocatalytic energy and environmental applications. *Renew. Sustain. Energy Rev.* **2023**, *173*, 113110. [CrossRef]
10. Balakrishnan, A.; Chinthala, M. Comprehensive review on advanced reusability of $g-C_3N_4$ based photocatalysts for the removal of organic pollutants. *Chemosphere* **2022**, *297*, 134190. [CrossRef]
11. Guo, R.T.; Wang, J.; Bi, Z.X.; Chen, X.; Hu, X.; Pan, W.G. Recent advances and perspectives of $g-C_3N_4$–based materials for photocatalytic dyes degradation. *Chemosphere* **2022**, *295*, 133834. [CrossRef] [PubMed]
12. Wang, T.; Nie, C.; Ao, Z.; Wang, S.; An, T. Recent progress in $g-C_3N_4$ quantum dots: Synthesis, properties and applications in photocatalytic degradation of organic pollutants. *J. Mater. Chem. A* **2019**, *8*, 485–502. [CrossRef]
13. Ma, P.; Zhang, X.; Wang, C.; Wang, Z.; Wang, K.; Feng, Y.; Wang, J.; Zhai, Y.; Deng, J.; Wang, L.; et al. Band alignment of homojunction by anchoring CN quantum dots on $g-C_3N_4$ (0D/2D) enhance photocatalytic hydrogen peroxide evolution. *Appl. Catal. B Environ.* **2021**, *300*, 120736. [CrossRef]
14. Xing, Y.; Wang, X.; Hao, S.; Zhang, X.; Ma, W.; Zhao, G.; Xu, X. Recent advances in the improvement of $g-C_3N_4$ based photocatalytic materials. *Chin. Chem. Lett.* **2021**, *32*, 13–20. [CrossRef]
15. Huang, R.; Wu, J.; Zhang, M.; Liu, B.; Zheng, Z.; Luo, D. Strategies to enhance photocatalytic activity of graphite carbon nitride-based photocatalysts. *Mater. Des.* **2021**, *210*, 110040. [CrossRef]
16. Zhu, F.; Ma, J.; Ji, Q.; Cheng, H.; Komarneni, S. Visible-light-driven activation of sodium persulfate for accelerating orange II degradation using $ZnMn_2O_4$ photocatalyst. *Chemosphere* **2021**, *278*, 130404. [CrossRef]
17. Al-Buriahi, A.K.; Al-Gheethi, A.A.; Kumar, P.S.; Mohamed, R.M.S.R.; Yusof, H.; Alshalif, A.F.; Khalifa, N.A. Elimination of rhodamine B from textile wastewater using nanoparticle photocatalysts: A review for sustainable approaches. *Chemosphere* **2022**, *287*, 132162. [CrossRef]
18. Li, S.-S.; Liu, M.; Wen, L.; Xu, Z.; Cheng, Y.-H.; Chen, M.-L. Exploration of long afterglow luminescent materials composited with graphitized carbon nitride for photocatalytic degradation of basic fuchsin. *Environ. Sci. Pollut. Res.* **2023**, *30*, 322–336. [CrossRef]
19. Tsvetkov, M.; Zaharieva, J. Milanova, Ferrites, modified with silver nanoparticles, for photocatalytic degradation of malachite green in aqueous solutions. *Catal. Today* **2020**, *357*, 453–459. [CrossRef]
20. Khan, I.; Saeed, K.; Zekker, I.; Zhang, B.; Hendi, A.H.; Ahmad, A.; Ahmad, S.; Zada, N.; Ahmad, H.; Shah, L.A.; et al. Review on Methylene Blue: Its Properties, Uses, Toxicity and Photodegradation. *Water* **2022**, *14*, 242. [CrossRef]

21. Li, Y.; Shu, S.; Huang, L.; Liu, J.; Liu, J.; Yao, J.; Liu, S.; Zhu, M.; Huang, L. Construction of a novel double S-scheme structure $WO_3/g-C_3N_4/BiOI$: Enhanced photocatalytic performance for antibacterial activity. *J. Colloid Interface Sci.* **2023**, *633*, 60–71. [CrossRef] [PubMed]
22. Jing, Y.; Chen, Z.; Ding, E.; Yuan, R.; Liu, B.; Xu, B.; Zhang, P. High-yield production of $g-C_3N_4$ quantum dots as photocatalysts for the degradation of organic pollutants and fluorescent probes for detection of Fe^{3+} ions with live cell application. *Appl. Surf. Sci.* **2022**, *586*, 152812. [CrossRef]
23. Basaleh, A.S.; Mohamed, R.M. Influence of doped silver nanoparticles on the photocatalytic performance of $ZnMn_2O_4$ in the production of methanol from CO_2 photocatalytic reduction. *Appl. Nanosci.* **2020**, *10*, 3865–3874. [CrossRef]
24. Wei, Y.; Li, X.; Zhang, Y.; Yan, Y.; Huo, P.; Wang, H. $G-C_3N_4$ quantum dots and Au nano particles co-modified CeO_2/Fe_3O_4 micro-flowers photocatalyst for enhanced CO_2 photoreduction. *Renew. Energy* **2021**, *179*, 756–765. [CrossRef]
25. Preeyanghaa, M.; Vinesh, V.; Sabarikirishwaran, P.; Rajkamal, A.; Ashokkumar, M.; Neppolian, B. Investigating the role of ultrasound in improving the photocatalytic ability of CQD decorated boron-doped $g-C_3N_4$ for tetracycline degradation and first-principles study of nitrogen-vacancy formation. *Carbon* **2022**, *192*, 405–417. [CrossRef]
26. He, L.; Fei, M.; Chen, J.; Tian, Y.; Jiang, Y.; Huang, Y.; Xu, K.; Hu, J.; Zhao, Z.; Zhang, Q.; et al. Graphitic C_3N_4 quantum dots for next-generation QLED displays. *Mater. Today* **2018**, *22*, 76–84. [CrossRef]
27. Xu, C.; Li, D.; Liu, X.; Ma, R.; Sakai, N.; Yang, Y.; Lin, S.; Yang, J.; Pan, H.; Huang, J.; et al. Direct Z-scheme construction of $g-C_3N_4$ quantum dots/TiO_2 nanoflakes for efficient photocatalysis. *Chem. Eng. J.* **2021**, *430*, 132861. [CrossRef]
28. Yan, S.; Yanlong, Y.; Cao, Y. Synthesis of porous $ZnMn_2O_4$ flower-like microspheres by using MOF as precursors and its application on photoreduction of CO_2 into CO. *Appl. Surf. Sci.* **2019**, *465*, 383–388. [CrossRef]
29. Olusegun, S.J.; Larrea, G.; Osial, M.; Jackowska, K.; Krysinski, P. Photocatalytic Degradation of Antibiotics by Superparamagnetic Iron Oxide Nanoparticles. Tetracycline Case. *Catalysts* **2021**, *11*, 1243. [CrossRef]
30. Papadas, I.T.; Galatopoulos, F.; Armatas, G.S.; Tessler, N.; Choulis, S.A. Nanoparticulate Metal Oxide Top Electrode Interface Modification Improves the Thermal Stability of Inverted Perovskite Photovoltaics. *Nanomaterials* **2019**, *9*, 1616. [CrossRef]
31. Alhaddad, M.; Mohamed, R.M. Synthesis and characterizations of ZnMn2O4-ZnO nanocomposite photocatalyst for enlarged photocatalytic oxidation of ciprofloxacin using visible light irradiation. *Appl. Nanosci.* **2020**, *10*, 2269–2278. [CrossRef]
32. Wu, X.; Xiang, Y.; Peng, Q.; Wu, X.; Li, Y.; Tang, F.; Song, R.; Liu, Z.; He, Z.; Wu, X. Green-low-cost rechargeable aqueous zinc-ion batteries using hollow porous spinel $ZnMn_2O_4$ as the cathode material. *J. Mater. Chem. A* **2017**, *5*, 17990–17997. [CrossRef]
33. Ni, Q.; Cheng, H.; Ma, J.; Kong, Y.; Komarneni, S. Efficient degradation of orange II by ZnMn2O4 in a novel photo-chemical catalysis system. *Front. Chem. Sci. Eng.* **2020**, *14*, 956–966. [CrossRef]
34. Iurilli, P.; Brivio, C.; Wood, V. On the use of electrochemical impedance spectroscopy to characterize and model the aging phenomena of lithium-ion batteries: A critical review. *J. Power Sources* **2021**, *505*, 229860. [CrossRef]
35. Schlumberger, C.; Thommes, M. Characterization of Hierarchically Ordered Porous Materials by Physisorption and Mercury Porosimetry—A Tutorial Review. *Adv. Mater. Interfaces* **2021**, *8*, 2002181. [CrossRef]
36. Deng, H.; Fei, X.; Yang, Y.; Fan, J.; Yu, J.; Cheng, B.; Zhang, L. S-scheme heterojunction based on p-type $ZnMn_2O_4$ and n-type ZnO with improved photocatalytic CO_2 reduction activity. *Chem. Eng. J.* **2020**, *409*, 127377. [CrossRef]
37. Wang, A.; Guo, S.; Zheng, Z.; Wang, H.; Song, X.; Zhu, H.; Zeng, Y.; Lam, J.; Qiu, R.; Yan, K. Highly dispersed Ag and $g-C_3N_4$ quantum dots co-decorated 3D hierarchical Fe_3O_4 hollow microspheres for solar-light-driven pharmaceutical pollutants degradation in natural water matrix. *J. Hazard. Mater.* **2022**, *434*, 128905. [CrossRef]
38. Khan, I.; Saeed, K.; Ali, N.; Khan, I.; Zhang, B.; Sadiq, M. Heterogeneous photodegradation of industrial dyes: An insight to different mechanisms and rate affecting parameters. *J. Environ. Chem. Eng.* **2020**, *8*, 104364. [CrossRef]
39. Ahmad, S.; Almehmadi, M.; Janjuhah, H.T.; Kontakiotis, G.; Abdulaziz, O.; Saeed, K.; Ahmad, H.; Allahyani, M.; Aljuaid, A.; Alsaiari, A.A.; et al. The Effect of Mineral Ions Present in Tap Water on Photodegradation of Organic Pollutants: Future Perspectives. *Water* **2023**, *15*, 175. [CrossRef]
40. Yin, Y.; Liu, M.; Shi, L.; Zhang, S.; Hirani, R.A.K.; Zhu, C.; Chen, C.; Yuan, A.; Duan, X.; Wang, S.; et al. Highly dispersive Ru confined in porous ultrathin $g-C_3N_4$ nanosheets as an efficient peroxymonosulfate activator for removal of organic pollutants. *J. Hazard. Mater.* **2022**, *435*, 128939. [CrossRef]

Disclaimer/Publisher's Note: The statements, opinions and data contained in all publications are solely those of the individual author(s) and contributor(s) and not of MDPI and/or the editor(s). MDPI and/or the editor(s) disclaim responsibility for any injury to people or property resulting from any ideas, methods, instructions or products referred to in the content.

Article

Synthesis of Nanocrystalline Metal Tungstate NiWO₄/CoWO₄ Heterojunction for UV-Light-Assisted Degradation of Paracetamol

Fahad Ahmed Alharthi *, Alanoud Abdullah Alshayiqi, Wedyan Saud Al-Nafaei, Adel El Marghany, Hamdah Saleh Alanazi and Imran Hasan *

Department of Chemistry, College of Science, King Saud University, Riyadh 11451, Saudi Arabia
* Correspondence: fharthi@ksu.edu.sa (F.A.A.); iabdulateef@ksu.edu.sa (I.H.)

Abstract: The discharge of pharma products such as paracetamol (PCT) into water has resulted in great harm to humans and emerged as a potential threat requiring a solution. Therefore, the development of smart and efficient materials as photocatalysts has become imperative in order to treat PCT in wastewater. The present study demonstrates the synthesis of pristine NiWO$_4$ and CoWO$_4$ and a heterojunction nanostructure, NiWO$_4$/CoWO$_4$, through a hydrothermal process using a Teflon-lined autoclave at 180 °C for 18 h. Various spectroscopic techniques, such as X-ray diffraction (XRD), Fourier transform infrared (FTIR), ultraviolet–visible (UV–Vis), transmission electron microscopy (TEM), scanning electron microscopy–energy dispersive X-ray (SEM–EDX), and X-ray photoelectron spectroscopy (XPS) were utilised to determine the lattice, structural, optical, and morphological information of the solid nanomaterial upon heterojunction formation. The synthesised nanomaterials were exploited for the photocatalytic degradation of paracetamol (PCT) under UV light irradiation. Photocatalytic experiments were performed for the optimization of various reaction parameters, such as irradiation time, pH, catalyst dose, and PCT concentration at room temperature. The results obtained suggested that the heterojunction nanocomposite NiWO$_4$/CoWO$_4$ exhibited enhanced photocatalytic efficiency (97.42%) with PCT as compared to its precursors—96.50% for NiWO$_4$ and 97.12% for CoWO$_4$. The photocatalytic data were best defined by the Langmuir–Hinshelwood (L–H) model of pseudo-first-order kinetics, with apparent rates constant at 0.015 min^{-1} for NiWO$_4$, 0.017 min^{-1} for CoWO$_4$, and 0.019 min^{-1} for NiWO$_4$/CoWO$_4$ NC. It was observed that NiWO$_4$/CoWO$_4$ NC with enhanced optical properties effected a higher rate of PCT degradation due to the improved bandgap energy upon heterojunction formation. The scavenger test revealed the involvement of •OH radicals as reactive oxidant species (ROS) in PCT degradation. The material was found to be highly stable and reusable for the degradation of PCT at optimized reaction conditions.

Keywords: heterojunction; nanocomposites; tungstate; photocatalytic degradation; pharmaceuticals

Citation: Alharthi, F.A.; Alshayiqi, A.A.; Al-Nafaei, W.S.; El Marghany, A.; Alanazi, H.S.; Hasan, I. Synthesis of Nanocrystalline Metal Tungstate NiWO₄/CoWO₄ Heterojunction for UV-Light-Assisted Degradation of Paracetamol. *Catalysts* **2023**, *13*, 152. https://doi.org/10.3390/catal13010152

Academic Editor: Meng Li

Received: 20 December 2022
Revised: 3 January 2023
Accepted: 5 January 2023
Published: 9 January 2023

Copyright: © 2023 by the authors. Licensee MDPI, Basel, Switzerland. This article is an open access article distributed under the terms and conditions of the Creative Commons Attribution (CC BY) license (https:// creativecommons.org/licenses/by/ 4.0/).

1. Introduction

In recent years, pharmaceutical products, personal care products (PPCs), and pesticides in water bodies have been categorized as contaminants of emerging concern (CECs) [1]. Among these emerging pollutants, pharma products have begun to gain researchers' attention owing to their toxic influence on human health [2]. One of these pharma products, paracetamol (PCT), also known as acetaminophen, is widely used in analgesic and antipyretic drugs for treating headache, fever, etc. [3]. The high consumption of the drug during COVID-19 has resulted in a harmful impact on ecology and human health through its presence in wastewater effluents [4]. The main sources of PCT inclusion into the aquatic system are the pharmaceutical industry and the excretory waste of both humans and animals who have received medical treatment [4,5]. With each successive year, PCT concentrations in lakes and rivers have significantly increased, which can cause diseases such as liver failure, hepatic necrosis, nephrite toxicity, and possibly death [6,7].

Therefore, to address this issue, there is a need to develop efficient methods and technology to treat PCT from wastewater streams.

There are numerous methods reported in the literature for removing these pollutants from the environment, such as electrochemical separation, liquid extraction, chemical oxidation, membrane separation, biodegradation, and adsorption technology [8–10]. However, sludge formation, slow procedures, and mass transfer properties from one phase to another make these methods less efficient. Therefore, an advanced oxidation process (AOP) in association with photocatalytic degradation was taken into consideration for the treatment of these PPCPs. Due to its lack of toxicity, affordability, lack of mass transfer limitations, chemical stability, and potential operation at room temperature, photocatalytic oxidation has emerged as one of the most efficient and innovative technologies to trigger pharma pollutants without causing secondary waste to enter the environment [11]. The technology utilizes the irradiation of a catalyst material using solar energy, and depending on its optical properties, generates reactive oxidant species (ROS), hydroxy ($^{\bullet}OH$), or superoxide ($^{\bullet}O_2^-$) radicals, which are responsible for mineralization process [12].

Based on their high aspect ratio, smaller size, and good optical properties, nanomaterials have been identified as the most effective and potentially useful materials to address this sustainability issue, offering some fundamental and practical approaches for wastewater treatment operations [13]. Among these nanomaterials, semiconductor metal oxides, such as zirconia (ZrO_2), zinc oxide (ZnO), titanium dioxide (TiO_2), iron (III) oxide (Fe_2O_3), tungsten trioxide (WO_3), and vanadium (V) oxide (V_2O_5), etc., have been proved to be promising photocatalysts for wastewater treatment processes due to their excellent physicochemical properties [8,14–17]. The nanostructured metal tungstates with the empirical formula MWO_4 (M = Co, Zn, Sn, Ni, Mn etc.) have been recognized as materials with advanced photocatalytic activity, chemical stability, low cost, and non-toxicity [18]. They are associated with a wolframite monoclinic structure and appropriate band gap energy, which classifies them as exemplary materials for scientific and engineering applications such as conventional catalysis, scintillator materials, photoluminescence, optical fibers, microwave technology, supercapacitors, and semiconductors [18,19]. $NiWO_4$ is one of the members of the tungstate family with intriguing structural properties, large surface area, and photoluminescent characteristics, with an energy bandgap value of 2.97 eV [20]. However, rapid recombination of electron–hole pairs in pure substances restricts their photocatalytic activities, and to address this issue, the strategy of heterojunction formation was taken into consideration [21]. The heterojunction formation results in spatial separation of photogenerated electron–hole pairs, and thus enhances the photocatalytic activities [22]. In the present study, $CoWO_4$ was used for heterojunction formation with $NiWO_4$ through the hydrothermal process. $CoWO4$, with an energy bandgap value of 2.8 eV, exhibits excellent photocatalytic activities [23]. The resultant nanostructured $NiWO_4/CoWO_4$ heterojunction was used as a catalyst for photocatalytic degradation of paracetamol under mercury lamp (Ultraviolet radiation) in photocatalytic reactor. Various reaction factors influencing the degradation process, such as PCT concentration, irradiation time, pH, and amount of catalyst, were optimized. The novelty of this work lies in the use of mixed metal tungstate heterojunction ($MWO_4/M'WO_4$) nanomaterials in photocatalytic degradation processes to remove pharmaceuticals from wastewater.

2. Results and Discussion
2.1. Material Characterization
2.1.1. FTIR Analysis

The FTIR spectrum of the prepared metals and mixed metal tungstate catalysts is shown in Figure 1. In the FTIR spectra of Figure 1a,b, 466 and 456 cm^{-1} belong to Ni–O and Co–O bonds, 537 and 523 cm^{-1} belong to Ni–O–W and Co–O–W bonds, 618 belongs to the W–O bonds, 700–880 cm^{-1} belongs to the W–O stretching mode of the WO_4 tetrahedron, W–O–W bond vibrational modes [24–26]. The –OH group stretching and bending vibrations as a result of moisture adsorption appeared at 3427–3446 cm^{-1} and

1635–1640 cm^{-1} [27]. The FTIR spectra of heterojunction NiWO$_4$/CoWO$_4$ NC Figure 1c exhibit most of the peaks from NiWO$_4$ and CoWO$_4$, but with some shifted values, which suggest that in the wolframite monoclinic structure, both Co and Ni took the lattice position in the solid structure.

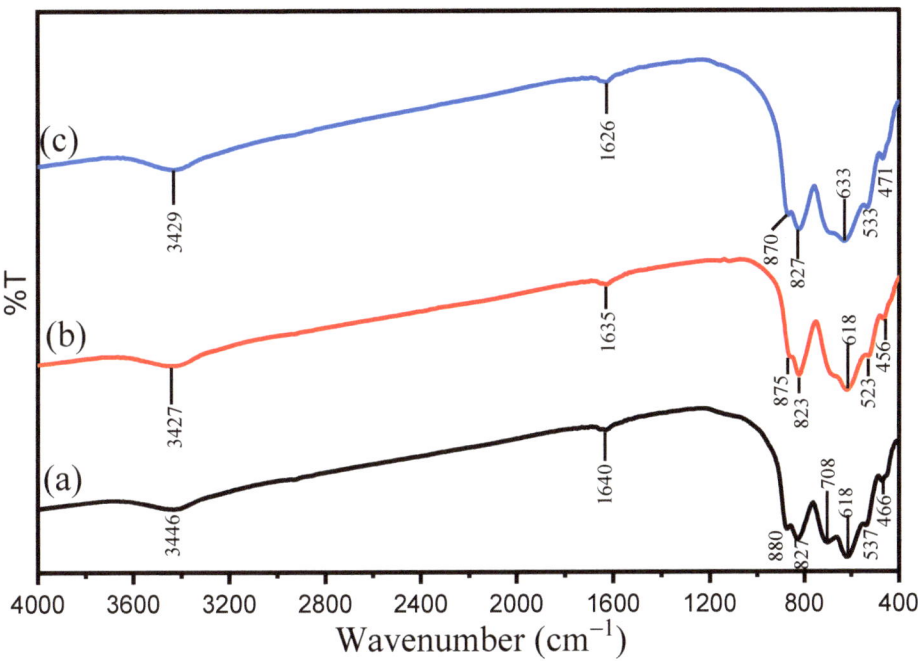

Figure 1. FTIR spectra of (**a**) NiWO$_4$ (**b**) CoWO$_4$ and (**c**) mix metal tungstate NiWO$_4$/CoWO$_4$ NC.

2.1.2. X-ray Diffraction (XRD)

The crystal structures of the prepared metals and mixed metal tungstate catalysts were identified by X-ray diffraction (XRD) patterns. Figure 2 shows the XRD pattern of the catalysts synthesized through a typical reaction. The XRD spectra of CoWO$_4$ show characteristic peaks at 2θ value of 2.37°, 24.91°, 24.91°, 30.75°, 36.78°, 37.93°, 48.66°, 54.53°, 61.91°, and 65.21°, which belong to miller indices (100), (011), (110), (020), (002), (200), (022), (031), (310), and (040) respectively. The XRD spectra of NiWO$_4$ show characteristic peaks at 2θ value of 6.51°, 19.35°, 24.03°, 24.99°, 31.02°, 36.65°, 39.24°, 41.76°, 44.84°, 46.52°, 49.09°, 52.42°, 54.74°, 62.41°, 65.82°, and 69.10°, which belong to miller indices (010), (100), (011), (110), (111), (021), (200), (121), (112), (211), (022), (220), (130), (202), (113), (311), and (041) respectively. All reflection peaks can be indexed as the monoclinic crystal systems of CoWO$_4$ (JCPDS Card No. (96–317)) and NiWO$_4$ (JCPDS Card No. (96–278)). Finally, the XRD pattern of NiWO$_4$/CoWO$_4$ NC shows peaks ascribed to the NiWO$_4$ at 16.91° (010), 30.63° (111), 41.37° (121), 44.43° (112), 46.02° (211), 52.10° (202), 65.32° (311), and 68.82° (041), and peaks ascribed to the CoWO$_4$ at 30.81° (020), 36.32° (002), 54.25° (031), and 61.95° (310), respectively. Moreover, some peaks ascribed to both NiWO$_4$ and CoWO$_4$ show at 18.97° (100), 24.62° (011), 24.62° (110), 38.72° (200), and 48.78° (022), which suggests the successful formation of NiWO$_4$/CoWO$_4$ heterojunctions. For further information about the crystallite size and dislocation density, the Scherrer's equation was taken into consideration. The crystallite size and interlayer spacing were calculated using the Scherrer's Equation:

$$D = \frac{0.9\lambda}{\beta \cos \theta} \quad (1)$$

$$\text{Dislocation Density }(\delta) = \frac{1}{D^2} \quad (2)$$

$$\text{Interlayer Spacing }(d_{111}) = \frac{n\lambda}{2\sin\theta} \quad (3)$$

$$\%\text{Crystallinity} = \frac{\text{Area Under Crystalline Peaks}}{\text{Total Area}} \times 100 \quad (4)$$

where D is the crystallite size, λ is the characteristic wavelength of the X-ray, β represents angular width in radian at intensity equal to half of its maximum of the peak, and θ is the diffraction angle. The average particle size of $CoWO_4$, $NiWO_4$, and $NiWO_4/CoWO_4$ NC were 27.16, 22.16, and 19.07 nm, respectively, calculated using Equation (2). Moreover, from Table 1, it can be seen that the crystallinity of the heterojunction nanocomposite increased to 38% after heterojunction formation, suggesting a successful occupation of lattice sites by Co in the $NIWO_4$ solid structure without producing any impurity. The process reflected a contraction in particle size and an increase in dislocation density due to deformation in the solid lattice of $CoWO_4$ and $NiWO_4$.

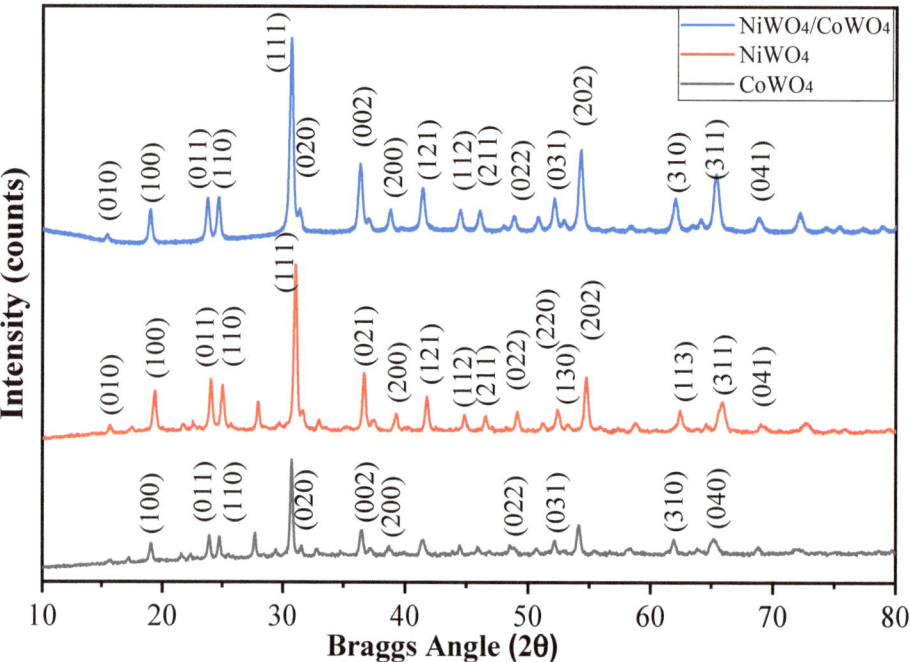

Figure 2. XRD pattern of ($CoWO_4$, $NiWO_4$, $NiWO_4/CoWO_4$ NC) nanomaterials.

Table 1. XRD parameter of $CoWO_4$, $NiWO_4$, and $NiWO_4/CoWO_4$ NC.

Component	2θ	FWHM (β_{hkl})	Interlayer Spacing (nm) at 2θ	Crystallite Size (nm) at 2θ	Dislocation Density (δ)×10^{15} Lines (m^{-2})	Crystallinity (%)
$CoWO_4$	30.748	0.26203	0.020414	27.16	1.35	23.60
$NiWO_4$	31.015	0.32392	0.020587	22.16	2.03	23.25
$NiWO_4/CoWO_4$ NC	30.64	0.36296	0.020344	19.07	2.75	38.14

2.1.3. Scanning Electron Microscopy (SEM)

The surface morphological changes in the material during the solid-state processes were observed using scanning electron microscopy (SEM). The SEM images of the sintered NiWO$_4$ taken at 10,000× magnification is shown in Figure 3a. The sample's morphology is porous. The complete coverage of the substrate is evident in the micrographs. These surfaces evince uneven surface morphology and irregular geometry quite clearly, while in the SEM image of CoWO$_4$ shown in the Figure 3b, the grain arrangements and sizes are irregular. Figure 3c presents a SEM image of NiWO$_4$/CoWO$_4$ NC, which shows large, irregularly sized grains. The EDX spectrum recorded for the sample, which is depicted in Figure 3d, was used to explore the elements present in the sample. The presence of the components Co, W, Ni, and O in the mixed metal tungstate NiWO$_4$/CoWO$_4$ NC was confirmed by the EDX analysis.

Figure 3. SEM images of the metal tungstate (**a**) NiWO$_4$; (**b**) CoWO$_4$, mix metal tungstate (**c**) NiWO$_4$/CoWO$_4$ NC; (**d**) EDX of NiWO$_4$/CoWO$_4$ NC.

2.1.4. Transmission Electron Microscopy (TEM)

The optimal diameter of the mix metal tungstate was determined using transmission electron microscopy (TEM), and further investigation was carried out to establish the crystallinity and morphology of the metal's tungstate. Figure 4a–c presents the TEM images of NiWO$_4$, CoWO$_4$, and NiWO$_4$/CoWO$_4$ NC synthesized nanoparticles, respectively, in which NiWO$_4$ shows rod-like morphology and CoWO$_4$ shows a mix of rod- and irregular-shaped particles, while after heterojunction, a new morphology appears for NiWO$_4$/CoWO$_4$ NC constituted of irregular-shaped particles with mitigated monoclinic structure (Figure 4d). The statistical gaussian distribution profile applied for average particle size distribution suggested an average of 20 nm particle size for NiWO$_4$/CoWO$_4$ NC shown in the Figure 4e, which has been found to be in close agreement with the Scherer formula.

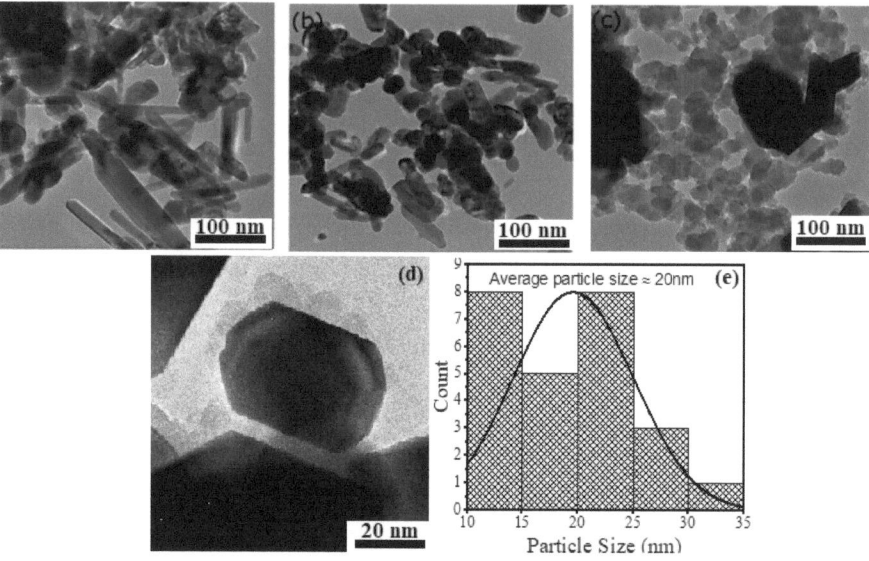

Figure 4. TEM images of the metal tungstates. (**a**) NiWO$_4$; (**b**) CoWO$_4$; (**c**) NiWO$_4$/CoWO$_4$ NC; (**d**) particle size distribution of NiWO$_4$/CoWO$_4$ NC; (**e**) statistical gaussian distribution profile for average particle size distribution.

2.1.5. UV/Vis Spectrophotometer

The produced catalyst's optical characteristics can be determined by UV–Vis spectroscopic studies, and the spectra are shown in Figure 5. The UV–vis spectrum has a broad line, which denotes absorption throughout the whole UV–Vis region. The existence of numerous consecutive energy levels in the photocatalyst was attributable to the broad absorption spectrum shown by Figure 5, in which the absorption maxima (λ_{max}) of NiWO$_4$, CoWO$_4$, and NiWO$_4$/CoWO$_4$ NC were observed around 263, 331, and 360 nm, respectively. The UV spectrum of NiWO$_4$ and CoWO$_4$ suggested the synthesized NPs were UV light-active only, while a broad spectrum from 200–650 nm in the case of NiWO$_4$/CoWO$_4$ NC suggested UV as well as visible light activity of the heterojunction. Thus, the formation of heterojunction results in enhancement of the light-absorption capacity of the material [28]. Tauc's relationship was used to determine the band gap using the absorption data:

$$(\alpha h \nu) = A(h\nu - E_g)^n \tag{5}$$

where α = absorption coefficient, h = plank constant, ν = frequency of radiations, A = constant, and n is a constant of transition variations depending on the type of electronic transition, i.e., n = 1/2 for direct transitions and n = 2 for the indirect transitions. Energy bandgap values were taken from the slope of a linear plot between $(\alpha h \nu)^2$ and E_g; the intercept give rise to the value of energy band gap. Tauc's plot specified the value of E_g as 3.04, 2.32, and 1.14 eV for the synthesized NiWO$_4$, CoWO$_4$, and NiWO$_4$/CoWO$_4$ NC, respectively. Thus, it is also quantitively proved that heterojunction formation results in contraction of the energy bandgap, and thus restricts the electron–hole pair recombination rate.

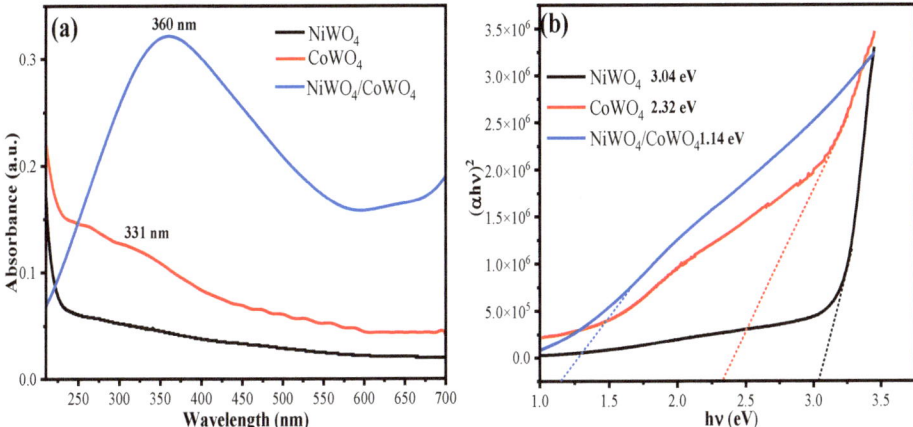

Figure 5. (a) UV–Vis plot of NiWO$_4$, CoWO$_4$, NiWO$_4$/CoWO$_4$ NC in wavelength range of 200–700 nm; (b) Tauc's plot for the band gap energy (E$_g$) of the materials.

To further investigate the chemical oxidation state of the elements in the synthesized NiWO$_4$/CoWO$_4$ NC material, XPS analysis was taken into consideration. Figure 6a represents the XPS survey spectra of NiWO$_4$/CoWO$_4$ NC, which confirm the presence of W4f, O1s, Ni2p, and Co2p in the material. In order to observe the change in the oxidation states of elements during heterojunction formation, the deconvoluted spectres were studied. Figure 6b represents the deconvoluted XPS spectra of W4f, which show two peaks at 46.48 eV and 49.49 eV, corresponding to the W4f$_{7/2}$ and W4f$_{5/2}$ states associated with W^{6+} chemical state [29]. Figure 6c represents three deconvoluted peaks for O1s, corresponding to 538.44, 539.44, and 541.44 eV, associated with O1 (Co–O), O2 (W–O), and O3 (Ni–O) bonds [30,31]. The XPS spectra in Figure 6d for Co2P show two deconvoluted peaks at 868.01 and 872.78 eV associated with two states Co2p$_{3/2}$ and Co2p$_{1/2}$ states of Co^{2+}, respectively [32]. Similarly, Figure 6e, corresponding to the XPS spectra of Ni2p, represents the two states Ni2p$_{3/2}$ and Ni2p$_{1/2}$ for the Ni^{2+} chemical state at 884.94 eV and 890.87 eV [33]. The outcomes of the XPS studies clearly confirm the formation of heterojunction NiWO$_4$/CoWO$_4$ NC by the validation of the chemical stability of the Ni^{2+} and Co^{2+} ions in the lattice.

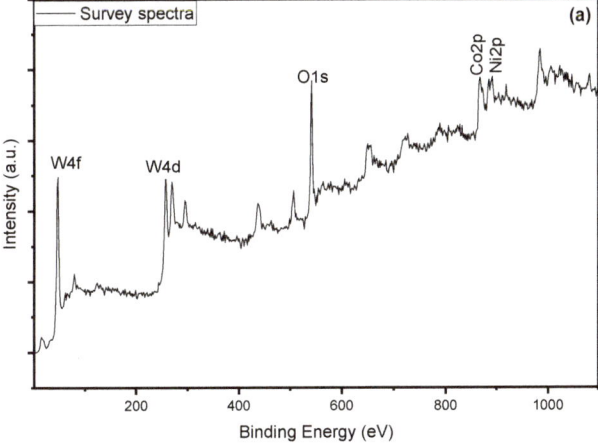

Figure 6. Cont.

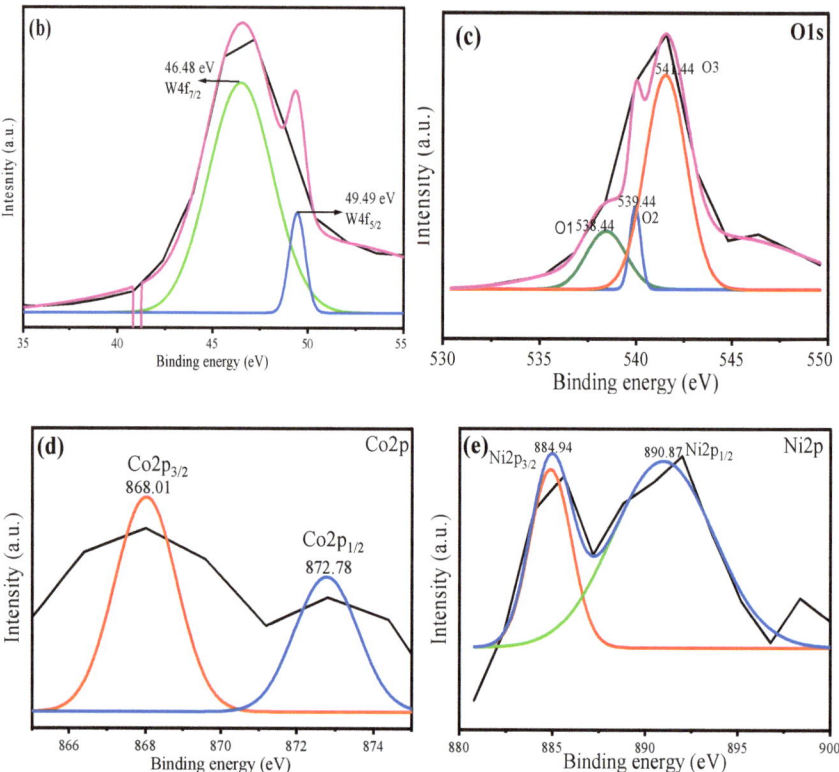

Figure 6. (a) XPS survey spectra of $NiWO_4/CoWO_4$ NC deconvoluted XPS spectra for (b) W4f, (c) O1s, (d) Co2p, and (e) Ni2P.

2.2. Photocatalysis and Optimization of Reaction Parameters

2.2.1. Photocatalysis with Variable Paracetamol Concentration

The photocatalytic experiments were performed at different PCT concentrations in the range of 5–60 ppm to observe the effect of PCT concentration on degradation rates using $NiWO_4$, $CoWO_4$, and $NiWO_4/CoWO_4$ NC. The results obtained are shown in Figure 7a–d, and suggest that over a span of 120 min of irradiation, the degradation efficiencies of the synthesized nanomaterials decreased with increasing paracetamol concentration. These findings were in close agreement with a number of studies, where the activation of the reaction by photon absorption was often the initial step [34]. Another explanation for the result shown in Figure 7 is that the formation of radicals and holes under irradiation remained consistent and sufficient for the breakdown of paracetamol at low concentrations, while at high concentrations, screening hindered their path towards the catalyst surface, which led to a decrease in the paracetamol degradation rate [35]. Acceptable degradation rates were obtained up to 50 ppm PCT concentration, which was used in further photocatalytic experiments.

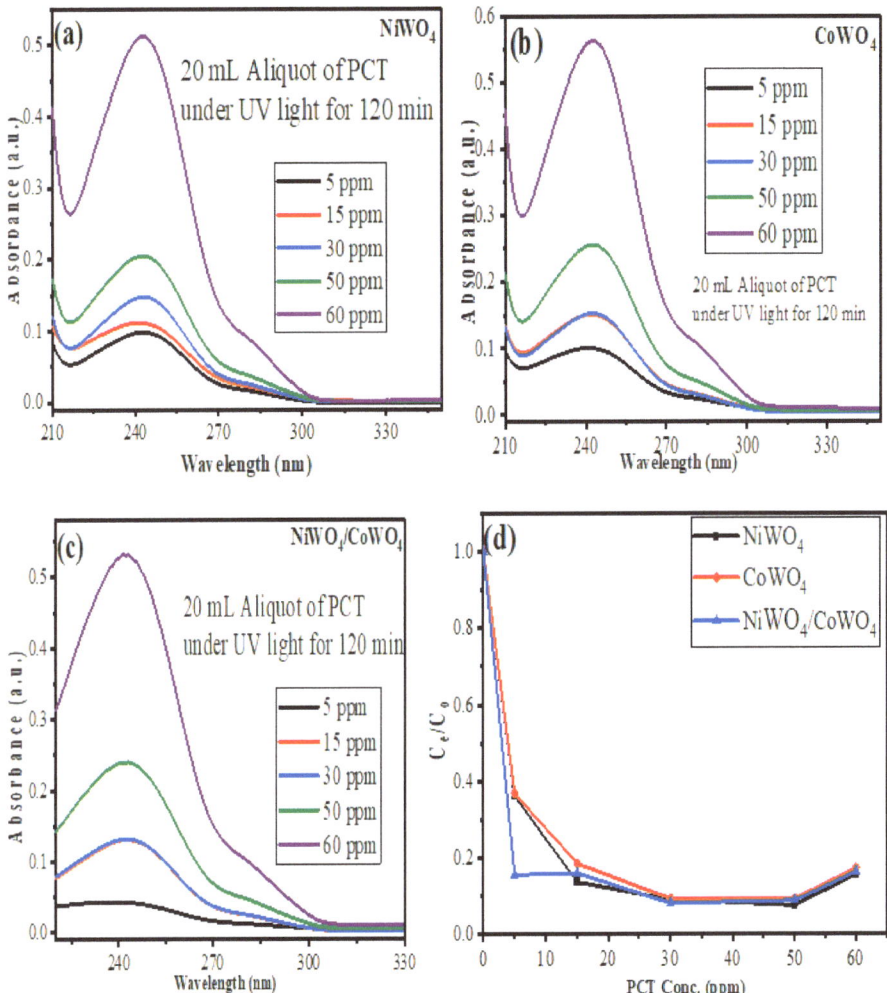

Figure 7. Effect of variable PCT concentration (5–60 ppm) using 5 mg prepared (**a**) NiWO$_4$, (**b**) CoWO$_4$, (**c**) NiWO$_4$/CoWO$_4$ NC, and (**d**) C$_e$/C$_o$ vs. PCT conc. Plot.

2.2.2. Photocatalysis with Variable Catalyst Dosage

The photodegradation of paracetamol under irradiation is shown in Figure 8a–d as a function of catalyst dosage at 5, 10, 15, and 20 mg. Additionally, the other reaction conditions remained the same. As observed in Figure 8, the results showed that the degradation rate decreased by 15–20 mg with the addition of more catalyst. The number of active sites that could absorb additional photons rose as the catalyst dosage was raised. However, the increase in catalyst dose appeared to result in an excess dosage, which produced suspension turbidity in the solution. The efficiency of the photocatalytic process also declined as a result of the lower light penetration [36,37]. After 2 h of irradiation with the 10 mg catalyst dosage was taken as optimum for further experiments.

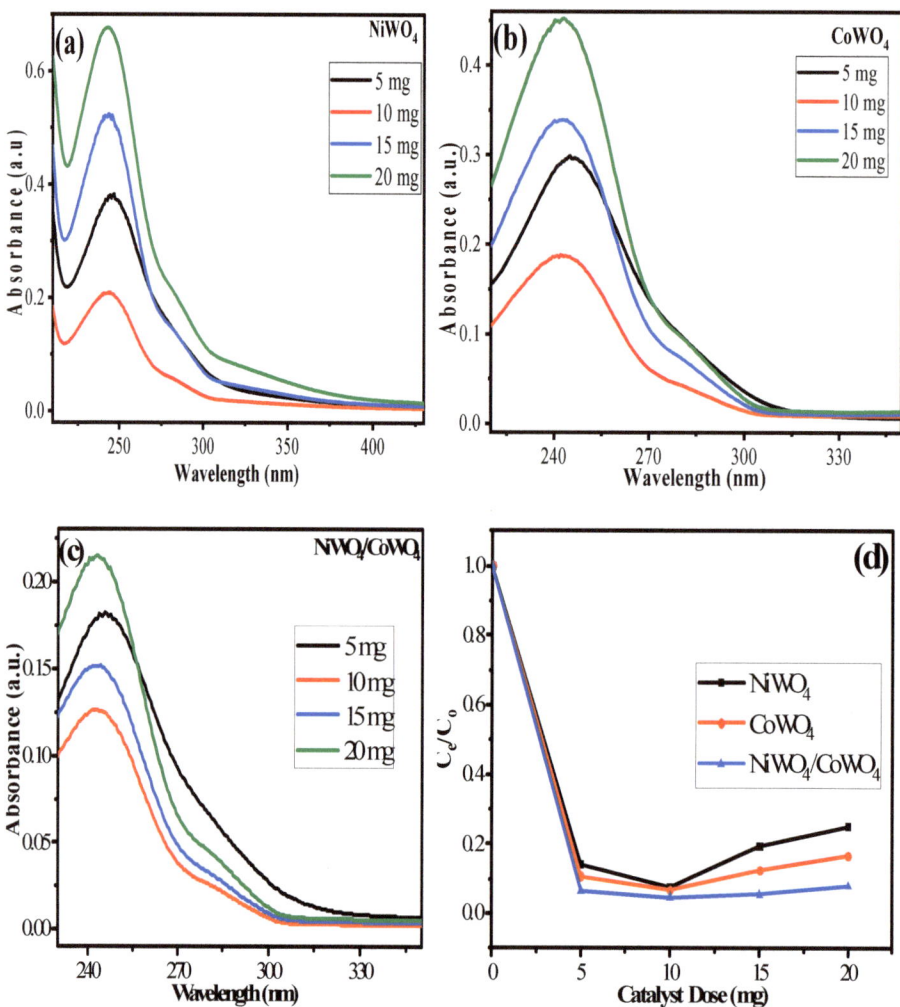

Figure 8. Effect of variable catalyst dose (5–20 mg) using 50 ppm PCT for (**a**) NiWO$_4$, (**b**) CoWO$_4$, (**c**) NiWO$_4$/CoWO$_4$ NC, and (**d**) C$_e$/C$_o$ vs. catalyst dose plot.

2.2.3. Photocatalysis with Variable Paracetamol Solution pH

The pH is an important parameter in photochemical reactions. The results, which are presented in Figure 9, demonstrate how the effects of pH on the paracetamol degradation were investigated at various pH ranges from 3 to 10 in order to determine the ideal pH for the degradation process. This study included three catalysts (NiWO$_4$, CoWO$_4$, and NiWO$_4$/CoWO$_4$ NC), which were studied separately. The particle surface was positively charged at pH levels below 5, and negatively charged at pH levels above 5. Additionally, paracetamol has a pKa value of 9.38, making it a weak base. In light of this, the adsorption–desorption characteristics of the catalyst surface are significantly influenced by the pH value [34]. The results presented in Figure 8 show that the degradation rate rose as the pH value rose. This can be ascribed to increased hydroxide ion creation, because at high pH, more hydroxide ions that are already present on the surface can readily be oxidized to create new hydroxyl radicals, which in turn boosts the efficiency of paracetamol degradation [38]. Finally, in this study, we registered that the degradation was maximal at pH 9.

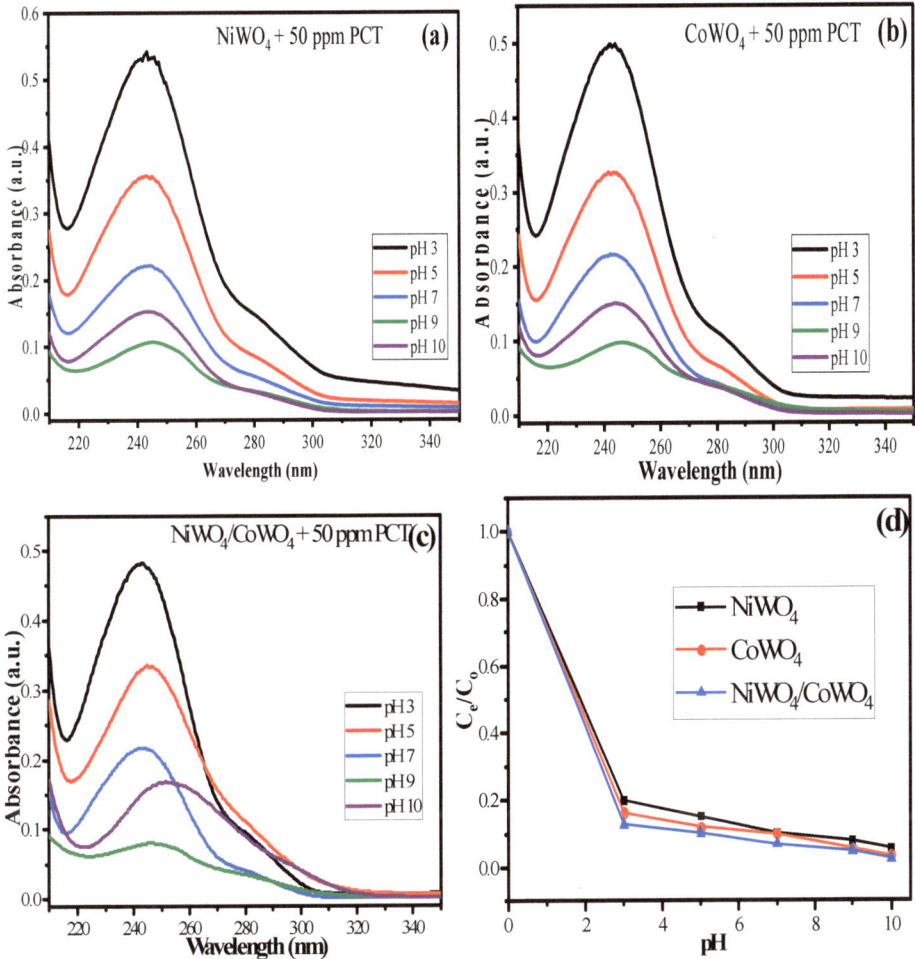

Figure 9. Effect of variable pH of the media (3–10) using 50 ppm PCT for 10 mg of (**a**) NiWO$_4$, (**b**) CoWO$_4$, (**c**) NiWO$_4$/CoWO$_4$ NC, and (**d**) C$_e$/C$_o$ vs. pH of the medium plot.

2.2.4. Photocatalysis with Variable Irradiation Time and Kinetics of the Reaction

Photocatalytic experiments were carried out using 50 ppm paracetamol concentration, pH 9, and 10 mg of catalyst under variable irradiation time from 10 to 120 min. The results of the experiments given in Figure 10a–c show that the photocatalytic degradation of the paracetamol similarly increased with a steady increase in the irradiation time from 10 to 120 min. This trend was caused by the valence electron of a catalyst being excited from its ground state to its excited state, which produced a photoelectron by the absorption of radiation. Since there was a high density of –OH groups on the catalyst surface, when these high energy electrons coupled with them, they produced •OH radicals, which were what caused the paracetamol to photodegrade [39]. The findings in Figure 10 show that as the irradiation period increased, the degradation rate also decreased continually, and typically followed the Langmuir–Hinshelwood first-order kinetic pattern, which assumes the decomposition of the pollutant molecule at the catalyst surface to be the determining step in heterogeneous catalysis processes. The kinetic constant linked to the decomposition and the adsorbate concentration are therefore considered to equal the product of the first-order reaction rate and the adsorbate concentration [40].

$$r = \frac{dC}{dt} = -\frac{kKC}{1+KC} \tag{6}$$

where k is the rate constant, which is affected by light intensity, K is the catalyst adsorption constant, and C is the paracetamol concentration. (KC < 1) applies to modest low adsorption magnitudes and concentrations. Equation (7) states that Equation (6) simplifies to the first-order kinetics.

$$\frac{dC}{dt} = -kKC \tag{7}$$

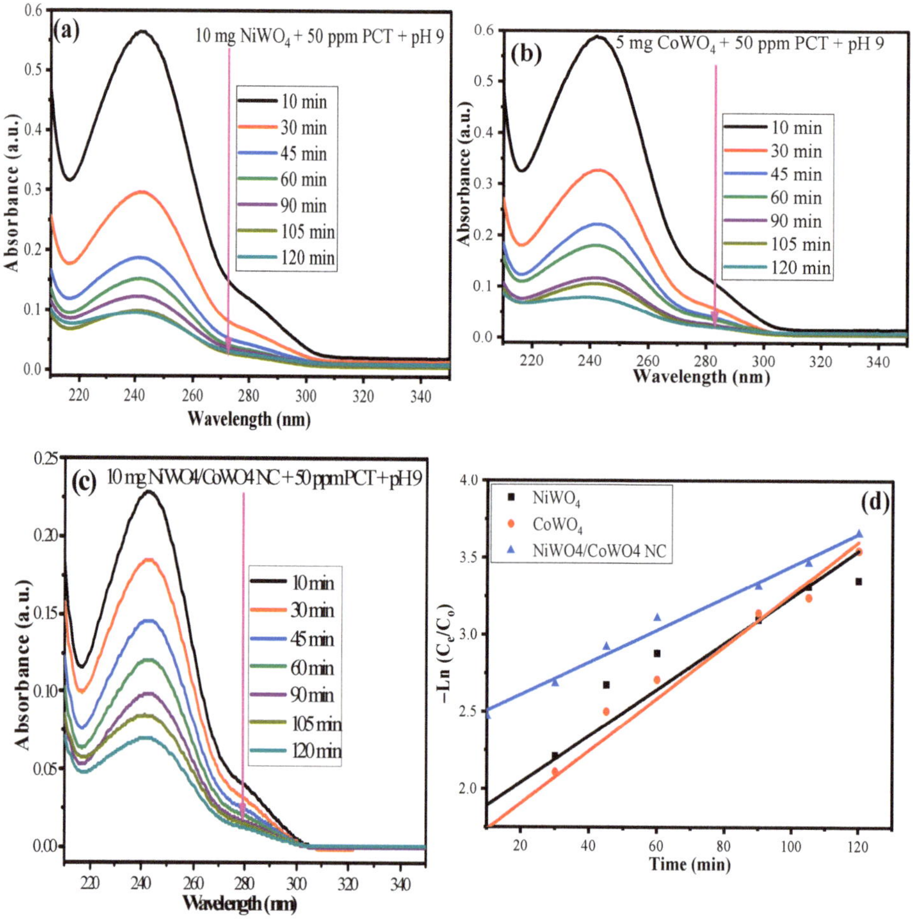

Figure 10. Effect of variable UV irradiation time (10–120 min) using 50 ppm PCT and 10 mg of (**a**) NiWO$_4$, (**b**) CoWO$_4$, (**c**) NiWO$_4$/CoWO$_4$ NC, and (**d**) −Ln (C$_e$/C$_o$) vs. time (min) pseudo first order plot.

Integrating and separating variables between the initial conditions t = 0 and C = C$_0$ at time t and if Kk = k′, where k′ is the apparent rate constant for the photocatalytic degradation.

$$\ln\left(\frac{C_0}{C}\right) = kKt = k't \tag{8}$$

Equation (3) can be plotted, and the slope represents the reaction rate constant, k′, expressed in units of min^{-1}. Given that C = C$_0$/2 in Equation (8), the half-life (t$_{1/2}$), which is an important factor in photocatalytic degradation, is the period of time needed to reduce the concentration of paracetamol by half, it may be calculated as follows by Equation (9):

$$t_{1/2} = \frac{\ln 2}{k'} \quad (9)$$

The obtained results are listed in Table 2, in which the value of apparent rate constant k′ was found to be 0.015 min^{-1} for NiWO$_4$, 0.017 min^{-1} for CoWO$_4$, and 0.019 min^{-1} for NiWO$_4$/CoWO$_4$ NC. The synthesized heterojunction nanocomposite exhibited a high rate of paracetamol degradation as compared to its precursors. The corresponding half-life time values were 46.20, 40.76, and 36.47 min for NiWO$_4$, CoWO$_4$, and NiWO$_4$/CoWO$_4$ NC. The outcomes of the kinetic studies suggested that the heterojunction formation resulted in an enhanced rate of PCT degradation as compared to pristine NiWO$_4$ and CoWO$_4$.

Table 2. Pseudo-first-order kinetic parameters for the photocatalytic degradation of paracetamol by synthesized NiWO$_4$, CoWO$_4$, and NiWO$_4$/CoWO$_4$ NC.

Material	k$_1$ (min^{-1})	R^2	t$_{1/2}$ (min)	Error
NiWO$_4$	0.015	0.95	46.20	2.31 × 10^{-3}
CoWO$_4$	0.017	0.98	40.76	1.46 × 10^{-3}
NiWO$_4$/CoWO$_4$	0.019	0.99	36.47	5.11 × 10^{-4}

2.3. Scavenging Study

The main reactive species involved in the photocatalytic process for paracetamol using NiWO$_4$/CoWO$_4$ NC were discovered using a variety of scavengers. Among the reactive species, the superoxide ($^\bullet$O$_2^-$), hydroxyl radical ($^\bullet$OH), valence band hole (h$^+$), and photogenerated electron (e$^-$) are key species on which the photodegradation of organic pollutants relies [41]. Benzoic acid (BA), acrylamide (AA), ethylenediaminetetraacetic acid (EDTA), and triphenylphosphine (TPP) were the scavengers for $^\bullet$OH, h$^+$, e$^-$, and $^\bullet$O$_2^-$, respectively. Figure 11 represent the results of the scavengers, which show that photogenerated (e$^-$) and superoxide ($^\bullet$O$_2^-$) radicals did not have any significant effect on the photocatalytic performance of the system; therefore, we can conclude that electrons do not play an important role in degradation. The degradation rate was dramatically hindered in ($^\bullet$OH scavenger); as a result, it was concluded that hydroxyl radicals are the main reactive species involved in the removal of paracetamol.

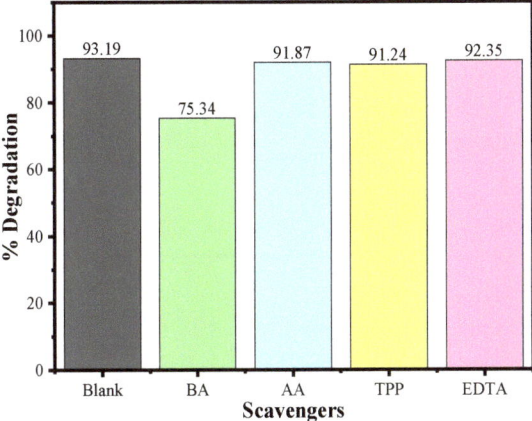

Figure 11. Photodegradation of paracetamol by NiWO$_4$/CoWO$_4$ NC in the presence of different scavengers.

2.4. Effect of Various Radiation and Reaction Process

Experiments were performed to observe the effects of various reaction process and radiations on the PCT degradation by $NiWO_4/CoWO_4$ NC, such as photolysis (PCT solution without catalyst under radiations), adsorption (PCT with catalyst agitated in dark), and photocatalytic reactions under UV and visible radiations. The results obtained are given in Figure 12a,b, and suggested that there is a negligible contribution from the photolysis process. PCT removal by adsorption was found to be 81%, while under visible light it was 69.27%. The maximum removal was obtained under UV light, i.e., 93.95%.

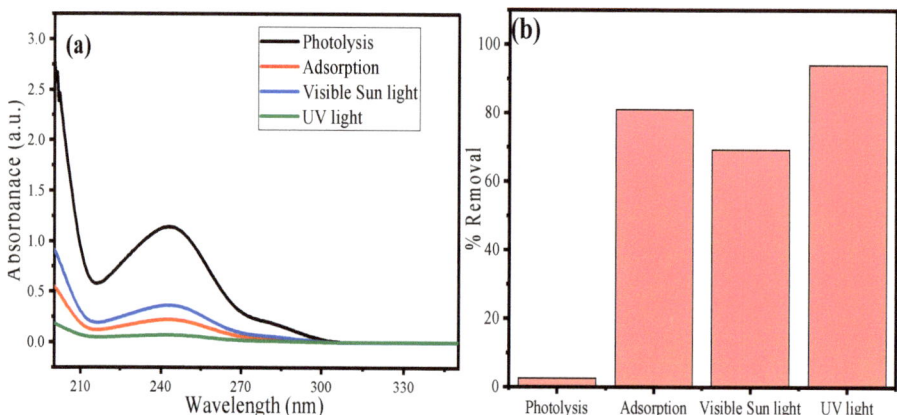

Figure 12. Effect of various reaction process and radiations on PCT degradation. (**a**) UV–Vis spectra, and (**b**) bar graph showing process vs. %PCT removal.

2.5. Comparison with Literature

Table 3 shows the photocatalytic data regarding the degradation of PCT reported in the literature as compared with the present study. It was found that no single study reported on the synthesized material in regard to PCT degradation, which confirms the novelty of this study.

Table 3. Comparison of present study with literature regarding PCT degradation.

Photocatalyst	Light Source	PCT Concentration (ppm)	Irradiation Time (min)	% Degradation	References
ZnO–Ag	Visible light	5	240	92	[42]
CuO@C	LED light	5	60	95	[43]
$g-C_3N_4/(101)-(001)-TiO_2$	300 W xenon lamp	10	300	69.49	[44]
$Cu_2O/WO_3/TiO_2$	Solar light	1	60	92.50	[45]
Bi_2WO_6-CNP-TiO_2	LCS-100 W solar simulator	5	180	84	[46]
$NiWO_4/CoWO_4$ NC	UV light	50	120	97.42	Present study

3. Methods and Materials

3.1. Chemicals and Reagents

Acetaminophen ($CH_3CONHC_6H_4OH$, Analytical grade, 99%), sodium tungstate dihydrate ($Na_2WO_4 \cdot 2H_2O$, 96%, laboratory reagent) were purchased from Merck, Darmstadt, Germany), Nickel nitrate hexahydrate (Ni $(NO_3)_2 \cdot 6H_2O$ 98%), cobalt nitrate hexahydrate (Co $(NO_3)_2 \cdot 6H_2O$, 98%) were supplied by Thermo Fisher Scientific, Dreieich, Germany.

General purpose reagent such as ammonia solution (NH₄OH, 25%, extra pure), and ethanol (C_2H_5OH, 96%, ACS grade) were supplied by Alfa Aesar, Kandel, Germany. The obtained chemicals were employed directly without any further processing, and the solutions were made using distilled water.

3.2. Synthesis of Nickel Tungstate (NiWO₄) Nanomaterials

The hydrothermal process was taken into consideration for synthesis of the nanomaterials reported elsewhere [20,23]. Ni $(NO_3)_2 \cdot 6H_2O$ (3 mmol) and $Na_2WO_4 \cdot 2H_2O$ (5 mmol) were taken individually in 25 mL of distilled water while being stirred magnetically for 15 min for complete dissolution. The two aqueous solutions above were then combined using magnetic stirring for 30 min at room temperature. The mixture was then transferred into 80 mL Teflon-lined steel autoclave and heated for 18 h at 180 °C in a convection oven. After the completion of the reaction, the autoclave was left to cool normally at room temperature and the precipitate was collected through centrifugation. The material was washed with distilled water and ethanol and then dehydrated in a vacuum oven at 80 °C for 5 h. The material was finally calcined at 600 °C for 4 h and stored in desiccator for further characterization and applications. In a similar fashion, CoWO₄ (3 mmol) was prepared, and the heterojunction nanocomposite was prepared by adding 3 mmol of Ni $(NO_3)_2 \cdot 6H_2O$ and 1.5 mmol of Co $(NO_3)_2 \cdot 6H_2O$ with 5 mmol of $Na_2WO_4 \cdot 2H_2O$.

3.3. Characterization of the Synthesized Nanomaterials

Fourier Transform Infrared Spectroscopy (FTIR) system spectrum BX (Perkin Elmer, Akron, OH, USA) which operates in the 4400–400 cm^{-1} range, was used to explore the various types of bonding and functional group contained in the material. The XRD diffractometer, (A Rigaku Ultima) (Rigaku, Austin, TX, USA) was employed to identify the crystalline structure, crystalline size and lattice phase of synthesized nanoparticles. SEM integrated with EDX (SEM; JEOL GSM 6510LV, Tokyo, Japan) was used to examine the surface morphology of the material in order to learn more about its elemental composition as well as the chemical composition and homogeneity of its manufactured NiWO₄/CoWO₄ NC. Through the use of a transmission electron microscope, the particle size and distribution were evaluated (TEM, TEM: JEM 2100, Tokyo, Japan). UV-1800 spectrophotometer (SHIMADZU, Kyoto, Japan), was used for the analysis of synthesized nanoparticles before and after the photocatalytic reaction in the range of 200–700 nm. The chemical state of the elements in the synthesized material was analyzed by X-ray photoelectron spectroscopy (XPS, PHI 5000 Versa Probe III, Physical Electronics, Chanhassen, MN, USA).

3.4. Photocatalysis Experiment

The photocatalytic technique was utilized to observe the degradation of paracetamol under ultraviolet radiation. Accordingly, 20 mL aliquots of variable concentration of paracetamol (5–60 ppm), solution pH (3–10), and (5–20 mg) of catalyst were subjected to magnetic stirring time (10–120 min) under mercury lamp irradiation using a photocatalytic reactor. The percentage of paracetamol (%) that was degraded in an aqueous solution was varied to determine the photocatalyst's effectiveness. The UV–Vis spectrophotometer was used to examine the paracetamol effluents following the photocatalytic reaction at the highest absorption wavelength (λ_{max} = 243 nm). Equation (10) was used to assess the percentage of paracetamol degradation.

$$\% \text{ Degradation} = \frac{C_o - C_t}{C_o} \times 100 \tag{10}$$

where C_o is the initial paracetamol concentration and C_t is the concentrations of paracetamol after the photocatalysis reaction.

4. Conclusions

The present study considers the use of a hydrothermal reaction to synthesize $NiWO_4$, $CoWO_4$, and their heterojunction $NiWO_4/CoWO_4$ nanocomposite materials. The synthesized materials were subjected to various structural, morphological, and optical tests to confirm their correct formation and to test the hypothesis of the research. The FTIR analysis confirmed the formation of the heterojunction through electrostatic interactions between CoO_6 and NiO_6 octahedra in the corners with a WO_6^{6-} frame in a wolframite monoclinic lattice. The outcomes of the FTIR studies were further supported by the XRD studies, with an increase in dislocation density to 2.75 in $NiWO_4/CoWO_4$ from 1.35 in $CoWO_4$ and 2.03 $NiWO_4$. The heterojunction formation also resulted in a contraction in Scherer particle size to 19.07 nm for $NiWO_4/CoWO_4$, which was also found to be in close agreement with the TEM analysis. The SEM–TEM analysis resulted in the formation of a mitigated wolframite monoclinic structure for the synthesized heterojunction nanocomposite, with an average particle size of around 20 nm. The EDX analysis confirmed the presence of Ni, Co, W, and O in the synthesized nanocomposite material. The UV–Vis spectroscopy revealed the values of the energy band gap for $NiWO_4$ (3.04 eV), $CoWO_4$ (2.32 eV), and $NiWO_4/CoWO_4$ (1.14 eV), which supported the conclusion that the formation of heterojunction results in a contraction in the energy bandgap. The synthesized materials were explored as photocatalysts for the degradation of the paracetamol drug under UV light radiation. The optimized operational parameters were found be to 120 min irradiation time, pH 9, and 10 mg of catalyst dose with 50 ppm of paracetamol. At optimized conditions, the photocatalytic efficiencies of the synthesized nanocatalysts were calculated as 96.50% for $NiWO_4$, 97.12% for $CoWO_4$, and 97.42% for $NiWO_4/CoWO_4$ NC. The values of the apparent rate constant obtained from the Langmuir–Hinshelwood model were found to be 0.015 min^{-1} for $NiWO_4$, 0.017 min^{-1} for $CoWO_4$, and 0.019 min^{-1} for $NiWO_4/CoWO_4$ NC. The high values of the rate constant suggested the high photocatalytic efficiency of the heterojunction NC $NiWO_4/CoWO_4$. The outcomes of this study suggest that the heterojunction $NiWO_4/CoWO_4$ NC can be used for photocatalytic degradation of other pharmaceutical pollutants with high efficiency.

Author Contributions: Conceptualization, F.A.A., H.S.A. and A.E.M.; methodology, I.H., F.A.A. and H.S.A.; software, I.H., A.E.M. and F.A.A.; validation, I.H., F.A.A. and H.S.A.; formal analysis, A.A.A. and W.S.A.-N.; investigation, A.A.A. and W.S.A.-N.; resources A.A.A., W.S.A.-N and I.H.; data curation, I.H., A.E.M. and H.S.A.; writing—original draft preparation, A.A.A. and W.S.A.-N.; writing—review and editing, I.H. and F.A.A.; visualization, H.S.A., I.H. and A.E.M.; supervision, F.A.A. and H.S.A.; project administration, F.A.A.; funding acquisition, F.A.A. All authors have read and agreed to the published version of the manuscript.

Funding: This research received no external funding.

Data Availability Statement: Data is contained within the article.

Acknowledgments: The authors extend their appreciation to the deputyship of Research and Innovation, Ministry of Education in Saudi Arabia for funding this research work through project number (IFKSURG-2-1329).

Conflicts of Interest: The authors declare that there is no conflict of interest related to this research, and the paper has not been submitted to any other journal simultaneously.

References

1. Jóźwiak-Bebenista, M.; Nowak, J.Z. Paracetamol: Mechanism of action, applications and safety concern. *Acta Pol. Pharm.* **2014**, *71*, 11–23. [PubMed]
2. Klotz, U. Paracetamol (acetaminophen)—A popular and widely used nonopioid analgesic. *Arzneimittelforschung* **2012**, *62*, 355–359. [CrossRef] [PubMed]
3. Wexler, P.; Anderson, B.D.; Gad, S.C.; Hakkinen, P.B.; Kamrin, M.; de Peyster, A.; Locey, B.; Pope, C.; Mehendale, H.M.; Shugart, L.R.; et al. *Encyclopedia of Toxicology*; Academic Press: Cambridge, MA, USA, 2005; Volume 1.

4. Galani, A.; Alygizakis, N.; Aalizadeh, R.; Kastritis, E.; Dimopoulos, M.A.; Thomaidis, N.S. Patterns of Pharmaceuticals Use during the First Wave of COVID-19 Pandemic in Athens, Greece as Revealed by Wastewater-Based Epidemiology. *Sci. Total Environ.* 2021, *798*, 149014. [CrossRef] [PubMed]
5. Karungamye, P.N. Methods used for removal of pharmaceuticals from wastewater: A review. *Appl. J. Environ. Eng. Sci.* 2020, *6*, 412–428.
6. Choina, J.; Kosslick, H.; Fischer, C.; Flechsig, G.U.; Frunza, L.; Schulz, A. Photocatalytic decomposition of pharmaceutical ibuprofen pollutions in water over titania catalyst. *Appl. Catal. B Environ.* 2013, *129*, 589–598. [CrossRef]
7. Shakir, M.; Faraz, M.; Sherwani, M.A.; Al-Resayes, S. Photocatalytic degradation of the Paracetamol drug using Lanthanum doped ZnO nanoparticles and their in-vitro cytotoxicity assay. *J. Lumin.* 2016, *176*, 159–167. [CrossRef]
8. Amrane, A.; Rajendran, S.; Nguyen, T.A.; Assadi, A.; Sharoba, A. *Nanotechnology in the Beverage Industry: Fundamentals and Applications*; Elsevier: Amsterdam, The Netherlands, 2020.
9. Bibi, S.; Ahmad, A.; Anjum, M.A.R.; Haleem, A.; Siddiq, M.; Shah, S.S.; Kahtani, A. Photocatalytic Degradation of Malachite Green and Methylene Blue over Reduced Graphene Oxide (RGO) Based Metal Oxides (RGO-Fe_3O_4/TiO_2) Nanocomposite under UV-Visible Light Irradiation. *J. Environ. Chem. Eng.* 2021, *9*, 105580. [CrossRef]
10. Shah, L.A.; Malik, T.; Siddiq, M.; Haleem, A.; Sayed, M.; Naeem, A. TiO_2 Nanotubes Doped Poly (Vinylidene Fluoride) Polymer Membranes (PVDF/TNT) for Efficient Photocatalytic Degradation of Brilliant Green Dye. *J. Environ. Chem. Eng.* 2019, *7*, 103291. [CrossRef]
11. Hayat, M.; Shah, A.; Hakeem, M.K.; Irfan, M.; Haleem, A.; Khan, S.B.; Shah, I. A Designed Miniature Sensor for the Trace Level Detection and Degradation Studies of the Toxic Dye Rhodamine B. *RSC Adv.* 2022, *12*, 15658–15669. [CrossRef]
12. Hayat, M.; Shah, A.; Nisar, J.; Shah, I.; Haleem, A.; Ashiq, M.N. A Novel Electrochemical Sensing Platform for the Sensitive Detection and Degradation Monitoring of Methylene Blue. *Catalysts* 2022, *12*, 306. [CrossRef]
13. Tahir, M.B.; Rafique, M.; Nabi, G.; Shafiq, F. Photocatalytic nanomaterials for the removal of pharmaceuticals. In *Nanotechnology and Photocatalysis for Environmental Applications*; Elsevier: Amsterdam, The Netherlands, 2020; pp. 191–202.
14. Ismail, W.N.W.; Mokhtar, S.U. Various methods for removal, treatment, and detection of emerging water contaminants. In *Emerging Contaminants*; IntechOpen: London, UK, 2020.
15. Mirzaei, A.; Chen, Z.; Haghighat, F.; Yerushalmi, L. Removal of pharmaceuticals and endocrine disrupting compounds from water by zinc oxide-based photocatalytic degradation: A review. *Sustain. Cities Soc.* 2016, *27*, 407–418. [CrossRef]
16. Massima Mouele, E.S.; Tijani, J.O.; Badmus, K.O.; Pereao, O.; Babajide, O.; Zhang, C.; Shao, T.; Sosnin, E.; Tarasenko, V.; Fatoba, O.O.; et al. Removal of pharmaceutical residues from water and wastewater using dielectric barrier discharge methods—A review. *Int. J. Environ. Res. Public Health* 2021, *18*, 1683. [CrossRef] [PubMed]
17. Nor, N.A.M.; Jaafar, J.; Othman, M.H.D.; Rahman, M.A. A review study of nanofibers in photocatalytic process for wastewater treatment. *J. Teknol.* 2013, *65*, 2335. [CrossRef]
18. Montini, T.; Gombac, V.; Hameed, A.; Felisari, L.; Adami, G.; Fornasiero, P. Synthesis, characterization and photocatalytic performance of transition metal tungstates. *Chem. Phys. Lett.* 2010, *498*, 113–119. [CrossRef]
19. Chukwuike, V.I.; Sankar, S.S.; Kundu, S.; Barik, R.C. Nanostructured cobalt tungstate ($CoWO_4$): A highly promising material for fabrication of protective oxide film on copper in chloride medium. *J. Electrochem. Soc.* 2019, *166*, C631. [CrossRef]
20. Zawawi, S.M.; Yahya, R.; Hassan, A.; Mahmud, H.N.M.; Daud, M.N. Structural and optical characterization of metal tungstates (MWO_4; M = Ni, Ba, Bi) synthesized by a sucrose-templated method. *Chem. Cent. J.* 2013, *7*, 80. [CrossRef]
21. Sun, Y.; Wang, X.; Fu, Q.; Pan, C. Construction of Direct Z-Scheme Heterojunction NiFe-Layered Double Hydroxide (LDH)/$Zn_{0.5}Cd_{0.5}S$ for Photocatalytic H_2 Evolution. *ACS Appl. Mater. Interfaces* 2021, *13*, 39331–39340. [CrossRef]
22. Liao, Y.; Wang, G.; Wang, J.; Wang, K.; Yan, S.; Su, Y. Nitrogen Vacancy Induced in Situ G-C_3N_4 p-n Homojunction for Boosting Visible Light-Driven Hydrogen Evolution. *J. Colloid Interface Sci.* 2021, *587*, 110–120. [CrossRef]
23. Güy, N.; Atacan, K.; Özacar, M. Rational Construction of P-n-p CuO/CdS/$CoWO_4$ S-Scheme Heterojunction with Influential Separation and Directional Transfer of Interfacial Photocarriers for Boosted Photocatalytic H_2 Evolution. *Renew. Energy* 2022, *195*, 107–120. [CrossRef]
24. Jothivenkatachalam, K.; Prabhu, S.; Nithya, A.; Chandra Mohan, S.; Jeganathan, K. Solar, visible and UV light photocatalytic activity of $CoWO_4$ for the decolourization of methyl orange. *Desalination Water Treat.* 2015, *54*, 3134–3145. [CrossRef]
25. Srirapu, V.K.V.P.; Kumar, A.; Srivastava, P.; Singh, R.N.; Sinha, A.S.K. Nanosized $CoWO_4$ and $NiWO_4$ as efficient oxygen-evolving electrocatalysts. *Electrochim. Acta* 2016, *209*, 75–84. [CrossRef]
26. Taneja, P.; Sharma, S.; Umar, A.; Mehta, S.K.; Ibhadon, A.O.; Kansal, S.K. Visible-light driven photocatalytic degradation of brilliant green dye based on cobalt tungstate ($CoWO_4$) nanoparticles. *Mater. Chem. Phys.* 2018, *211*, 335–342. [CrossRef]
27. Kumaresan, A.; Arun, A.; Kalpana, V.; Vinupritha, P.; Sundaravadivel, E. Polymer-supported $NiWO_4$ nanocomposites for visible light degradation of toxic dyes. *J. Mater. Sci. Mater. Electron.* 2022, *33*, 9660–9668. [CrossRef]
28. Ahmadi, F.; Rahimi-Nasrabadi, M.; Fosooni, A.; Daneshmand, M. Synthesis and application of $CoWO_4$ nanoparticles for degradation of methyl orange. *J. Mater. Sci. Mater. Electron.* 2016, *27*, 9514–9519. [CrossRef]
29. Li, X.; Li, X.; Li, Z.; Wang, J.; Zhang, J. WS_2 Nanoflakes Based Selective Ammonia Sensors at Room Temperature. *Sens. Actuators B Chem.* 2017, *240*, 273–277. [CrossRef]
30. Wang, G.; Ling, Y.; Wang, H.; Yang, X.; Wang, C.; Zhang, J.Z.; Li, Y. Hydrogen-Treated WO_3 Nanoflakes Show Enhanced Photostability. *Energy Environ. Sci.* 2012, *5*, 6180–6187. [CrossRef]

31. Zhu, J.; Li, W.; Li, J.; Li, Y.; Hu, H.; Yang, Y. Photoelectrochemical Activity of NiWO$_4$/WO$_3$ Heterojunction Photoanode under Visible Light Irradiation. *Electrochim. Acta* **2013**, *112*, 191–198. [CrossRef]
32. Xu, S.; Li, X.; Yang, Z.; Wang, T.; Jiang, W.; Yang, C.; Wang, S.; Hu, N.; Wei, H.; Zhang, Y.; et al. Nanofoaming to Boost the Electrochemical Performance of Ni@Ni(OH)$_2$ Nanowires for Ultrahigh Volumetric Supercapacitors. *ACS Appl. Mater. Interfaces* **2016**, *8*, 27868–27876. [CrossRef]
33. Alshehri, S.M.; Ahmed, J.; Ahamad, T.; Arunachalam, P.; Ahmad, T.; Khan, A. Bifunctional Electro-Catalytic Performances of CoWO$_4$ Nanocubes for Water Redox Reactions (OER/ORR). *RSC Adv.* **2017**, *7*, 45615–45623. [CrossRef]
34. Liu, Y.; Wan, K.; Deng, N.; Wu, F. Photodegradation of paracetamol in montmorillonite KSF suspension. *React. Kinet. Mech. Catal.* **2010**, *99*, 493–502. [CrossRef]
35. Vaiano, V.; Sacco, O.; Matarangolo, M. Photocatalytic degradation of paracetamol under UV irradiation using TiO$_2$-graphite composites. *Catal. Today* **2018**, *315*, 230–236. [CrossRef]
36. Hu, C.; Xu, J.; Zhu, Y.; Chen, A.; Bian, Z.; Wang, H. Morphological effect of BiVO$_4$ catalysts on degradation of aqueous paracetamol under visible light irradiation. *Environ. Sci. Pollut. Res.* **2016**, *23*, 18421–18428. [CrossRef]
37. Harimisa, G.E.; Mustapha, M.H.; Masudi, A.; Jusoh, N.C.; Tan, L.S. March. Enhanced Degradation Rates of Paracetamol in Aqueous Solution using Silver Doped Durio Zibethinus Husk Catalyst. In *IOP Conference Series: Materials Science and Engineering*; IOP Publishing: Bristol, UK, 2020; Volume 808, p. 012014.
38. Borges, M.E.; García, D.M.; Hernández, T.; Ruiz-Morales, J.C.; Esparza, P. Supported photocatalyst for removal of emerging contaminants from wastewater in a continuous packed-bed photoreactor configuration. *Catalysts* **2015**, *5*, 77–87. [CrossRef]
39. Hasan, I.; Bassi, A.; Alharbi, K.H.; BinSharfan, I.I.; Khan, R.A.; Alslame, A. Sonophotocatalytic degradation of malachite green by nanocrystalline chitosan-ascorbic Acid@ NiFe$_2$O$_4$ spinel ferrite. *Coatings* **2020**, *10*, 1200. [CrossRef]
40. Lozano-Morales, S.A.; Morales, G.; López Zavala, M.Á.; Arce-Sarria, A.; Machuca-Martínez, F. Photocatalytic treatment of paracetamol using TiO$_2$ nanotubes: Effect of pH. *Processes* **2019**, *7*, 319. [CrossRef]
41. Mao, S.; Bao, R.; Fang, D.; Yi, J. Facile synthesis of Ag/AgX (X = Cl, Br) with enhanced visible-light-induced photocatalytic activity by ultrasonic spray pyrolysis method. *Adv. Powder Technol.* **2018**, *29*, 2670–2677. [CrossRef]
42. Al-Gharibi, M.A.; Kyaw, H.H.; Al-Sabahi, J.N.; Zar Myint, M.T.; Al-Sharji, Z.A.; Al-Abri, M.Z. Silver Nanoparticles Decorated Zinc Oxide Nanorods Supported Catalyst for Photocatalytic Degradation of Paracetamol. *Mater. Sci. Semicond. Process.* **2021**, *134*, 105994. [CrossRef]
43. Abdelhaleem, A.; Abdelhamid, H.N.; Ibrahim, M.G.; Chu, W. Photocatalytic Degradation of Paracetamol Using Photo-Fenton-like Metal-Organic Framework-Derived CuO@C under Visible LED. *J. Clean. Prod.* **2022**, *379*, 134571. [CrossRef]
44. Sun, J.; Deng, L.; Sun, J.; Shen, T.; Wang, X.; Zhao, R.; Zhang, Y.; Wang, B. Construction of a Double Heterojunction between Graphite Carbon Nitride and Anatase TiO$_2$ with Co-Exposed (101) and (001) Faces for Enhanced Photocatalytic Degradation. *RSC Adv.* **2022**, *12*, 20206–20216. [CrossRef]
45. Chau, J.H.F.; Lai, C.W.; Leo, B.F.; Juan, J.C.; Johan, M.R. Advanced Photocatalytic Degradation of Acetaminophen Using Cu$_2$O/WO$_3$/TiO$_2$ Ternary Composite under Solar Irradiation. *Catal. Commun.* **2022**, *163*, 106396. [CrossRef]
46. Mahhumane, N.; Cele, L.M.; Muzenda, C.; Nkwachukwu, O.V.; Koiki, B.A.; Arotiba, O.A. Enhanced Visible Light-Driven Photoelectrocatalytic Degradation of Paracetamol at a Ternary z-Scheme Heterojunction of Bi$_2$WO$_6$ with Carbon Nanoparticles and TiO$_2$ Nanotube Arrays Electrode. *Nanomaterials* **2022**, *12*, 2467. [CrossRef] [PubMed]

Disclaimer/Publisher's Note: The statements, opinions and data contained in all publications are solely those of the individual author(s) and contributor(s) and not of MDPI and/or the editor(s). MDPI and/or the editor(s) disclaim responsibility for any injury to people or property resulting from any ideas, methods, instructions or products referred to in the content.

Article

Removal of Persistent Acid Pharmaceuticals by a Biological-Photocatalytic Sequential Process: Clofibric Acid, Diclofenac, and Indomethacin

María J. Cruz-Carrillo [1], Rosa M. Melgoza-Alemán [1,*], Cecilia Cuevas-Arteaga [2] and José B. Proal-Nájera [3,*]

[1] Facultad de Ciencias Químicas e Ingeniería, Universidad Autónoma del Estado de Morelos, Av. Universidad 1001, Col. Chamilpa, Cuernavaca 62209, Mexico
[2] Centro de Investigación en Ingenierías y Ciencias Aplicadas, Universidad Autónoma del Estado de Morelos, Av. Universidad 1001, Col. Chamilpa, Cuernavaca 62209, Mexico
[3] Instituto Politécnico Nacional, CIIDIR-Unidad Durango, Calle Sigma 119, Fracc. 20 de Noviembre II, Durango 34220, Mexico
* Correspondence: rmelgoza@uaem.mx (R.M.M.-A.); jproal@ipn.mx (J.B.P-N.)

Abstract: The removal of three acid pharmaceuticals—clofibric acid (CLA), diclofenac (DCL), and indomethacin (IND)—by a biological-photocatalytic sequential system was studied. These pharmaceutical active compounds (PhACs) are considered to persist in the environment and have been found in water and sewage, producing adverse effects on the aquatic environment. For the biological process, in batch experiments, a fixed bed bioreactor and activated sludge (hybrid bioreactor), under aerobic conditions, was used as pretreatment. The pretreated effluent was exposed to a photocatalytic process employing TiO_2 nanotubular films ($NTF\text{-}TiO_2$) with the following characteristics: an internal diameter of 112 nm, a wall thickness of 26 nm, nanotube length of 15 μm, a roughness factor of 1840 points, and an anatase-rutile crystalline structure. In the hybrid bioreactor, 39% IND and 50% ACL and DCL were removed. The biological-photocatalysis sequential system achieved the degradation of up to 90% of the initial concentrations of the three acid pharmaceuticals studied. This approach appears to be a viable alternative for the treatment of these non-biodegradable effluents.

Keywords: persistent acid pharmaceuticals; sequential system; biological-photocatalytic processes; $NTF\text{-}TiO_2$; fixed bed bioreactor

1. Introduction

Pharmaceutical active compounds (PhACs), many of which are pervasive, recalcitrant, and biologically active substances, have incited growing concern, mainly because no legislation has been established for their discharge into surface water bodies [1,2]. Vast quantities of PhACs are used all over the world and are able to reach the aquatic environment via urinary excretion and the unsuitable disposal of medication. Hence, these compounds have been found in wastewaters and in surface waters, as they are currently not completely eliminated in wastewater treatment plants (WWTPs) [3,4]. Numerous studies have shown that levels of PhACs are high in aquatic environments and that there is a risk of adverse effects on aquatic organisms. Pharmaceuticals have been found in freshwater environments at concentrations above the threshold for pharmacological effects on fish [5,6].

Among PhACs, a group that has recently received a lot of attention due to its persistent occurrence in different water sources is that of analgesic and non-steroidal anti-inflammatory drugs (NSAIDs). This group currently comprises more than one hundred compounds that are widely used around the world. Another important and regularly observed group of PhACs is lipid regulators. Lipid regulators are widely used and are one of the most common PhAC groups, according to consumption and prescriptions [7,8]. Among PhACs, clofibric acid (CLA), diclofenac (DCL), and indomethacin (IND) are commonly detected in aquatic environments due to their incomplete removal during wastewater treatment processes. As such, these compounds are receiving increasing attention as recalcitrant environmental contaminants [9–12]. For some PhACs, considerably higher removal rates have been achieved through the use of biofilm carriers compared to activated sludge [13,14], suggesting that the further optimization of the biological micropollutant removal processes is within the realm of possibility. Hybrid suspended/attached growth processes combine biofilm carriers and activated sludge into a single treatment procedure. The probability of achieving effective removal using an attached growth biomass is high, as it comprises a diverse bacterial community with aerobic, anoxic, and anaerobic actions [15].

In addition, advanced oxidation processes (AOPs) have been proposed as tertiary treatments for WWTP effluents due to their versatility and ability to increase the biodegradability and detoxification of effluent streams containing polar and hydrophilic chemicals [16]. The principle of AOPs is based on the in situ generation of highly reactive transitory species (i.e., H_2O_2, OH•, $O_2^{•-}$, and O_3) for the mineralization of refractory organic compounds [17]. Within the most extensively used AOPs, heterogeneous photocatalysis (HP) occurs. The principle of HP is based on the excitation of a semiconductor (i.e., TiO_2) by light (UV or visible) [18]. Titanium dioxide is a wide-band gap semiconductor that is frequently used in photocatalysis, as it is easily available, stable to photochemical corrosion, cheap, non-toxic, and has excellent photocatalytic activity [19–22].

TiO_2 powders have been used for many years in the area of photocatalysis and their benefits have accelerated progress in the field. Nevertheless, this material has several shortcomings: it is difficult to disperse and to reuse, it is easily agglomerative, it has low light response properties, and, finally, it causes environmental contamination when used inadequately. The introduction of TiO_2 nanotube arrays (NT-TiO_2), a novel form of TiO_2, has resolved these problems. The structural parameters of NT-TiO_2, such as its specific surface area, wall thickness, tube length, and crystalline phase, have important effects on the photocatalytic activity of nanotube arrays [23,24]. NT-TiO_2 have been extensively applied for the degradation of various organic microcontaminants, such as azo dyes, phenols, and PhACs, among others [25–29].

The coupling of a biological process to an AOP is applied to treat wastewater containing refractory compounds and to limit the high treatment costs associated with chemical oxidation. In these coupled processes, chemical oxidation can be used as a pre-treatment to decrease toxicity or as a post-treatment for the final cleansing of the wastewater [30]. The measurements of the efficiency of the combined process depend on the purpose of the treatment; however, it is usually necessary to independently optimize each of the two stages, i.e., biological and chemical [31]. To augment the elimination of pharmaceutical products, conventional and alternative wastewater treatment processes and their combinations have been researched [32–34]. Nonetheless, the literature on the elimination of CLA, DCL, and IND using coupled processes–AOP systems is scarce.

Some research has been carried out to study the removal of various pharmaceutical products using biological methods combined with POA. Wilt et al. (2020) [35] used a combined process, i.e., photocatalysis UV/TiO_2–biological, to efficiently remove eight out of nine studied compounds (atenolol, atorvastatin, caffeine, carbamazepine, diclofenac, gemfibrozil, fluoxetine, ibuprofen, and naproxen). They determined that the degree of biodegradation following photocatalysis was improved for some drugs (caffeine, diclofenac, gemfibrozil, and ibuprofen) while for others, such as carbamazepine, the approach was ineffective. In 2019, Wang et al. [36] investigated the mineralization of amoxicillin via an intimately coupled photocatalysis (Ag-doped TiO_2) and biodegradation (biofilm) process. The results they obtained using the combined process showed removal rates 40% greater than when using photocatalysis alone, and 65% higher than when using biodegradation alone. The degree of mineralization was around 35% with the combined process, while in the separate processes, it was minimal. Zupanc et al. (2013) [37] studied the removal efficiencies for clofibric acid, ibuprofen, naproxen, ketoprofen, carbamazepine, and diclofenac by two treatment processes: a suspended, activated sludge and moving bed biofilm process

(MBBR) and hydrodynamic cavitation (HC) with the addition of H_2O_2 and UV irradiation. The highest average removals of all the investigated compounds were achieved when the biological treatment (MBBR), HC/H_2O_2 process, and UV treatment were applied consecutively.

The main objective of this study was to assess the feasibility of a hybrid biological process (i.e., a fixed bed bioreactor and activated sludge) sequenced to a photocatalysis process employing TiO_2 nanotubular films (NTF-TiO_2) for the removal of three acidic PhACs (clofibric acid, diclofenac, and indomethacin).

2. Results and Discussion

2.1. Biological Process

The hybrid bioreactor (HBR) was operated as a hybrid reactor using two processes: activated sludge and adhered biomass supported in biomedia. The HBR bioreactor operated for 86 days (28 cycles). During this period, the dissolved oxygen (DO) level varied in the range of <1 and 5 mg L^{-1}, with a pH of 7.5 ± 0.2 and a temperature inside the reactor of 22 °C ± 2 °C.

2.1.1. Biomass Acclimatization

The duration of the HBR acclimatization phase was 46 days (cycle 1 to 7). The acclimation began with a concentration of 2 mg L^{-1} of the PhACs dissolved in methanol (MeOH). During this phase, the removal efficiencies fluctuated between 3% and 17% for the CLA, 8% and 18% for the DCL, and 5 and 15% for the IND. The reaction time was 192 h to 72 h; this could have been the consequence of a non-acclimated biomass and an immature biofilm since biofilms need a long maturation time [38]. Regarding the total organic carbon (TOC) concentration, the removal of TOC was observed to be unstable, as it fluctuated between 20% and 50%. The low removal of TOC during the first days is a normal process in the start-up phase, because the biomass is not acclimated to the removal of PhACs. To consider an acclimatized biomass, 80% PhAC removal was set; however, due to drug recalcitrance, which involves complex structures that bacteria find difficult to degrade, fixed removal efficiencies were not achieved, so it was considered to take as an adaptation criterion the removal of 80% of the TOC. This adaptation was reached in cycle 7. According to the stoichiometric reaction, it was observed that the greatest contribution of carbon was due to MeOH, which acted as a co-substrate, providing an easily degradable carbon source for the bacteria. Table 1 summarizes the HBR's operating parameters during the acclimatization phase.

Table 1. Operating parameters of the HBR bioreactor during the acclimatization phase: cycle, reaction time, TOC concentration, and removal % of TOC, ACL, DCL, and IND.

Cycle	Reaction Time (h)	mg L^{-1} TOC Influent	% Removal TOC	% Removal		
				ACL	DCL	IND
1	192	155.57	21.50	18.0	18.8	31.8
2	192	143.43	29.74	4.5	8.4	10.2
3	192	148.52	51.12	6.8	17.3	7.3
4	148	174.24	55.18	11.8	7.4	12.9
5	148	188.15	74.35	17.6	11.9	5.0
6	148	144.44	52.33	2.9	8.1	14.6
7	72	166.15	80.19	3.0	18.2	11.9

2.1.2. Behavior of Biomass and Total Suspended Solids

The biomass was determined as volatile suspended solids in the mixed liquor (MLVSS). Figure 1 shows the evolution of the biomass that oscillated between 1800 and 1300 mg L^{-1}. During the first 40 days of the reactor's operation, a decrease in biomass was observed, causing a loss in bacterial activity, which was a product of the process of the biomass' adaptation to the degradation of the PhACs. To strengthen the biomass, for 12 days (cycle

8 to 11), only MeOH (0.02%) was added as an easily degradable substrate. At 58 days, the amount of biomass gradually increased until reaching 1800 mg L^{-1}. The results showed that there was a sufficient degree of biomass recovery in order to synthesize new cells.

The total suspended solids (TSS) in the HBR effluent decreased from 216 mg L^{-1} to 25 ± 3 mg L^{-1} (Figure 1) as a result of the reactor's stabilization. Some authors [39–42] have reported poor sludge settleability and high solids in the effluent when operating at low DO concentrations. In this work, no negative impacts on the sludge's settleability were observed when the reactor operated at DO concentrations < 1 mg/L. The operating conditions of the reactor favored the growth of bacteria capable of forming favorable flocs for a good settleability and development of biomass.

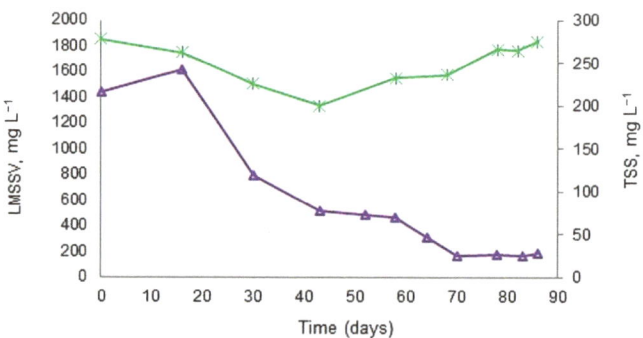

Figure 1. Behavior of biomass ( ) and TSS (Δ).

2.1.3. Total Organic Carbon

After the acclimatization phase, a decrease in TOC removal was observed due to biomass inhibition. As a strategy to strengthen the biomass, the addition of the PhACs was suspended (cycle 8 to 11), adding only MeOH (0.02%) to the HBR. At this phase, the average TOC removal was 79% with reaction times of 72 h. After this recovery phase, the PhACs were again added to the HBR. From this phase (cycle 12 to 28), the removal efficiencies gradually increased up to 90% ± 5%. The reaction time was optimized from 192 h to 24 h. Figure 2 presents the total organic carbon (TOC) concentration in the influent and effluent and the removal efficiency during all the operation periods of the HBR.

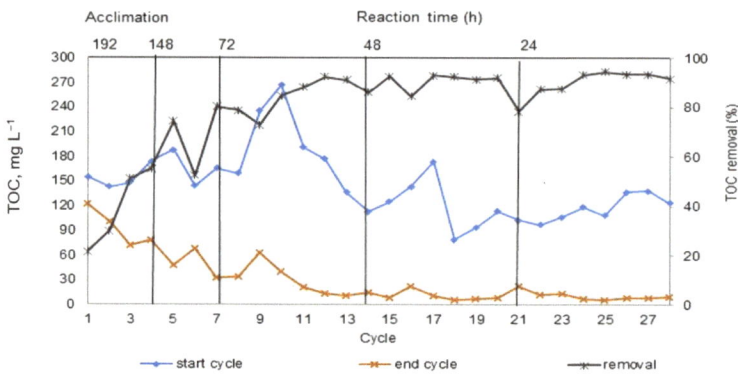

Figure 2. Behavior of influent and effluent TOC concentrations and TOC removal.

2.1.4. PhACs Biodegradation

After the biomass recovery phase, removal efficiencies of 39% were obtained for the IND and 50% for the ACL and DCL. The evolution of the removal efficiencies and reaction times during the operation of the reactor are presented in Figure 3a–c. The removal of the PhACs investigated in the present study fell into the range reported in [43–45].

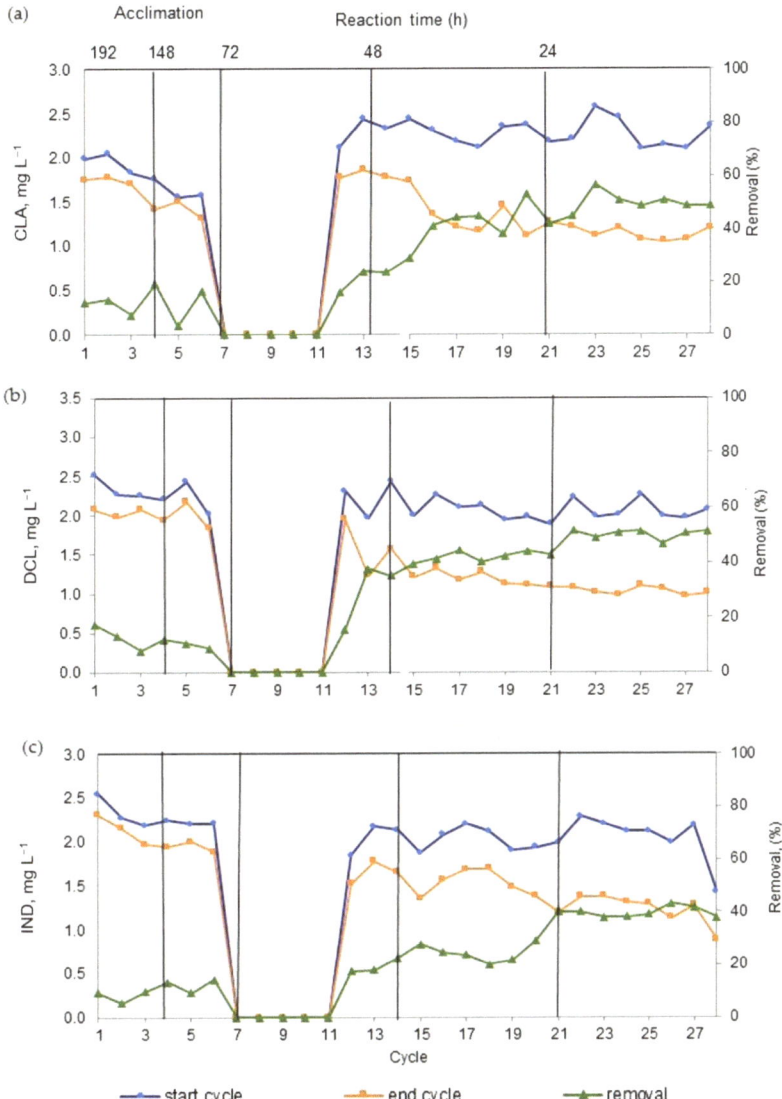

Figure 3. Evolution of the removal efficiencies and reaction times. (a) CLA; (b) DCL; (c) IND.

Some authors attribute high PhAC removal to the substrate and redox gradients within the biofilm, which may induce vastly stratified microbial communities, with microorganisms adapted to easily degradable organic substrates in the outer part of the biofilm and microorganisms adapted to the remaining and hardly degradable organic substrates in the inner part of the biofilm [14,38].

On the other hand, the addition of MeOH as a source of external carbon can lead to the development of a diverse microbial community that favors the growth of nitrifying bacteria associated with the biodegradation and biotransformation of PhACs [15,38,46–48].

2.2. Biological-Photocatalytic Sequential System

Since the fixed efficiencies criterion (80% PhACs removal) was not achieved and, consequently, there were still remnants of the PhACs not removed in the HBR bioreactor, coupling to photocatalysis was performed. To determine the operating conditions of the photocatalysis process, the effluent of the HBR bioreactor was used. The parameters evaluated to determine the optimal operating conditions in the photocatalytic process were the area of the NTF-TiO$_2$ catalyst, the wavelength of UV radiation (UV-A and UV-C), and pH.

2.2.1. NTF-TiO$_2$ Catalyst

Some factors that influence the efficiency of photocatalysis using NTF-TiO$_2$ are the morphology and structure (including the thickness of the tube wall), diameter, and length of the nanotubes [27]. In our study, NTF-TiO$_2$ with a wall thickness of 26 nm, a length of 65 μm, and a diameter of 112 nm were used. In comparison with other reported works, they are four times larger with respect to length and up to 12 times larger with respect to diameter [49]. Smith et al. 2009 [50] observed that as the diameter, length, and thickness of the NTF-TiO$_2$ increase, higher photocatalytic activity is observed. Figure 4 shows some Field Emission Scanning Electron Microscopy (FE-SEM) images of the morphological characteristics. Figure 4a shows the open top view, whereas Figure 4b,c show the cross section of the nanofilm.

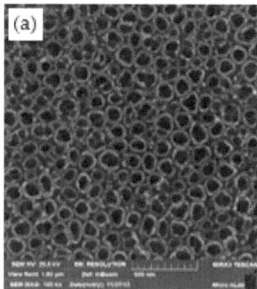 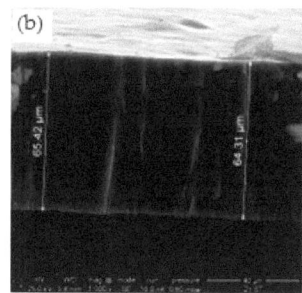 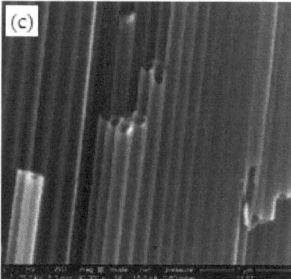

Figure 4. FE-SEM images of the NTF-TiO$_2$. (**a**) Top view; (**b**) and (**c**) cross sectional views.

With respect to the crystalline structure, Figure 5 presents the X-ray Diffraction (XRD) spectrum of the NTF-TiO$_2$. The XRD spectrum shows the main presence of anatase (JCPDS Card # 21-1272) in coexistence with rutile phases (JCPDS card # 21-1276), observing a very intensive peak of anatase together with two other medium intensive anatase peaks. From the XRD spectrum, it can be observed that the major peak corresponds to the anatase phase. As reported, the crystalline structure plays an important role in the photoactivity of the NTF-TiO$_2$; for instance, with the crystalline structure anatase as the major phase in combination with rutile, the best photocatalytic performance was determined, followed by the rutile phase and amorphous structures [51]. In this study, the presence of the anatase phase in a greater proportion was a factor that favored the degradation of the PhACs, which has been reported in studies on the influence of crystal structure on the removal of diverse molecules such as PhACs [52,53], phenols [27], and dyes [25].

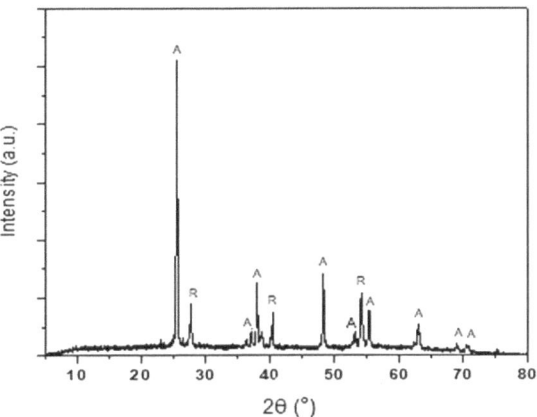

Figure 5. XRD spectrum of the NTF-TiO$_2$. A (anatase phase); R (rutile phase).

2.2.2. pH Effect

In the photocatalytic process, the generation of hydroxyl radicals is a function of pH. Thus, pH is an important parameter in photocatalytic reactions [54]. In this study, we investigated the effect of pH on photocatalytic performance. The experiments were carried out employing a 4 cm^2 area of NTF-TiO$_2$ and UV-A radiation. The studied pH levels were 3.5 ± 0.2 and 7.3 ± 0.2. After 48 h of reaction, the best performance was evidenced at pH 3.5 with removal efficiencies of 97% for the CLA and 100% for the IND and DCL. In the case of pH 7.3, after 72 h of reaction, the removals were 38%, 61%, and 82% for the CLA, DCL, and IND respectively. This influence of pH on the photodegradation of PhACs, using TiO$_2$ catalyst, is consistent with that reported by Molinari et al. in 2006 [55]. The previous results may be due to the fact that when working at pH 3.5 and 7.2, on the one hand, the ionization of PhACs is favored, since their pKa are 3.18, 4.10, and 4.50 for CLA, DCL, and IND, respectively, while on the other hand, the isoelectric point (PZC) reported for TiO$_2$ [17] is in the pH range of 4.5 and 7.0; thus, at pH lower than PZC, the catalyst acquires a positive charge and gradually exerts a electrostatic attraction force towards the ionized species of the PhACs favoring adsorption on the TiO$_2$ surface for subsequent photocatalytic reactions.

Several authors [50,56,57] have already reported the influence of pH on TiO$_2$ through the photocatalytic degradation of PhACs and found minimal influence in the near neutral pH range. In our study, as shown in Table 2, the best removal efficiencies were achieved at pH 3.5; however, the average pH of the biological treatment effluents was 7.3 ± 0.2, so the high removal efficiencies achieved at pH 7.3 are an important finding, since the biological treatment effluent can be treated directly without treatment prior to photocatalysis. On the other hand, there is an advantage of showing that the removal efficiency through a sequenced biological process to photocatalysis in acidic and near neutral pH conditions can be an alternative treatment for wastewaters with different pH conditions. Nevertheless, it is better to evaluate the effect of pH on photodegradation in a way that considers the contaminant properties, the type of photocatalyst, and the effluent to treat.

2.2.3. Effect of the NTF-TiO$_2$ Area and UV Radiation

Experiments were carried out using a film of NTF-TiO$_2$ (area 2 cm^2) and with two films of NTF-TiO$_2$ (area 4 cm^2), UV-A radiation, and pH 7.5. The results of the biological-photocatalytic sequential system under different operating conditions are presented in Table 2. In the experiments using an area of 2 cm^2, the degree of photodegradation was 26% for the ACL, 62% for the DCL, and 84% for the IND after 72 h of reaction (Table 2). Whereas when the catalyst area was increased to 4 cm^2, no changes in photodegradation were observed. The removals were 38%, 61%, and 82% for CLA, DCL, and IND, respectively. Subsequently, experiments with 4 and 8 cm^2 of NTF-TiO$_2$ area, applying UV-C radiation, and at pH 7.4 were carried out. When UV-C radiation and the 4 cm^2 area of NT-TiO$_2$ were used, an increase in the removal was observed for the three PhACs. The removals were 100%, 77%, and 64% for the CLA, DCL, and IND, respectively. The reaction time was 5 h. When increasing the area of the NTF-TiO$_2$ to 8 cm^2 under the same conditions of pH and UV radiation, a further removal was favored, reaching 100% degradation for the DCL and IND. The photodegradation times were 5 h for IND and 2 h for DCL. The CLA was photodegraded up to 90% over a reaction time of 3 h (Table 2).

Table 2. Results of the biological–photocatalytic sequential system under different operating conditions.

Experiments	UV-365 nm (UV-A)			UV-254 nm (UV-C)	
	2 cm^2 NTF-TiO$_2$	4 cm^2 NTF-TiO$_2$		4 cm^2 NTF-TiO$_2$	8 cm^2 NTF-TiO$_2$
	pH 7.5	pH 3.5	pH 7.3	pH 7.4	
	RT = 72 h	RT = 48 h	RT = 72 h	RT = 5 h	RT
PhACs	Removal Efficiencies (%)				
CLA	26	97	38	100	90 (3 h)
DCL	62	100	61	77	100 (2 h)
IND	84	100	82	64	100 (5 h)

RT = reaction time.

The experiments carried out without the nanotubes (control photolytic experiments) were performed only at a pH of 7.3, since it is the average pH of the biological treatment effluent. The removal efficiencies were 3%, 5%, and 25% for CLA, DCL, and IND, respectively, after 4 h of reaction and UV-C radiation (254 nm), so the effect of photolysis alone is not sufficient to achieve the total removal of the PhACs.

2.3. Kinetic Study

The reaction kinetics of the photocatalytic process were determined. The concentration was determined by UV–Vis spectroscopy and HPLC. The reaction kinetics of the PhAC mixture were determined by applying UV-A and UV-C radiation, pH 7.4, and areas of 4 cm^2 and 8 cm^2. The NTF-TiO$_2$ films were first immersed in the PhACs mixture (effluent of HBR bioreactor) in the dark for 30 min to reach the adsorption equilibrium. Then, the light was turned on (t = 0 min) and photocatalytic degradation occurred. Figure 6a shows the absorption spectrum of the CLA, DCL, and IND mixtures obtained from the monitoring of the reaction kinetics using UV–Vis, an area of 8 cm^2 of NTF-TiO$_2$, UV-C radiation, and pH 7.4. Figure 6b shows the absorption spectra of CLA, DCL, and IND.

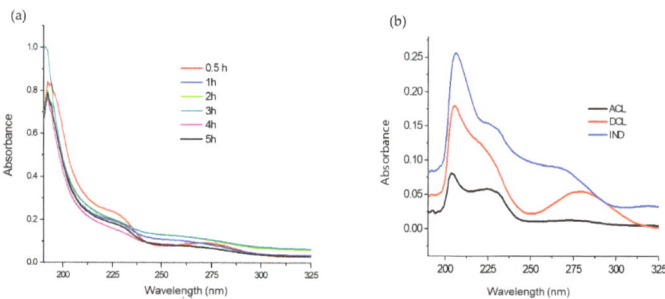

Figure 6. (a) Monitoring of reaction kinetics of CLA, DCL, and IND mixture, using area of 8 cm^2 of NTF-TiO$_2$, UV-C radiation, and pH 7.4 (b) Absorption spectrum of CLA, DCL, and IND.

Figure 7 shows the results of the photocatalytic degradation of the CLA, DCL, and IND mixtures, which follow a first order reaction model. Figure 7a shows the degradation kinetics using UV-A radiation and a 4 cm^2 area of the NTF-TiO$_2$, Figure 7b shows the degradation kinetics using UV-C radiation and a 4 cm^2 area of the NTF-TiO$_2$, and Figure 7c shows the degradation kinetics using UV-C radiation and an area of 8 cm^2 of NTF-TiO$_2$. The results in Figure 7 confirm the data presented in Table 2.

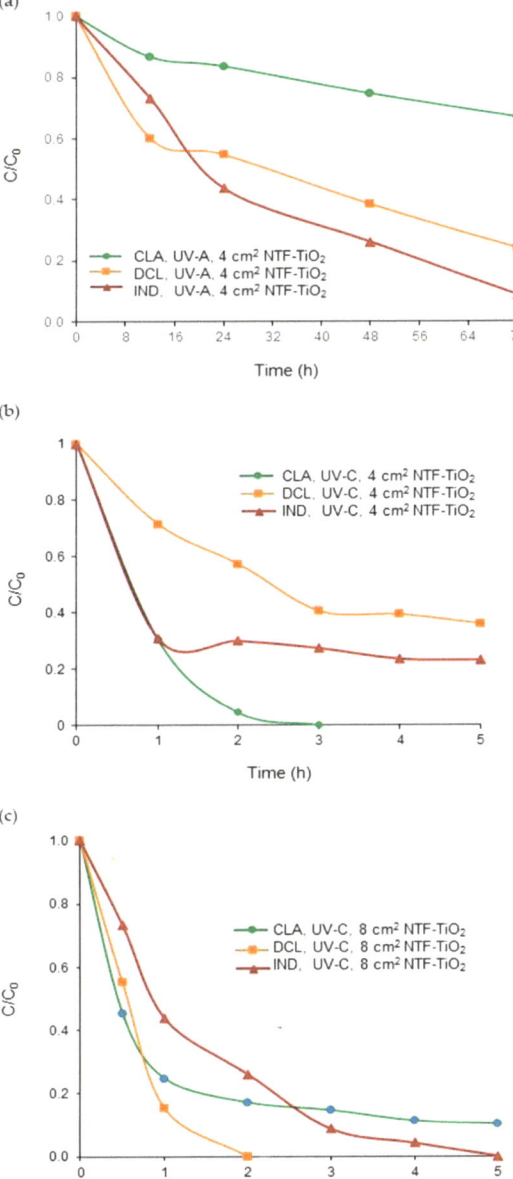

Figure 7. Photocatalytic degradations of CLA, DCL, and IND mixture. (**a**) UV-A radiation and 4 cm^2 of NTF-TiO$_2$; (**b**) UV-C radiation and 4 cm^2 of NTF-TiO$_2$; (**c**) UV-C radiation and 8 cm^2 of NTF-TiO$_2$.

The concentrations of the PhACs mixture, the results of the kinetic constants, the correlation coefficient, and the percentage removal (%) of the kinetic study of the biological-photocatalysis sequential system are summarized in Table 3. Based on the results, it can be observed that the photoactivity was the lowest—with rate constants of 0.0042, 0.0091, and 0.0121 for the CLA, DCL, and IND, respectively, and with removals of 33.27, 76.07, and 83.97% after 72 h—when UV-A radiation was used. By increasing the area of NTF-TiO$_2$ and applying UV-C radiation, the photoactivity increased significantly, showing better results when using an area of 8 cm^2 with the rate constants of 0.1304, 0.4855, and 0.1871 for CLA, DCL, and IND, respectively, and removals of 89.95% for CLA and 100% for the DCL and IND after 5 h, 2 h, and 4 h, respectively. Figure 8 shows the comparison of the achieved rate constants (Figure 8a) and removals (Figure 8b).

Table 3. Kinetic constants, correlation coefficients, and PhACs removal percentage (%) from kinetic study of the biological-photocatalysis sequential system.

PhACs	mg L^{-1}	UV-A/4 cm^2 K (h^{-1})	Removal (%)	mg L^{-1}	UV-C/4 cm^2 K (h^{-1})	Removal (%)	mg L^{-1}	UV-C/8 cm^2 K (h^{-1})	Removal (%)
CLA	1.93	0.0042 (R^2 = 0.9358)	33.27	1.75	0.3255 (R^2 = 0.8299)	99.88	2.40	0.1304 (R^2 = 0.5651)	89.95
DCL	1.88	0.0091 (R^2 = 0.8444)	76.07	1.27	0.1240 (R^2 = 8698)	76.80	2.57	0.4855 (R^2 = 0.8588)	100
IND	2.63	0.0121 (R^2 = 0.9156)	83.97	2.69	0.1167 (R^2 = 0.5311)	64.10	1.24	0.1871 (R^2 = 0.8413)	100

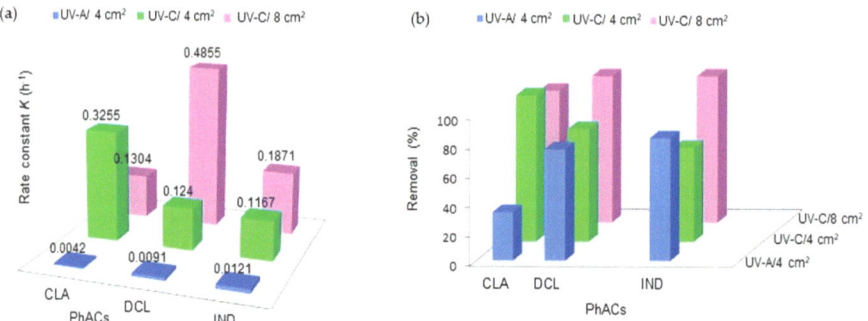

Figure 8. Comparison of rate constants (a) and the achieved removals (b) from kinetic study of the biologica-photocatalysis sequential system.

The results presented in Figure 8 correspond to the performance of NTF-TiO$_2$ as a catalyst when irradiated with UV radiation (being initially in their basal state), wherein they absorb the energy of the light and thereby form an excited state that can lead to photochemical reactions, such as the breaking of bonds [58]. The energy at a wavelength of 254 nm (UV-C) is greater that at a UV-A wavelength, which favors photodegradation. In addition, the absorption spectrum of the PhACs (Figure 6b) shows maximum absorption in the UV-C region. These results are consistent with previous studies, where the photodegradation of organic molecules (PhACs and azo-type dyes) was investigated using UV irradiation at various wavelengths [58–62]. With respect to the area of NTF-TiO$_2$, the results showed that increasing the area of the nanofilms increased removal due to a larger quantity of nanotubes; therefore, the ability to adsorb light and generate photo-induced electron-hole pairs inactive sites was greater, which led to an increase in photodegradation efficiency.

3. Materials and Methods

3.1. Reagents

PhACs of standard grade ≥ 98% purity, namely, clofibric acid (CLA), diclofenac (DCL), and indomethacin (IND); substances of nutrient mineral medium [63]; ethylene glycol; and NH_4F were acquired from Sigma Aldrich. Figure 9 shows the chemical structures of the utilized pharmaceuticals.

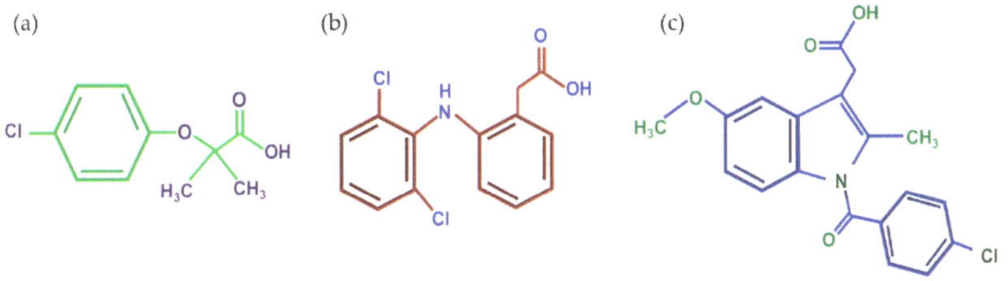

Figure 9. Chemical structures: (**a**) clofibric acid, (**b**) diclofenac, and (**c**) indomethacin.

3.2. Hybrid Biological Process

For the inoculation of the HBR, the activated sludge was obtained from the wastewater treatment plant ECCACIV, located in Jiutepec-Morelos, México.

3.2.1. Biomass Acclimatization and Operation Strategy

The biomass was acclimatized to the pharmaceuticals by feeding nutrient mineral medium (Table 4) and CLA, DCL, and IND dissolved in MeOH. Inlet concentrations of PhACs were about 2 mg L^{-1}. The adaptation of microorganisms for treating wastewater was carried out with variable reaction cycles to reach the acclimatization of biomass for degradation of PhACs using the criteria-fixed efficiencies [64], which consisted of allowing biomass to adapt to degradation of 80% of PhACs. The experiments were performed in batches. These batch experiments were carried out in 4 stages: filling, reaction, sedimentation, and emptying.

Table 4. Composition of nutrient mineral medium.

Substances	mg L^{-1}
K_2HPO_4	65.25
$Na_2HPO_4 \cdot 2H_2O$	100.2
KH_2PO_4	25.5
NH_4Cl	7.5
$MgSO_4 \cdot 7H_2O$	22.5
$CaCl_2 \cdot 2H_2O$	27.5
$FeCl_3 \cdot 6H_2O$	0.25
H_3BO_3	0.06
$MnSO_4 \cdot H_2O$	0.04
$ZnSO_4 \cdot 7H_2O$	0.04
$(NH_4)_6 \cdot Mo_7O_{24}$	0.03
EDTA	0.1

3.2.2. Experimental System (HBR Bioreactor)

The experimental set up consisted of an acrylic reactor with a total volume of 6.5 L and a useful volume of 4 L. For the control of loading, recirculating, and discharge, two peristaltic pumps were used (Master Flex, Cole Palmer, Vernon Hills, Chicago, IL, USA).

The air inside the reactor was distributed from the bottom of the reactor through a porous diffuser with an air pump. For HBR configurations, 50% of the useful volume of reactor was filled with high-density polypropylene plastic carriers (biomedia AMB, DYNAMIC AQUA SCIENCE, Kirkland, AZ, USA).

The control parameters of the HBR were the concentrations of biomass measured as MLVSS, TSS, TOC and PhACs, DO, pH, and temperature.

3.3. Photocatalytic System

The effect of pH, area of NTF-TiO$_2$, and UV radiation, on the photocatalytic performance of NTF-TiO$_2$ were investigated.

3.3.1. TiO$_2$ Nanostructures Films

NTF-TiO$_2$ were synthesized by the electrochemical anodization technique. Electrochemical anodization experiments were performed in a nonaqueous electrolytic solution of ethylene glycol (3 vol% H$_2$O) + 0.25 wt% NH$_4$F. The anodization process lasted 11.5 h, maintaining a constant voltage of 60 V at room temperature. After anodization, the samples were rinsed with deionized water and dried in a N$_2$ stream. The total area anodized was 2 cm^2. The morphological and structural characterization of NTF-TiO$_2$ was carried out by means of FE-SEM and an XRD.

3.3.2. Biological-Photocatalytic Sequential System

Photocatalysis experiments were performed in a glass cell with a water recirculation jacket to control the temperature, using 60 mL of the effluent from the HBR, which contained CLA, DCL, and IND not removed by the biological system. The sample for photocatalysis was taken immediately after finishing the cycle of the biological process.

The glass cell was irradiated with UV-A (365 nm) or UV-C (254 nm) lamps and in dark conditions. For the optimization of the process, experiments were carried out with two different pH values: 7.3 ± 0.2 and 3.5 ± 0.2, employing 2, 4, and 8 cm^2 areas of NTF-TiO$_2$. Once the optimal conditions of operation were determined, reaction kinetics were performed. The scheme of biological-photocatalytic sequential system is shown in Figure 10.

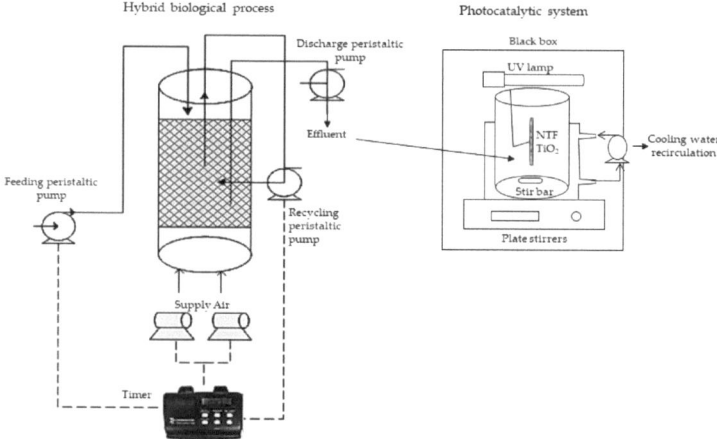

Figure 10. Biological-photocatalytic sequential system.

3.3.3. Analytical Methods

PhACs concentrations were determined by high performance liquid chromatograph HPLC 1100 (Agilent, Mexico City, Mexico), which was equipped with Zorbax Eclipse XDB C-18 (250 mm × 4.6 mm × 5 µm) column (LEACSA SA de CV, Mexico City, Mexico) at

35 °C and a UV detector with a wavelength of 230 nm. The mobile phase, with a flow rate of 0.8 mL/min, was composed of acetonitrile:1% acetic acid water mixture with a volumetric ratio of 65:35. The volume of the injected sample was 10 µL. The samples to determine the concentration of the PhACs were previously centrifuged at 3000 rpm for 15 min and then filtered through a membrane of 45 µm. PhACs absorption spectra were determined by a spectrophotometer UV–VIS Lambda 25, Perkin Elmer (Perkin Elmer de México, Mexico City, Mexico). TOC of samples was analyzed using a TOC analyzer (Torch TELEDYNE Tekmar, MAC Analítica SA de CV, Estado de México, Mexico). Samples were centrifuged at 3000 rpm for 15 min, and then they were filtered through fiberglass paper. The control parameters—DO, temperature, and pH—were determined by APHA Methods 2005 [65].

3.4. Kinetic Study

Kinetic analysis was performed considering a first order reaction (Equation (1)) [66]. The rate constant (k) of the photocatalytic reaction for the PhACs degradation process was determined through Equation (1).

$$\ln(C/C_0) = -k \times t \quad (1)$$

where C corresponds to PhACs concentration at time t, C_0 represents the PhACs initial concentration of the sample, and t is the time at which the sample was taken.

The kinetic curves for PhACs degradation were obtained by a non-linear relationship (C/C_0 vs. t) for each treatment [67].

4. Conclusions

In this study, a biological process utilizing a bioreactor (HBR) sequenced to perform photocatalysis using NTF-TiO$_2$ for the degradation of CLA, DCL, and IND was optimized.

In the hybrid bioreactor (HBR), the overall removal for IND was 39%, whereas for CLA and DCL the overall removals were 50% over a reaction time of 24 h. The degradation of organic carbon measured as TOC was 92%.

In the biological-photocatalysis sequential system, the overall removal efficiency for the CLA was 90%, whereas for DCL and the IND it was 100% (below the LOD) after 29 h of treatment (24 h of biological process and 5 h of photocatalysis).

It was shown that the photocatalytic degradation reaction of PhACs follows a first-order kinetics pattern, which was dependent on the UV radiation and the area of NTF-TiO$_2$. The optimal operating parameters were applying UV-C radiation and using 8 cm^2 of NTF-TiO$_2$.

The results indicated that the biological-photocatalysis sequential system is a promising method for treating water contaminated with CLA, DCL, and IND.

Author Contributions: Conceptualization, R.M.M.-A.; methodology, M.J.C.-C., R.M.M.-A., C.C.-A. and J.B.P.-N.; formal analysis, M.J.C.-C.; investigation, M.J.C.-C.; resources, M.J.C.-C. and C.C.-A.; data curation, M.J.C.-C., R.M.M.-A., C.C.-A. and J.B.P.-N., writing—original draft preparation, M.J.C.-C.; writing—review and editing, M.J.C.-C., R.M.M.-A., C.C.-A. and J.B.P.-N.; visualization, R.M.M.-A.; supervision, R.M.M.-A.; project administration, R.M.M.-A., C.C.-A. and J.B.P.-N.; funding acquisition, M.J.C.-C., R.M.M.-A., C.C.-A. and J.B.P.-N. All authors have read and agreed to the published version of the manuscript.

Funding: This research received no external funding.

Data Availability Statement: Data are contained within the article.

Acknowledgments: We thank the Facultad de Ciencias Químicas e Ingeniería of Universidad Autónoma del Estado de Morelos (UAEMorelos) and Centro de Investigación en Ingenierías (UAEMorelos) for their help with accomplishing the experiments. We also thank the National Science and Technology Council of Mexico (CONACYT) for the scholarship granted to the first author.

Conflicts of Interest: The authors declare no conflict of interest.

References

1. Verlicchi, P.; Al Aukidy, M.; Zambello, E. Occurrence of Pharmaceutical Compounds in Urban Wastewater: Removal, Mass Load and Environmental Risk after a Secondary Treatment—A Review. *Sci. Total Environ.* **2012**, *429*, 123–155. [CrossRef] [PubMed]
2. Mendoza, A.; Aceña, J.; Pérez, S.; López de Alda, M.; Barceló, D.; Gil, A.; Valcárcel, Y. Pharmaceuticals and Iodinated Contrast Media in a Hospital Wastewater: A Case Study to Analyse Their Presence and Characterise Their Environmental Risk and Hazard. *Environ. Res.* **2015**, *140*, 225–241. [CrossRef] [PubMed]
3. Boix, C.; Ibáñez, M.; Sancho, J.v.; Parsons, J.R.; Voogt, P.d.; Hernández, F. Biotransformation of Pharmaceuticals in Surface Water and during Waste Water Treatment: Identification and Occurrence of Transformation Products. *J. Hazard. Mater.* **2016**, *302*, 175–187. [CrossRef]
4. Chonova, T.; Keck, F.; Labanowski, J.; Montuelle, B. Science of the Total Environment Separate Treatment of Hospital and Urban Wastewaters: A Real Scale Comparison of Ef Fl Uents and Their Effect on Microbial Communities. *Sci. Total Environ.* **2016**, *542*, 965–975. [CrossRef]
5. Grabicova, K.; Lindberg, R.H.; Östman, M.; Grabic, R.; Randak, T.; Joakim Larsson, D.G.; Fick, J. Tissue-Specific Bioconcentration of Antidepressants in Fish Exposed to Effluent from a Municipal Sewage Treatment Plant. *Sci. Total Environ.* **2014**, *488–489*, 46–50. [CrossRef] [PubMed]
6. Ejhed, H.; Fång, J.; Hansen, K.; Graae, L.; Rahmberg, M.; Magnér, J.; Dorgeloh, E.; Plaza, G. The Effect of Hydraulic Retention Time in Onsite Wastewater Treatment and Removal of Pharmaceuticals, Hormones and Phenolic Utility Substances. *Sci. Total Environ.* **2018**, *618*, 250–261. [CrossRef] [PubMed]
7. Nikolaou, A.; Meric, S.; Fatta, D. Occurrence Patterns of Pharmaceuticals in Water and Wastewater Environments. *Anal. Bioanal. Chem.* **2007**, *387*, 1225–1234. [CrossRef]
8. Feng, L.; van Hullebusch, E.D.; Rodrigo, M.A.; Esposito, G.; Oturan, M.A. Removal of Residual Anti-Inflammatory and Analgesic Pharmaceuticals from Aqueous Systems by Electrochemical Advanced Oxidation Processes. A Review. *Chem. Eng. J.* **2013**, *228*, 944–964. [CrossRef]
9. Dordio, A.v.; Duarte, C.; Barreiros, M.; Carvalho, A.J.P.; Pinto, A.P.; Teixeira, C. Bioresource Technology Toxicity and Removal Efficiency of Pharmaceutical Metabolite Clofibric Acid by *Typha* spp.—Potential Use for Phytoremediation? *Bioresour. Technol.* **2009**, *100*, 1156–1161. [CrossRef]
10. Basha, S.; Keane, D.; Morrissey, A.; Nolan, K.; Oelgemöller, M.; Tobin, J. Studies on the Adsorption and Kinetics of Photodegradation of Pharmaceutical Compound, Indomethacin Using Novel Photocatalytic Adsorbents (IPCAs). *Ind. Eng. Chem. Res.* **2010**, *49*, 11302–11309. [CrossRef]
11. Kim, I.; Yamashita, N.; Tanaka, H. Chemosphere Photodegradation of Pharmaceuticals and Personal Care Products during UV and UV/H_2O_2 Treatments. *Chemosphere* **2009**, *77*, 518–525. [CrossRef]
12. Zhao, Y.; Kuang, J.; Zhang, S.; Li, X.; Wang, B.; Huang, J.; Deng, S.; Wang, Y.; Yu, G. Ozonation of Indomethacin: Kinetics, Mechanisms and Toxicity. *J. Hazard. Mater.* **2017**, *323*, 460–470. [CrossRef]
13. Yu, T.H.; Lin, A.Y.C.; Lateef, S.K.; Lin, C.F.; Yang, P.Y. Removal of Antibiotics and Non-Steroidal Anti-Inflammatory Drugs by Extended Sludge Age Biological Process. *Chemosphere* **2009**, *77*, 175–181. [CrossRef] [PubMed]
14. Falås, P.; Baillon-Dhumez, A.; Andersen, H.R.; Ledin, A.; La Cour Jansen, J. Suspended Biofilm Carrier and Activated Sludge Removal of Acidic Pharmaceuticals. *Water Res.* **2012**, *46*, 1167–1175. [CrossRef] [PubMed]
15. Arya, V.; Philip, L.; Murty Bhallamudi, S. Performance of Suspended and Attached Growth Bioreactors for the Removal of Cationic and Anionic Pharmaceuticals. *Chem. Eng. J.* **2016**, *284*, 1295–1307. [CrossRef]
16. Prieto-Rodríguez, L.; Oller, I.; Klamerth, N.; Agüera, A.; Rodríguez, E.M.; Malato, S. Application of Solar AOPs and Ozonation for Elimination of Micropollutants in Municipal Wastewater Treatment Plant Effluents. *Water Res.* **2013**, *47*, 1521–1528. [CrossRef]
17. Chong, M.N.; Jin, B.; Chow, C.W.; Saint, C. Recent Developments in Photocatalytic Water Treatment Technology: A Review. *Water Res.* **2010**, *44*, 2997–3027. [CrossRef]
18. Andronic, L.; Enesca, A.; Cazan, C.; Visa, M. TiO_2-Active Carbon Composites for Wastewater Photocatalysis. *J. Solgel. Sci. Technol.* **2014**, *71*, 396–405. [CrossRef]
19. Sakkas, V.A.; Calza, P.; Medana, C.; Villioti, A.E.; Baiocchi, C.; Pelizzetti, E.; Albanis, T. Heterogeneous Photocatalytic Degradation of the Pharmaceutical Agent Salbutamol in Aqueous Titanium Dioxide Suspensions. *Appl. Catal. B* **2007**, *77*, 135–144. [CrossRef]
20. Hu, L.; Flanders, P.M.; Miller, P.L.; Strathmann, T.J. Oxidation of Sulfamethoxazole and Related Antimicrobial Agents by TiO_2 Photocatalysis. *Water Res.* **2007**, *41*, 2612–2626. [CrossRef] [PubMed]
21. Wang, J.; Yang, C.; Wang, C.; Han, W.; Zhu, W. Photolytic and Photocatalytic Degradation of Micro Pollutants in a Tubular Reactor and the Reaction Kinetic Models. *Sep. Purif. Technol.* **2014**, *122*, 105–111. [CrossRef]
22. Dai, J.; Yang, J.; Wang, X.; Zhang, L.; Li, Y. Enhanced Visible-Light Photocatalytic Activity for Selective Oxidation of Amines into Imines over TiO_2(B)/Anatase Mixed-Phase Nanowires. *Appl. Surf. Sci.* **2015**, *349*, 343–352. [CrossRef]
23. Nakata, K.; Ochiai, T.; Murakami, T.; Fujishima, A. Photoenergy Conversion with TiO_2 photocatalysis: New Materials and Recent Applications. *Electrochim. Acta* **2012**, *84*, 103–111. [CrossRef]
24. Zhou, Q.; Fang, Z.; Li, J.; Wang, M. Applications of TiO_2 Nanotube Arrays in Environmental and Energy Fields: A Review. *Microporous Mesoporous Mater.* **2015**, *202*, 22–35. [CrossRef]
25. Lai, Y.; Sun, L.; Chen, Y.; Zhuang, H.; Lin, C.; Chin, J.W. Effects of the Structure of TiO_2 Nanotube Array on Ti Substrate on Its Photocatalytic Activity. *J. Electrochem. Soc.* **2006**, *153*, D123. [CrossRef]

26. Zhuang, H.; Lin, C. Some Critical Structure Factors of Titanium Oxide Nanotube Array in Its Photocatalytic Activity. *Environ. Sci. Technol.* **2007**, *41*, 4735–4740. [CrossRef]
27. Liang, H.; Li, X. Effects of Structure of Anodic TiO$_2$ Nanotube Arrays on Photocatalytic Activity for the Degradation of 2,3-Dichlorophenol in Aqueous Solution. *J. Hazard. Mater.* **2009**, *162*, 1415–1422. [CrossRef] [PubMed]
28. Camposeco, R.; Castillo, S.; Navarrete, J.; Gomez, R. Synthesis, Characterization and Photocatalytic Activity of TiO$_2$ nanostructures: Nanotubes, Nanofibers, Nanowires and Nanoparticles. *Catal. Today* **2016**, *266*, 90–101. [CrossRef]
29. Marien, C.B.D.; Cottineau, T.; Robert, D.; Drogui, P. Applied Catalysis B: Environmental TiO$_2$ Nanotube Arrays: Influence of Tube Length on the Photocatalytic Degradation of Paraquat. *Appl. Catal. B* **2016**, *194*, 1–6. [CrossRef]
30. Laera, G.; Chong, M.N.; Jin, B.; Lopez, A. An Integrated MBR-TiO$_2$ Photocatalysis Process for the Removal of Carbamazepine from Simulated Pharmaceutical Industrial Effluent. *Bioresour. Technol.* **2011**, *102*, 7012–7015. [CrossRef]
31. Oller, I.; Malato, S.; Sánchez-Pérez, J.A. Combination of Advanced Oxidation Processes and Biological Treatments for Wastewater Decontamination—A Review. *Sci. Total Environ.* **2011**, *409*, 4141–4166. [CrossRef] [PubMed]
32. Leyva-Díaz, J.C.; López-López, C.; Martín-Pascual, J.; Muñío, M.M.; Poyatos, J.M. Kinetic Study of the Combined Processes of a Membrane Bioreactor and a Hybrid Moving Bed Biofilm Reactor-Membrane Bioreactor with Advanced Oxidation Processes as a Post-Treatment Stage for Wastewater Treatment. *Chem. Eng. Process. Process Intensif.* **2015**, *91*, 57–66. [CrossRef]
33. Lester, Y.; Aga, D.S.; Love, N.G.; Singh, R.R.; Morrissey, I.; Linden, K.G. Integrative Advanced Oxidation and Biofiltration for Treating Pharmaceuticals in Wastewater. *Water Environ. Res.* **2016**, *88*, 1985–1993. [CrossRef]
34. Ganzenko, O.; Trellu, C.; Papirio, S.; Oturan, N.; Huguenot, D.; van Hullebusch, E.D.; Esposito, G.; Oturan, M.A. Bioelectro-Fenton: Evaluation of a Combined Biological—Advanced Oxidation Treatment for Pharmaceutical Wastewater. *Environ. Sci. Pollut. Res.* **2018**, *25*, 20283–20292. [CrossRef] [PubMed]
35. De Wilt, A.; Arlos, M.J.; Servos, M.R.; Rijnaarts, H.H.M.; Langenhoff, A.A.M.; Parker, W.J. Improved Biodegradation of Pharmaceuticals after Mild Photocatalytic Pretreatment. *Water Environ. J.* **2020**, *34*, 704–714. [CrossRef]
36. Wang, Y.; Chen, C.; Zhou, D.; Xiong, H.; Zhou, Y.; Dong, S.; Rittmann, B.E. Eliminating Partial-Transformation Products and Mitigating Residual Toxicity of Amoxicillin through Intimately Coupled Photocatalysis and Biodegradation. *Chemosphere* **2019**, *237*, 124491. [CrossRef] [PubMed]
37. Zupanc, M.; Kosjek, T.; Petkovšek, M.; Dular, M.; Kompare, B.; Širok, B.; Blažeka, Ž.; Heath, E. Removal of Pharmaceuticals from Wastewater by Biological Processes, Hydrodynamic Cavitation and UV Treatment. *Ultrason. Sonochem.* **2013**, *20*, 1104–1112. [CrossRef] [PubMed]
38. De la Torre, T.; Alonso, E.; Santos, J.L.; Rodríguez, C.; Gómez, M.A.; Malfeito, J.J. Trace Organics Removal Using Three Membrane Bioreactor Configurations: MBR, IFAS-MBR and MBMBR. *Water Sci. Technol.* **2015**, *71*, 761–768. [CrossRef]
39. Zeng, A.P.; Deckwer, W.D. Bioreaction Techniques under Microaerobic Conditions: From Molecular Level to Pilot Plant Reactors. *Chem. Eng. Sci.* **1996**, *51*, 2305–2314. [CrossRef]
40. Chu, L.; Zhang, X.; Yang, F.; Li, X. Treatment of Domestic Wastewater by Using a Microaerobic Membrane Bioreactor. *Desalination* **2006**, *189*, 181–192. [CrossRef]
41. Martins, A.M.P.; Heijnen, J.J.; Van Loosdrecht, M.C.M. Effect of Dissolved Oxygen Concentration on Sludge Settleability. *Appl. Microbiol. Biotechnol.* **2003**, *62*, 586–593. [CrossRef]
42. Stadler, L.B.; Su, L.; Moline, C.J.; Ernstoff, A.S.; Aga, D.S.; Love, N.G. Effect of Redox Conditions on Pharmaceutical Loss during Biological Wastewater Treatment Using Sequencing Batch Reactors. *J. Hazard. Mater.* **2015**, *282*, 106–115. [CrossRef] [PubMed]
43. Bo, L.; Urase, T.; Wang, X. Biodegradation of Trace Pharmaceutical Substances in Wastewater by a Membrane Bioreactor. *Front. Environ. Sci. Eng. China* **2009**, *3*, 236–240. [CrossRef]
44. Xue, W.; Wu, C.; Xiao, K.; Huang, X.; Zhou, H. Elimination and Fate of Selected Micro-Organic Pollutants in a Full-Scale Anaerobic/Anoxic/Aerobic Process Combined with Membrane Bioreactor for Municipal Wastewater Reclamation. *Water Res.* **2010**, *44*, 5999–6010. [CrossRef] [PubMed]
45. Casas, M.E.; Chhetri, R.K.; Ooi, G.; Hansen, K.M.S.; Litty, K.; Christensson, M.; Kragelund, C.; Andersen, H.R.; Bester, K. Biodegradation of Pharmaceuticals in Hospital Wastewater by Staged Moving Bed Biofilm Reactors (MBBR). *Water Res.* **2015**, *83*, 293–302. [CrossRef] [PubMed]
46. Krkosek, W.H.; Payne, S.J.; Gagnon, G.A. Removal of Acidic Pharmaceuticals within a Nitrifying Recirculating Biofilter. *J. Hazard. Mater.* **2014**, *273*, 85–93. [CrossRef] [PubMed]
47. Stadler, L.B.; Love, N.G. Impact of Microbial Physiology and Microbial Community Structure on Pharmaceutical Fate Driven by Dissolved Oxygen Concentration in Nitrifying Bioreactors. *Water Res.* **2016**, *104*, 189–199. [CrossRef]
48. Torresi, E.; Casas, M.E.; Polesel, F.; Plósz, B.G.; Christensson, M.; Bester, K. Impact of External Carbon Dose on the Removal of Micropollutants Using Methanol and Ethanol in post-denitrifying Moving Bed Biofilm Reactors. *Water Res.* **2017**, *108*, 95–105. [CrossRef]
49. Zhang, A.; Zhou, M.; Han, L.; Zhou, Q. The Combination of Rotating Disk Photocatalytic Reactor and TiO$_2$ Nanotube Arrays for Environmental Pollutants Removal. *J. Hazard. Mater.* **2011**, *186*, 1374–1383. [CrossRef]
50. Smith, Y.R.; Kar, A.; Subramanian, V.R. Kinetics, Catalysis, and Reaction Engineering Investigation of Physicochemical Parameters That Influence Photocatalytic Degradation of Methyl Orange over TiO$_2$ Nanotubes. *Ind. Eng. Chem. Res.* **2009**, *48*, 10268–10276. [CrossRef]

51. Macak, J.M.; Zlamal, M.; Krysa, J.; Schmuki, P. Self-Organized TiO$_2$ Nanotube Layers as Highly Efficient Photocatalysts. *Small* **2007**, *3*, 300–304. [CrossRef]
52. Nie, X.; Chen, J.; Li, G.; Shi, H.; Zhao, H.; Wong, P.K.; An, T. Synthesis and Characterization of TiO$_2$ Nanotube Photoanode and Its Application in Photoelectrocatalytic Degradation of Model Environmental Pharmaceuticals. *J. Chem. Technol. Biotechnol.* **2013**, *88*, 1488–1497. [CrossRef]
53. Ye, Y.; Feng, Y.; Bruning, H.; Yntema, D.; Rijnaarts, H.H.M. Applied Catalysis B: Environmental Photocatalytic Degradation of Metoprolol by TiO$_2$ Nanotube Arrays and UV- LED: EFfects of Catalyst Properties, Operational Parameters, Commonly Present Water Constituents, and Photo-Induced Reactive Species. *Appl. Catal. B Environ.* **2018**, *220*, 171–181. [CrossRef]
54. Rezaei, M.; Royaee, S.J.; Jafarikojour, M. Performance Evaluation of a Continuous Flow Photocatalytic Reactor for Wastewater Treatment. *Environ. Sci. Pollut. Res.* **2014**, *21*, 12505–12517. [CrossRef]
55. Molinari, R.; Pirillo, F.; Loddo, V.; Palmisano, L. Heterogeneous Photocatalytic Degradation of Pharmaceuticals in Water by Using Polycrystalline TiO$_2$ and a Nanofiltration Membrane Reactor. *Catal. Today* **2006**, *118*, 205–213. [CrossRef]
56. Gar, M.; Taw, A.; Ookawara, S. Enhancement of Photocatalytic Activity Journal of Environmental Chemical Engineering Enhancement of Photocatalytic Activity of TiO$_2$ by Immobilization on Activated Carbon for Degradation of Pharmaceuticals. *J. Environ. Chem. Eng.* **2016**, *4*, 1929–1937. [CrossRef]
57. Burak Ozkal Can, M.S. A Comparative Heterogeneous Photocatalytic Removal Study on Amoxicillin and Clarithyromicin Antibiotics in Aqueous Solutions Can. *J. Water Technol. Treat. Methods* **2018**, *1*, 4–8.
58. Kawabata, K.; Sugihara, K.; Sanoh, S.; Kitamura, S.; Ohta, S. Photodegradation of Pharmaceuticals in the Aquatic Environment by Sunlight and UV-A, -B and -C Irradiation. *J. Toxicol. Sci.* **2013**, *38*, 215–223. [CrossRef]
59. Chang, M.T.; Wu, N.; Faqing, Z. A Kinetic Model for Photocatalytic Degradation of Organic Contaminants in a thin-film TiO$_2$ catalyst. *Water Res.* **2000**, *34*, 407–416. [CrossRef]
60. Cortés, J.A.; Alarcón-Herrera, M.T.; Villicaña-Méndez, M.; González-Hernández, J.; Pérez-Robles, J.F. Impact of the Kind of Ultraviolet Light on the Photocatalytic Degradation Kinetics of the TiO$_2$/UV Process. *Environ. Prog. Sustain. Energy* **2011**, *30*, 318–325. [CrossRef]
61. Choi, J.; Lee, H.; Choi, Y.; Kim, S.; Lee, S.; Lee, S.; Choi, W.; Lee, J. Heterogeneous Photocatalytic Treatment of Pharmaceutical Micropollutants: Effects of Wastewater Effluent Matrix and Catalyst Modifications. *Appl. Catal. B* **2014**, *147*, 8–16. [CrossRef]
62. Finčur, N.L.; Krstić, J.B.; Šibul, F.S.; Šojić, D.V.; Despotović, V.N.; Banić, N.D.; Agbaba, J.R.; Abramović, B. Removal of Alprazolam from Aqueous Solutions by Heterogeneous Photocatalysis: Influencing Factors, Intermediates, and Products. *Chem. Eng. J.* **2017**, *307*, 1105–1115. [CrossRef]
63. AFNOR. Evaluation En Milieu Aqueux de La Biodégradabilité Aérobie 'Ultime' Des Produits Organiques Solubles. Méthode Par Analyse de Dioxyde Dégagé. 1994, T 90-306. Available online: https://www.iso.org/fr/standard/42155.html (accessed on 6 May 2022).
64. Melgoza, R.M.; Chew, M.; Buitrón, G. Start–up of a Sequential Anaerobic/Aerobic Batch Reactor for the Mineralization of p–Nitrophenol. *Water Sci. Technol.* **2000**, *42*, 289–292. [CrossRef]
65. American Public Health Association; American Water Works Association; Water Environment Federation (Eds.) *Standard Methods for Examination of Water and Wastewater*, 21st ed.; APHA-AWWA-WEF Publisher: Washington, DC, USA, 2005.
66. Kuhn, H.; Fösterling, H. Kinetics of Chemical Reactions. In *Principles of Physical Chemistry*; Wiley: West Sussex, UK, 2000.
67. GáborLente Facts and Alternative Facts in Chemical Kinetics: Remarks about the Kinetic Use of Activities, Termolecular Processes, and Linearization Techniques. *Curr. Opin. Chem. Eng.* **2018**, *21*, 76–83. [CrossRef]

Review

A Systematic Review on Solar Heterogeneous Photocatalytic Water Disinfection: Advances over Time, Operation Trends, and Prospects

Felipe de J. Silerio-Vázquez [1], Cynthia M. Núñez-Núñez [2], José B. Proal-Nájera [3,*] and María T. Alarcón-Herrera [1,*]

1. Departamento de Ingeniería Sustentable, Centro de Investigación en Materiales Avanzados, S.C. Calle CIMAV 110, Colonia 15 de Mayo, Durango 34147, Mexico
2. Ingeniería en Tecnología Ambiental, Universidad Politécnica de Durango, Carretera Durango-México km 9.5, Durango 34300, Mexico
3. CIIDIR-Durango, Instituto Politécnico Nacional, Calle Sigma 119, Fraccionamiento 20 de Noviembre II, Durango 34220, Mexico
* Correspondence: jproal@ipn.mx (J.B.P.-N.); teresa.alarcon@cimav.edu.mx (M.T.A.-H.); Tel.: +52-618-1341781 (J.B.P.-N.); +52-614-4394896 (M.T.A.-H.)

Citation: Silerio-Vázquez, F.d.J.; Núñez-Núñez, C.M.; Proal-Nájera, J.B.; Alarcón-Herrera, M.T. A Systematic Review on Solar Heterogeneous Photocatalytic Water Disinfection: Advances over Time, Operation Trends, and Prospects. *Catalysts* **2022**, *12*, 1314. https://doi.org/10.3390/catal12111314

Academic Editor: Meng Li

Received: 30 September 2022
Accepted: 24 October 2022
Published: 26 October 2022

Publisher's Note: MDPI stays neutral with regard to jurisdictional claims in published maps and institutional affiliations.

Copyright: © 2022 by the authors. Licensee MDPI, Basel, Switzerland. This article is an open access article distributed under the terms and conditions of the Creative Commons Attribution (CC BY) license (https:// creativecommons.org/licenses/by/ 4.0/).

Abstract: Access to drinking water is a human right recognized by the United Nations. It is estimated that more than 2.1 billion people lack access to drinking water with an adequate microbiological quality, which is associated to 80% of all diseases, as well as with millions of deaths caused by infections, especially in children. Water disinfection technologies need a continuous improvement approach to meet the growing demand caused by population growth and climate change. Heterogeneous photocatalysis with semiconductors, which is an advanced oxidation process, has been proposed as a sustainable technology for water disinfection, as it does not need addition of any chemical substance and it can make use of solar light. Nevertheless, the technology has not been deployed industrially and commercially yet, mainly because of the lack of efficient reactor designs to treat large volumes of water, as most research focus on lab-scale experimentation. Additionally, very few applications are often tested employing actual sunlight. The present work provide a perspective on the operation trends and advances of solar heterogeneous photocatalytic reactors for water disinfection by systematically analyzing pertaining literature that made actual use of sunlight, with only 60 reports found out of the initially 1044 papers detected. These reports were discussed in terms of reactor employed, photocatalyst used, microorganism type, overall disinfection efficiency, and location. General prospects for the progression of the technology are provided as well.

Keywords: sunlight; reactor design; advanced oxidation processes; water treatment; water potabilization

1. Introduction

As of 2010, the access to water has been recognized as a human right by the United Nations. Water purposed for personal and domestic use should comply with sufficiency, physical availability, safeness, and affordability [1]. Nevertheless, the World Health Organization (WHO) estimates that 30% of the global population, which accounts for 2.1 billion people, lack access to water sources which meet guidelines for safe drinking water [2]. Water which does not observe those set guidelines cannot be considered as drinking water and its consumption can be hazardous; it is estimated than the intake of unsafe water is at fault for 80% of all of the world diseases [3]. Among these illnesses, infectious diseases which are caused by pathogens, mainly bacteria and virus, are recurrent [4]. Some of them are typhoid, cholera, dysentery, parasitic infections [3] or viral infections [4]. Water-borne diseases can become lethal, especially if patients do not receive medical attention; these diseases cause 1.8 million deaths each year, children being the most vulnerable group. Water quality improvement reduces this morbidity [3].

The WHO has also projected, that by year 2025, half of the world population will live in water-stressed areas. To provide safe drinking water has become one of the biggest challenges for mankind in the present century [5].

Among the WHO guidelines for drinking water, it is stated that no microorganism known to be pathogenic should be contained within water, hence, drinking water should be disinfected to ensure this guideline [6]. Water supply systems may use one or several disinfection technologies in order to ensure no pathogenic microorganisms are present, the selected technology depends on a plethora of factors, such as availability, water initial microbial quality, and final quality intended or needed, cost effectiveness, volume required or even level of automation and local level costs [7]. Table 1 lists some of the most used water disinfection processes, along with some of their advantages and disadvantages.

Table 1. Advantages and disadvantages of some of the most used water disinfection technologies.

Water Disinfection Process	Advantages	Disadvantages
Chlorination	Low cost; effective at low concentrations; residual effect, widely available [8]	Formation of toxic byproducts; modified taste and odor; ineffective against biofouling; cannot kill parasite eggs [9,10]
Ozonation	High biocidal efficacy over a wide antimicrobial spectrum; color, odor, and taste control [11]	Lack of residual effect as ozone is unstable in water; byproducts formation from bromide and natural organic matter [12]
Disinfection with colloidal silver	Well-known biocidal and disinfection properties; able to remove organic compounds; lacks the adverse effects of chlorination and ozonation [13]	Relatively high cost and time inversion; loss of effectivity over time [14]
Peracetic acid treatment	Low dependence on pH; high sterilization ability; reduced toxic byproducts' formation; easy implementation [15]	Costly activation (UV light or metal catalysis); not proven technical feasibility to inactivate fungi, algae of microorganisms on biofilms; scarce pilot plant applications which limit economic feasibility evaluation [16]
Ultraviolet radiation	No chemical addition, reduce disinfection byproduct (DBP) formation, high efficiency in inactivating chlorine-resistant organisms [17]	Energy intensive; unsuitable for places without stable energy supply, lack of a residual effect [18,19]
Solar disinfection (SODIS)	Non-energy intensive as it harvests solar energy; effective for several microorganisms; point of use technology [20]	Low efficiency in solar energy conversion; long exposure time; microorganism regrowth might happen if UV exposure is not high enough [21]

Considering the increasing water stress context and the drawbacks of the known disinfection water technologies, the need to re-design or improve these processes to obtain technologies which are robust, simple to use, chemical-free, and inexpensive arises [22]. Some emerging water disinfection technologies include electrodisinfection [23], water cavitation [24] or heterogeneous photocatalysis (HP) with semiconductors, also known as photocatalysts (PC).

HP is an advanced oxidation process (AOP) which was first reported in 1972 when water splitting was observed on the surface of a titanium electrode, namely, over titanium dioxide (TiO_2) due to the effect of light irradiation [25]. When a PC is exposed to radiation (hv) with energy higher than its band-gap (space between the molecule conduction band and valence band) level energy, an electron from the valence band migrates to the conduction band, creating a hole with a positive charge in the valence band (h^+) and an extra electron with a negative charge in the conduction band (e^-) [26,27]. The photo-generated charges move up to the PC's surface, which then give place to redox reactions when oxygen and water are present, generating reactive oxidizing species (ROS), mainly hydroxyl radical ($HO^\bullet$) and superoxide radical ($O_2^{\bullet-}$), but also hydrogen peroxide (H_2O_2) [28].

HP is a potentially sustainable technology for water disinfection, as PC can be activated employing solar light, and moreover, no additional chemical substance is needed in the process, which minimizes the potential formation of DBP and environmental harmful effects [29,30]. HP has also been researched for inorganic pollutants' removal from water, such as hexavalent chromium [31] or trivalent arsenic [32], and also for recalcitrant organic compound degradation, such as dyes, phenolic compounds, pesticides and pharmaceutical active compounds [33], posing HP as potential technology for comprehensive water treatment.

TiO_2 has remained as the most studied PC, despite the fact it is only active under UV irradiation (which accounts for nearly 5% of the total Sun spectral irradiation), whose wavelengths are in the range of 200–400 nm (TiO_2 peak absorbance is around 387 nm). It also has a relative high recombination rate of the photo-generated charges. Several approaches have been researched to address these issues, such as doping TiO_2 with other elements, coupling it with other PC to form heterojunctions, or even with itself to form homojunctions; these strategies generally improved TiO_2 photocatalytic activity to some extent, although most studies have only been carried out at lab-scale [25,34–37].

Other materials have also been researched, such as the bismuth oxyhalides, which are a group of layered materials with narrow band-gaps [38,39]. It is still unclear if a narrow band-gap is the answer to use solar light more efficiently. Black TiO_2, a variation of TiO_2 exhibiting dark coloration instead of white, and sometimes also referred to as reduced, hydrogenated or oxygen-vacant TiO_2, was first reported in 2011 and its ability to show photocatalytic activity even under infrared light was noticed; although activity under visible light has been reported and some studies show it outperforms pristine TiO_2, it is believed that this happens due to an improved use of UV irradiation rather than an effective use of visible or infrared light [40,41].

Water disinfection via HP has been broadly researched for at least the last twenty years, although disinfection via UV or heat might happen simultaneously [42]; in disinfection via HP, the generated ROS cause oxidative stress to a wide variety of microorganisms, including Gram-positive and Gram-negative bacteria, DNA viruses, and RNA viruses amongst others [43]. The microorganism cell integrity is compromised as the ROS cause damage to the cell membrane, resulting in cytoplasm leakage; ROS can also obstruct cell vital functions like protein synthesis or break biomolecules covalent bonds [44–46]. A schematic representation of ROS generation at molecular level (depicting TiO_2) and microorganism disinfection via HP is shown in Figure 1.

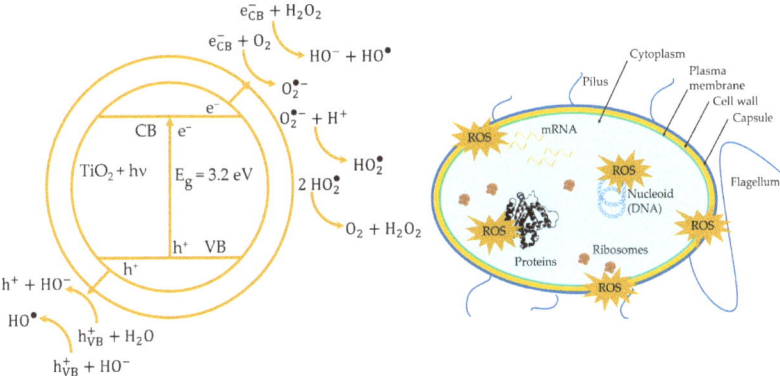

Figure 1. Schematic representation of ROS generation and disinfection via heterogeneous photocatalysis.

Water disinfection via solar HP depends on several factors, to name a few: nature and initial concentration of microorganisms, process duration, irradiation intensity, composition of the water matrix, turbidity, water layer depth, sunlight angle of incidence, water pH, and PC properties [45,47,48].

Despite its potential advantages, solar HP has not been deployed as a full-scale process yet; several limitations have been addressed such as low photoconversion efficiency, scarce information on energy consumption, and yields on catalyst preparation processes amongst others, but the lack of knowledge on the field of solar HP reactor design, as well as the absence of a consensus on proper design methodologies, are recognized as the main ones [49]. Several reactor designs have been proposed based on devices designed for solar thermal applications, such as the parabolic trough reactor (PTR) based on the parabolic trough solar collector, the compound parabolic reactor (CPC) based on the compound parabolic solar collector or the flat plate reactor (FPR), based on flat devices for solar energy collection [50,51].

The number of papers published on the topic of disinfection via HP has been increasing year by year during the last two decades, which reflects the interest of the scientific community in these technologies [52]. However, the amount of research focusing on reactors is generally scarce [53], and even though one of the most promoted characteristics of HP is its ability to make use of solar energy, plenty of research is performed using solar simulators instead of actual sunlight, as the controlled conditions allowed by these devices provide accurate and reproducible results [54]. PC efficiency differs between simulated sunlight and actual sunlight [55], hence, carrying out research employing actual sunlight is also relevant and needed.

The objective of the present work is to analyze scientific works focusing on the use of solar HP reactors for water disinfection. A systematic literature search was conducted following the preferred reporting items for systematic reviews and meta-analyses (PRISMA) protocol to discriminate non-pertaining works. The operations have been discussed in function of reactor type used, PC properties, type of the microorganism treated, disinfection performance, and location.

2. Methods for Literature Search, Inclusion Criteria, and Review

The literature review method was performed following the PRISMA four-step procedure [56,57]. The four steps are:
1. Identification of relevant papers indexed by databases.
2. Screen the papers for the determined criteria.
3. Verify papers' eligibility.
4. Incorporate the eligible papers in the systematic review.

An electronic search of articles was performed on Scopus and Web of Knowledge using the terms "photocataly*", "disinfection", and "solar", searching on title, abstract, and keywords in Scopus, and on all fields in Web of Knowledge. As it has been reported that research focused on reactors is scarce, comprising less than 2% of all papers related to HP [58], it was decided to include any work involving water disinfection regardless of the objective or publication year.

Books, book chapters, review articles, conference proceedings, and articles not published in English were excluded. Following the first exclusion, papers were screened for the inclusion criteria. Papers not focused on water disinfection (i.e., energy generation or pollutants' degradation), papers that did not make use of actual sunlight (i.e., solar simulators or UV-lamps), and papers that did not use a photocatalytic reactor (i.e., test tubes of bakers) were not included. Figure 2 illustrates the PRISMA steps taken for papers' eligibility.

Eligible papers were then analyzed for data extraction including the type of reactor employed, the PC used, type of microorganisms, microorganism concentration, disinfection efficiency, operation duration, volume of water treated, operation timeframe, and experimentation location.

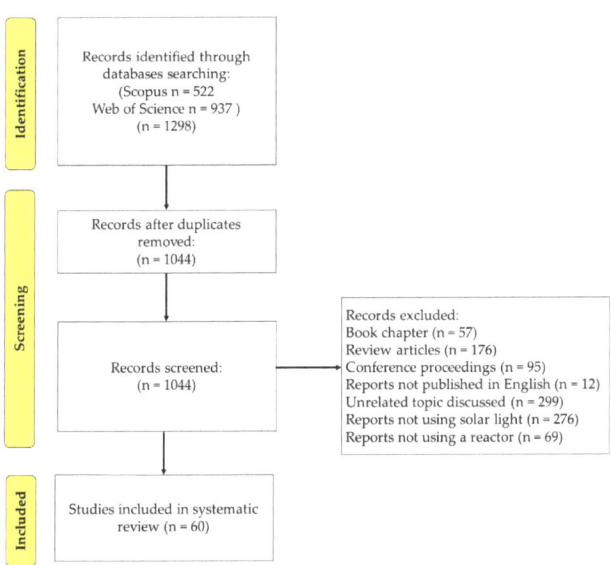

Figure 2. Flowchart for selection of literature.

3. Results and Discussion
3.1. Data Synthesis

A total of 60 reports were deemed as appropriate for screening in this systematic review out of the 1043 records originally found. Table 2 shows the extracted data.

Table 2. Research reports focused on water disinfection via heterogeneous photocatalysis employing actual sunlight carried out in photocatalytic reactors.

Reactor	Photocatalyst	Microorganism Treated	Initial Concentration (CFU/mL)	Log Cycle Reduction	Process Time (min)	Volume Capacity (L)	Operation Timeframe	Location	Reference
CPC type	TiO_2	Escherichia coli	10^5	−5	35	150	Not reported	Not reported	[59]
CPC type	TiO_2	Escherichia coli	10^6	−6	120	Not reported	14:00–16:00	Seoul, Korea (38° S, 127° E)	[60]
CPC type	TiO_2	Escherichia coli	10^5	−4	30	35	Not reported	Almeria, Spain (37° N, 2° W)	[61][a]
CPC type	TiO_2	Escherichia coli	10^6	−4	12	11	Not reported	Almeria, Spain (37° N, 2° W)	[62][a]
CPC type	TiO_2	Escherichia coli	10^6	−6	90	70	12:00–16:00	Lausanne, Switzerland (46° N, 6° E)	[63][a]
CPC type	TiO_2	Escherichia coli	10^4	−3	45	11	Not reported	Almeria, Spain (37° N, 2° W)	[64][a]
CPC type	TiO_2	Escherichia coli	10^6	−6	60	1	Not reported	Dublin, Ireland (53° N, 15° W)	[65][a]
CPC type	TiO_2	Fecal coliforms	10^6	−6	360	20	Not reported	Tucumán, Argentina (26° S, 64° W)	[66]
CPC type	TiO_2	Escherichia coli	10^6	−6	150	37	Not reported	Lausanne, Switzerland (46° N, 6° E)	[67][a]
CPC type	TiO_2	Escherichia coli	10^6	−6	180	35	Not reported	Lausanne, Switzerland (46° N, 6° E)	[68]
CPC type	TiO_2	Escherichia coli	10^6	−6	30	35	Not reported	Lausanne, Switzerland (46° N, 6° E)	[69][a]
FPR type	TiO_2	Escherichia coli	10^4	−4	60	1	Not reported	Not reported	[70]
CPC type	TiO_2	Escherichia coli	10^7	−5	90	14	Not reported	Almeria, Spain (37° N, 2° W)	[71][a]
CPC type	TiO_2	Escherichia coli	10^6	−4	90	14	09:00–10:30	Almeria, Spain (37° N, 2° W)	[72][a]
Rectangular type	TiO_2	Coliforms	10^6	−5	120	12.8	13:00–15:00	Not reported	[73]
CPC type	TiO_2	Fusarium solani	10^3	−3	300	14	11:00–16:00	Almeria, Spain (37° N, 2° W)	[74][a]
CPC type	TiO_2	Escherichia coli	10^6	−6	50	20	Not reported	Porto, Portugal (41° N, 8° W)	[75][a]
CPC type	TiO_2	Fusarium spp.	10^3	−3	240	60	11:00–16:00	Almeria, Spain (37° N, 2° W)	[76][a]
CPC type	TiO_2	Escherichia coli	10^5	−5	240	12	10:00–14:00	Lares, Peru (13° S, 72° W)	[77][a]
CPC type	TiO_2	Escherichia coli	10^6	−6	Not reported	10	Not reported	Almeria, Spain (37° N, 2° W)	[78][a]
Offset tubular type	TiO_2	Coliforms	10^3	−3	240	300	11:00–15:00	Not reported	[79]
Offset tubular type	TiO_2	Escherichia coli	10^6	−6	300	7	10:30–15:30	Almeria, Spain (37° N, 2° W)	[80]
FPR type	TiO_2	Aeromonas hydrophila	10^5	−1.2	2.5	0.2	Not reported	Queensland, Australia (20° S, 142° E)	[81]

Table 2. *Cont.*

Reactor	Photocatalyst	Microorganism Treated	Initial Concentration (CFU/mL)	Log Cycle Reduction	Process Time (min)	Volume Capacity (L)	Operation Timeframe	Location	Reference
FPR type	TiO_2	*Aeromonas hydrophila*	10^5	−1.38	2.5	0.2	Not reported	Queensland, Australia (20° S, 142° E)	[82]
CPC type	TiO_2	Phage ΦX174	Not reported	−3	30	Not reported	Not reported	Dublin, Ireland (53° N, 6° W)	[83] [a]
CPC type	TiO_2	*Microcystis aeruginosa*	10^7	−7	20	20	Not reported	Porto, Portugal (41° N, 8° W)	[84] [a]
CPC type	TiO_2	*Escherichia coli*	10^5	−4	300	10	11:00–15:00	Almeria, Spain (37° N, 2° W)	[85] [a]
PTR type	TiO_2	Coliforms	Not reported	−2	300	5	Not reported	Indore, India (22° N, 75° E)	[86]
FPR type	N-TiO_2	Coliforms	10^3	−3	180	1.332	10:00–16:00	Not reported	[87]
CPC type	TiO_2	Fecal coliforms	10^3	−1	120	1	Not reported	Medellin, Colombia (6° N, 75° W)	[88] [a]
CPC type	TiO_2	*Fusarium solani*	10^2	−2	120	60	10:00–16:00	Almeria, Spain (37° N, 2° W)	[89] [a]
Staircase reactor	TiO_2	*Escherichia coli*	10^6	−2	140	Not reported	Not reported	Not reported	[90] [a]
CPC type	TiO_2	*Escherichia coli*	10^6	−6	300	8.5	10:00–15:00	Almeria, Spain (37° N, 2° W)	[91] [a]
CPC type	TiO_2	*Escherichia coli*	10^6	−6	Not reported	60	Not reported	Almeria, Spain (37° N, 2° W)	[92] [a]
CPC type	TiO_2	*Escherichia coli*	10^7	−4	300	15	Not reported	Perpignan, France (42° N, 2° E)	[93]
CPC type	TiO_2	*Escherichia coli*	10^7	−2	90	6.4	Not reported	Bangkok, Thailand (13° N, 100° E)	[94] [a]
PTR type	TiO_2	Fecal coliforms	10^6	−6	360	Not reported	09:00–17:00	Madhya Pradesh, India (22° N, 75° E)	[95]
CPC type	TiO_2	*Escherichia coli*	10^5	−5	80	27	Not reported	Cadiz, Spain (36° N, 6° W)	[96] [a]
Box type	TiO_2	*Escherichia coli*	10^7	−2	480	0.66	08:30–16:30	Not reported	[97]
CPC type	TiO_2	*Curvularia* sp.	10^3	−3	120	20	Not reported	Almeria, Spain (37° N, 2° W)	[98] [a]
CPC type	Ag-$BiVO_4$	*Escherichia coli*	10^6	−6	40	10	10:00–15:00	Almeria, Spain (37° N, 2° W)	[99] [a]
CPC type	TiO_2	*Escherichia coli*	10^6	−4	81	15	Not reported	Dublin, Ireland (53° N, 6° W)	[100] [a]
Concave dish type	TiO_2	Heterotrophic bacteria	10^3	−4	240	1	10:30–14:30	Gonabad, Iran (34° N, 58° E)	[101]
CPC type	TiO_2	*Enterobacter cloacae*	10^9	−4.9	180	3	Not reported	Medellin, Colombia (6° N, 75° W)	[102] [a]
CPC type	TiO_2	Fecal coliforms	10^5	−5	300	20	Not reported	Almeria, Spain (37° N, 2° W)	[103] [a]
CPC type	TiO_2	*Escherichia coli*	10^7	−7	105	0.6	10:00–13:00	Nsukka, Nigeria (6° N, 7° E)	[104]

Table 2. Cont.

Reactor	Photocatalyst	Microorganism Treated	Initial Concentration (CFU/mL)	Log Cycle Reduction	Process Time (min)	Volume Capacity (L)	Operation Timeframe	Location	Reference
Compound Triangular collector type	N-TiO$_2$	*Escherichia coli*	10^3	−1	180	Not reported	10:00–13:00	Salerno, Italy (37° N, 14° E)	[105] [a]
Offset tubular type	Ag-TiO$_2$	*Escherichia coli*	10^5	−5	30	20	10:00–14:00	Pondicherry, India (12° N, 79° E)	[106]
FPR type	TiO$_2$	*Escherichia coli*	10^6	−6	60	1	Not reported	Not reported	[107] [a]
PTR Type	Ag-TiO$_2$	*Escherichia coli*	10^6	−6	180	Not reported	10:00–15:00	Not reported	[108]
Offset tubular type	TiO$_2$	Coliforms	10^2	−2	560	Not reported	08:00–17:00	Mae Salong Nok, Thailand (20° N, 99° E)	[109]
CPC type	ZnO	Fecal coliforms	10^7	−7	15	1.1	11:00–15:00	Tehran, Iran (35° N, 51° E)	[110]
FPR type	TiO$_2$	Fecal Coliform	10^6	−4	45	2	Not reported	Durango City, Mexico (23° N, 104° W)	[111] [a]
CPC type	TiO$_2$	*Escherichia coli*	10^5	−3.5	360	2.2	Not reported	Cadiz, Spain (36° N, 6° W)	[112]
CPC type	rGO-TiO$_2$	*Klebsiella pneumonia*	10^9	−9.3	210	0.39	Not reported	Stellenbosch, South Africa (33° S, 18° E)	[113]
CPC type	TiO$_2$	*Escherichia coli*	10^7	−4	300	15	11:00–15:00	Perpignan, France (42° N, 2° E)	[114]
Optofluidic capillary type	Red phosphorous	*Escherichia coli*	10^8	−8	28	0.004	Not reported	Not reported	[5]
Linear Fresnel type	TiO$_2$	*Escherichia coli*	10^4	−3	300	1	08:00–13:00	Tsukuba, Japan (36° N, 140° E)	[115]
Through reactor	Fe-TiO$_2$	*Escherichia coli*	10^6	−6	120	6	12:00–14:00	Patiala, India (30° N, 76° E)	[116]
Offset tubular type	TiO$_2$	*Escherichia coli*	Not reported	Not reported	420	120	Not reported	Boyacá, Colombia (5° N, 72° O)	[117]

[a] Cumulative UV dose (Q_{UV}) reported.

3.2. Type of Reactor

CPC commonly is composed of tubes made of borosilicate glass placed above parabolic reflectors made of polished aluminum, installed on an inclinable stand tilted at local latitude; water reservoirs and water pumps are also needed, and the most sophisticated ones count with radiometers, flow meters, and sensors for temperature, pH and dissolved oxygen measuring [118]. Due to its optical efficiency and its ability to use both direct and diffuse solar UV light, it has been considered as the most ideal solar reactor design available at the moment, especially for environmental applications [119], hence, its use on the majority of papers reviewed was within expectations. Figure 3 depicts a scheme of a CPC and its main components, as well as one of a FPR and one of a PTR.

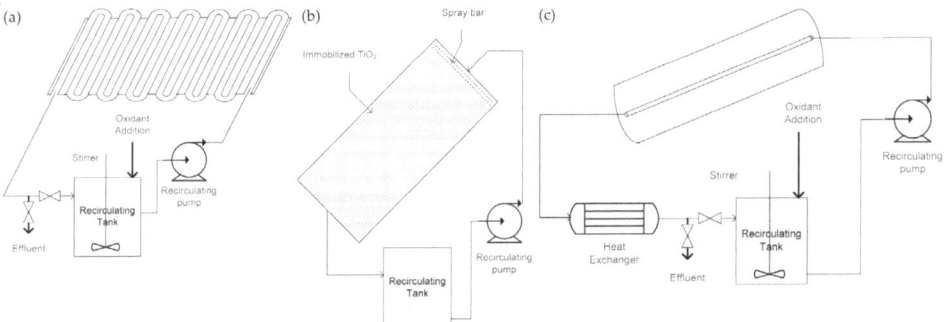

Figure 3. Schematic drawing of: (**a**) slurry CPC (Compound Parabolic Concentrator) system; (**b**) FPR (Flat plate reactor) system; (**c**) PTR (Parabolic Trough reactor) system (Reprinted from Ref. [120]).

FPR consists of a flat or corrugated surface over which a thin film of water flows (in most cases, in the laminar flow regime); it can be a proper approach for small-scale applications, as it offers a large surface area where the PC can be immobilized and its design is simple [121]. Although the reports reviewed in this paper employing this design were a lot less in number than those using a CPC design, it is worth mentioning that two of the papers reviewed reported steady-state operation rather than batch operation, which is one of the sought-after characteristics for reactors. An example on the operation of an FPR reactor can be seen in Figure 3 above.

Offset tubular reactors were employed in four studies. This reactor is similar to the CPC, with the only difference that it does not count with a reflector, which makes it less expensive than the CPC. As it does not need a reflector, more tubular sections could be positioned within the same space a CPC occupies, allowing the treatment of a higher volume of water, as a work which compared both reactor designs suggested [122]. A potential disadvantage in comparison to the CPC is the lower temperature the water might reach. Temperature in the range of 20–80 °C does not affect TiO_2 photo-excitation, and although dissolved oxygen concentration in water decreases as water temperature increases, ionic products of water (OH^- and H_3O^+) do increase in the temperature range of 20–80 °C, which promotes $HO^\bullet$ generation [92]. In addition to damage caused by ROS to the microorganism, inactivation also happens due to UV light effect alone and temperature increase [123].

Three more papers reported the use of a PTR, in which main components include a parabolic reflector mounted over a rotating platform to concentrate sunlight; its use has dropped as it has numerous insuperable disadvantages: its components are relatively expensive, the high sunlight concentration leads to an excessive heating which hinders photocatalytic activity, and it is only able to use direct sunlight [115]. Likewise, a scheme showing an example of the operation of a PTR can be seen above in Figure 3. Table 3 shows the summarized advantages and disadvantages of the most used photocatalytic reactor types.

Table 3. Summarized advantages and disadvantages of the most used photocatalytic reactor types.

Type of Reactor	Brief Description	Advantages	Disadvantages
Compound parabolic concentrator (CPC)	Transparent cylindrical receptors are placed onto two joined half-parabola shape reflectors [120]	Uniform light irradiation for cylindrical receptors; able to use both direct and diffuse solar irradiation; able to work at low solar concentration rates [50,120,124]	Aluminum reflector is imperative, which implies a higher cost; limited optical efficiency due to light being reflected multiple times [50,125]
Flat plate reactor (FPR)	Water flows over a tilted plate whence the PC has been immobilized [48]	Relatively lower price as no reflectors are needed [50]	Atmosphere can prevent the system to work in an appropriate way; high pressure is needed to pump water onto large surfaces [124]
Parabolic trough reactor (PTR)	A parabolic light-reflecting surface concentrates solar irradiation in a transparent receptor; usually operated in the turbulent flow regime; equipped with a sun tracking system [120]	Easily adaptable and developed technology; able to use both direct and diffuse solar irradiation [50,124]	Sun tracking systems implies a higher cost; aluminum reflector is imperative, which implies a higher cost; solar concentration factor above one sun inhibits photocatalytic activity; increase in temperature reduces dissolved O_2 concentration, which slows down photocatalytic activity; unable to use diffuse solar irradiation, rendering it impractical on overcast or cloudy days; a relatively large area needed for installation [50,120]

The remaining eight reactors can be considered as empirical approaches (as no design parameters are included), that due to relative youngness of the field, results in research teams frequently employing their own singular design, as pointed out more than a decade ago [48].

It has already been stated that the biggest challenge to translate HP to commercial applications remains as the lack of an efficient reactor design suitable for treating large volumes of water [126]. CPC remains as the most studied reactor, with no update, which again points out the need of more research focused on reactors. Additionally, reactor innovation can be made in diverse ways; some process intensification attempts have been reported which involve combining solar HP with other processes, such as solar pasteurization [42] or ozonation [127]. Additionally, more research is needed to understand the interesting synergy between the different microorganism inactivation mechanisms which take place within a solar HP reactor (UV light inactivation and thermal inactivation).

3.3. Photocatalyst Used

TiO_2 was the most used PC in the papers analyzed, with a total of 51 studies. Around 1600 papers examining its usage in disinfection have been published over the last 20 years, with an increase from 5 papers in 2000 to 165 papers in 2019, which is an indicator of the growing research interest in heterogeneous photocatalytic disinfection, leading to scaled-up applications [30], just as the ones analyzed in the present work. TiO_2 is known for its noteworthy photocatalytic activity, due in part to its specific surface area of around 46.06 m^2 g^{-1} and a surface energy of 80 mJ·m^{-2}, but also for its optical and electronic properties, high chemical stability, low cost, non-toxicity, and environmental friendliness [128–130].

On the other hand, TiO_2 has limitations, including fast electron-hole recombination even though adding H_2O_2 results in an almost inhibited recombination by scavenging e$^-$ [131], and other drawbacks include slow charge carrier transfer, elevated recycling cost, and photocatalytic activity under UV irradiation only, due to its wide band-gap of 3.2 V, limiting its efficiency for solar applications, as UV irradiation account for less than 5% of the received solar energy [132]. ZnO, a PC used in one of the papers reviewed, also has a band-gap value of 3.2 V, and its drawbacks are almost identical to those of TiO_2 [47].

To overcome these issues, TiO_2 modification by doping and coupling has been researched, i.e., $Ag-TiO_2$, $Fe-TiO_2$, $N-TiO_2$, and $rGO-TiO_2$, as used in some of the papers reviewed. Research towards PC with narrower band-gap and photocatalytic activity under visible light irradiation, such as the ones used in the papers reviewed, Ag/BiVO4 and red phosphorus, is the other relevant research trend [133].

When TiO_2 surface is modified by metal doping, a Schottky's barrier arises when irradiated with UV light, causing the metal Fermi levels to be lower than those of TiO_2 CB, increasing the metal ability to accept electrons, inhibiting charge recombination. When doping with a noble metal, surface plasmon resonance happens, which allows for electrons to be transferred directly to TiO_2 CB [134].

$Fe-TiO_2$ has shown a better performance than TiO_2, which is attributed to Fe atoms acting as electron-hole traps, slowing down charge recombination and enhancing photocatalytic activity as a result; it also enhances specific surface area [135,136].

$Ag-TiO_2$ has been researched since 1984 and it has several advantages over conventional TiO_2, such as a narrower band-gap of 2.77 eV, a higher specific surface area (239 $m^2 \, g^{-1}$) and the plasmonic effect which increase visible light response. $Ag-TiO_2$ has been gaining attention from the scientific community, which is also aimed at overcoming its share of drawbacks, such as photocatalytic activity gradual loss and Ag leaching [132].

Nitrogen is the most common non-metal used as a doping agent for TiO_2, predominantly because of its small ionization energy and its atomic size comparable with that of oxygen. Doping with oxygen confers TiO_2 photocatalytic activity under visible light, although the exact mechanism for this enhancement is still elusive and not totally understood; $N-TiO_2$ has been used for several applications [137,138], including solar disinfection in a photocatalytic reactor.

Graphene-based and TiO_2 composites have also been researched, which has resulted in an increased photocatalytic performance [139]. Composited of TiO_2 and reduced graphene oxide (rGO) higher photocatalytic activity is attributed to a synergism involving a higher number of photocatalytic active sites, a superior light collection (due to the hierarchical structural interface between unidimensional TiO_2 and bidimensional rGO) and an enhanced charge separation rate [140].

PC based on bismuth have been widely researched for water disinfection applications, as they are non-toxic, chemically stable, visible light active, synthesized with ease, reusable, and relatively economic. Bismuth-based PC disadvantages include low light absorption, high charge recombination, and a slow charge migration [141]. One of the most promising PC is bismuth vanadate ($BiVO_4$), whose properties can be improved when doped with metals (such as in one of the works reviewed in this paper) [142]. Although 15% Ag doped $BiVO_4$ was effective for water disinfection, it was not able to outperform conventional TiO_2 [99].

Red phosphorous belongs to a different kind of PC, which are elemental PC; it is an allotrope inert to chemical reactions and of elevated thermodynamic stability; its band-gap in the range of 1.4–2.0 allows it to be active under visible light, its production is non-expensive as the raw material is of low cost, and it is non-toxic; it is a promising material for photocatalytic disinfection, although there is still non-consensus about its resistance to oxidation [143,144].

3.4. Microorganism Type

The vast majority of works reviewed in this paper (a total of 38) analyzed water disinfection employing *E. coli*, which is within expectations due to the interest of the scientific community in it, being a research subject in evolutionary, biological genetic, and molecular studies [145]. *E. coli* is regarded as an indicator microorganism for the presence of bacteria by the WHO, as it has been characterized extensively and its presence, which indicates fecal contamination, is common in untreated water sources; however, it is inactivated with more ease than other microorganisms, hence, its absence does not guarantee that any other fecal coliform or microorganism are absent as well [146].

Among the reviewed works, five and six papers analyzed disinfection of total coliforms and fecal coliforms, respectively, which is completely appropriate within the interest of performing experiments in conditions as close to reality as possible, as it is well-known that disinfection rate is different between microorganisms of different species, and might even vary between strains of the same species [147]. Methods to quantify coliforms in the works reviewed included test paper and plate count [88,109].

One of the reviewed papers analyzed total heterotrophic bacteria disinfection, and four more analyzed disinfection employing a determined species. Several bacteria genres are included within the heterotrophic bacteria, such as *Aeromonas*, *Citrobacter*, *Enterobacter*, *Helicobacter*, *Klebsiella*, and *Serratia* among others. Some of these bacteria can cause health issues to humans with suppressed immunologic response, and are also associated with a low organoleptic water quality [148,149].

The last paper analyzing bacteria disinfection focused on *Microcystis aeruginosa*, which is a cyanobacteria that causes immense harm to ecosystems due to the release of cyanobacterial toxins and algal organic matter; its rapid growth greatly affects the efficiency of drinking water treatments [149]. Previous work has reported that *Microcystis aeruginosa* regrowth after photocatalytic is inhibited, as cell density is less than 85% compared to control experiments [150]. Since HP is also able to degrade organic compounds, it can offer a comprehensive alternative for cyanobacteria disinfection, as the detrimental cyanobacterial toxins can be degraded as well [126,151].

In the case of fungi disinfection, three papers focused on the genre *Fusarium*, which is a fungus that can cause a condition called fusariosis. Fusariosis symptoms depend on the affected area and the host's immunological response, but they can include nail, skin, and eye infection [152]. According to estimations, the species *Fusarium solani* is associated with 50% of all the fusariosis reported cases [153].

One of the papers examined *Curvularia* spp. disinfection; although it is infrequent, this genre can cause several types of human mycoses, including: fungal keratitis, onychomycosis, peritonitis, invasive sinusitis, subcutaneous disease, and systemic infections among others [154].

Viral disinfection was also reported. A paper analyzed disinfection of the ΦX174 virus, which is a phage that commonly infects *E. coli*, hence, its presence in water is also considered an indicator of fecal pollution [155].

Summarizing, solar HP reactors have been studied for water disinfection involving several types of microorganisms, and although the exact mechanism does surely differ from one microorganism to another, the oxidation caused by the ROS surely plays an important role.

3.5. Disinfection Performance

Most of the papers reviewed reported a high disinfection rate of several orders of magnitude. However, this data alone is not enough to properly assess process efficiency. It is well known that there are no established and unanimous figures of merit to evaluate the efficiency of HP processes, as the amount of diverse studies is vast, which include emerging technologies and processes combinations, giving as a result a massive challenge to critically assess HP efficiency, which is also multidimensional, as operational costs, sustainability, feasibility and yields, among other parameters, such as microorganism nature, initial concentration, water matrix and pH, to name a few, should, preferably, be considered [47,156].

Proposed benchmarks to evaluate HP efficiency include disinfection rate constant, photocatalytic space-time yield, photonic yield, quantum yield, and population log reduction among others; however, the methodologies to determine these figures of merit can differ from one study to another, hence, the IUPAC recommends to treat these figures of merit as only apparent, as a consensus on how to properly determine a benchmark for comparison among different reports is still lacking [30,53,157]. This issue has been known for decades and in 2001, the IUPAC published a technical report which suggested the use of several figures of merit to evaluate AOPs (including solar HP) [158]. Although the use of some of these benchmarks (i.e., collector area per order) is still reported on in recent scientific literature, none of the reports reviewed in the present work made use of any of them.

Thirty-two papers reported the use of Q_{UV}, which is the cumulative UV energy during an irradiation time per unit of volume of water; as this figure of merit considers solar irradiation intermittencies throughout the day, it can be considered accurate for efficiency assessments for the time being [159], considering the complex and variable composition of the water microbial consortium.

Ten out of the 69 analyzed papers reported an order of magnitude reduction smaller than 3, which indicates that in the vast majority of the reviewed works, a disinfection rate of at less 99.9% was achieved, which complies with the minimal health risk standard set by the WHO [160].

3.6. Volume Treated

In 2011, the WHO estimated that a person needs between 50 and 100 L of water per day to meet their basic needs which include, but are not limited to, drinking water, food preparation, and sanitation. However, in water scarcity scenarios, the WHO recommends a minimum of 7.5 L of drinking water per capita per day [161–164]. The lack of reactor designs able to treat large volumes of water is one of the main reasons which restrains HP commercial and industrial application [165]; nevertheless, some of the papers reviewed in this present work report on treating water volumes in the range of 100–300 L, which could very well meet drinking water requirements of households or small public facilities [166], although actual application depends on more factors, such as level of automation or cost effectiveness, to mention some [7].

Future research should also focus on exploring solar HP reactors or systems able to disinfect water in a steady-state operation rather than batch operation; the volume output still needs to be increased, and even though scaling-up might pose a challenge due to light distribution, scaling-out or numbering-up might offer a feasible alternative, providing land for installation is available; finally, any system should be tested in conditions as close to reality as possible [167,168].

3.7. Experimentation Location

The reviewed works were performed around the globe in latitudes as northernmost as 53° N and southernmost as 38° S, as well as in longitudes as westernmost as 104° W and easternmost as 142° E, which indicates water disinfection via HP could be used in varied locations. However, UV radiation reaching Earth's surface varies around the world and through time, and is dependent on many factors, such as presence of clouds, atmosphere ozone concentration, sunlight reaching the surface oblique angle, aerosol particles concentration, sun elevation, and surface reflectivity, among others [169].

Disinfection via solar HP is related to SODIS, which is being globally promoted. Guidance has been made public to facilitate the worldwide implementation of a standardized procedure [170,171], although, at the moment, there are no available predictive approaches for SODIS expected efficiency worldwide [172], hence, the scenario is analogous for disinfection via solar HP.

3.8. Discussion, Considerations, and Prospects

Based on the present literature review and considering the relatively small number of works that make use of real solar irradiation, there has not been any considerable advance in reactor design in more than two decades, with batch-operated CPC reactor being the most common operation. One of the papers reported coupling solar HP with an electrochemical process [114], which can be considered a process intensification approach [173], rather than an improvement in reactor design by itself.

HP reactor design is challenging, as it involves complex interactions between the PC and microorganisms with the light [174], which can explain the minimal innovation in solar HP reactor designs. It is worth mentioning that recent research on reactor modeling has generated important information regarding reactor design in function of the radiant energy absorbed, shedding light on the relevance of PC chemical properties, PC loading and the high relevance of the Damköhler number (the ratio of the rate of a chemical reaction to diffusive mass transfer rate) [175,176].

HP has also shown efficiency in antibiotic resistant microorganism disinfection, which is not always achieved by conventional technologies such as chlorination [177–179].

Another important reason for HP research on water treatment is that the presence of emerging pollutants (such as: active pharmaceutical ingredients or personal care products) has been reported in groundwater, surface water, and tap water; even if they are commonly found in trace concentrations, its occurrence poses a threat to human health and HP-based technologies can degrade the pollutants or their organic compounds' precursors, mitigating their formation [180–182]. In addition to the intrinsic effect these substances can cause on their own account, when water undergoes disinfection by chlorination, organic compounds can react with chlorine and give rise to the formation of DBP; drinking water with trace concentrations of disinfection byproducts can have a chronic adverse effect on human health [183]. As HP is theoretically able to mineralize organic compounds [184], its implementation within a water potabilization process could potentially mitigate disinfection byproducts' formation [185].

It has been reported than there is a disproportion on the amount of studies focusing on fundamental science regarding HP compared to that regarding applied science; more than 129,000 papers have been published on the HP topic, although usually these studies focus on the application of photocatalytic reactions employing a benchmark PC to a specific process, or in the performance of a new PC applied to a benchmark process [186]. More multidisciplinary endeavors are needed to keep improving HP efficiency.

4. Conclusions

The present work provides a systematic review making use of the PRISMA methodology, which served the purpose of discriminating non-relevant works, including only papers focused on water disinfection employing HP making use of actual sunlight, with a total of 60 papers found. This information sheds light on the need of performing research employing real sunlight as well, as photocatalysis efficiency differs when simulated sunlight is used.

The found papers were analyzed in terms of several operational parameters, identifying the following trends:

1. CPC reactor is the most used type of reactor and its design has not received any major modification in decades.
2. TiO_2 remains the most researched PC despite being unable to use visible light. The use of modified TiO_2 for allowing its visible light activity was more reported than the use of other PCs.
3. The reports indicated good disinfection efficiency, but the use of proper benchmarks is not a standardized practice.
4. Most of the works reported the working volume, with some of them treating enough water for households or small public buildings.
5. Water disinfection via solar HP has been performed in many places around the globe, but proper models to predict disinfection efficiency in different locations are still lacking.

More research is needed in several disciplines to keep improving HP efficiency, aiming to its large-scale application in the future.

Author Contributions: Conceptualization, F.d.J.S.-V., J.B.P.-N. and M.T.A.-H.; Formal analysis, F.d.J.S.-V.; Investigation, F.d.J.S.-V.; Methodology, F.d.J.S.-V.; Resources, M.T.A.-H. and J.B.P.-N.; Supervision, M.T.A.-H. and J.B.P.-N.; Visualization, C.M.N.-N., M.T.A.-H. and J.B.P.-N.; Writing—original draft, F.d.J.S.-V.; Writing—review and editing, C.M.N.-N., M.T.A.-H. and J.B.P.-N. All authors have read and agreed to the published version of the manuscript.

Funding: This research was internally financed by Centro de Investigación en Materiales Avanzados and Instituto Politécnico Nacional. The first author is grateful to the Consejo Nacional de Ciencia y Tecnología (CONACyT) for the doctorate scholarship granted to him, student identification DCTA1001002.

Data Availability Statement: All data and materials have been provided within the manuscript.

Acknowledgments: The authors would like to express their solemn appreciation to the original researchers who performed different kinds of studies on water disinfection via solar heterogeneous photocatalysis, which provided the foundations of this review article. The authors gratefully acknowledge the Consorcio Nacional de Recursos de Información Científica y Tecnológica (CONRICyT) for its ongoing support to facilitate access to scientific literature. The first author wishes to thank Isidro Rocha Núñez for his unceasing morale and emotional support.

Conflicts of Interest: The authors declare no conflict of interest to disclose. The funders played no role in study design; data collection, analysis, and interpretation; manuscript writing or in the publication of results.

References

1. López-Ruiz, S.; Tortajada, C.; González-Gómez, F. Is the Human Right to Water Sufficiently Protected in Spain? Affordability and Governance Concerns. *Util. Policy* 2020, *63*, 101003. [CrossRef]
2. Porley, V.; Chatzisymeon, E.; Meikap, B.C.; Ghosal, S.; Robertson, N. Field Testing of Low-Cost Titania-Based Photocatalysts for Enhanced Solar Disinfection (SODIS) in Rural India. *Environ. Sci. Water Res. Technol.* 2020, *6*, 809–816. [CrossRef]
3. Gebrewahd, A.; Adhanom, G.; Gebremichail, G.; Kahsay, T.; Berhe, B.; Asfaw, Z.; Tadesse, S.; Gebremedhin, H.; Negash, H.; Tesfanchal, B.; et al. Bacteriological Quality and Associated Risk Factors of Drinking Water in Eastern Zone, Tigrai, Ethiopia, 2019. *Trop. Dis. Travel Med. Vaccines* 2020, *6*, 1–7. [CrossRef] [PubMed]
4. Salvador, D.; Caeiro, M.F.; Serejo, F.; Nogueira, P.; Carneiro, R.N.; Neto, C. Monitoring Waterborne Pathogens in Surface and Drinking Waters. Are Water Treatment Plants (WTPs) Simultaneously Efficient in the Elimination of Enteric Viruses and Fecal Indicator Bacteria (FIB)? *Water* 2020, *12*, 2824. [CrossRef]
5. Roshith, M.; Pathak, A.; Nanda Kumar, A.K.; Anantharaj, G.; Saranyan, V.; Ramasubramanian, S.; Satheesh Babu, T.G.; Ravi Kumar, D.V. Continuous Flow Solar Photocatalytic Disinfection of *E. coli* Using Red Phosphorus Immobilized Capillaries as Optofluidic Reactors. *Appl. Surf. Sci.* 2021, *540*, 148398. [CrossRef]
6. Wen, X.; Chen, F.; Lin, Y.; Zhu, H.; Yuan, F.; Kuang, D.; Jia, Z.; Yuan, Z. Microbial Indicators and Their Use for Monitoring Drinking Water Quality—A Review. *Sustainability* 2020, *12*, 2249. [CrossRef]
7. Van Haute, S.; Sampers, I.; Jacxsens, L.; Uyttendaele, M. Selection Criteria for Water Disinfection Techniques in Agricultural Practices. *Crit. Rev. Food Sci. Nutr.* 2015, *55*, 1529–1551. [CrossRef]
8. Gelete, G.; Gokcekus, H.; Ozsahin, D.U.; Uzun, B.; Gichamo, T. Evaluating Disinfection Techniques of Water Treatment. *Desalin. Water Treat.* 2020, *177*, 408–415. [CrossRef]
9. Goncharuk, V.V.; Terletskaya, A.V.; Zuy, O.V.; Pshinko, G.N.; Saprykina, M.N. Principally New Technology of Preparing Drinking Water of Sanitary-Hygienic Purpose. *J. Water Chem. Technol.* 2018, *40*, 11–15. [CrossRef]
10. Al-Abri, M.; Al-Ghafri, B.; Bora, T.; Dobretsov, S.; Dutta, J.; Castelletto, S.; Rosa, L.; Boretti, A. Chlorination Disadvantages and Alternative Routes for Biofouling Control in Reverse Osmosis Desalination. *NPJ Clean Water* 2019, *2*, 2. [CrossRef]
11. Shi, Q.; Chen, Z.; Liu, H.; Lu, Y.; Li, K.; Shi, Y.; Mao, Y.; Hu, H. Efficient Synergistic Disinfection by Ozone, Ultraviolet Irradiation and Chlorine in Secondary Effluents. *Sci. Total Environ.* 2020, *758*, 143641. [CrossRef] [PubMed]
12. Laflamme, O.; Sérodes, J.-B.; Simard, S.; Legay, C.; Dorea, C.; Rodriguez, M.J. Occurrence and Fate of Ozonation Disinfection By-Products in Two Canadian Drinking Water Systems. *Chemosphere* 2020, *260*, 127660. [CrossRef] [PubMed]
13. Khan, S.A.; Jain, M.; Pandey, A.; Pant, K.K.; Ziora, Z.M.; Blaskovich, M.A.T.; Shetti, N.P.; Aminabhavi, T.M. Leveraging the Potential of Silver Nanoparticles-Based Materials towards Sustainable Water Treatment. *J. Environ. Manag.* 2022, *319*, 115675. [CrossRef] [PubMed]
14. Shi, Y.; Ma, J.; Chen, Y.; Qian, Y.; Xu, B.; Chu, W.; An, D. Recent Progress of Silver-Containing Photocatalysts for Water Disinfection under Visible Light Irradiation: A Review. *Sci. Total Environ.* 2022, *804*, 150024. [CrossRef] [PubMed]
15. Domínguez Henao, L.; Turolla, A.; Antonelli, M. Disinfection By-Products Formation and Ecotoxicological Effects of Effluents Treated with Peracetic Acid: A Review. *Chemosphere* 2018, *213*, 25–40. [CrossRef] [PubMed]
16. Ao, X.W.; Eloranta, J.; Huang, C.H.; Santoro, D.; Sun, W.J.; Lu, Z.D.; Li, C. Peracetic Acid-Based Advanced Oxidation Processes for Decontamination and Disinfection of Water: A Review. *Water Res.* 2021, *188*, 116479. [CrossRef]

17. Ao, X.; Chen, Z.; Li, S.; Li, C.; Lu, Z.; Sun, W. The Impact of UV Treatment on Microbial Control and DBPs Formation in Full-Scale Drinking Water Systems in Northern China. *J. Environ. Sci.* **2020**, *87*, 398–410. [CrossRef]
18. Heidarinejad, G.; Bozorgmehr, N.; Safarzadeh, M. Effect of Highly Reflective Material on the Performance of Water Ultraviolet Disinfection Reactor. *J. Water Process Eng.* **2020**, *36*, 101375. [CrossRef]
19. Mazhar, M.A.; Khan, N.A.; Ahmed, S.; Khan, A.H.; Hussain, A.; Rahisuddin; Changani, F.; Yousefi, M.; Ahmadi, S.; Vambol, V. Chlorination Disinfection By-Products in Municipal Drinking Water—A Review. *J. Clean. Prod.* **2020**, *273*, 123159. [CrossRef]
20. Cowie, B.E.; Porley, V.; Robertson, N. Solar Disinfection (SODIS) Provides a Much Underexploited Opportunity for Researchers in Photocatalytic Water Treatment (PWT). *ACS Catal.* **2020**, *10*, 11779–11782. [CrossRef]
21. Parsa, S.M.; Rahbar, A.; Koleini, M.H.; Davoud Javadi, Y.; Afrand, M.; Rostami, S.; Amidpour, M. First Approach on Nanofluid-Based Solar Still in High Altitude for Water Desalination and Solar Water Disinfection (SODIS). *Desalination* **2020**, *491*, 114592. [CrossRef]
22. Burzio, E.; Bersani, F.; Caridi, G.C.A.; Vesipa, R.; Ridolfi, L.; Manes, C. Water Disinfection by Orifice-Induced Hydrodynamic Cavitation. *Ultrason. Sonochem.* **2020**, *60*, 104740. [CrossRef] [PubMed]
23. Rahmani, A.R.; Nematollahi, D.; Poormohammadi, A.; Azarian, G.; Zamani, F. Electrodisinfection of Bacteria-Laden in Surface Water Using Modified Ti Electrode by Antimony-and Nickel-Doped Tin Oxide Composite. *Chemosphere* **2021**, *263*, 127761. [CrossRef] [PubMed]
24. Sun, X.; Liu, J.; Ji, L.; Wang, G.; Zhao, S.; Yoon, J.Y.; Chen, S. A Review on Hydrodynamic Cavitation Disinfection: The Current State of Knowledge. *Sci. Total Environ.* **2020**, *737*, 139606. [CrossRef] [PubMed]
25. Fujishima, A.; Honda, K. Electrochemical Photolysis of Water at a Semicondcutor Electrode. *Nature* **1972**, *238*, 37–38. [CrossRef] [PubMed]
26. You, J. Review of visible light-active photocatalysts for water disinfection: F. and prospectsnhua; Guo, Y.; Guo, R.; Liu, X. A Review of Visible Light-Active Photocatalysts for Water Disinfection: Features and Prospects. *Chem. Eng. J.* **2019**, *373*, 624–641. [CrossRef]
27. Litter, M.I. Mechanisms of Removal of Heavy Metals and Arsenic from Water by TiO_2-Heterogeneous Photocatalysis. *Pure Appl. Chem.* **2015**, *87*, 557–567. [CrossRef]
28. Litter, M.I. Last Advances on TiO_2-Photocatalytic Removal of Chromium, Uranium and Arsenic. *Curr. Opin. Green Sustain. Chem.* **2017**, *6*, 150–158. [CrossRef]
29. Dodoo-Arhin, D.; Bowen-Dodoo, E.; Agyei-Tuffour, B.; Nyankson, E.; Obayemi, J.D.; Salifu, A.A.; Yaya, A.; Agbe, H.; Soboyejo, W.O. Modified Nanostructured Titania Photocatalysts for Aquatic Disinfection Applications. *Mater. Today Proc.* **2020**, *38*, 1183–1190. [CrossRef]
30. He, J.; Kumar, A.; Khan, M.; Lo, I.M.C. Critical Review of Photocatalytic Disinfection of Bacteria: From Noble Metals- and Carbon Nanomaterials-TiO_2 Composites to Challenges of Water Characteristics and Strategic Solutions. *Sci. Total Environ.* **2021**, *758*, 143953. [CrossRef]
31. Bashir, M.S.; Ramzan, N.; Najam, T.; Abbas, G.; Gu, X.; Arif, M.; Qasim, M.; Bashir, H.; Shah, S.S.A.; Sillanpää, M. Metallic Nanoparticles for Catalytic Reduction of Toxic Hexavalent Chromium from Aqueous Medium: A State-of-the-Art Review. *Sci. Total Environ.* **2022**, *829*, 154475. [CrossRef] [PubMed]
32. García, A.; Rosales, M.; Thomas, M.; Golemme, G. Arsenic Photocatalytic Oxidation over TiO_2-Loaded SBA-15. *J. Environ. Chem. Eng.* **2021**, *9*, 106443. [CrossRef]
33. Zhang, Y.; Hawboldt, K.; Zhang, L.; Lu, J.; Chang, L.; Dwyer, A. Carbonaceous Nanomaterial-TiO_2 Heterojunctions for Visible-Light-Driven Photocatalytic Degradation of Aqueous Organic Pollutants. *Appl. Catal. A Gen.* **2022**, *630*, 118460. [CrossRef]
34. Sarngan, P.P.; Lakshmanan, A.; Sarkar, D. Influence of Anatase-Rutile Ratio on Band Edge Position and Defect States of TiO_2 Homojunction Catalyst. *Chemosphere* **2022**, *286*, 131692. [CrossRef] [PubMed]
35. Cerrato, E.; Gaggero, E.; Calza, P.; Paganini, M.C. The Role of Cerium, Europium and Erbium Doped TiO_2 Photocatalysts in Water Treatment: A Mini-Review. *Chem. Eng. J. Adv.* **2022**, *10*, 100268. [CrossRef]
36. Hu, L.; Wang, R.; Wang, M.; Wang, C.; Xu, Y.; Wang, Y.; Gao, P.; Liu, C.; Song, Y.; Ding, N.; et al. The Inactivation Effects and Mechanisms of *Karenia Mikimotoi* by Non-Metallic Elements Modified TiO_2 (SNP-TiO_2) under Visible Light. *Sci. Total Environ.* **2022**, *820*, 153346. [CrossRef]
37. Barrocas, B.T.; Ambrožová, N.; Kočí, K. Photocatalytic Reduction of Carbon Dioxide on TiO_2 Heterojunction Photocatalysts—A Review. *Materials* **2022**, *15*, 967. [CrossRef]
38. Wang, L.; Wang, L.; Du, Y.; Xu, X.; Dou, S.X. Progress and Perspectives of Bismuth Oxyhalides in Catalytic Applications. *Mater. Today Phys.* **2021**, *16*, 100294. [CrossRef]
39. Arumugam, M.; Natarajan, T.S.; Saelee, T.; Praserthdam, S.; Ashokkumar, M.; Praserthdam, P. Recent Developments on Bismuth Oxyhalides (BiOX; X = Cl, Br, I) Based Ternary Nanocomposite Photocatalysts for Environmental Applications. *Chemosphere* **2021**, *282*, 131054. [CrossRef]
40. Ullattil, S.G.; Narendranath, S.B.; Pillai, S.C.; Periyat, P. Black TiO_2 Nanomaterials: A Review of Recent Advances. *Chem. Eng. J.* **2018**, *343*, 708–736. [CrossRef]
41. Rajaraman, T.S.; Parikh, S.P.; Gandhi, V.G. Black TiO_2: A Review of Its Properties and Conflicting Trends. *Chem. Eng. J.* **2020**, *389*, 123918. [CrossRef]

42. Monteagudo, J.M.; Durán, A.; Martín, I.S.; Acevedo, A.M. A Novel Combined Solar Pasteurizer/TiO$_2$ Continuous-Flow Reactor for Decontamination and Disinfection of Drinking Water. *Chemosphere* **2017**, *168*, 1447–1456. [CrossRef] [PubMed]
43. Hosseini, F.; Assadi, A.A.; Nguyen-Tri, P.; Ali, I.; Rtimi, S. Titanium-Based Photocatalytic Coatings for Bacterial Disinfection: The Shift from Suspended Powders to Catalytic Interfaces. *Surf. Interfaces* **2022**, *32*, 102078. [CrossRef]
44. John, D.; Jose, J.; Bhat, S.G.; Achari, V.S. Integration of Heterogeneous Photocatalysis and Persulfate Based Oxidation Using TiO$_2$-Reduced Graphene Oxide for Water Decontamination and Disinfection. *Heliyon* **2021**, *7*, e07451. [CrossRef]
45. Byrne, J.; Dunlop, P.; Hamilton, J.; Fernández-Ibáñez, P.; Polo-López, I.; Sharma, P.; Vennard, A. A Review of Heterogeneous Photocatalysis for Water and Surface Disinfection. *Molecules* **2015**, *20*, 5574–5615. [CrossRef]
46. Xu, Y.; Liu, Q.; Liu, C.; Zhai, Y.; Xie, M.; Huang, L.; Xu, H.; Li, H.; Jing, J. Visible-Light-Driven Ag/AgBr/ZnFe$_2$O$_4$ Composites with Excellent Photocatalytic Activity for *E. coli* Disinfection and Organic Pollutant Degradation. *J. Colloid Interface Sci.* **2018**, *512*, 555–566. [CrossRef]
47. Jabbar, Z.H.; Esmail Ebrahim, S. Recent Advances in Nano-Semiconductors Photocatalysis for Degrading Organic Contaminants and Microbial Disinfection in Wastewater: A Comprehensive Review. *Environ. Nanotechnol. Monit. Manag.* **2022**, *17*, 100666. [CrossRef]
48. Braham, R.J.; Harris, A.T. Review of Major Design and Scale-up Considerations for Solar Photocatalytic Reactors. *Ind. Eng. Chem. Res.* **2009**, *48*, 8890–8905. [CrossRef]
49. Iervolino, G.; Zammit, I.; Vaiano, V.; Rizzo, L. Limitations and Prospects for Wastewater Treatment by UV and Visible-Light-Active Heterogeneous Photocatalysis: A Critical Review. *Top. Curr. Chem.* **2020**, *378*, 7. [CrossRef]
50. Malato, S.; Maldonado, M.I.; Fernández-Ibáñez, P.; Oller, I.; Polo, I.; Sánchez-Moreno, R. Decontamination and Disinfection of Water by Solar Photocatalysis: The Pilot Plants of the Plataforma Solar de Almeria. *Mater. Sci. Semicond. Process.* **2016**, *42*, 15–23. [CrossRef]
51. Malato, S.; Blanco, J.; Alarcón, D.C.; Maldonado, M.I.; Fernández-Ibáñez, P.; Gernjak, W. Photocatalytic Decontamination and Disinfection of Water with Solar Collectors. *Catal. Today* **2007**, *122*, 137–149. [CrossRef]
52. Wang, H.; Li, X.; Zhao, X.; Li, C.; Song, X.; Zhang, P.; Huo, P.; Li, X. A Review on Heterogeneous Photocatalysis for Environmental Remediation: From Semiconductors to Modification Strategies. *Chin. J. Catal.* **2022**, *43*, 178–214. [CrossRef]
53. Silerio-Vázquez, F.; Nájera, J.B.P.; Bundschuh, J.; Alarcon-Herrera, M.T. Photocatalysis for Arsenic Removal from Water: Considerations for Solar Photocatalytic Reactors. *Environ. Sci. Pollut. Res.* **2021**, *29*, 61594–61607. [CrossRef] [PubMed]
54. Galushchinskiy, A.; González-Gómez, R.; McCarthy, K.; Farràs, P.; Savateev, A. Progress in Development of Photocatalytic Processes for Synthesis of Fuels and Organic Compounds under Outdoor Solar Light. *Energy Fuels* **2022**, *36*, 4625–4639. [CrossRef]
55. Liu, M.; Xing, Z.; Li, Z.; Zhou, W. Recent Advances in Core–Shell Metal Organic Frame-Based Photocatalysts for Solar Energy Conversion. *Coord. Chem. Rev.* **2021**, *446*, 214123. [CrossRef]
56. Ansari, M.; Sharifian, M.; Ehrampoush, M.H.; Mahvi, A.H.; Salmani, M.H.; Fallahzadeh, H. Dielectric Barrier Discharge Plasma with Photocatalysts as a Hybrid Emerging Technology for Degradation of Synthetic Organic Compounds in Aqueous Environments: A Critical Review. *Chemosphere* **2021**, *263*, 128065. [CrossRef]
57. Moher, D. Preferred Reporting Items for Systematic Reviews and Meta-Analyses: The PRISMA Statement. *Ann. Intern. Med.* **2009**, *151*, 264. [CrossRef]
58. Veréb, G.; Hernádi, K.; Baia, L.; Rákhely, G.; Pap, Z. *Pilot-Plant Scaled Water Treatment Technologies, Standards for the Removal of Contaminants of Emerging Concern Based on Photocatalytic Materials*; Elsevier: Amsterdam, The Netherlands, 2020; ISBN 9780128158821.
59. Vidal, A.; Díaz, A.I.; El Hraiki, A.; Romero, M.; Muguruza, I.; Senhaji, F.; González, J. Solar Photocatalysis for Detoxification and Disinfection of Contaminated Water: Pilot Plant Studies. *Catal. Today* **1999**, *54*, 283–290. [CrossRef]
60. Cho, I.-H.; Moon, I.-Y.; Chung, M.-H.; Lee, H.-K.; Zoh, K.-D. Disinfection Effects on *E. coli* Using TiO$_2$/UV and Solar Light System. *Water Supply* **2002**, *2*, 181–190. [CrossRef]
61. McLoughlin, O.A.; Ibáñez, P.F.; Gernjak, W.; Rodríguez, S.M.; Gill, L.W. Photocatalytic Disinfection of Water Using Low Cost Compound Parabolic Collectors. *Sol. Energy* **2004**, *77*, 625–633. [CrossRef]
62. McLoughlin, O.A.; Kehoe, S.C.; McGuigan, K.G.; Duffy, E.F.; Al Touati, F.; Gernjak, W.; Alberola, I.O.; Rodríguez, S.M.; Gill, L.W. Solar Disinfection of Contaminated Water: A Comparison of Three Small-Scale Reactors. *Sol. Energy* **2004**, *77*, 657–664. [CrossRef]
63. Rincón, A.-G.; Pulgarin, C. Field Solar *E. coli* Inactivation in the Absence and Presence of TiO$_2$: Is UV Solar Dose an Appropriate Parameter for Standardization of Water Solar Disinfection? *Sol. Energy* **2004**, *77*, 635–648. [CrossRef]
64. Fernández, P.; Blanco, J.; Sichel, C.; Malato, S. Water Disinfection by Solar Photocatalysis Using Compound Parabolic Collectors. *Catal. Today* **2005**, *101*, 345–352. [CrossRef]
65. Gill, L.W.; McLoughlin, O.A. Solar Disinfection Kinetic Design Parameters for Continuous Flow Reactors. *J. Sol. Energy Eng.* **2007**, *129*, 111–118. [CrossRef]
66. Navntoft, C.; Araujo, P.; Litter, M.I.; Apella, M.C.; Fernández, D.; Puchulu, M.E.; del V. Hidalgo, M.; Blesa, M.A. Field Tests of the Solar Water Detoxification SOLWATER Reactor in Los Pereyra, Tucumán, Argentina. *J. Sol. Energy Eng.* **2007**, *129*, 127–134. [CrossRef]
67. Rincón, A.-G.; Pulgarin, C. Absence of *E. coli* Regrowth after Fe^{3+} and TiO$_2$ Solar Photoassisted Disinfection of Water in CPC Solar Photoreactor. *Catal. Today* **2007**, *124*, 204–214. [CrossRef]

68. Rincón, A.-G.; Pulgarin, C. Fe^{3+} and TiO_2 Solar-Light-Assisted Inactivation of *E. coli* at Field Scale. *Catal. Today* **2007**, *122*, 128–136. [CrossRef]
69. Rincón, A.-G.; Pulgarin, C. Solar Photolytic and Photocatalytic Disinfection of Water at Laboratory and Field Scale. Effect of the Chemical Composition of Water and Study of the Postirradiation Events. *J. Sol. Energy Eng.* **2007**, *129*, 100–110. [CrossRef]
70. Rodrigues, C.P.; Ziolli, R.L.; Guimarães, J.R. Inactivation of *Escherichia Coli* in Water by TiO_2-Assisted Disinfection Using Solar Light. *J. Braz. Chem. Soc.* **2007**, *18*, 126–134. [CrossRef]
71. Sichel, C.; Blanco, J.; Malato, S.; Fernández-Ibáñez, P. Effects of Experimental Conditions on *E. coli* Survival during Solar Photocatalytic Water Disinfection. *J. Photochem. Photobiol. A Chem.* **2007**, *189*, 239–246. [CrossRef]
72. Sichel, C.; Tello, J.; de Cara, M.; Fernández-Ibáñez, P. Effect of UV Solar Intensity and Dose on the Photocatalytic Disinfection of Bacteria and Fungi. *Catal. Today* **2007**, *129*, 152–160. [CrossRef]
73. Hernández-García, H.; López-Arjona, H.; Rodríguez, J.F.; Enríquez, R. Preliminary Study of the Disinfection of Secondary Wastewater Using a Solar Photolytic-Photocatalytic Reactor. *J. Sol. Energy Eng.* **2008**, *130*, 0410041–0410045. [CrossRef]
74. Fernández-Ibáñez, P.; Sichel, C.; Polo-López, M.I.; de Cara-García, M.; Tello, J.C. Photocatalytic Disinfection of Natural Well Water Contaminated by *Fusarium solani* Using TiO_2 Slurry in Solar CPC Photo-Reactors. *Catal. Today* **2009**, *144*, 62–68. [CrossRef]
75. Gomes, A.I.; Santos, J.C.; Vilar, V.J.P.; Boaventura, R.A.R. Inactivation of Bacteria *E. coli* and Photodegradation of Humic Acids Using Natural Sunlight. *Appl. Catal. B Environ.* **2009**, *88*, 283–291. [CrossRef]
76. Polo-López, M.I.; Fernández-Ibáñez, P.; García-Fernández, I.; Oller, I.; Salgado-Tránsito, I.; Sichel, C. Resistance of *Fusarium* sp. Spores to Solar TiO_2 Photocatalysis: Influence of Spore Type and Water (Scaling-up Results). *J. Chem. Technol. Biotechnol.* **2010**, *85*, 1038–1048. [CrossRef]
77. Rodríguez, J.; Jorge, C.; Zúñiga, P.; Palomino, J.; Zanabria, P.; Ponce, S.; Solís, J.L.; Estrada, W. Solar Water Disinfection Studies with Supported TiO_2 and Polymer-Supported Ru(II) Sensitizer in a Compound Parabolic Collector. *J. Sol. Energy Eng.* **2010**, *132*, 0110011–0110015. [CrossRef]
78. Sordo, C.; Van Grieken, R.; Marugán, J.; Fernández-Ibáñez, P. Solar Photocatalytic Disinfection with Immobilised TiO_2 at Pilot-Plant Scale. *Water Sci. Technol.* **2010**, *61*, 507–512. [CrossRef]
79. Mehrabadi, A.R.; Kardani, N.; Fazeli, M.; Hamidian, L.; Mousavi, A.; Salmani, N. Investigation of Water Disinfection Efficiency Using Titanium Dioxide (TiO_2) in Permeable to Sunlight Tubes. *Desalin. Water Treat.* **2011**, *28*, 17–22. [CrossRef]
80. Alrousan, D.M.A.; Polo-López, M.I.; Dunlop, P.S.M.; Fernández-Ibáñez, P.; Byrne, J.A. Solar Photocatalytic Disinfection of Water with Immobilised Titanium Dioxide in Re-Circulating Flow CPC Reactors. *Appl. Catal. B Environ.* **2012**, *128*, 126–134. [CrossRef]
81. Khan, S.J.; Reed, R.H.; Rasul, M.G. Thin-Film Fixed-Bed Reactor for Solar Photocatalytic Inactivation of *Aeromonas hydrophila*: Influence of Water Quality. *BMC Microbiol.* **2012**, *12*, 285. [CrossRef]
82. Khan, S.J.; Reed, R.H.; Rasul, M.G. Thin-Film Fixed-Bed Reactor (TFFBR) for Solar Photocatalytic Inactivation of Aquaculture Pathogen *Aeromonas hydrophila*. *BMC Microbiol.* **2012**, *12*, 5. [CrossRef] [PubMed]
83. Misstear, D.B.; Gill, L.W. The Inactivation of Phages MS2, ΦX174 and PR772 Using UV and Solar Photocatalysis. *J. Photochem. Photobiol. B Biol.* **2012**, *107*, 1–8. [CrossRef] [PubMed]
84. Pinho, L.X.; Azevedo, J.; Vasconcelos, V.M.; Vilar, V.J.P.; Boaventura, R.A.R. Decomposition of *Microcystis aeruginosa* and Microcystin-LR by TiO_2 Oxidation Using Artificial UV Light or Natural Sunlight. *J. Adv. Oxid. Technol.* **2012**, *15*, 98–106. [CrossRef]
85. Agulló-Barceló, M.; Polo-López, M.I.; Lucena, F.; Jofre, J.; Fernández-Ibáñez, P. Solar Advanced Oxidation Processes as Disinfection Tertiary Treatments for Real Wastewater: Implications for Water Reclamation. *Appl. Catal. B Environ.* **2013**, *136*, 341–350. [CrossRef]
86. Singh, C.; Chaudhary, R.; Gandhi, K. Solar Photocatalytic Oxidation and Disinfection of Municipal Wastewater Using Advanced Oxidation Processes Based on PH, Catalyst Dose, and Oxidant. *J. Renew. Sustain. Energy* **2013**, *5*, 023124. [CrossRef]
87. Arya, V.; Philip, L. Visible and Solar Light Photocatalytic Disinfection of Bacteria by N-Doped TiO_2. *Water Supply* **2014**, *14*, 924–930. [CrossRef]
88. GilPavas, E.; Acevedo, J.; López, L.F.; Dobrosz-Gómez, I.; Gómez-García, M.Á. Solar and Artificial UV Inactivation of Bacterial Microbes by Ca-Alginate Immobilized TiO_2 Assisted by H_2O_2 Using Fluidized Bed Photoreactors. *J. Adv. Oxid. Technol.* **2014**, *17*, 343–351. [CrossRef]
89. Polo-López, M.I.; Castro-Alférez, M.; Oller, I.; Fernández-Ibáñez, P. Assessment of Solar Photo-Fenton, Photocatalysis, and H_2O_2 for Removal of Phytopathogen Fungi Spores in Synthetic and Real Effluents of Urban Wastewater. *Chem. Eng. J.* **2014**, *257*, 122–130. [CrossRef]
90. Ahmad, N.; Gondal, M.A.; Sheikh, A.K. Comparative Study of Different Solar-Based Photo Catalytic Reactors for Disinfection of Contaminated Water. *Desalin. Water Treat.* **2015**, *57*, 1–8. [CrossRef]
91. Ferro, G.; Fiorentino, A.; Alferez, M.C.; Polo-López, M.I.; Rizzo, L.; Fernández-Ibáñez, P. Urban Wastewater Disinfection for Agricultural Reuse: Effect of Solar Driven AOPs in the Inactivation of a Multidrug Resistant *E. coli* Strain. *Appl. Catal. B Environ.* **2015**, *178*, 65–73. [CrossRef]
92. García-Fernández, I.; Fernández-Calderero, I.; Inmaculada Polo-López, M.; Fernández-Ibáñez, P.; Polo-López, M.I.; Fernández-Ibáñez, P. Disinfection of Urban Effluents Using Solar TiO_2 Photocatalysis: A Study of Significance of Dissolved Oxygen, Temperature, Type of Microorganism and Water Matrix. *Catal. Today* **2015**, *240*, 30–38. [CrossRef]
93. Kacem, M.; Goetz, V.; Plantard, G.; Wery, N. Modeling Heterogeneous Photocatalytic Inactivation of *E. coli* Using Suspended and Immobilized TiO_2 Reactors. *AIChE J.* **2015**, *61*, 2532–2542. [CrossRef]

94. Nararom, M.; Thepa, S.; Kongkiattikajorn, J.; Songprakorp, R. Disinfection of Water Containing *Escherichia Coli* by Use of a Compound Parabolic Concentrator: Effect of Global Solar Radiation and Reactor Surface Treatment. *Res. Chem. Intermed.* **2015**, *41*, 6543–6558. [CrossRef]
95. Barwal, A.; Chaudhary, R. Feasibility Study for the Treatment of Municipal Wastewater by Using a Hybrid Bio-Solar Process. *J. Environ. Manag.* **2016**, *177*, 271–277. [CrossRef]
96. Gutiérrez-Alfaro, S.; Acevedo, A.; Rodríguez, J.; Carpio, E.A.; Manzano, M.A. Solar Photocatalytic Water Disinfection of *Escherichia Coli*, *Enterococcus* Spp. and *Clostridium Perfringens* Using Different Low-Cost Devices. *J. Chem. Technol. Biotechnol.* **2016**, *91*, 2026–2037. [CrossRef]
97. Yoriya, S.; Chumphu, A.; Pookmanee, P.; Laithong, W.; Thepa, S.; Songprakorp, R. Multi-Layered TiO_2 Films towards Enhancement of *Escherichia Coli* Inactivation. *Mater.* **2016**, *9*, 808. [CrossRef] [PubMed]
98. Aguas, Y.; Hincapie, M.; Fernández-Ibáñez, P.; Polo-López, M.I. Solar Photocatalytic Disinfection of Agricultural Pathogenic Fungi (*Curvularia* Sp.) in Real Urban Wastewater. *Sci. Total Environ.* **2017**, *607*, 1213–1224. [CrossRef]
99. Booshehri, A.Y.; Polo-Lopez, M.I.I.; Castro-Alférez, M.; He, P.; Xu, R.; Rong, W.; Malato, S.; Fernández-Ibáñez, P.; Yoosefi, A.; Polo-Lopez, M.I.I.; et al. Assessment of Solar Photocatalysis Using $Ag/BiVO_4$ at Pilot Solar Compound Parabolic Collector for Inactivation of Pathogens in Well Water and Secondary Effluents. *Catal. Today* **2017**, *281*, 124–134. [CrossRef]
100. Mac Mahon, J.; Pillai, S.C.; Kelly, J.M.; Gill, L.W. Solar Photocatalytic Disinfection of *E. coli* and Bacteriophages MS2, ΦX174 and PR772 Using TiO_2, ZnO and Ruthenium Based Complexes in a Continuous Flow System. *J. Photochem. Photobiol. B Biol.* **2017**, *170*, 79–90. [CrossRef]
101. Afsharnia, M.; Kianmehr, M.; Biglari, H.; Dargahi, A.; Karimi, A. Disinfection of Dairy Wastewater Effluent through Solar Photocatalysis Processes. *Water Sci. Eng.* **2018**, *11*, 214–219. [CrossRef]
102. Aguas, Y.; Hincapié, M.; Sánchez, C.; Botero, L.; Fernández-Ibañez, P. Photocatalytic Inactivation of *Enterobacter cloacae* and *Escherichia Coli* Using Titanium Dioxide Supported on Two Substrates. *Processes* **2018**, *6*, 137. [CrossRef]
103. Moreira, N.F.F.; Narciso-da-Rocha, C.; Polo-López, M.I.; Pastrana-Martínez, L.M.; Faria, J.L.; Manaia, C.M.; Fernández-Ibáñez, P.; Nunes, O.C.; Silva, A.M.T. Solar Treatment (H_2O_2, TiO_2-P25 and $GO-TiO_2$ Photocatalysis, Photo-Fenton) of Organic Micropollutants, Human Pathogen Indicators, Antibiotic Resistant Bacteria and Related Genes in Urban Wastewater. *Water Res.* **2018**, *135*, 195–206. [CrossRef] [PubMed]
104. Nwoke, O.O.; Ezema, F.I.; Chigor, N.V.; Anoliefo, C.E.; Mbajiorgu, C.C. Disinfection of a Farmstead Roof Harvested Rainwater for Potable Purposes Using an Automated Solar Photocatalytic Reactor. *Agric. Eng. Int. CIGR J.* **2018**, *20*, 52–60.
105. Sacco, O.; Vaiano, V.; Rizzo, L.; Sannino, D. Photocatalytic Activity of a Visible Light Active Structured Photocatalyst Developed for Municipal Wastewater Treatment. *J. Clean. Prod.* **2018**, *175*, 38–49. [CrossRef]
106. Saran, S.; Arunkumar, P.; Devipriya, S.P. Disinfection of Roof Harvested Rainwater for Potable Purpose Using Pilot-Scale Solar Photocatalytic Fixed Bed Tubular Reactor. *Water Supply* **2018**, *18*, 49–59. [CrossRef]
107. Achouri, F.; BenSaid, M.; Bousselmi, L.; Corbel, S.; Schneider, R.; Ghrabi, A. Comparative Study of Gram-Negative Bacteria Response to Solar Photocatalytic Inactivation. *Environ. Sci. Pollut. Res.* **2019**, *26*, 18961–18970. [CrossRef]
108. Mecha, A.C.; Onyango, M.S.; Ochieng, A.; Momba, M.N.B. UV and Solar Photocatalytic Disinfection of Municipal Wastewater: Inactivation, Reactivation and Regrowth of Bacterial Pathogens. *Int. J. Environ. Sci. Technol.* **2019**, *16*, 3687–3696. [CrossRef]
109. Negishi, N.; Chawengkijwanich, C.; Pimpha, N.; Larpkiattaworn, S.; Charinpanitkul, T. Performance Verification of the Photocatalytic Solar Water Purification System for Sterilization Using Actual Drinking Water in Thailand. *J. Water Process Eng.* **2019**, *31*, 100835. [CrossRef]
110. Yazdanbakhsh, A.; Rahmani, K.; Rahmani, H.; Sarafraz, M.; Tahmasebizadeh, M.; Rahmani, A. Inactivation of Fecal Coliforms during Solar and Photocatalytic Disinfection by Zinc Oxide (ZnO) Nanoparticles in Compound Parabolic Concentrators (CPCs). *Iran. J. Catal.* **2019**, *9*, 339–346. [CrossRef]
111. Núñez-Núñez, C.M.; Osorio-Revilla, G.I.; Villanueva-Fierro, I.; Antileo, C.; Proal-Nájera, J.B. Solar Fecal Coliform Disinfection in a Wastewater Treatment Plant by Oxidation Processes: Kinetic Analysis as a Function of Solar Radiation. *Water* **2020**, *12*, 639. [CrossRef]
112. Rueda-Márquez, J.J.; Palacios-Villarreal, C.; Manzano, M.; Blanco, E.; Ramírez del Solar, M.; Levchuk, I. Photocatalytic Degradation of Pharmaceutically Active Compounds (PhACs) in Urban Wastewater Treatment Plants Effluents under Controlled and Natural Solar Irradiation Using Immobilized TiO_2. *Sol. Energy* **2020**, *208*, 480–492. [CrossRef]
113. Waso, M.; Khan, S.; Singh, A.; McMichael, S.; Ahmed, W.; Fernández-Ibáñez, P.; Byrne, J.A.; Khan, W. Predatory Bacteria in Combination with Solar Disinfection and Solar Photocatalysis for the Treatment of Rainwater. *Water Res.* **2020**, *169*, 115281. [CrossRef] [PubMed]
114. McMichael, S.; Waso, M.; Reyneke, B.; Khan, W.; Byrne, J.A.; Fernandez-Ibanez, P. Electrochemically Assisted Photocatalysis for the Disinfection of Rainwater under Solar Irradiation. *Appl. Catal. B Environ.* **2021**, *281*, 119485. [CrossRef]
115. Zhang, C.; Liu, N.; Ming, J.; Sharma, A.; Ma, Q.; Liu, Z.; Chen, G.; Yang, Y. Development of a Novel Solar Energy Controllable Linear Fresnel Photoreactor (LFP) for High-Efficiency Photocatalytic Wastewater Treatment under Actual Weather. *Water Res.* **2022**, *208*, 117880. [CrossRef]
116. Thakur, I.; Verma, A.; Örmeci, B. Visibly Active $Fe-TiO_2$ Composite: A Stable and Efficient Catalyst for the Catalytic Disinfection of Water Using a Once-through Reactor. *J. Environ. Chem. Eng.* **2021**, *9*, 106322. [CrossRef]

117. Murcia Mesa, J.J.; Hernández Niño, J.S.; González, W.; Rojas, H.; Hidalgo, M.C.; Navío, J.A. Photocatalytic Treatment of Stained Wastewater Coming from Handicraft Factories. A Case Study at the Pilot Plant Level. *Water* **2021**, *13*, 2705. [CrossRef]
118. Durán, A.; Monteagudo, J.M.; San Martín, I. Operation Costs of the Solar Photo-Catalytic Degradation of Pharmaceuticals in Water: A Mini-Review. *Chemosphere* **2018**, *211*, 482–488. [CrossRef]
119. Colina-Márquez, J.; Machuca-Martínez, F.; Puma, G.L. Radiation Absorption and Optimization of Solar Photocatalytic Reactors for Environmental Applications. *Environ. Sci. Technol.* **2010**, *44*, 5112–5120. [CrossRef]
120. Abdel-Maksoud, Y.; Imam, E.; Ramadan, A. TiO$_2$ Solar Photocatalytic Reactor Systems: Selection of Reactor Design for Scale-up and Commercialization—Analytical Review. *Catalysts* **2016**, *6*, 138. [CrossRef]
121. Fendrich, M.; Quaranta, A.; Orlandi, M.; Bettonte, M.; Miotello, A. Solar Concentration for Wastewaters Remediation: A Review of Materials and Technologies. *Appl. Sci.* **2018**, *9*, 118. [CrossRef]
122. Ochoa-Gutiérrez, K.S.; Tabares-Aguilar, E.; Mueses, M.Á.; Machuca-Martínez, F.; Li Puma, G. A Novel Prototype Offset Multi Tubular Photoreactor (OMTP) for Solar Photocatalytic Degradation of Water Contaminants. *Chem. Eng. J.* **2018**, *341*, 628–638. [CrossRef]
123. García-Gil, Á.; Feng, L.; Moreno-SanSegundo, J.; Giannakis, S.; Pulgarín, C.; Marugán, J. Mechanistic Modelling of Solar Disinfection (SODIS) Kinetics of *Escherichia Coli*, Enhanced with H$_2$O$_2$—Part 1: The Dark Side of Peroxide. *Chem. Eng. J.* **2022**, *439*, 135709. [CrossRef]
124. Ghosh, S. *Visible-Light-Active Photocatalysis: Nanostructured Catalyst Design, Mechanisms, and Applications*; Wiley & Sons: New York, NY, USA, 2018; ISBN 9783527342938.
125. Tanveer, M.; Tezcanli Guyer, G. Solar Assisted Photo Degradation of Wastewater by Compound Parabolic Collectors: Review of Design and Operational Parameters. *Renew. Sustain. Energy Rev.* **2013**, *24*, 534–543. [CrossRef]
126. Serrà, A.; Philippe, L.; Perreault, F.; Garcia-Segura, S. Photocatalytic Treatment of Natural Waters. Reality or Hype? The Case of Cyanotoxins Remediation. *Water Res.* **2021**, *188*, 116543. [CrossRef] [PubMed]
127. Mecha, A.C.; Onyango, M.S.; Ochieng, A.; Momba, M.N.B. Evaluation of Synergy and Bacterial Regrowth in Photocatalytic Ozonation Disinfection of Municipal Wastewater. *Sci. Total Environ.* **2017**, *601*, 626–635. [CrossRef] [PubMed]
128. Intisar, A.; Ramzan, A.; Sawaira, T.; Kareem, A.T.; Hussain, N.; Din, M.I.; Bilal, M.; Iqbal, H.M.N. Occurrence, Toxic Effects, and Mitigation of Pesticides as Emerging Environmental Pollutants Using Robust Nanomaterials—A Review. *Chemosphere* **2022**, *293*, 133538. [CrossRef]
129. Janczarek, M.; Klapiszewski, Ł.; Jędrzejczak, P.; Klapiszewska, I.; Ślosarczyk, A.; Jesionowski, T. Progress of Functionalized TiO$_2$-Based Nanomaterials in the Construction Industry: A Comprehensive Review. *Chem. Eng. J.* **2022**, *430*, 132062. [CrossRef]
130. Dharma, H.N.C.; Jaafar, J.; Widiastuti, N.; Matsuyama, H.; Rajabsadeh, S.; Othman, M.H.D.; Rahman, M.A.; Jafri, N.N.M.; Suhaimin, N.S.; Nasir, A.M.; et al. A Review of Titanium Dioxide (TiO$_2$)-Based Photocatalyst for Oilfield-Produced Water Treatment. *Membranes* **2022**, *12*, 345. [CrossRef]
131. Malato, S.; Fernández-Ibáñez, P.; Maldonado, M.I.; Blanco, J.; Gernjak, W. Decontamination and Disinfection of Water by Solar Photocatalysis: Recent Overview and Trends. *Catal. Today* **2009**, *147*, 1–59. [CrossRef]
132. Kanakaraju, D.; anak Kutiang, F.D.; Lim, Y.C.; Goh, P.S. Recent Progress of Ag/TiO$_2$ Photocatalyst for Wastewater Treatment: Doping, Co-Doping, and Green Materials Functionalization. *Appl. Mater. Today* **2022**, *27*, 101500. [CrossRef]
133. Karim, A.V.; Krishnan, S.; Shriwastav, A. An Overview of Heterogeneous Photocatalysis for the Degradation of Organic Compounds: A Special Emphasis on Photocorrosion and Reusability. *J. Indian Chem. Soc.* **2022**, *99*, 100480. [CrossRef]
134. Ibrahim, N.S.; Leaw, W.L.; Mohamad, D.; Alias, S.H.; Nur, H. A Critical Review of Metal-Doped TiO$_2$ and Its Structure–Physical Properties–Photocatalytic Activity Relationship in Hydrogen Production. *Int. J. Hydrogen Energy* **2020**, *45*, 28553–28565. [CrossRef]
135. Dutta, V.; Devasia, J.; Chauhan, A.; Jayalakshmi, M.; Vasantha, V.L.; Jha, A.; Nizam, A.; Lin, K.-Y.A.; Ghotekar, S. Photocatalytic Nanomaterials: Applications for Remediation of Toxic Polycyclic Aromatic Hydrocarbons and Green Management. *Chem. Eng. J. Adv.* **2022**, *11*, 100353. [CrossRef]
136. Peiris, S.; Silva, H.B.; Ranasinghe, K.N.; Bandara, S.V.; Perera, I.R. Recent Development and Future Prospects of TiO$_2$ Photocatalysis. *J. Chin. Chem. Soc.* **2021**, *68*, 738–769. [CrossRef]
137. Piątkowska, A.; Janus, M.; Szymański, K.; Mozia, S. C-,N- and S-Doped TiO$_2$ Photocatalysts: A Review. *Catalysts* **2021**, *11*, 144. [CrossRef]
138. Sacco, O.; Venditto, V.; Pragliola, S.; Vaiano, V. Catalytic Composite Systems Based on N-Doped TiO$_2$/Polymeric Materials for Visible-Light-Driven Pollutant Degradation: A Mini Review. *Photochem* **2021**, *1*, 330–344. [CrossRef]
139. Purabgola, A.; Mayilswamy, N.; Kandasubramanian, B. Graphene-Based TiO$_2$ Composites for Photocatalysis & Environmental Remediation: Synthesis and Progress. *Environ. Sci. Pollut. Res.* **2022**, *29*, 32305–32325. [CrossRef]
140. Padmanabhan, N.T.; Thomas, N.; Louis, J.; Mathew, D.T.; Ganguly, P.; John, H.; Pillai, S.C. Graphene Coupled TiO$_2$ Photocatalysts for Environmental Applications: A Review. *Chemosphere* **2021**, *271*, 129506. [CrossRef]
141. Kumar, R.; Raizada, P.; Verma, N.; Hosseini-Bandegharaei, A.; Thakur, V.K.; Van Le, Q.; Nguyen, V.-H.; Selvasembian, R.; Singh, P. Recent Advances on Water Disinfection Using Bismuth Based Modified Photocatalysts: Strategies and Challenges. *J. Clean. Prod.* **2021**, *297*, 126617. [CrossRef]
142. Tayebi, M.; Lee, B.-K. Recent Advances in BiVO$_4$ Semiconductor Materials for Hydrogen Production Using Photoelectrochemical Water Splitting. *Renew. Sustain. Energy Rev.* **2019**, *111*, 332–343. [CrossRef]

143. Singh, S.; Kansal, S.K. Recent Progress in Red Phosphorus-Based Photocatalysts for Photocatalytic Water Remediation and Hydrogen Production. *Appl. Mater. Today* **2022**, *26*, 101345. [CrossRef]
144. Ren, X.; Philo, D.; Li, Y.; Shi, L.; Chang, K.; Ye, J. Recent Advances of Low-Dimensional Phosphorus-Based Nanomaterials for Solar-Driven Photocatalytic Reactions. *Coord. Chem. Rev.* **2020**, *424*, 213516. [CrossRef]
145. Ajiboye, T.O.; Babalola, S.O.; Onwudiwe, D.C. Photocatalytic Inactivation as a Method of Elimination of *E. coli* from Drinking Water. *Appl. Sci.* **2021**, *11*, 1313. [CrossRef]
146. World Health Organization. *Results of Round I of the WHO International Scheme to Evaluate Eousehold Water Treatment Technologies*; World Health Organization: Geneva, Switzerland, 2016.
147. Žvab, U.; Lavrenčič Štangar, U.; Bergant Marušič, M. Methodologies for the Analysis of Antimicrobial Effects of Immobilized Photocatalytic Materials. *Appl. Microbiol. Biotechnol.* **2014**, *98*, 1925–1936. [CrossRef]
148. Chowdhury, S. Heterotrophic Bacteria in Drinking Water Distribution System: A Review. *Environ. Monit. Assess.* **2012**, *184*, 6087–6137. [CrossRef]
149. Gomes Gradíssimo, D.; Pereira Xavier, L.; Valadares Santos, A. Cyanobacterial Polyhydroxyalkanoates: A Sustainable Alternative in Circular Economy. *Molecules* **2020**, *25*, 4331. [CrossRef]
150. Menezes, I.; Capelo-Neto, J.; Pestana, C.J.; Clemente, A.; Hui, J.; Irvine, J.T.S.; Nimal Gunaratne, H.Q.; Robertson, P.K.J.; Edwards, C.; Gillanders, R.N.; et al. Comparison of UV-A Photolytic and UV/TiO$_2$ Photocatalytic Effects on *Microcystis aeruginosa* PCC7813 and Four Microcystin Analogues: A Pilot Scale Study. *J. Environ. Manag.* **2021**, *298*, 113519. [CrossRef] [PubMed]
151. Munoz, M.; Cirés, S.; de Pedro, Z.M.; Colina, J.Á.; Velásquez-Figueroa, Y.; Carmona-Jiménez, J.; Caro-Borrero, A.; Salazar, A.; Santa María Fuster, M.-C.; Contreras, D.; et al. Overview of Toxic Cyanobacteria and Cyanotoxins in Ibero-American Freshwaters: Challenges for Risk Management and Opportunities for Removal by Advanced Technologies. *Sci. Total Environ.* **2021**, *761*, 143197. [CrossRef] [PubMed]
152. Al Yazidi, L.S.; Al-Hatmi, A.M.S. Fusariosis: An Update on Therapeutic Options for Management. *Expert Opin. Orphan Drugs* **2021**, *9*, 95–103. [CrossRef]
153. Sáenz, V.; Alvarez-Moreno, C.; Pape, P.L.; Restrepo, S.; Guarro, J.; Ramírez, A.M.C. A One Health Perspective to Recognize *Fusarium* as Important in Clinical Practice. *J. Fungi* **2020**, *6*, 235. [CrossRef]
154. Revankar, S.G.; Sutton, D.A. Melanized Fungi in Human Disease. *Clin. Microbiol. Rev.* **2010**, *23*, 884–928. [CrossRef] [PubMed]
155. Kitajima, M.; Sassi, H.P.; Torrey, J.R. Pepper Mild Mottle Virus as a Water Quality Indicator. *npj Clean Water* **2018**, *1*, 19. [CrossRef]
156. Miklos, D.B.; Remy, C.; Jekel, M.; Linden, K.G.; Drewes, J.E.; Hübner, U. Evaluation of Advanced Oxidation Processes for Water and Wastewater Treatment—A Critical Review. *Water Res.* **2018**, *139*, 118–131. [CrossRef] [PubMed]
157. Habibi-Yangjeh, A.; Asadzadeh-Khaneghah, S.; Feizpoor, S.; Rouhi, A. Review on Heterogeneous Photocatalytic Disinfection of Waterborne, Airborne, and Foodborne Viruses: Can We Win against Pathogenic Viruses? *J. Colloid Interface Sci.* **2020**, *580*, 503–514. [CrossRef] [PubMed]
158. Bolton, J.R.; Bircher, K.G.; Tumas, W.; Tolman, C.A. Figures-of-Merit for the Technical Development and Application of Advanced Oxidation Technologies for Both Electric- and Solar-Driven Systems (IUPAC Technical Report). *Pure Appl. Chem.* **2001**, *73*, 627–637. [CrossRef]
159. Berruti, I.; Nahim-Granados, S.; Abeledo-Lameiro, M.J.; Oller, I.; Polo-López, M.I. Recent Advances in Solar Photochemical Processes for Water and Wastewater Disinfection. *Chem. Eng. J. Adv.* **2022**, *10*, 100248. [CrossRef]
160. Komba, F.E.; Fabian, C.; Elimbinzi, E.; Shao, G.N. Efficiency of Common Filters for Water Treatment in Tanzania. *Bull. Natl. Res. Cent.* **2022**, *46*, 208. [CrossRef]
161. Hernández Aguilar, B.; Lerner, A.M.; Manuel-Navarrete, D.; Siqueiros-García, J.M. Persisting Narratives Undermine Potential Water Scarcity Solutions for Informal Areas of Mexico City: The Case of Two Settlements in Xochimilco. *Water Int.* **2021**, *46*, 919–937. [CrossRef]
162. Jegede, A.; Shikwambane, P. Water 'Apartheid' and the Significance of Human Rights Principles of Affirmative Action in South Africa. *Water* **2021**, *13*, 1104. [CrossRef]
163. Benarroch, A.; Rodríguez-Serrano, M.; Ramírez-Segado, A. New Water Culture versus the Traditional Design and Validation of a Questionnaire to Discriminate between Both. *Sustainability* **2021**, *13*, 2174. [CrossRef]
164. World Health Organization. *Guidelines for Drinking-Water Quality: Fourth Edition Incorporating the First Addendum*; World Health Organization: Geneva, Switzerland, 2017.
165. Antonopoulou, M.; Kosma, C.; Albanis, T.; Konstantinou, I. An Overview of Homogeneous and Heterogeneous Photocatalysis Applications for the Removal of Pharmaceutical Compounds from Real or Synthetic Hospital Wastewaters under Lab or Pilot Scale. *Sci. Total Environ.* **2021**, *765*, 144163. [CrossRef]
166. Siwila, S.; Brink, I.C. Comparison of Five Point-of-Use Drinking Water Technologies Using a Specialized Comparison Framework. *J. Water Health* **2019**, *17*, 568–586. [CrossRef] [PubMed]
167. Chaúque, B.J.M.; Brandão, F.G.; Rott, M.B. Development of Solar Water Disinfection Systems for Large-Scale Public Supply, State of the Art, Improvements and Paths to the Future—A Systematic Review. *J. Environ. Chem. Eng.* **2022**, *10*, 107887. [CrossRef]
168. Zhang, J.; Mo, Y. A Scalable Light-Diffusing Photochemical Reactor for Continuous Processing of Photoredox Reactions. *Chem. Eng. J.* **2022**, *435*, 134889. [CrossRef]
169. Reboredo, F.; Lidon, F.J.C. UV-B Radiation Effects on Terrestrial Plants—A Perspective. *Emir. J. Food Agric.* **2012**, *24*, 502–509. [CrossRef]

170. García-Gil, Á.; García-Muñoz, R.A.; McGuigan, K.G.; Marugán, J. Solar Water Disinfection to Produce Safe Drinking Water: A Review of Parameters, Enhancements, and Modelling Approaches to Make SODIS Faster and Safer. *Molecules* **2021**, *26*, 3431. [CrossRef] [PubMed]
171. Kataki, S.; Chatterjee, S.; Vairale, M.G.; Sharma, S.; Dwivedi, S.K. Concerns and Strategies for Wastewater Treatment during COVID-19 Pandemic to Stop Plausible Transmission. *Resour. Conserv. Recycl.* **2021**, *164*, 105156. [CrossRef]
172. Azamzam, A.A.; Rafatullah, M.; Yahya, E.B.; Ahmad, M.I.; Lalung, J.; Alharthi, S.; Alosaimi, A.M.; Hussein, M.A. Insights into Solar Disinfection Enhancements for Drinking Water Treatment Applications. *Sustainability* **2021**, *13*, 10570. [CrossRef]
173. Constantino, D.S.M.; Dias, M.M.; Silva, A.M.T.; Faria, J.L.; Silva, C.G. Intensification Strategies for Improving the Performance of Photocatalytic Processes: A Review. *J. Clean. Prod.* **2022**, *340*, 130800. [CrossRef]
174. Sundar, K.P.; Kanmani, S. Progression of Photocatalytic Reactors and It's Comparison: A Review. *Chem. Eng. Res. Des.* **2020**, *154*, 135–150. [CrossRef]
175. Mueses, M.A.; Colina-Márquez, J.; Machuca-Martínez, F.; Li Puma, G. Recent Advances on Modeling of Solar Heterogeneous Photocatalytic Reactors Applied for Degradation of Pharmaceuticals and Emerging Organic Contaminants in Water. *Curr. Opin. Green Sustain. Chem.* **2021**, *30*, 100486. [CrossRef]
176. Russo, D. Kinetic Modeling of Advanced Oxidation Processes Using Microreactors: Challenges and Opportunities for Scale-Up. *Appl. Sci.* **2021**, *11*, 1042. [CrossRef]
177. Öncü, N.B.; Menceloğlu, Y.Z.; Akmehmet Balcıoğlu, I. Comparison of the Effectiveness of Chlorine, Ozone, and Photocatalytic Disinfection in Reducing the Risk of Antibiotic Resistance Pollution. *J. Adv. Oxid. Technol.* **2011**, *14*, 196–203. [CrossRef]
178. Oncu, N.B.; Balcioglu, I.A. Antimicrobial Contamination Removal from Environmentally Relevant Matrices: A Literature Review and a Comparison of Three Processes for Drinking Water Treatment. *Ozone Sci. Eng.* **2013**, *35*, 73–85. [CrossRef]
179. Miranda, A.C.; Lepretti, M.; Rizzo, L.; Caputo, I.; Vaiano, V.; Sacco, O.; Lopes, W.S.; Sannino, D. Surface Water Disinfection by Chlorination and Advanced Oxidation Processes: Inactivation of an Antibiotic Resistant *E. coli* Strain and Cytotoxicity Evaluation. *Sci. Total Environ.* **2016**, *554*, 1–6. [CrossRef] [PubMed]
180. O'Flynn, D.; Lawler, J.; Yusuf, A.; Parle-McDermott, A.; Harold, D.; Mc Cloughlin, T.; Holland, L.; Regan, F.; White, B. A Review of Pharmaceutical Occurrence and Pathways in the Aquatic Environment in the Context of a Changing Climate and the COVID-19 Pandemic. *Anal. Methods* **2021**, *13*, 575–594. [CrossRef]
181. Plattard, N.; Dupuis, A.; Migeot, V.; Haddad, S.; Venisse, N. An Overview of the Literature on Emerging Pollutants: Chlorinated Derivatives of Bisphenol A (ClxBPA). *Environ. Int.* **2021**, *153*, 106547. [CrossRef]
182. Gowland, D.C.A.; Robertson, N.; Chatzisymeon, E. Photocatalytic Oxidation of Natural Organic Matter in Water. *Water* **2021**, *13*, 288. [CrossRef]
183. Yang, M.; Liberatore, H.K.; Zhang, X. Current Methods for Analyzing Drinking Water Disinfection Byproducts. *Curr. Opin. Environ. Sci. Health* **2019**, *7*, 98–107. [CrossRef]
184. Sibhatu, A.K.; Weldegebrieal, G.K.; Sagadevan, S.; Tran, N.N.; Hessel, V. Photocatalytic Activity of CuO Nanoparticles for Organic and Inorganic Pollutants Removal in Wastewater Remediation. *Chemosphere* **2022**, *300*, 134623. [CrossRef]
185. Dong, F.; Lin, Q.; Li, C.; He, G.; Deng, Y. Impacts of Pre-Oxidation on the Formation of Disinfection Byproducts from Algal Organic Matter in Subsequent Chlor(Am)Ination: A Review. *Sci. Total Environ.* **2021**, *754*, 141955. [CrossRef] [PubMed]
186. Giamello, E.; Pacchioni, G. Applied vs Fundamental Research in Heterogeneous Photocatalysis: Problems and Perspectives. An Introduction to 'Physical Principles of Photocatalysis. *J. Phys. Condens. Matter* **2020**, *32*, 360301. [CrossRef] [PubMed]

Article

Effect of the Active Metal on the NO$_x$ Formation during Catalytic Combustion of Ammonia SOFC Off-Gas

Tobias Weissenberger *, Ralf Zapf, Helmut Pennemann and Gunther Kolb

Fraunhofer Institute for Microengineering and Microsystems IMM, Carl-Zeiss-Straße 18-20, 55129 Mainz, Germany
* Correspondence: tobias.weissenberger@imm.fraunhofer.de

Abstract: Catalytic combustion of hydrogen and ammonia containing off-gas surrogate from an ammonia solid oxide fuel cell (SOFC) was studied with a focus on nitrogen oxides (NO$_x$) mitigation. Noble and transition metals (Pt, Pd, Ir, Ru, Rh, Cu, Fe, Ni) supported on Al$_2$O$_3$ were tested in the range of 100 to 800 °C. The tested catalysts were able to completely convert hydrogen and ammonia present in the off-gas. The selectivity to NO$_x$ increased with reaction temperature and stagnated at temperatures of 600 °C and higher. At low temperatures, the formation of N$_2$O was evident, which declined with increasing temperature until no N$_2$O was observed at temperatures exceeding 400 °C. Over nickel and iridium-based catalysts, the NO$_x$ formation was reduced drastically, especially at 300 to 400 °C. To the best knowledge of the authors, the current paper is the first study about catalytic combustion of hydrogen-ammonia mixtures as a surrogate of an ammonia-fed SOFC off-gas.

Keywords: ammonia; selective catalytic oxidation; SCO; SOFC; solid oxide fuel cell; catalytic combustion; nitrogen oxides; microreactor

Citation: Weissenberger, T.; Zapf, R.; Pennemann, H.; Kolb, G. Effect of the Active Metal on the NO$_x$ Formation during Catalytic Combustion of Ammonia SOFC Off-Gas. Catalysts 2022, 12, 1186. https://doi.org/10.3390/catal12101186

Academic Editor: Antonio Vita

Received: 30 August 2022
Accepted: 28 September 2022
Published: 7 October 2022

Publisher's Note: MDPI stays neutral with regard to jurisdictional claims in published maps and institutional affiliations.

Copyright: © 2022 by the authors. Licensee MDPI, Basel, Switzerland. This article is an open access article distributed under the terms and conditions of the Creative Commons Attribution (CC BY) license (https://creativecommons.org/licenses/by/4.0/).

1. Introduction

The maritime transport of goods emits less CO$_2$ per ton of cargo and km compared to other modes of transport. However, since a large part of international trade is transported by ship, the CO$_2$ emissions caused by international maritime shipping are immense [1,2].

Thus, global shipping is responsible for about 3% of the global greenhouse gas emissions (GHG) [3]. To reduce the negative impact of the shipping industry, the International Maritime Organization has set the goal to reduce the global CO$_2$ emission of the maritime shipping sector by at least 50%, compared to the emissions in the year 2008 [4].

To achieve this massive reduction in GHG emissions associated with shipping, environmentally friendly fuels and power systems must be introduced.

Fuel cells convert chemical energy directly into electrical energy, which results in a much higher efficiency compared to internal combustion engines [5]. The most common energy carrier used in fuels cells is hydrogen, which allows the operation of the fuel cell without emissions of CO$_2$ or other pollutants. Therefore, fuel cells are a very promising technology for the green energy generation, especially in transport applications such as maritime shipping.

However, when used as transportation fuel, the low energy density and high flammability of hydrogen represents major challenges. To increase the energy density, hydrogen must be stored at very high pressures, often exceeding 300 bar, or in liquid state at cryogenic temperatures [6]. High pressure and cryogenic hydrogen storage come with much higher energy demand for compression and cooling, respectively. Further, the necessary high-pressure tanks and compressors as well as the cooling system increase the investment costs drastically.

An attractive alternative for the direct hydrogen storage is the use of hydrogen carriers, which can be handled and stored under atmospheric or near atmospheric pressures and ambient temperatures [7].

One promising hydrogen carrier is ammonia due to advantages in terms of energy density, flammability and its possible direct use in solid oxide fuel cells (SOFC) [8]. Ammonia can be liquefied at 20 °C at a moderate pressure of 8.6 bar [9,10]. The lower storage pressure makes the tanks for liquefied ammonia much cheaper than the high-pressure tanks necessary for storage of compressed hydrogen or the cryogenic equipment necessary for storage of liquefied hydrogen. Therefore, the cost of ammonia storage per kWh would be much cheaper compared to storage of pure hydrogen [11].

Another advantage of ammonia as hydrogen carrier is the possibility of its direct conversion in SOFCs without the requirement for cracking reactor. The ammonia is decomposed into hydrogen and nitrogen directly in the SOFC due to its high operation temperature and the nickel catalyst present. Therefore, ammonia-fed SOFCs have attracted much interest for energy generation [12–14].

Recently, the European Union awarded funding for the ShipFC project to convert the world's first offshore vessel to run on ammonia-powered fuel cells under its Fuel Cells and Hydrogen Joint Undertaking (FCH JU) of the Research and Innovation program Horizon 2020. The ShipFC project consists of a consortium of 14 European companies and institutions and aims to install a 2 MW ammonia solid oxide fuel cell system on a shipping vessel to demonstrate that zero emission large-scale shipping is feasible. Thus, current SOFC systems will be scaled up to 2 MW and installed on the vessel Viking Energy in 2023 and operated for 3000 h during a one-year period. In addition to shipping, ammonia-powered SOFCs are an emerging technology for other mobile and stationary applications and can help to store and distribute energy form renewable sources such as wind and solar energy [15].

SOFCs and all other fuel cells cannot convert the fuel completely. Therefore, the fuel cell anode off-gas contains small quantities of unreacted hydrogen, even when a recycle of the off-gas back into the fuel cell is installed. The remaining hydrogen in the off-gas must be removed before releasing it to the atmosphere, which is commonly performed by combustion in a catalytic afterburner which also generates heat [14]. The heat generated in the afterburner can be used to preheat the ammonia and air feed before the fuel cell and excess heat can be used for heating on the vessel itself, increasing the overall efficiency of the SOFC system. In addition to hydrogen, the off-gas also contains traces of unconverted ammonia which must be removed in the catalytic afterburner too. Combustion of the ammonia traces would lead to relatively low NO_x emissions, due to the low NH_3 concentration in the feed. Nevertheless, selective catalytic oxidation of the ammonia to water and nitrogen is much preferred. Hence, the catalytic afterburner must be able to completely remove hydrogen and ammonia via combustion, while minimizing the formation of NO_x species.

In the literature, different studies about the catalytic ammonia combustion can be found. The studies could demonstrate that different catalysts are effective for the selective catalytic oxidation (SCO) of ammonia into nitrogen and water rather than NO_x, with nitrogen selectivities of up to 99% [16]. The tested catalysts include supported noble metals such as platinum [17], palladium [18], iridium [19], ruthenium [20], rhodium [21], silver [22] and gold [23]. Platinum-based catalysts showed relative low nitrogen selectivities while rhodium, palladium and silver reached selectivities of up to 97%. Another class of highly active NH_3 SCO catalysts are transition metals, e.g., iron [24], copper [25,26], cobalt [27], nickel [28], manganese [29], molybdenum [30] and vanadium [31]. Copper-based catalysts showed by far the highest nitrogen selectivities of the transition metal-based catalysts, converting up to 99% of the ammonia into nitrogen [32]. Bimetallic catalysts based on Pt/Rh and Pt/Pd have also been studied and have proven to be effective NH_3 SCO catalysts [33,34]. Further, catalysts based on mixed oxides such as Cu-Ce-Zn, Fe-Mg-Al or Cu-Mg-Al have also been studied for NH_3 SCO [35–37]. Common support materials for NH_3 SCO catalysts are Al_2O_3 [38], SiO_2 [25], TiO_2 [39], Nb_2O_5 [40], and zeolites such as CHA and ZSM-5 [41–43].

However, all of the NH$_3$ SCO studies found in the literature are dedicated to catalytic combustion of ammonia. No reports can be found about the catalytic combustion of hydrogen ammonia mixtures or anything reassembling an ammonia SOFC off-gas. Thus, the results might not apply to the SOFC off-gas combustion.

Here, we report investigations about the combustion of ammonia SOFC off-gas surrogate over different catalysts with focus on mitigating NOx formation. Microstructured reactors coated with different supported noble and transition metal catalysts were used for the off-gas combustion and their performance in terms of NOx formation was evaluated and compared at different temperatures.

2. Results

2.1. Catalyst Characterisation

The metal-loadings of the calcined catalysts are close to the values targeted during preparation, as it gets evident from the XRF data summarized in Table 1. Only the rhodium and ruthenium-based catalysts contain higher amounts of the noble metals than the desired loading of 5 wt.-% active metal, with 5.79 and 6.50 wt.-%, respectively.

Table 1. Characterization results (XRD and TEM) for the prepared catalysts.

Catalyst	Metal Loading (XRF)/wt.-%	Particle Size (TEM)/nm
Pt/Al$_2$O$_3$	5.28	2.35
Pd/Al$_2$O$_3$	4.90	4.30
Rh/Al$_2$O$_3$	5.79	1.02
Ru/Al$_2$O$_3$	6.50	50.3
Ir/Al$_2$O$_3$	4.80	3.72
Ni/Al$_2$O$_3$	5.01	>30 nm
Fe/Al$_2$O$_3$	4.87	>25 nm
Cu/Al$_2$O$_3$	10.35	not measured

The XRD pattern (see Figures S1 and S2 in the supporting information) reveals no differences in the lines observed for alumina, thus it is evident that the aluminium support was not altered by the impregnation with the corresponding metals. The metal phases display distinctive diffraction lines in the XRD patterns. The identification of the metal phases according to the XRD pattern reveal that other that platinum, which is in elemental state, all other metals are present as oxides (palladium oxide PdO, rhodium oxide Rh$_2$O$_3$, ruthenium oxide RuO$_2$, iridium oxide IrO$_2$, nickel oxide (bunsenite) NiO, iron oxide (hematite) Fe$_2$O$_3$ and copper oxide (tenorite) CuO.

The nitrogen sorption isotherms of the different catalysts and the pure support Al$_2$O$_3$, shown in Figure S5 in the supporting information, are of IUPAC type II typical for mesoporous materials. The metal deposition did not alter the isotherms and thus did not change the textural properties significantly. This is also evident in the BET surface areas (see Supporting Information Table S1) of 147 to 154 m^2 g^{-1}, which are slightly lower than the surface area of the pure Al$_2$O$_3$ support of 156 m$^2 \cdot$g^{-1}. The small reduction in surface area is likely caused by the lower specific surface area of the added metals. The surface area of the copper sample is further reduced to 140 m$^2 \cdot$g^{-1} due to higher metal content.

The transmission electron microscopy (TEM) micrographs are shown in Figure 1 (higher magnification micrographs can be found in Figure S3 in the supporting information). The TEM images of all samples display amorphous Al$_2$O$_3$ particles as well as rod shaped Al$_2$O$_3$ crystallites. The metal phases as well as support material were identified using selected area electron diffraction (SAED). The SAED pattern and TEM images with the interplanar distances are shown in Figure S4 in the supporting information. Pt and Rh particles were determined to be in metallic state while all other metals are present as oxides (PdO, RuO$_2$, IrO$_2$, NiO, Fe$_2$O$_3$ and CuO). These results are in good accordance to the XRD data. Only exemption is rhodium which was identified as Rh$_2$O$_3$ by XRD and metallic rhodium by TEM.

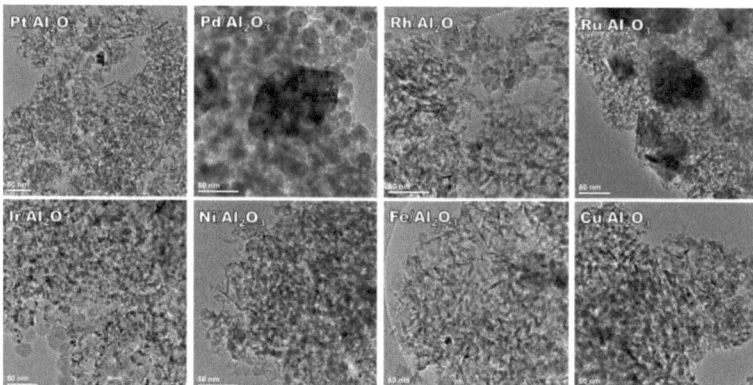

Figure 1. TEM micrographs of the different catalysts.

In case of Pt, Pd, Rh, Ru and Ir-based catalysts, the particles of the active metal phase are well dispersed over the support and can be seen easily due to their high molecular mass and thus high contrast. The average crystallite diameters can be found in Table 1. The CuO particles are hard to see and distinguish from the Al_2O_3 support, which made the measurement of a particle size distribution impractical. For the Ni and Fe-based catalysts, only a small numbers of large particles of the active phase could be seen on the TEM images.

2.2. Catalytic Tests

The evaluation of the catalytic activity was carried out under identical conditions applying a surrogate of ammonia SOFC off-gas with a composition as shown in Table 2. For the presented studies, the off-gas surrogate composition was chosen according to simulation results. For the simulation, the presumption was made that the off-gas will not be recycled in the SOFC, thus still contains a relative high concentration of hydrogen. Since the off-gas is not recycled, no separation of the water in the off-gas by condensation is necessary. Hence, the water content in the off-gas surrogate is very high. The model off-gas was mixed with air using an air-to-fuel ratio $\lambda = 4$ and then fed into the reactor.

Table 2. Composition of the model off-gas surrogate and feed gas used for the catalytic tests.

Component	SOFC Off-Gas Surrogate	Reactor Feed
N_2	25%	56.8%
H_2	15%	5.9%
NH_3	100 ppm	35.8 ppm
H_2O	60%	23.9%
O_2	-	13.2%

2.2.1. Noble Metal Catalysts

The noble metals platinum, palladium, rhodium, ruthenium and iridium supported on Al_2O_3 were tested as catalysts for the ammonia SOFC off-gas combustion. The hydrogen and ammonia concentrations as observed for different reaction temperatures are shown in Figure 2. Differences in low temperature activity of the catalysts tested are obvious.

For all tested noble metal-based catalysts, the hydrogen and ammonia conversions increase with increasing reaction temperature and reach complete conversion at about 400 °C latest. The lowest light off temperatures for hydrogen and ammonia are evident for the platinum catalyst, with full conversion of hydrogen at 150 °C and ammonia at 200 °C, respectively. The palladium-based catalyst shows a slightly reduced low temperature activity, reaching full conversion of hydrogen at 200 °C and ammonia 400 °C. The rhodium and iridium-based catalysts display comparable low temperature activity and reach full conversion of hydrogen and ammonia at 400 °C. For the ruthenium-based catalyst, the lowest activity can be observed, with complete ammonia conversion at 400 °C and complete hydrogen conversion at 500 °C, respectively. Thus, the observed light off temperatures for hydrogen and ammonia increase in the order Pt < Pd < Rh < Ir < Ru.

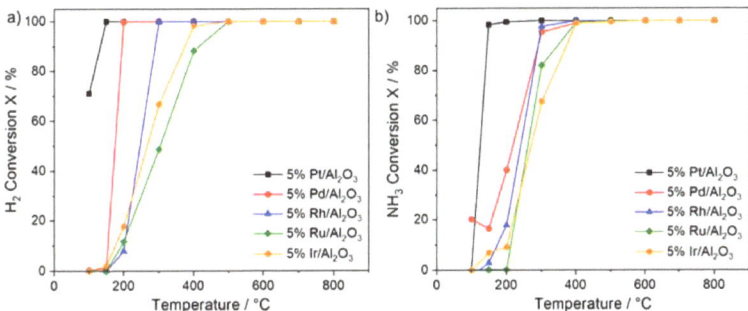

Figure 2. (a) Hydrogen and (b) ammonia conversion vs. reaction temperature for different noble metal catalysts, WHSV = 600 L/g h.

The concentrations of the nitrogen oxides (nitric oxide NO and nitrogen dioxide NO_2), shown in Figure 3a, reveal that ammonia is oxidized and forms nitrogen oxides (NO_x). Generally, once the reaction temperature is high enough for catalytic ammonia combustion, NO_x can be detected at the reactor outlet. Therefore, at low temperatures the tested catalysts show first NO_x formation according to their activity for ammonia combustion. With increasing reaction temperature, the observed NO_x concentrations increase and then stagnate at reaction temperatures exceeding 500 °C for all tested noble metal catalysts. At high temperatures in the range of 700 °C–800 °C, the NO_x concentrations increase in the order Ru < Rh < Ir < Pd < Pt.

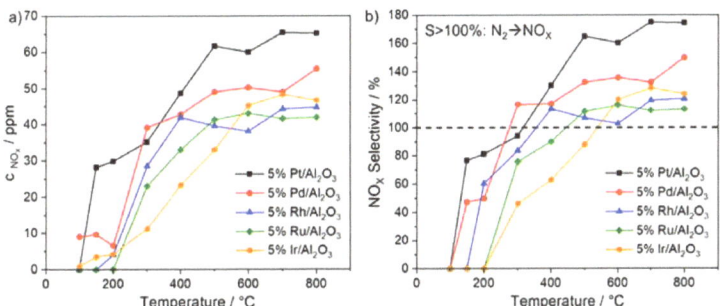

Figure 3. (a) NO_X concentration and (b) NO_X selectivities vs. reaction temperature for different noble metal catalysts, WHSV = 600 L/g h.

The different noble metal catalysts display differences in NO_X selectivity as summarized in Figure 3b. Interestingly, the measured NO_x concentrations at high temperatures exceed the ammonia concentration present in the off-gas feed of the reactor resulting in NO_x selectivities exceeding 100% if referred only to the NH_3 present in the feed. This indicates that the ammonia is completely converted into NO_x and a small fraction of the nitrogen is oxidized to NO_x as well, resulting in formation of additional NO_x. Especially the platinum catalyst displays a high additional NO_x formation, giving rise to a NO_x selectivity of about 170%. For the other noble metal catalysts high temperature NO_X selectivities between 115% and 140% were observed.

At lower temperatures, the observed NO_X molar flows are often lower than the converted ammonia resulting in NO_X selectivities lower than 100%. Therefore, it can be assumed that at lower reaction temperatures a certain fraction of the ammonia is not converted into NO_x, but very likely into nitrogen. The selectivities to NO_x at 300 and 400 °C increase in order Ir<Ru<Rh<Pd<Pt. The lowest NO_x selectivity at full ammonia conversion of the tested noble metal catalysts was observed for the iridium catalyst with a NO_x selectivity of 63% at 400 °C.

The formation of nitrous oxide (N_2O) shows a different behaviour compared to the NO_x formation (see Figure 4). The N_2O concentrations have a maxima at 150 or 200 °C and with further increase in reaction temperature the N_2O concentration decreases. At 500 °C almost no N_2O is detectable at the reactor outlet for all tested noble metal catalysts.

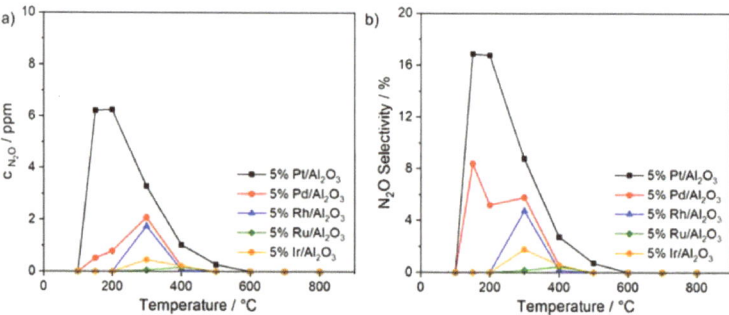

Figure 4. (a) N_2O concentration and (b) N_2O selectivities over reaction temperature for different noble metal catalysts, WHSV = 600 L/g h.

The highest N_2O concentrations are evident for the platinum catalyst, followed by palladium and rhodium. For the ruthenium and iridium catalysts almost no N_2O formation is observable, which can be traced back to the low activity of these catalysts at temperatures below 300 °C.

The selectivity to N_2O (see Figure 4b) follows the same trend as the N_2O concentrations, with exception of the palladium catalyst which shows an increased N_2O selectivity at 150 °C. The highest observed N_2O selectivity is 17% for the platinum catalyst. For all other noble metal catalyst N_2O selectivities below 10% can be observed.

Among the tested noble metals, iridium displays the most promising catalytic properties for the off-gas combustion. At a temperature of 400 °C, the iridium catalyst obtains complete hydrogen and ammonia conversion, low N_2O formation and comparatively low selectivity to NO_x.

2.2.2. Transition Metal Catalysts

In addition to noble metals, the transition metals iron, nickel and copper supported on alumina were tested as catalysts for the SOFC off-gas combustion and compared to the platinum catalyst. Both hydrogen and ammonia conversion follow the same trend as observed for the noble metal-based catalysts (see Figure 5). The differences in light off temperatures, however, are more pronounced for the transition metal catalysts. Again, with increasing reaction temperature both hydrogen and ammonia conversion increase for all tested transition metal-based catalysts. The nickel-based catalyst displays the highest activity of the tested transition metal catalyst, with complete hydrogen and ammonia conversion at 400 °C. The iron-based catalyst is slightly less active with full hydrogen and ammonia conversion at 400 °C and 600 °C, respectively. The copper-based catalyst shows a very low activity and thus high light off temperatures for both hydrogen and ammonia, despite its higher active metal loading of 10% compared to 5% for all other tested metals. To reach complete hydrogen and ammonia conversion over the copper-based catalyst a reaction temperature of about 600 °C is necessary.

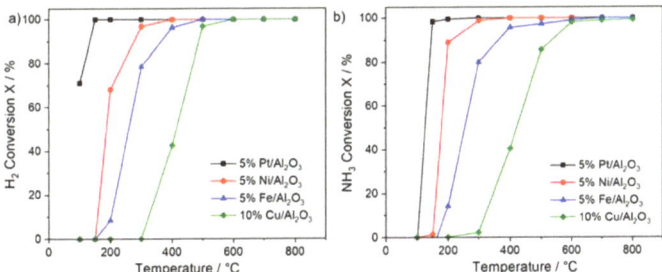

Figure 5. (a) Hydrogen and (b) ammonia conversion vs. reaction temperature for different transition metal catalysts, the platinum-based catalyst is shown for comparison, WHSV = 600 L/g h.

The NO_x concentrations observed for the tested transition metal catalysts are much lower compared to the platinum catalyst (see Figure 6a). The trend however is the same as for the noble metal catalysts. Once the reaction temperature is sufficiently high for catalytic ammonia combustion, NO_x can be detected at the reactor outlet. The NO_x concentrations and selectivities observed for the transition metal-based catalysts increase with reaction temperature and then stagnate at temperatures of 500 °C and above.

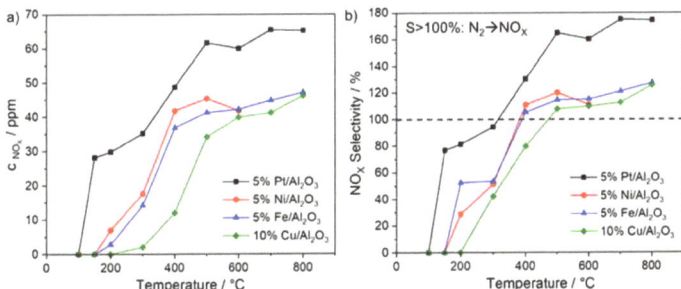

Figure 6. (a) NO_x concentration and (b) NO_x selectivity vs. reaction temperature for different transition metal catalysts, platinum-based catalyst shown for comparison, WHSV = 600 L/g h.

The nickel and iron-based catalysts display similar NO_x concentrations. However, since nickel is more active at low temperatures compared to iron, the nickel catalyst has a lower selectivity to NO_x below 500 °C as summarised in Figure 6b. The copper-based catalyst shows fairly low NO_x concentrations too, but the NO_x selectivity of the copper catalyst is higher than observed for iron and nickel. Again, at higher temperatures the NO_x concentrations surpass the ammonia concentration present in the feed gas and thus the NO_x selectivities are larger than 100%. This is a clear indication that at high temperatures NO_x is formed by ammonia combustion as well as by oxidation of nitrogen present in the off-gas.

Different studies have reported that copper oxides are good catalysts for ammonia combustion in terms of low NO_x selectivity. In our study the copper catalyst showed a much higher selectivity to NO_x compared to literature. However, the literature results were obtained for combustion in the absence of hydrogen. The presence of hydrogen could for example alter the oxidation state of copper in the catalyst which was proven to play an important role [26].

The formation of N_2O follows the same trend as observed for the noble metal catalysts (see Figure 7). At low temperatures the concentrations and selectivities increase with increasing ammonia conversions and reach a maximum at 200 °C to 300 °C. With further temperature increase the N_2O formation decreases and no N_2O can be overserved at temperatures exceeding 500 °C for all tested transition metal-based catalysts.

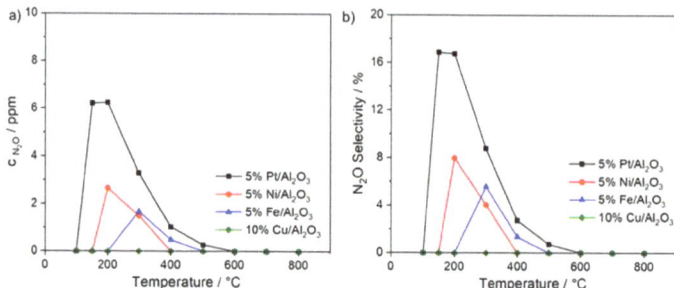

Figure 7. (a) N_2O concentration and (b) N_2O selectivity over reaction temperature for different transition metal catalysts, platinum-based catalyst shown for comparison, WHSV = 600 L/g h.

The nickel-based catalyst shows lower N_2O concentrations and selectivities compared to the platinum-based catalyst despite its relative high activity at 200 and 300 °C. At 400 °C no N_2O was observed for the nickel catalyst. The iron catalyst displays even lower N_2O concentrations but also lower activity at low temperatures resulting in slightly higher N_2O selectivity at 300 and 400 °C. No N_2O can be observed for the copper catalyst, likely because it is almost inactive at temperatures below 400°C, thus preventing N_2O formation as well.

Under the tested transitions of metal-based catalysts, the nickel-based catalyst displays the best performance. At a temperature of 300 °C, the nickel catalyst shows complete hydrogen and ammonia conversion with a rather low NO_x selectivity of 51%.

2.2.3. Catalyst Stability

Since the catalyst will be operated for prolonged times, catalytic stability was tested. A platinum catalyst was chosen for the experiments. The reaction was carried out at 800 °C with frequent cold starts from 100 to 800 °C.

The hydrogen conversion over time on stream in Figure 8a reveals that the catalyst fully converts the hydrogen even after prolonged reaction time at 800 °C. However, the hydrogen conversion during the cold start experiments (see Figure 8b) reveals catalyst deactivation. Initially during the first 24 h of reaction the catalyst loses activity as evident in the reduced hydrogen conversion at lower temperatures compared to the fresh catalyst.

With prolonged time on stream, no further reduction in hydrogen conversions can be observed. This indicated that the catalyst deactivation has stopped after the initial deactivation. Due to the observed catalyst deactivation the light-off temperature for hydrogen increases from below 100 °C to about 270 °C. Hence the preheating temperature necessary to start up the afterburner would be around 300 °C, which is still feasible.

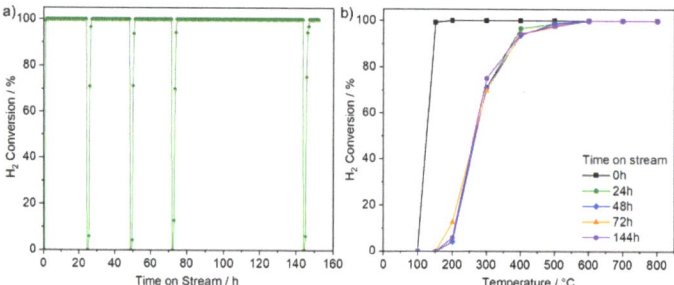

Figure 8. (a) Hydrogen conversion over time on stream with frequent cold start experiments and (b) hydrogen conversion over temperature for the corresponding cold start experiments after different times, catalyst 5% Pt/Al_2O_3, $T_{Reactor}$ = 800 °C (cold starts from 100 to 800 °C), WHSV = 600 L/g h.

3. Discussion

The light off temperatures for all tested catalysts, calculated at a 50% conversion level, for hydrogen and ammonia are summarized in Table 3. Despite the different homogenous light off temperatures for ammonia and hydrogen, the differences observed between hydrogen and ammonia light off are relatively small over all tested catalysts.

Table 3. Comparison of the catalytic performance of different metal-based catalysts tested in this study.

Catalyst	H_2 Light-Off $T_{X=50\%}$	NH_3 Light-Off $T_{X=50\%}$	Lowest S_{NOx} at X_{NH3} > 98%	S NOx at 600 °C
Pt/Al_2O_3	<100	125	81.1 (200 °C)	160.3
Pd/Al_2O_3	178	214	117.2 (400 °C)	135.9
Rh/Al_2O_3	243	244	113.6 (400 °C)	102.9
Ru/Al_2O_3	303	266	90.2 (400 °C)	116.4
Ir/Al_2O_3	267	275	63.1 (400 °C)	120.3
Ni/Al_2O_3	183	177	51.4 (300 °C)	110.7
Fe/Al_2O_3	257	254	115.0 (600 °C)	115.0
Cu/Al_2O_3	412	419	109.6 (600 °C)	109.6

Overall, platinum, nickel and palladium show low light-off temperatures, below 200 °C. For iridium, rhodium, iron and ruthenium the light–off temperatures are around 200 to 300 °C. For the copper-based catalyst the highest light-off temperature exceeding 400 °C can be observed, despite its higher loading of active metal on the catalyst support.

The selectivity to NO_x at 600 °C is close to the maximum, since the NO_x formation usually reaches a plateau at 600 °C. Here, the selectivity increases in the order Rh < Cu < Ni < Fe < Ru < Ir < Pd < Pt. The lowest selectivity to NO_x at full ammonia conversion (X_{NH3} > 98%) also corresponds to the lowest temperature necessary for full ammonia conversion since the NO_x selectivities increased with reaction temperature. The lowest value can be found for nickel and increases in the order Ni < Ir < Pt < Ru < Cu < Rh < Fe < Pd.

Comparison to Internal Combustion Engines

Compared to conventional internal combustion engines used for shipping, the presented NO_x emissions of the SOFC system are much lower.

NO_x emissions of maritime diesel engines are limited by regulation 13 of the MARPOL Annex VI of the International Maritime Organization (IMO). The current Tier II regulations set the NO_x limit at 14.4 g_{NOx}/kWh for slow running engines with rotational speeds of under 130 rpm and 7.7 g_{NOx}/kWh for engines running at over 2000 rpm.

The 2016 Tier III regulations by the IMO set more stringent NO_x limits in defined Emission Control Areas. In this areas, which include the Baltic and northern sea, the NO_x emission limits are 3.4 and 1.96 g_{NOx}/kWh, for engines operating at <130 and >2000 revolutions per minute, respectively. NO_x emission according to Tier II can be archived by modern diesel engines, Tier III however makes a deNO_x process such as ammonia-SCR necessary.

The presented system of NH_3 SOFC and afterburner emits 0.18 g NO_x per kWh (electric) in the worst case of complete NH_3 conversion to NO_2 at full load. Thus, the SOFC system only emits 9% of the NO_x emitted by Tier III conventional diesel engines. For the iridium and nickel-based catalysts developed in this work, values of about 5% of the specific NO_x emission of Tier III diesel engines can be obtained. The NO_x emission of the SOFC system were calculated at full load. Emissions are measured according to ISO 8178 at different loads which would lower the emissions even further.

Direct use of ammonia in internal combustion engines is possible too. Ammonia is partially split to hydrogen, since pure ammonia is not suitable for use in internal combustion engines. The ammonia/hydrogen mixture can be combusted at lower temperatures compared to diesel engines (CO_2 formation) and thus NO_x formation can be reduced.

Studies have shown that engines fuelled by hydrogen ammonia mixtures can achieve NO_x concentrations as low as 1–2 g NO_x per kWh depending of the ammonia concentration in the feed gas, which is still at least 10 times higher than the NO_x emissions of the NH_3 SOFC system.

Additionally, the tested catalysts were able to obtain complete conversion of ammonia at feasible operation temperatures for the afterburner, thus preventing ammonia slip from the SOFC system which can be a problem for ammonia-fuelled internal combustion engines.

4. Materials and Methods

4.1. Catalyst Preparation and Reactor Assembly

The investigations were carried out using microreactors containing microchannels coated with catalyst.

The reactors consisted of two plates with 14 channels each. The channels are 500 μm in width, 250 μm deep and 25 mm long. Once the two plates were welded together, channels with a depth of 500 μm were obtained.

The preparation and assembly consisted of three main steps: preparation of the catalyst powder, wash coating of the catalyst in the microchannels and finally assembly of the reactor via laser welding.

Catalyst powder preparation: The catalyst powders were prepared by impregnation of the Al_2O_3 support (Puralox, Sasol) with the calculated amount of an aqueous solution containing the corresponding metal precursor to archive the desired catalyst composition. After impregnation the samples were calcined at 450 °C for 6 h and the powders were milled and used for the wash coating. The metal precursors used were: $Pt(NH_3)_4(NO_3)_2$ (Alfa Aesar), $Ni(NO_3)_2 \cdot 6H_2O$ (Sigma-Aldrich), $Rh(NO_3)_3$ solution (chemPUR), $IrCl_3 \cdot xH_2O$ (Alfa Aesar), $Fe(NO_3)_3 \cdot 9H_2O$ (Sigma-Aldrich), $Cu(NO_3)_2 \cdot 3H_2O$ (Honeywell / Fluka), $Pd(NO_3)_2$ solution (Sigma-Aldrich), $Ru(NO)(NO_3)_3$ (Alfa Aesar).

Wash coating: The microchannels were coated by using a suspension containing the catalyst powder, polyvinyl alcohol as binder, acetic acid and deionised water. The polyvinyl alcohol was dissolved in water at 65 °C for 3 h before 1 wt.-% acetic acid and the catalyst powder were added. The suspension was then stirred at 65 °C for additional 3 h followed by stirring at room temperature for 2–3 days to obtain a homogeneous suspension.

The microchannels were filled with the suspension and excess suspension was removed using a blade. Then the plates were dried at room temperature and calcined at 450 °C. The wash coating process was described in greater detail in previous publications [44].

Reactor assembly: The coated plates as well as the inlet and outlet capillaries were assembled via laser welding to form the microreactor. More details can be found in a previous publication [45,46].

4.2. Catalytic Testing

The catalyst coated microstructured reactors were placed in a steal heating block equipped with two heating cartridges and thermocouples and connected to the test rig.

The gas mixture consisting of nitrogen, hydrogen and 250 ppm ammonia was mixed with air and steam before it was fed into the reactor. The gases were dosed via Bronkhorst mass flow controllers. Steam was delivered by dosing liquid water using a Bronkhorst Cori-flow into an evaporator. All lines were heated to 180 °C to avoid condensation.

Ammonia concentrations were measured using a MKS FTIR spectrometer. The hydrogen concentration was measured using an Agilent μGC.

To obtain a weight hourly space velocity (WHSV) of 600 L/g h a catalyst mass of 20 to 25 mg was used and the flow rate adjusted accordingly. For example, for a catalyst mass of 20 mg the flow rates were set to 120.1 mL/min air, 31.98 mL/min off-gas surrogate and 2.31 g/h water, resulting in a total flow rate of 200 mL/min.

4.3. Catalyst Characterization

X-ray diffraction (XRD): X-ray diffraction measurements were carried out using a D8-Advance diffractometer (Bruker) equipped with Cu-K$_\alpha$ radiation source and LYNXEYE XT-E detector in Bragg-Brentano geometry from 5–90° 2Theta with a step size of 0.02°.

X-ray fluorescence (XRF): The elemental composition of the samples was measured by XRF spectroscopy on an ED-XRF spectrometer model 1510 (Canberra Packard, USA). As radiation source Cd-109 (22 keV) and Am-241 (60 keV) were used.

Transmission electron microscopy (TEM): the transmission electron microscopy images were taken on a Zeiss Libra 120 instrument operating at an accelerating voltage of 120 kV. Before the measurements, the catalyst samples were suspended in ethanol, dropped onto the TEM grids and dried.

Nitrogen physisorption: nitrogen sorption experiments were carried out at −273 °C using an Anton Parr Autosorb iQ. The samples were degassed for 12 h at 250 °C under vacuum before measurement.

5. Conclusions

The present study shows that it is possible to use an afterburner catalyst to completely combust both hydrogen and ammonia present in an ammonia SOFC off-gas stream. However, all tested catalysts oxidized at least parts of the ammonia to nitrogen oxides. The selectivity to NO_x increases with reaction temperature and reaches a plateau at about 600 °C. The formation of N_2O increases initially with increasing reaction temperature and reaches a maximum at about 200 to 300 °C. With further temperature increases, the N_2O formation decreases and no N_2O is detectable at temperatures between 400 and 500 °C, depending on the catalyst.

Compared to other studies reported in the literature, much higher NO_x selectivities were observed for the catalysts tested in this study. However, the studies in literature were dedicated to ammonia combustion without any other fuel present in the feed gas. The relative high concentration of hydrogen in the SOFC off-gas clearly alters the nitrogen selectivities of the catalysts.

Another difference is the WHSV used for the catalytic tests. The afterburner for the Ship-FC project needs to be cost effective and relatively small and light for use onboard a shipping vessel. Thus, the used WHSV of 600 L/g_{cat} h is high compared to the WHSV reported in literature.

Nevertheless, by selecting a suitable catalyst it is feasible to reduce the NO_X formation drastically, especially at low temperatures. Nevertheless, a complete mitigation of NO_X formation was not found possible using the presented catalyst systems.

Very promising candidates are nickel and iridium-based catalysts. The nickel-based catalyst shows the lowest NO_x selectivity in the low temperature range, as well as good cold start performance. However, operating the afterburner at low temperatures results in the formation of small quantities of N_2O. The observed N_2O concentration at 300 °C is below 2 ppm, and could be reduced by further catalyst optimization. For complete N_2O mitigation, iridium is the most promising active metal tested in this study.

The iridium-based catalyst has the smallest NO_x selectivities at 400 and 500 °C of all catalysts tested in this work. The selectivity to NO_x at 400 °C is higher compared to the nickel-based catalyst at 300 °C, but due to higher reaction temperature, the N_2O formation is suppressed completely. One drawback is the reduced cold start performance of iridium, with higher light off temperatures for both hydrogen and ammonia, compared to nickel, palladium and platinum-based catalysts.

In future, further reduction in NO_x emissions could be achieved by reducing the ammonia concentration in the off-gas, by further advancements in catalyst development or by removal of the formed NO_x. Studies to determine the effect of different catalyst supports, other active metals and bi-metallic catalysts could be beneficial to further reduce the NO_x formation in the afterburner. For the off-gas cleaning, well-established processes such as catalytic selective reduction (SCR) of NO_x to N_2 with ammonia are available and could be implemented to lower NO_x emissions further.

Supplementary Materials: The following supporting information can be downloaded at: https://www.mdpi.com/article/10.3390/catal12101186/s1, Figure S1: Powder XRD pattern of noble metal-based catalysts; Figure S2: Powder XRD pattern of transition metal-based catalysts; Figure S3: TEM micrographs of the used catalysts with two different magnifications; Figure S4: TEM micrographs with interplanar distances and selected area electron beam diffraction images used for identification, Figure S5: Nitrogen sorption isotherm of Al_2O_3 support; Figure S6: Nitrogen sorption isotherms of the used catalysts, Figure S7: Particle size distribution of different catalysts determined by TEM, Table S1: Specific surface area (BET method) of the used catalysts.

Author Contributions: Conceptualization, G.K., H.P. and T.W.; investigation, T.W and R.Z.; writing—original draft preparation, T.W.; writing—review and editing, T.W., R.Z., H.P. and G.K.; visualization, T.W.; supervision, G.K.; project administration, H.P. and G.K.; funding acquisition, G.K. All authors have read and agreed to the published version of the manuscript.

Funding: This research was funded by the ShipFC project which has received funding from the Fuel Cells and Hydrogen Joint Undertaking under grant agreement No 875156. This Joint Undertaking receives support from the European Union's Horizon 2020 research and innovation program and from Hydrogen Europe.

Data Availability Statement: Not applicable.

Conflicts of Interest: The authors declare no conflict of interest.

References

1. Miola, A.; Ciuffo, B. Estimating air emissions from ships: Meta-analysis of modelling approaches and available data sources. *Atmos. Environ.* **2011**, *45*, 2242–2251. [CrossRef]
2. Balcombe, P.; Brierley, J.; Lewis, C.; Skatvedt, L.; Speirs, J.; Hawkes, A.; Staffell, I. How to decarbonise international shipping: Options for fuels, technologies and policies. *Energy Convers. Manag.* **2019**, *182*, 72–88. [CrossRef]
3. CAIT. *World Resources Institute. Historical Emissions Data*; World Resources Institute: Washington, DC, USA, 2017.
4. International Maritime Organization. *Initial IMO Strategy on Reduction of GHG Emissions from Ships*; International Maritime Organization: London, UK, 2018.
5. Stambouli, A.B.; Traversa, E. Fuel cells, an alternative to standard sources of energy. *Renew. Sustain. Energy Rev.* **2002**, *6*, 295–304. [CrossRef]
6. Xu, Q.; Kobayashi, T. *Advanced Materials for Clean Energy*; CRC Press: Boca Raton, FL, USA, 2019; ISBN 1482205807.
7. He, T.; Pachfule, P.; Wu, H.; Xu, Q.; Chen, P. Hydrogen carriers. *Nat. Rev. Mater.* **2016**, *1*, 16059. [CrossRef]
8. Wojcik, A.; Middleton, H.; Damopoulos, I. Ammonia as a fuel in solid oxide fuel cells. *J. Power Sources* **2003**, *118*, 342–348. [CrossRef]
9. Lan, R.; Irvine, J.T.S.; Tao, S. Ammonia and related chemicals as potential indirect hydrogen storage materials. *Int. J. Hydrogen Energy* **2012**, *37*, 1482–1494. [CrossRef]
10. Valera-Medina, A.; Xiao, H.; Owen-Jones, M.; David, W.I.F.; Bowen, P.J. Ammonia for power. *Prog. Energy Combust. Sci.* **2018**, *69*, 63–102. [CrossRef]
11. Minutillo, M.; Perna, A.; Di Trolio, P.; Di Micco, S.; Jannelli, E. Techno-economics of novel refueling stations based on ammonia-to-hydrogen route and SOFC technology. *Int. J. Hydrogen Energy* **2021**, *46*, 10059–10071. [CrossRef]
12. Afif, A.; Radenahmad, N.; Cheok, Q.; Shams, S.; Kim, J.H.; Azad, A.K. Ammonia-fed fuel cells: A comprehensive review. *Renew. Sustain. Energy Rev.* **2016**, *60*, 822–835. [CrossRef]
13. Jeerh, G.; Zhang, M.; Tao, S. Recent progress in ammonia fuel cells and their potential applications. *J. Mater. Chem. A* **2021**, *9*, 727–752. [CrossRef]
14. Siddiqui, O.; Dincer, I. A review and comparative assessment of direct ammonia fuel cells. *Therm. Sci. Eng. Prog.* **2018**, *5*, 568–578. [CrossRef]
15. Jiao, F.; Xu, B. Electrochemical ammonia synthesis and ammonia fuel cells. *Adv. Mater.* **2019**, *31*, 1805173. [CrossRef] [PubMed]
16. Chmielarz, L.; Jabłońska, M. Advances in selective catalytic oxidation of ammonia to dinitrogen: A review. *RSC Adv.* **2015**, *5*, 43408–43431. [CrossRef]
17. Sobczyk, D.P.; van Grondelle, J.; Thüne, P.C.; Kieft, I.E.; de Jong, A.; van Santen, R.A. Low-temperature ammonia oxidation on platinum sponge studied with positron emission profiling. *J. Catal.* **2004**, *225*, 466–478. [CrossRef]
18. Jabłońska, M.; Król, A.; Kukulska-Zajac, E.; Tarach, K.; Chmielarz, L.; Góra-Marek, K. Zeolite Y modified with palladium as effective catalyst for selective catalytic oxidation of ammonia to nitrogen. *J. Catal.* **2014**, *316*, 36–46. [CrossRef]

19. Chen, W.; Qu, Z.; Huang, W.; Hu, X.; Yan, N. Novel effect of SO$_2$ on selective catalytic oxidation of slip ammonia from coal-fired flue gas over IrO$_2$ modified Ce–Zr solid solution and the mechanism investigation. *Fuel* **2016**, *166*, 179–187. [CrossRef]
20. Shin, J.H.; Kim, G.J.; Hong, S.C. Reaction properties of ruthenium over Ru/TiO$_2$ for selective catalytic oxidation of ammonia to nitrogen. *Appl. Surf. Sci.* **2020**, *506*, 144906. [CrossRef]
21. Long, R.Q.; Yang, R.T. Noble metal (Pt, Rh, Pd) promoted Fe-ZSM-5 for selective catalytic oxidation of ammonia to N$_2$ at low temperatures. *Catal. Lett.* **2002**, *78*, 353–357. [CrossRef]
22. Zhang, L.; He, H. Mechanism of selective catalytic oxidation of ammonia to nitrogen over Ag/Al$_2$O$_3$. *J. Catal.* **2009**, *268*, 18–25. [CrossRef]
23. Lin, M.; An, B.; Niimi, N.; Jikihara, Y.; Nakayama, T.; Honma, T.; Takei, T.; Shishido, T.; Ishida, T.; Haruta, M. Role of the acid site for selective catalytic oxidation of NH$_3$ over Au/Nb$_2$O$_5$. *ACS Catal.* **2019**, *9*, 1753–1756. [CrossRef]
24. Long, R.Q.; Yang, R.T. Selective catalytic oxidation of ammonia to nitrogen over Fe$_2$O$_3$–TiO$_2$ prepared with a sol–gel method. *J. Catal.* **2002**, *207*, 158–165. [CrossRef]
25. Hinokuma, S.; Kiritoshi, S.; Kawabata, Y.; Araki, K.; Matsuki, S.; Sato, T.; Machida, M. Catalytic ammonia combustion properties and operando characterization of copper oxides supported on aluminum silicates and silicon oxides. *J. Catal.* **2018**, *361*, 267–277. [CrossRef]
26. Hirabayashi, S.; Ichihashi, M. Gas-phase reactions of copper oxide cluster cations with ammonia: Selective catalytic oxidation to nitrogen and water molecules. *J. Phys. Chem. A* **2018**, *122*, 4801–4807. [CrossRef] [PubMed]
27. Hinokuma, S.; Araki, K.; Iwasa, T.; Kiritoshi, S.; Kawabata, Y.; Taketsugu, T.; Machida, M. Ammonia-rich combustion and ammonia combustive decomposition properties of various supported catalysts. *Catal. Commun.* **2019**, *123*, 64–68. [CrossRef]
28. Amblard, M.; Burch, R.; Southward, B.W.L. A study of the mechanism of selective conversion of ammonia to nitrogen on Ni/γ-Al$_2$O$_3$ under strongly oxidising conditions. *Catal. Today* **2000**, *59*, 365–371. [CrossRef]
29. Lee, J.Y.; Kim, S.B.; Hong, S.C. Characterization and reactivity of natural manganese ore catalysts in the selective catalytic oxidation of ammonia to nitrogen. *Chemosphere* **2003**, *50*, 1115–1122. [CrossRef]
30. De Boer, M.; Huisman, H.M.; Mos, R.J.M.; Leliveld, R.G.; van Dillen, A.J.; Geus, J.W. Selective oxidation of ammonia to nitrogen over SiO$_2$-supported MoO$_3$ catalysts. *Catal. Today* **1993**, *17*, 189–200. [CrossRef]
31. Lee, S.M.; Hong, S.C. Promotional effect of vanadium on the selective catalytic oxidation of NH3 to N2 over Ce/V/TiO2 catalyst. *Appl. Catal. B Environ.* **2015**, *163*, 30–39. [CrossRef]
32. Song, S.; Jiang, S. Selective catalytic oxidation of ammonia to nitrogen over CuO/CNTs: The promoting effect of the defects of CNTs on the catalytic activity and selectivity. *Appl. Catal. B Environ.* **2012**, *117*, 346–350. [CrossRef]
33. Hung, C.-M.; Lai, W.-L.; Lin, J.-L. Removal of gaseous ammonia in Pt-Rh binary catalytic oxidation. *Aerosol Air Qual. Res.* **2012**, *12*, 583–591. [CrossRef]
34. Zhou, M.; Wang, Z.; Sun, Q.; Wang, J.; Zhang, C.; Chen, D.; Li, X. High-performance Ag–Cu nanoalloy catalyst for the selective catalytic oxidation of ammonia. *ACS Appl. Mater. Interfaces* **2019**, *11*, 46875–46885. [CrossRef] [PubMed]
35. Chmielarz, L.; Jabłońska, M.; Strumiński, A.; Piwowarska, Z.; Węgrzyn, A.; Witkowski, S.; Michalik, M. Selective catalytic oxidation of ammonia to nitrogen over Mg-Al, Cu-Mg-Al and Fe-Mg-Al mixed metal oxides doped with noble metals. *Appl. Catal. B Environ.* **2013**, *130*, 152–162. [CrossRef]
36. Chmielarz, L.; Kuśtrowski, P.; Rafalska-Łasocha, A.; Dziembaj, R. Selective oxidation of ammonia to nitrogen on transition metal containing mixed metal oxides. *Appl. Catal. B Environ.* **2005**, *58*, 235–244. [CrossRef]
37. Qu, Z.; Wang, Z.; Zhang, X.; Wang, H. Role of different coordinated Cu and reactive oxygen species on the highly active Cu–Ce–Zr mixed oxides in NH$_3$-SCO: A combined in situ EPR and O$_2$-TPD approach. *Catal. Sci. Technol.* **2016**, *6*, 4491–4502. [CrossRef]
38. Sun, M.; Liu, J.; Song, C.; Ogata, Y.; Rao, H.; Zhao, X.; Xu, H.; Chen, Y. Different reaction mechanisms of ammonia oxidation reaction on Pt/Al$_2$O$_3$ and Pt/CeZrO$_2$ with various Pt states. *ACS Appl. Mater. Interfaces* **2019**, *11*, 23102–23111. [CrossRef]
39. Jabłońska, M. TPR study and catalytic performance of noble metals modified Al$_2$O$_3$, TiO$_2$ and ZrO$_2$ for low-temperature NH$_3$-SCO. *Catal. Commun.* **2015**, *70*, 66–71. [CrossRef]
40. Lin, M.; An, B.; Takei, T.; Shishido, T.; Ishida, T.; Haruta, M.; Murayama, T. Features of Nb$_2$O$_5$ as a metal oxide support of Pt and Pd catalysts for selective catalytic oxidation of NH$_3$ with high N$_2$ selectivity. *J. Catal.* **2020**, *389*, 366–374. [CrossRef]
41. Akah, A.; Cundy, C.; Garforth, A. The selective catalytic oxidation of NH3 over Fe-ZSM-5. *Appl. Catal. B Environ.* **2005**, *59*, 221–226. [CrossRef]
42. Guo, J.; Yang, W.; Zhang, Y.; Gan, L.; Fan, C.; Chen, J.; Peng, Y.; Li, J. A multiple-active-site Cu/SSZ-13 for NH$_3$-SCO: Influence of Si/Al ratio on the catalytic performance. *Catal. Commun.* **2020**, *135*, 105751. [CrossRef]
43. Yue, Y.; Liu, B.; Qin, P.; Lv, N.; Wang, T.; Bi, X.; Zhu, H.; Yuan, P.; Bai, Z.; Cui, Q. One-pot synthesis of FeCu-SSZ-13 zeolite with superior performance in selective catalytic reduction of NO by NH$_3$ from natural aluminosilicates. *Chem. Eng. J.* **2020**, *398*, 125515. [CrossRef]
44. Zapf, R.; Thiele, R.; Wichert, M.; O'Connell, M.; Ziogas, A.; Kolb, G. Application of rhodium nanoparticles for steam reforming of propane in microchannels. *Catal. Commun.* **2013**, *41*, 140–145. [CrossRef]

45. Kolb, G.; Zapf, R.; Hessel, V.; Löwe, H. Propane steam reforming in micro-channels—results from catalyst screening and optimisation. *Appl. Catal. A Gen.* **2004**, *277*, 155–166. [CrossRef]
46. Zapf, R.; Becker-Willinger, C.; Berresheim, K.; Bolz, H.; Gnaser, H.; Hessel, V.; Kolb, G.; Löb, P.; Pannwitt, A.-K.; Ziogas, A. Detailed characterization of various porous alumina-based catalyst coatings within microchannels and their testing for methanol steam reforming. *Chem. Eng. Res. Des.* **2003**, *81*, 721–729. [CrossRef]

Article

Effective Removal of Refractory Pollutants through Cinnamic Acid-Modified Wheat Husk Biochar: Experimental and DFT-Based Analysis

Umme Habiba [1], Sadaf Mutahir [1,2,*], Muhammad Asim Khan [1,2,*], Muhammad Humayun [3], Moamen S. Refat [4] and Khurram Shahzad Munawar [5]

1. Department of Chemistry, University of Sialkot, Sialkot 51300, Pakistan
2. School of Chemistry and Chemical Engineering, Linyi University, Linyi 276000, China
3. Wuhan National Laboratory for Optoelectronics, School of Optical and Electronics Information, Huazhong University of Science and Technology, Wuhan 430074, China
4. Department of Chemistry, College of Science, Taif University, Taif 21944, Saudi Arabia
5. Department of Chemistry, University of Mianwali, Mianwali 42200, Pakistan
* Correspondence: sadafmutahir@hotmail.com (S.M.); khanabdali@hotmail.com (M.A.K.)

Citation: Habiba, U.; Mutahir, S.; Khan, M.A.; Humayun, M.; Refat, M.S.; Munawar, K.S. Effective Removal of Refractory Pollutants through Cinnamic Acid-Modified Wheat Husk Biochar: Experimental and DFT-Based Analysis. *Catalysts* 2022, 12, 1063. https://doi.org/10.3390/catal12091063

Academic Editor: Meng Li

Received: 21 August 2022
Accepted: 14 September 2022
Published: 17 September 2022

Publisher's Note: MDPI stays neutral with regard to jurisdictional claims in published maps and institutional affiliations.

Copyright: © 2022 by the authors. Licensee MDPI, Basel, Switzerland. This article is an open access article distributed under the terms and conditions of the Creative Commons Attribution (CC BY) license (https://creativecommons.org/licenses/by/4.0/).

Abstract: The removal of refractory pollutants, i.e., methylene blue (MB) and ciprofloxacin (CIP), relies heavily on sorption technologies to address global demands for ongoing access to clean water. Because of the poor adsorbent–pollutant contact, traditional sorption procedures are inefficient. To accomplish this, a wheat husk biochar (WHB), loaded with cinnamic acid, was created using a simple intercalation approach to collect dangerous organic pollutants from an aqueous solution. Batch experiments, detecting technologies, and density functional theory (DFT) calculations were used to investigate the interactions at the wheat husk biochar modified with cinnamic acid (WHB/CA) and water interface to learn more about the removal mechanisms. With MB (96.52%) and CIP (94.03%), the functionalized WHB exhibited outstanding adsorption capabilities, with model fitting results revealing that the adsorption process was chemisorption and monolayer contact. Furthermore, DFT studies were performed to evaluate the interfacial interaction between MB and CIP with the WHB/CA surface. The orbital interaction diagram provided a visual representation of the interaction mechanism. These findings open up a new avenue for researchers to better understand adsorption behavior for the utilization of WHB on an industrial scale.

Keywords: wheat husk biochar; cinnamic acid; wastewater treatment; adsorption; DFT

1. Introduction

Rapid industrialization and massive growth in the human population have resulted in major environmental contamination in the last few decades. Increased concentrations of a wide variety of contaminants or pollutants, such as toxic heavy metal ions, inorganic anions, micropollutants, and organic compounds such as dyes, phenols, pesticides, humic substances, detergents, and other persistent organic pollutants, have been widely reported in recent decades in various parts of the world. The release of these harmful chemicals into natural water bodies has devastated flora and fauna and has disrupted the ecological balance. Many of these pollutants are not only chemically or biologically resistant, but they also have a high level of environmental mobility and a high potential for bioaccumulation in the food chain [1]. Water contaminants include inorganic dangerous elements, organic compounds, and microorganisms. Inorganic hazardous elements include mercury, cadmium, lead, chromium, copper, and other inorganic metallic elements. Common organic contaminants in water include pharmaceuticals, personal care products, endocrine disruptors, pesticides, organic dyes, detergents, and common industrial organic wastes, such as phenolics, halogens, and aromatics [2–4].

Dyes are an important class of aromatic hydrocarbons that pollute the environment. These are colored substances used in various industries [5]. Each year, 5–10% of the 7×10^5 metric tons of dyes produced are discharged as waste into water [6]. Although these are used in a variety of industries for a variety of purposes, their presence in water is extremely hazardous due to their poor degradability, unpleasant odor, high toxicity level, and retardation of light penetration into the water, causing the photosynthesis process to halt, thereby disrupting the entire ecosystem [7].

There are various classes of dyes; among these, methylene blue is an important dye. It is a heterocyclic compound with the chemical formula $C_{16}H_{18}N_3SCl\cdot3H_2O$. Methylene blue (MB), a biological staining agent, is the most commonly used cationic dye in the wool, silk, leather, calico, cotton, and tannin industries. MB, which can inflict eye burns on humans and aquatic critters and permanently harm their eyes. Furthermore, depending on the length of time that people are exposed to MB, it can be hazardous to their health. Long-term MB exposure can result in a burning sensation, mental disorientation, nausea, vomiting, methemoglobinemia, cyanosis, convulsions, dyspnea, anemia, and hypertension, whereas short-term exposure might result in quick or difficult breathing [8–10].

Ciprofloxacin is a second-generation fluoroquinolone antibiotic that targets Gram-negative and Gram-positive bacteria and has a fluorine atom at position 6 in the quinoline group. Its superior properties, such as excellent tissue penetration, good bioavailability, and fewer side effects, led doctors and veterinarians all over the world to prescribe it to treat a variety of bacterial infections, including infections of the respiratory system, skin, bone, joints, urinary tract, and gastrointestinal tract [11]. According to many studies from around the world, ciprofloxacin has been identified in alarming proportions in a range of environmental systems, including agricultural soils, groundwater, surface water, and effluents from various wastewater treatment plants (i.e., domestic, hospital, and industrial) [12,13]. Even though ciprofloxacin concentrations in aquatic environments range from ngL^{-1} to gL^{-1}, these levels are adequate to cause and accelerate antimicrobial resistance gene (AMR) proliferation in water. These genes can infect the body (human or animal) with fatal diseases, while tolerating or competing with any antibiotic, posing a severe threat to the world's limited water supply. If reasonable measures to lower ciprofloxacin content and related resistant genes in water are not performed, AMRs could become a future pandemic similar to COVID-19 [14].

As a result, contaminated water and wastewater must be cleaned before being discharged into the natural environment. To remove organic contaminants from wastewater, many approaches have been tried, including electrochemical treatment [15], biological treatment [16], photodegradation [17], ozonation [18], electrochemical treatment [19], membrane filtration, oxidation [20], adsorption [21], reverse osmosis [22], physical nanofiltration [23,24], chemical oxidation [25,26], ion exchange, and electrocoagulation [27]. Among these methods, adsorption is considered one of the best methods, because it is low-cost, the nontoxic adsorbent can be regenerated, it is easy to operate, and has no chance of high toxicity [28]. However, the surface area, porosity, and pore diameter of the adsorbent and adsorbate that are utilized determine the efficacy of the adsorption process. This approach has used a variety of adsorbents, including carbon-based adsorbents [29], polymers and resins [30,31], metal-organic frameworks, clays and minerals [32,33], nanomaterials [34] [35,36] such as PTA/Zr-MCM-41, and metals and their oxides [37], which have been used for the adsorption of antibiotics/dyes. Activated carbon is the most important and frequently used cost-effective adsorbent; it can be regenerated with some mass loss. The production of activated carbon from different sources results in a difference in porosity, adsorption, and other properties. Moreover, the method of preparation of activated carbon and parameters used during this method affect the acidic and basic nature of activated carbon [38].

One of the serious problems is that the direct usage of these adsorbents creates many problems, such as poor adsorption capacity, emission of secondary pollutants, high COD, BOD, and TOC [39], while wheat husk biochar modified with cinnamic acid has no such problems. It does not create sludge formation, nor does it emit secondary pollutants, and it has no effects on BOD, COD, or TOC.

Biochar derived from wheat husk can be used as an adsorbent because it has a high surface area, insolubility in water, high mechanical strength, and various functional groups. Moreover, it has some alkali metal content that also enhances its adsorption capacity. The chemical composition of wheat husk includes cellulose, 33.7–40%; hemicellulose, 21–26%; lignin, 11–22.9%; and extractives that include alkali metal contents and silica, which account for 15.3% [40].

Tons of wheat husk are produced every year, with very little usage in feedstock, energy production, and the adsorption process. A large amount of wheat husk is thrown out as waste material. The use of wheat husk as an adsorbent for the removal of organic pollutants not only provides a solution for the usage of this waste material in a useful form, but also plays an effective role in wastewater treatment. It is the cheapest and most renewable source, but the problem is that natural wheat husk exhibits very poor adsorption capacity. Although wheat straw biochar has been reported in the literature for the removal of MB, it has a very low adsorption capacity of 46.6 mg/g in the absence of a magnetic field, and 62.5 mg/g in the presence of a magnetic field. This poor adsorption capacity is not sufficient for today's threatening levels of pollutants [41]. Other modification techniques can be used to improve the properties of the wheat husk [42]. The modification increases surface area and adsorption properties [43].

This paper reports the modification of wheat husk-derived biochar with cinnamon acid, as it is not a harsh acid, is less toxic, and is environmentally friendly, while the previously reported modifying agents are harsh and have some toxic effects. There is currently no literature available on the adsorption of organic pollutants using cinnamic acid-modified wheat husk biochar. A study has been carried out on wheat husk biochar that has very little affinity toward the removal of MB only, but no study has been carried out on wheat husk biochar loaded with cinnamic acid for the removal of antibiotics. As a result, the goal of this study was to fill a research gap by analyzing organic pollutant removal utilizing cinnamic acid-modified wheat husk biochar and batch adsorption. Methylene blue, a well-known thiazine dye, was chosen to represent cationic dyes, while ciprofloxacin, a fluoroquinolone antibiotic with a fluorine atom at position 6 in the quinoline group, was chosen to represent non-cationic colors. As part of the research, the adsorbent was characterized based on its surface appearance and pore characteristics. In batch mode, the impacts of key adsorption factors, such as adsorbent dosage, beginning pH, contact time, and initial concentration, were also investigated. Finally, the adsorption mechanisms on both MB and ciprofloxacin, with cinnamic acid-modified wheat husk biochar as the adsorbent, were investigated using the fitting of experimental data with adsorption equilibrium isotherm, kinetic, and mechanism models, as well as theoretical studies (Scheme 1).

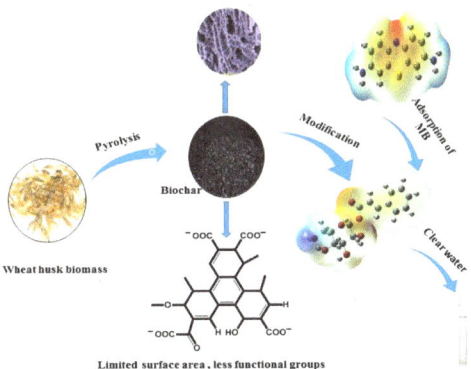

Scheme 1. Schematic representation of modified wheat husk-based wastewater treatment.

Hence, this study signifies that by preparing WHB/CA, we can remove refractory pollutants from wastewater in heavy amounts. So, we can not only overcome water pollution, but also land pollution.

2. Results and Discussion

2.1. SEM Analysis

The adsorbent's surface micromorphology has a significant impact on its adsorption performance. The micromorphology of biochar may be seen easily and intuitively using scanning electron microscopy. The morphological changes of materials before and after alteration can be better examined using electron microscopy. Figure 1 shows the SEM-EDS pictures of the biochar before and after modification. The micromorphology of both WHB and 7.5%WHB/CA is an inhomogeneous porous structure with a roughened surface, as seen by SEM images (Figure 1). It can be seen from Figure 1a that the surface of WHB has many prominent cracks that are lessened in the surface of 7.5%WHB/CA (Figure 1b), and macropores are converted into microspores. Hence, the modifier aggregated on the surface of WHB, which is very effective for the adsorption of organic pollutants on its surface, while it is shown in Figure 1c that the surface of WHB has a small amount of debris on its surface, with some smoother contents, while the surface of the 7.5%WHB/CA (Figure 1d) is rough and was fully covered with the modifier. This aggregation on the surface of 7.5%WHB/CA was an indication of successful modification [44].

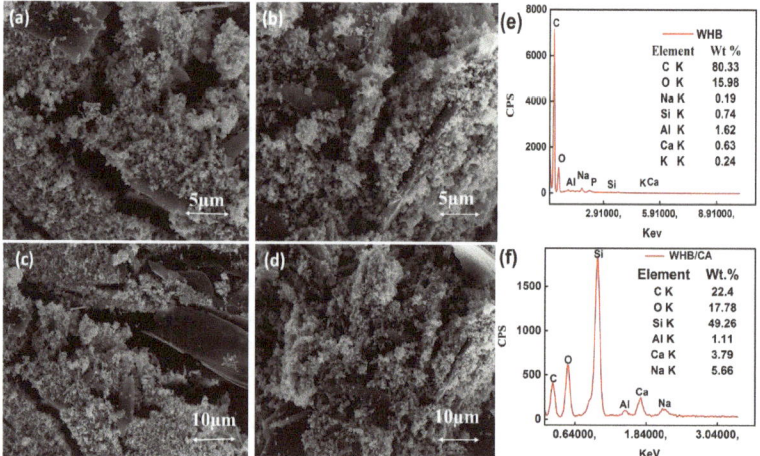

Figure 1. SEM of WHB (**a**,**c**) and WHB/CA (**b**,**d**), EDS of WHB (**e**), and EDS of MWH/CA (**f**).

2.2. XRD Analysis

XRD analysis was performed to investigate the crystallinity of the material. As shown in Figure 2, a characteristic peak at around 22° appeared, which corresponded to cellulose for WHB, while the same diffraction for 7.5%WHB/CA was very sharp, which is an indication of preferred orientation. It shows the removal of amorphous constituents, such as lignin and hemicellulose, and the insertion of cinnamic acid into the layers of WHB [45].

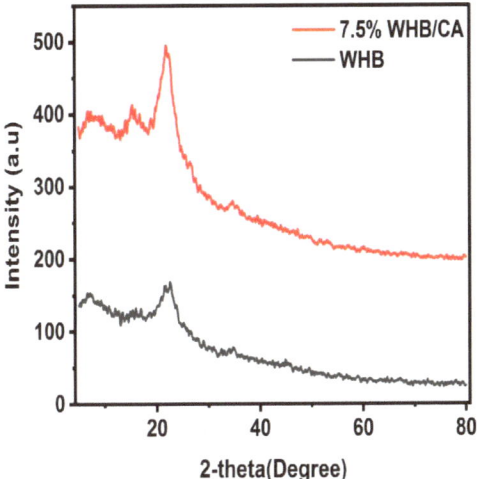

Figure 2. XRD of WHB and 7.5%WHB/CA.

2.3. FTIR Analysis

FTIR spectra of WHB, WHB/CA, and MB-WHB/CA samples were analyzed (Figure 3). The peak at 2350 cm^{-1} gives evidence of CH stretch, and a peak at 1640 cm^{-1} corresponded to the C=C of the aromatic ring [46]. The vibrations of the carboxylate (COO-) group were represented by a peak at 1450 cm^{-1}, whereas the OCC stretch of the acetate group was represented by a peak at 1160 cm^{-1}, indicating that WHB was esterified. The intensity of both peaks (C=O and COO-) increased after modification, which indicates that modification was performed, and some sort of rearrangement may have been performed in the biochar [47]. A band at 1070 cm^{-1} was due to COC stretch [46], which was suppressed in the treated WH sample. Furthermore, the broadening of the peak at 1160 cm^{-1} and the broadband of hydrogen-bonded OH at 3400 cm^{-1} also indicates the adsorption of dye onto the modified sample. Hence, it is inferred that WHB/CA can adsorb MB effectively.

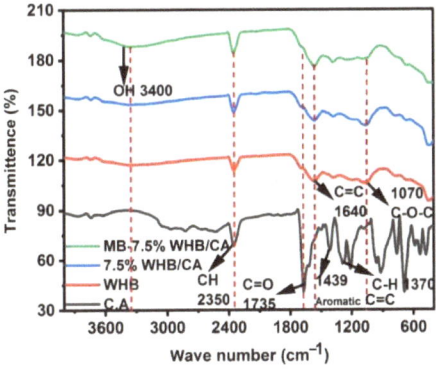

Figure 3. FTIR spectra of WHB, WHB/CA, and MB-WHB/CA.

2.4. Adsorption Experiments

Adsorption experiments were carried out by selecting the optimum impregnation ratio of WHB and CA. Mixing 0.1 g of adsorbent with 100 mL of dye solution and 100 mL (100 mg/L) of ciprofloxacin solution at room temperature for 160 minutes at 25 °C on an orbital shaker at 100 RPM yielded the best impregnation ratio. It was seen from the results that the modified sample of 7.5% WHB/CA (wheat husk biochar/cinnamic acid) showed maximum adsorption capacity against both adsorbates. Figure 4a shows the plot of impregnation ratios of the samples vs. adsorption capacity of MB, while Figure 4b shows the plot of impregnation ratios of the samples vs. adsorption capacity of Ciprofloxacin. It is inferred from the results that a sample of 7.5% WHB/CA showed the highest adsorption capacity for both adsorbates. So, the whole of the work was performed on a 7.5% WHB/CA sample.

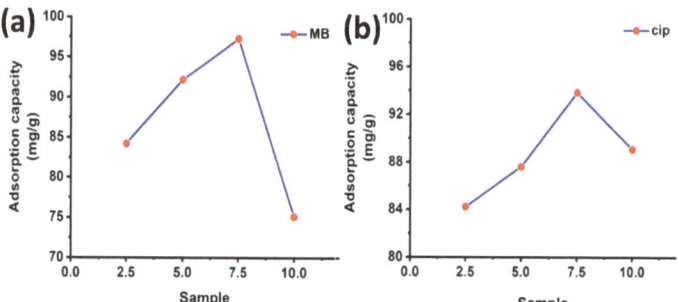

Figure 4. Plot of impregnation ratio vs. adsorption capacity of MB (**a**) and adsorption capacity of ciprofloxacin (**b**).

2.4.1. Effect of Adsorbent Dosage

The effect of the adsorbent dosage was evaluated by changing the adsorbent dosage from 50 mg to 200 mg in 100 mL of dye solution. The adsorption percentage was enhanced by increasing the dosage of the adsorbent from 83.155 to 99%, after which it became constant, which indicates that 100 mg of adsorbent dosage is sufficient for the maximum removal of pollutants; so, in this work, 100 mg of adsorbent dosage was used in all experiments. This increase in adsorption percentage might be due to an increase in adsorption sites. The adsorbent dose surface area increased, and thus, the adsorption rate also increased. However, further enhancement or poor increment in the adsorption rate was due to the unavailability of MB binding sites (Figure 5a) [48].

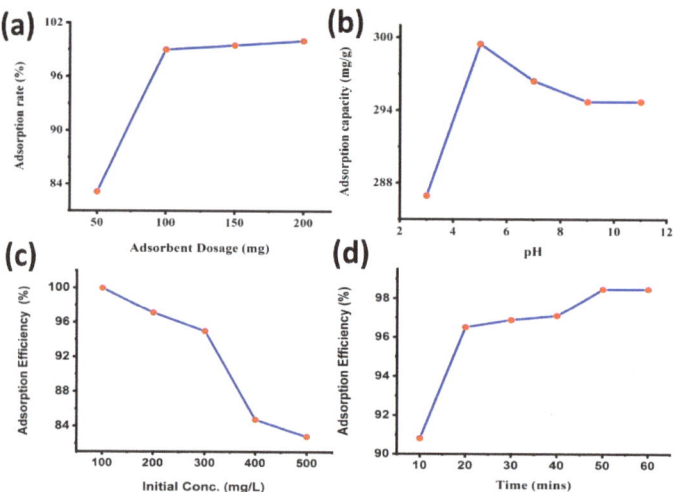

Figure 5. Effect of parameters on MB sorption, effect of adsorbent dosage (**a**), effect of pH (**b**), effect of initial conc. (**c**), and effect of contact time (**d**).

2.4.2. Effect of pH

The pH of wastewater is a very important parameter that may affect the removal rate of the pollutant, so the effect of pH on the adsorption rate was monitored by changing the pH from 3 to 11. The MB removal rate was lowest at pH 3, and the adsorption percentage increased from 95.658% to 99.6% as the pH increased from 3 to 5. So, the maximum removal ratio was obtained at pH 5, which decreased, and then became almost constant. This indicates that sorption is mainly due to hydrogen bonding functional groups (-NH$^-$, SO^{-3}) in MB and hydroxyl groups present in the adsorbent. For this reason, all experiments were performed at pH 5. Such results were reported by Jasmin S et al. (Figure 5b) [49].

2.4.3. Effect of Initial Concentration of Dye

The effect of the initial dye concentration was studied by adjusting the dye concentration from 100 to 500 mg/L, while keeping the other parameters constant. It was discovered from the data that the sorption rate was initially high, which could be related to the presence of a significant number of active sites on the adsorbent. The sorption rate gradually decreased after fast MB diffusion from the aqueous solution to the adsorbent surface. This could be attributed to the saturation of all active sites on the adsorbent's surface, as well as dye molecules penetrating the adsorbent's pores, lowering the rate of sorption (Figure 5c) [50].

2.4.4. Effect of Contact Time

An adsorption process is mainly affected by contact time. To elucidate the effect of time on the adsorption process, the contact time was varied in the range of 10–60 min at 100 rpm on an orbital shaker, while other parameters were kept constant. The results showed that a maximum adsorption of up to 96.52% was achieved in the early 20 min. Initially, the adsorption rate was very high, which might be due to the presence of enough active sites on the surface of the adsorbent, but as time increased, these sites were gradually filled, clogging the sorption sites, and hence, the adsorption rate slowly increased, and after forty minutes, it became constant. Therefore, the contact time was optimized as 120 min (Figure 5d) [51].

2.4.5. Adsorption Isotherms

Adsorption isotherms are important to the understanding of the mechanism of interaction. The adsorption mechanism of MB on the adsorbent was determined by using the Langmuir and Freundlich models, and their respective parameters are shown in Table 1. It was shown that the R^2 value was 0.97627 and 0.94287 for the Langmuir and Freundlich isotherms, respectively. This showed that adsorption isotherm properties are consistent with the Langmuir isotherm. The Langmuir isotherm is based on a monolayer adsorption mechanism on a homogenous surface. Figure 6a shows the isothermal plot for the Langmuir isotherm, while Figure 6b shows the isotherm plot for the Freundlich isotherm. These two Figures demonstrate that the adsorption mechanism was based on the Langmuir isotherm.

Table 1. Parameters of Langmuir and Freundlich isotherm.

Adsorption Model	Isotherm Parameters	R^2
Freundlich Model	Kf = 111.9026 mg/g 1/n = 3.34314	0.94287
Langmuir Model	Q_{max} = 427.35 mg/g K_L = 27721.8 m^{-1}·g^{-1} R_L = 0.003314	0.97627

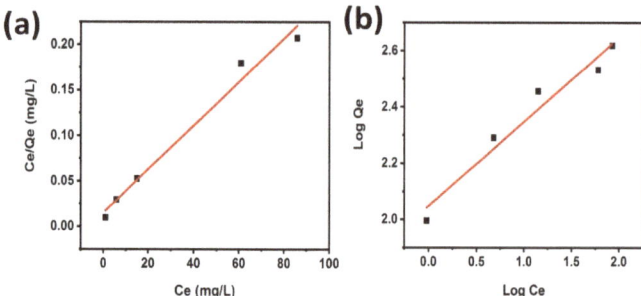

Figure 6. Langmuir graph (**a**) and Freundlich graph (**b**) MB adsorption on WHB/CA at pH = 5, 0.1 g biochar, T = 25 °C, and initial concentration of 300 mg/L.

2.4.6. Adsorption Kinetics

Adsorption kinetics is very important for the investigation of adsorption efficiency and the adsorption mechanism. So, kinetic parameters were evaluated by using two models, i.e., pseudo-first-order kinetics and pseudo-second-order kinetics. Their parameters are shown in Table 2, and their plots are shown in Figure 7a,b. The correlation coefficients (R^2) for pseudo-second-order and pseudo-first-order were (R^2) 0.9999 and 0.83843, respectively, and the adsorption capacity from pseudo-second-order was consistent with the experimental value.

Table 2. Parameters of pseudo-first- and pseudo-second-order.

Pseudo-First-Order Model	Co (mg/L)	Qe (mg/g) (exp)	Qe (mg/g) (cal)	K_1 (min^{-1})	R^2
	300	299.495	2162.46	−0.0005745	0.8383
Pseudo-Second-Order Model	Co (mg/L)	Qe (mg/g) (exp)	Qe (mg/g) (cal)	K_2 (g.mg^{-1} min^{-1})	R^2
	200	299.495	300.300	321.54	0.9999

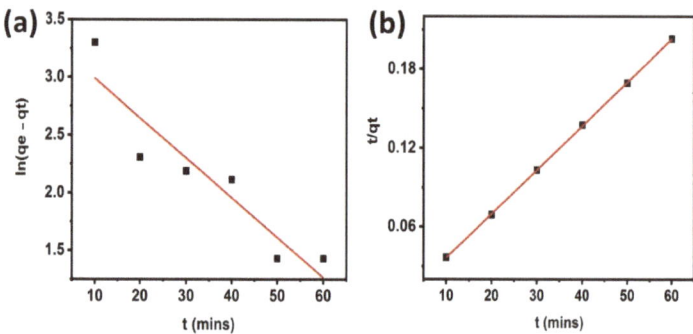

Figure 7. Plot of pseudo-first-order (**a**) and plot of pseudo-second-order (**b**).

2.5. Investigation of Adsorption Mechanism Based on DFT Analysis

To explore molecular communications among adapted biochar and MB, batch adsorption experiments were combined with density functional theory (DFT). Modified wheat husk biochar has a much-enhanced ability to adsorb MB and CIP. To investigate the adsorption mechanism and possible sites of adsorption, a DFT study was performed by using the 6–31G Hartree–Fock method. The structures of the adsorbates and adsorbents were optimized, and their molecular electrostatic potentials (MEP) were calculated. The MEP analysis showed the region of upper and lower potential, as designated by blue and red colors, respectively.

Figure 8a shows the MEP diagram of cellulose cinnamate, representing possible sites of attack. The hydrogen of the hydroxyl group is shown in blue color, having maximum potential. Hence, it was a possible site of attack by the nucleophile. While the red color indicates that the oxygen of the carbonyl group had the least potential, it was a possible site of attack for an electrophile. While the MEP of MB is shown in Figure 8b, the phenothiazine nitrogen atom was at a lower potential, represented by red, and could be a target for an electrophilic attack, whereas the atoms of dimethylamine benzene were at a higher potential, represented by blue, and could be a target for a nucleophilic assault. As a result of the MEP analysis, electrophilic interaction in the nitrogen of methylene blue and nucleophilic interaction in the dimethylamine group were discovered. The MEP diagram of CIP in neutral form is shown in Figure 8c. It shows that oxygen attached to the benzene ring and oxygen of the carbonyl group showed the least potential. Hence, these were the possible sites for the attack of the electrophile, while hydrogen of the carbonyl group showed maximum potential, so it was a possible site for the attack of the nucleophile. MEP of CIP in zwitterion form is shown in Figure 8d; here, the oxygen of the carbonyl group and oxygen-bearing negative charge showed the least potential, so these were the possible sites for the attack of the electrophile, while the nitrogen of the amino group showed maximum potential, and hence, it was a possible site for the attack of the nucleophile. In this way, MB and CIP were adsorbed on the surface of WHB/CA and effectively improved the quality of wastewater. [52].

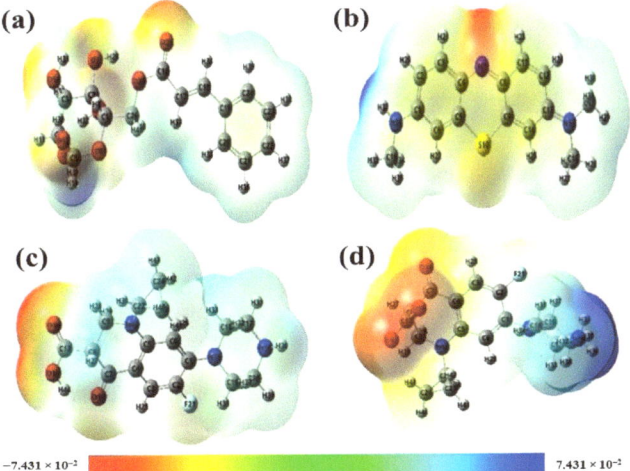

Figure 8. MEP of cellulose cinnamate (**a**), MEP of MB (**b**), MEP of CIP in neutral form (**c**), and MEP diagram of CIP in zwitterion form (**d**).

2.5.1. Possible Interactions of CIP with Modified Biochar

pH plays an important role during the adsorption of ciprofloxacin, as CIP exists in protonated, deprotonated, neutral, and zwitterion forms at different pH. Hence, the mechanism of adsorption of ciprofloxacin is highly dependent on the pH. So, pH is a crucial parameter for the adsorption of ciprofloxacin on modified biochar. CIP exists in protonated form at pH < 6.09 and deprotonated at pH > 8.64, while it exists in neutral and zwitterion form at 6.09 > pH < 8.64, but here, the zwitterion form is more dominated than the neutral form. At pH < 6.09, the amine group of ciprofloxacin gets protonated; hence, the protonated form exists at this pH, while at pH > 8.64, the amine group of CIP gets deprotonated, so at this pH, the deprotonated form exists. During the 6.09 < pH < 8.64, the amine group of CIP gets protonated, and the carboxyl group is deprotonated. So, at this pH, both positive and negative charges exist in CIP. Hence, at this pH, CIP exists in form zwitterion. All forms of CIP have a great attraction to modified biochar. This attraction is based on H-bonding, pi–pi interaction, pore diffusion, and EDA (electron donor–acceptor) interaction. CIP has electronegative and electropositive functional groups that form a hydrogen bond with modified biochar. In zwitterion form, CIP has positively charged NH^{2+} and negatively charged $C-O^-$; both of these make a hydrogen bond with modified biochar. Moreover, the presence of a highly electronegative fluorine atom also provides an extra site for electron acceptance from modified biochar. Hence, EDA interaction also plays a part during the adsorption of CIP on WHB/CA. Figure 9 shows various forms of CIP at different pH, while Figures 2 and 3 show the interaction of modified biochar with the neutral form of CIP and with the zwitterion form of CIP [53].

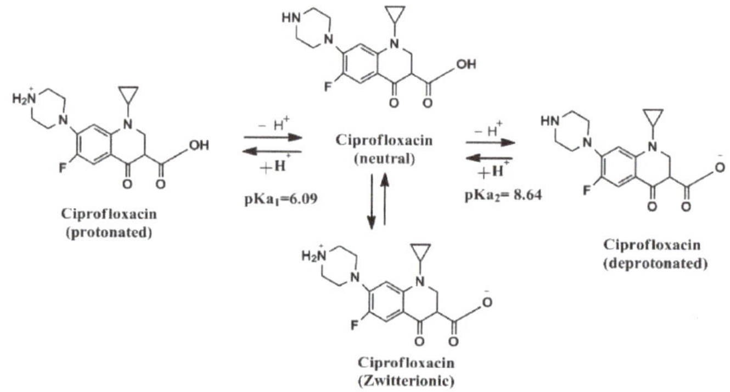

Figure 9. Forms of ciprofloxacin at different pH.

2.5.2. Interaction Mechanism Based upon HOMO and LUMO

The interaction mechanism was explored by calculating the HOMO and LUMO of ciprofloxacin in neutral form and zwitterion form (Figure 10). The calculated energy gap between HOMO and LUMO of CIP in neutral form was 0.12974, while for zwitterion form, it was 0.04452, which showed that more energy is required for the transition from LUMO to HOMO for CIP in neutral form, while less energy is required for the transition in the case of CIP in zwitterion form, suggesting that it is more reactive than neutral form. A possible reaction mechanism for both neutral and zwitterion forms is given in Figures 11 and 12, respectively, and their parameters are shown in Table 3 [54]. The WHB/CA for the adsorption of MB and CIP was compared with other adsorbents in terms of the maximum uptake of different adsorbents, as reported in Table 4.

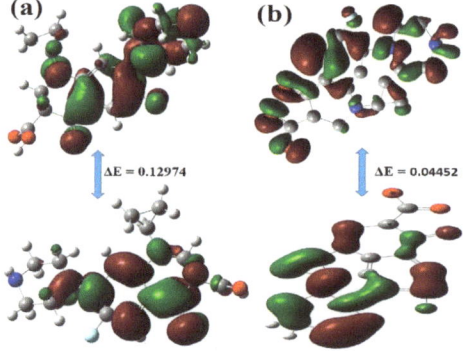

Figure 10. Interaction mechanism of CIP in neutral form (**a**) and zwitterion form (**b**) with modified biochar based on HOMO and LUMO.

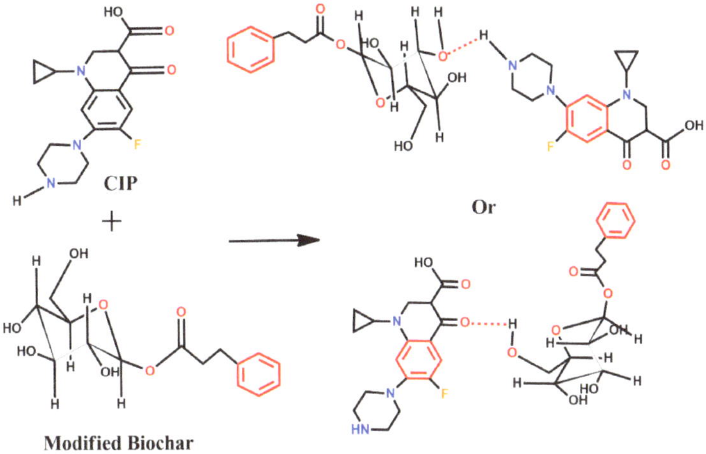

Figure 11. Interaction of CIP in neutral form with modified biochar.

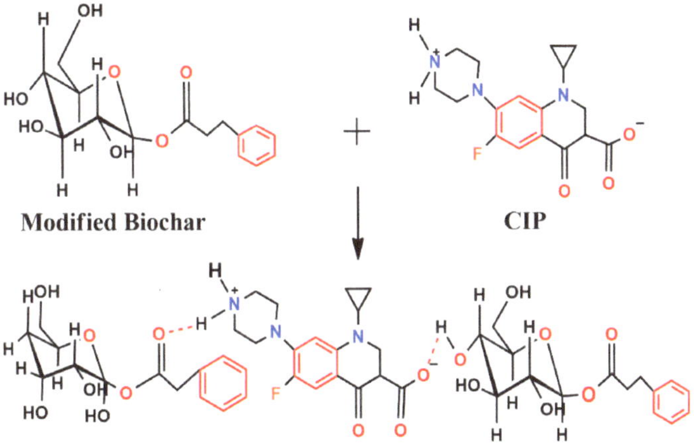

Figure 12. Interaction mechanism of CIP in zwitterion form with modified biochar.

Table 3. HOMO and LUMO energy values and other parameters.

Parameters (eV)	CIP (Neutral)	CIP (Zwitterion)
E_{LUMO}	−0.21903	−0.10424
E_{HOMO}	−0.08929	−0.05972
Energy gap (E_{HOMO}-E_{LUMO})	0.12974	0.04452
Ionization potential (I = −E_{HOMO})	0.08929	0.05972
Electron affinity (A = −E_{LUMO})	0.21903	0.10424
Chemical hardness (η = (I−A)/2)	−0.06487	−0.02226
Chemical softness (ζ = 1/2η)	−7.7077	−1.34141
Electronegativity (χ = (I + A)/2)	0.15416	0.08198
Chemical potential (μ = −(I + A)/2)	−0.15416	−0.08198
Electrophilicity index (ω = μ^2/2η)	−0.183135	−0.15094

Table 4. Comparison of maximum adsorption capacities of biochar prepared from different sources.

S. N.	Adsorbent	Adsorption Capacity (mg/g)	Reference
1	Wheat bran sawdust/Fe_3O_4 composite	51.28	[55]
2	Biochar from mixed municipal discarded material (MMDM)	7.2	[56]
3	Wodyetia bifurcate biochar	149.3	[57]
4	Mangolia Gradiflora Linn Leaf biochar (MGB)	101.27	[58]
5	Sour lemon saw dust	52.4	[59]
6	Date palm sawdust	54	[59]
7	Eucalyptus sawdust	53.5	[59]
8	Reed	53.23	[60]
9	Ficus Carica Bast Activated carbon	47.62	[61]
10	Biowaste sawdust	58.14	[62]
11	Peanut shell biochar	208	[49]
13	WHB/CA	427.35	This study

3. Materials and Methods

The wheat husk used in this work was collected from the Alnoor flour mill in Pasrur, Punjab, Pakistan. All reagents, including sodium hydroxide, nitric acid, cinnamic acid, hydrochloric acid, and methylene blue, were acquired from Sigma Aldrich and used as received. Throughout the experiment, distilled water was used. The stock solutions of methylene blue (500 mg/L) and ciprofloxacin (500 mg/L) were prepared in distilled water, while other required solutions were prepared by making dilutions from stock solutions. In this experiment, pH and temperature were measured using a digital meter. Concentrations of MB and ciprofloxacin were determined at 698 nm and 276 nm, respectively, via a UV-Visible spectrophotometer.

3.1. Preparation of Biochar

Firstly, the washing of wheat husk (WH) was performed with tap water and then with distilled water several times to remove surface impurities. The washed WH was dried for 24 h at 353 K and then chilled to room temperature before being crushed [63]. WH was transformed into biochar by slow pyrolysis in an electrical furnace at a heating rate of 10 °C/min and held for 2 h at 800 °C. Biochar was obtained under a limited supply of oxygen [64,65]. It was then crushed with a pestle and mortar to generate a fine powder, which was sieved to produce particles with a particle size of 40–60 mesh and stored in a glass bottle with the words "Wheat Husk Biochar" on it (WHB).

3.2. Modification of WHB

The modification of WHB was performed by the method reported by Gong et al., with slight modifications [66]. WHB was modified with cinnamic acid (C.A) at different impregnation ratios of 2.5%, 5%, 7.5%, and 10% (WHB/CA, W/W), followed by magnetic stirring for 2 h at room temperature. Esterification was achieved by adding 2–3 drops of nitric acid. It was then dried in a drying oven at 100 °C for 24 h, before being stored in a glass bottle. The optimum impregnation ratio was determined by dissolving 100 mg of adsorbent in 100 mL of dye solution at room temperature and shaking it for 160 min at 25 °C on an orbital shaker at 100 RPM. It was seen from the results that a modified sample of 7.5% WHB/CA (wheat husk biochar/cinnamic acid) showed maximum adsorption. So, the whole of the work was performed on this sample and named 7.5% WHB/CA.

3.3. Characterization

To investigate the surface topology and elemental content of the sample, scanning electron microscopy (SEM), equipped with energy dispersive spectroscopy (EDS) (FEI Nova 450 Nano SEM), was performed. Functional groups present in the samples were analyzed by using Fourier transform infrared spectroscopy (IRSpirit-T, Shimadzu). For this, FTIR spectra of all impregnation ratios, WHB, and samples after adsorbing MB (TWH) were recorded and analyzed. The concentration of pollutants was determined via a UV-Vis spectrophotometer (Cecil 7400S). The crystallinity and phase of the adsorbent were evaluated by X-ray diffraction (XRD) measurements in the range of 10–80° (2θ).

3.4. Batch Adsorption Study

The experiments were performed in batch mode in a 100 mL Erlenmeyer flask, shaken in an orbital shaker (orbital shaker incubator ES 20) at 120 rpm for 120 min. NaOH and HCl were used to adjust the initial pH. The effect of various parameters was investigated by performing adsorption of MB at various initial concentrations (100–500 mg/L), adsorbent dosages (0.5 g–0.2 g), times (10–60 min), and pH from 3–11. For reproducibility and accuracy, each experiment was repeated thrice. The concentrations of dye and ciprofloxacin were determined using a Cecil 7400S UV-Vis spectrophotometer at the wavelength of maximum absorbance, λ_{max} = 664 nm and 278 nm, respectively. Adsorption ability q_e (mg/g) and removal efficiency (%) were computed using Equations (1) and (2), respectively.

$$q_e = \frac{(C_0 - C_t)V}{W} \quad (1)$$

$$\% = \frac{C_0 - C_e}{C_0} \quad (2)$$

Here, C_e and C_0 represent the initial and equilibrium concentrations (mg/g) of the adsorbate, while V represents the volume of the solution (L), and W represents the weight of the adsorbent (g) used.

3.5. Adsorption Isotherm

Adsorption isotherms depict the relationship between adsorbents and adsorbates. They depict the relationship between the amount of adsorbate adsorbed on the adsorbent and amount present in the solution at equilibrium. The adsorption mechanism can also be inferred from these models. In this work, two adsorption isothermal models were employed to analyze adsorption equilibrium data, i.e., the Freundlich and Langmuir models. The Langmuir model is based on the monolayer adsorption mechanism on the homogenous surface of the adsorbent [52]. The Langmuir isotherm can be given by the following equation.

$$\frac{C_e}{Q_e} = \frac{1}{K_L q_m} + \frac{1}{q_m} C_e \quad (3)$$

Here, C_e represents the equilibrium concentration (mg/L), and Q_e represents the equilibrium adsorption capacity (mg/g), while q_m is the maximum adsorption capacity calculated by the Langmuir isotherm, and K_L is the Langmuir constant. Table 1 shows that the data obtained by the Langmuir isotherm, q_e, K_L, and q_m are computed by the slope and intercept of the fitted line of the Langmuir isotherm.

The Freundlich isotherm relies on the assumption that adsorption occurs on the heterogeneous surface through a multi-layer adsorption mechanism [49]. The Freundlich isotherm is given by the following equation.

$$\ln q_e = \frac{1}{n}\ln C_e + \ln K_F \quad (4)$$

K_F is the Freundlich constant (L/mg), n is a constant number that is related to adsorption strength, and 1/n shows adsorption favorability, which is related to the degree of heterogeneity of the surface [67].

3.6. Adsorption Kinetics

To explain adsorption characteristics and the adsorption mechanism, kinetic models were studied. For the adsorption process, two commonly used models were employed:

pseudo-first-order and pseudo-second-order. At various time intervals, the concentration of the solution was determined, and the adsorption capacity at that time is defined as q_t, calculated by the following equation.

Pseudo-first-order is given by Equation (5).

$$\ln\left(\frac{q_e}{q_e - q_t}\right) = \frac{K_1 t}{2.303} \tag{5}$$

K_1 (1/min) shows the adsorption rate constant for pseudo-first-order, while q_t and q_e show adsorption capacities at time t and equilibrium, respectively.

A pseudo-second-order kinetic model is given by Equation (6).

$$\frac{t}{q_t} = \frac{t}{q_e} + \frac{1}{K_2 q_e^2} \tag{6}$$

q_t (mg/g) and q_e (mg/g) are the adsorption capacities at time t and equilibrium, respectively, and K_2 is a pseudo-second-order constant.

4. Conclusions

The WH-based adsorbent was prepared through acid modification. Optimal conditions were pH 5, a dosage of 0.1 g, an initial dye concentration of 300 mg/L, and a contact time of 2 h. Results showed that the maximum adsorption capacity was 299.493 mg/g for MB and 94.03% for CIP. The best fit for the adsorption process was pseudo-second-order kinetics combined with the Langmuir isotherm. Hence, their values also proved that WHB/CA is the most acceptable adsorbent material for the removal of MB from wastewater. This study also showed that WHB/CA is an efficient, novel, and cost-effective adsorbent for wastewater treatment. Its synthesis also provides a new way for the utilization of wheat husk.

Author Contributions: U.H.: writing—original draft. S.M.: supervision, funding acquisition, and writing—review and editing. M.A.K.: writing—editing, conceptualization, software, and resources. M.S.R., K.S.M., and M.H.: editing and reviewing. All authors have read and agreed to the published version of the manuscript.

Funding: This research was funded by the Higher Education Commission (HEC) in Pakistan (NRPU-Project No. 16132). Taif University Researchers Supporting Project number (TURSP-2020/01), Taif University, Taif, Saudi Arabia.

Data Availability Statement: Not applicable.

Acknowledgments: The authors are thankful to IRCBM, Comsats University Islamabad Lahore campus, and the University of Mianwali for providing characterization facilities. Taif University Researchers Supporting Project number (TURSP-2020/01), Taif University, Taif, Saudi Arabia.

Conflicts of Interest: The authors declare no conflict of interest.

Sample Availability: Not available.

References

1. Hokkanen, S.; Bhatnagar, A.; Sillanpää, M. A review on modification methods to cellulose-based adsorbents to improve adsorption capacity. *Water Res.* **2016**, *91*, 156–173. [CrossRef]
2. Lu, F.; Astruc, D. Nanocatalysts and other nanomaterials for water remediation from organic pollutants. *Coord. Chem. Rev.* **2020**, *408*, 213180. [CrossRef]
3. Li, G.; Li, J.; Tan, W.; Yang, M.; Wang, H.; Wang, X. Effectiveness and mechanisms of the adsorption of carbendazim from wastewater onto commercial activated carbon. *Chemosphere* **2022**, *304*, 135231. [CrossRef]
4. Lan, T.; Cao, F.; Cao, L.; Wang, T.; Yu, C.; Wang, F. A comparative study on the adsorption behavior and mechanism of pesticides on agricultural film microplastics and straw degradation products. *Chemosphere* **2022**, *303*, 135058. [CrossRef] [PubMed]
5. Mohanraj, J.; Durgalakshmi, D.; Balakumar, S.; Aruna, P.; Ganesan, S.; Rajendran, S.; Naushad, M. Low cost and quick time absorption of organic dye pollutants under ambient condition using partially exfoliated graphite. *J. Water Process Eng.* **2020**, *34*, 101078. [CrossRef]
6. Sen, T.K.; Afroze, S.; Ang, H.M. Equilibrium, Kinetics and Mechanism of Removal of Methylene Blue from Aqueous Solution by Adsorption onto Pine Cone Biomass of *Pinus radiata*. *Water Air Soil Pollut.* **2011**, *218*, 499–515. [CrossRef]

7. Kafshgari, L.A.; Ghorbani, M.; Azizi, A. Fabrication and investigation of MnFe$_2$O$_4$/MWCNTs nanocomposite by hydrothermal technique and adsorption of cationic and anionic dyes. *Appl. Surf. Sci.* **2017**, *419*, 70–83. [CrossRef]
8. Ghosh, D.; Bhattacharyya, K.G. Adsorption of methylene blue on kaolinite. *Appl. Clay Sci.* **2002**, *20*, 295–300. [CrossRef]
9. Senthilkumaar, S.; Varadarajan, P.; Porkodi, K.; Subbhuraam, C. Adsorption of methylene blue onto jute fiber carbon: Kinetics and equilibrium studies. *J. Colloid Interface Sci.* **2005**, *284*, 78–82. [CrossRef]
10. Durrani, W.Z.; Nasrullah, A.; Khan, A.S.; Fagieh, T.M.; Bakhsh, E.M.; Akhtar, K.; Khan, S.B.; Din, I.U.; Khan, M.A.; Bokhari, A. Adsorption efficiency of date palm based activated carbon-alginate membrane for methylene blue. *Chemosphere* **2022**, *302*, 134793. [CrossRef]
11. Adefurin, A.; Sammons, H.; Jacqz-Aigrain, E.; Choonara, I. Ciprofloxacin safety in paediatrics: A systematic review. *Arch. Dis. Child.* **2011**, *96*, 874–880. [CrossRef]
12. Ashfaq, M.; Li, Y.; Rehman, M.S.U.; Zubair, M.; Mustafa, G.; Nazar, M.F.; Yu, C.-P.; Sun, Q. Occurrence, spatial variation and risk assessment of pharmaceuticals and personal care products in urban wastewater, canal surface water, and their sediments: A case study of Lahore, Pakistan. *Sci. Total Environ.* **2019**, *688*, 653–663. [CrossRef] [PubMed]
13. Arsand, J.B.; Hoff, R.B.; Jank, L.; Bussamara, R.; Dallegrave, A.; Bento, F.M.; Kmetzsch, L.; Falção, D.A.; Peralba, M.D.C.R.; Gomes, A.D.A.; et al. Presence of antibiotic resistance genes and its association with antibiotic occurrence in Dilúvio River in southern Brazil. *Sci. Total Environ.* **2020**, *738*, 139781. [CrossRef] [PubMed]
14. Falyouna, O.; Maamoun, I.; Bensaida, K.; Tahara, A.; Sugihara, Y.; Eljamal, O. Encapsulation of iron nanoparticles with magnesium hydroxide shell for remarkable removal of ciprofloxacin from contaminated water. *J. Colloid Interface Sci.* **2022**, *605*, 813–827. [CrossRef] [PubMed]
15. Jafari, K.; Heidari, M.; Rahmanian, O. Wastewater treatment for Amoxicillin removal using magnetic adsorbent synthesized by ultrasound process. *Ultrason. Sonochemistry* **2018**, *45*, 248–256. [CrossRef]
16. Arikan, O.A. Degradation and metabolization of chlortetracycline during the anaerobic digestion of manure from medicated calves. *J. Hazard. Mater.* **2008**, *158*, 485–490. [CrossRef]
17. Sturini, M.; Speltini, A.; Maraschi, F.; Rivagli, E.; Pretali, L.; Malavasi, L.; Profumo, A.; Fasani, E.; Albini, A. Sunlight photodegradation of marbofloxacin and enrofloxacin adsorbed on clay minerals. *J. Photochem. Photobiol. A Chem.* **2015**, *299*, 103–109. [CrossRef]
18. Gomes, J.; Costa, R.; Quinta-Ferreira, R.M.; Martins, R.C. Application of ozonation for pharmaceuticals and personal care products removal from water. *Sci. Total Environ.* **2017**, *586*, 265–283. [CrossRef]
19. Hirose, J.; Kondo, F.; Nakano, T.; Kobayashi, T.; Hiro, N.; Ando, Y.; Takenaka, H.; Sano, K. Inactivation of antineoplastics in clinical wastewater by electrolysis. *Chemosphere* **2005**, *60*, 1018–1024. [CrossRef]
20. Zhang, X.; Guo, W.; Ngo, H.H.; Wen, H.; Li, N.; Wu, W. Performance evaluation of powdered activated carbon for removing 28 types of antibiotics from water. *J. Environ. Manag.* **2016**, *172*, 193–200. [CrossRef]
21. Wang, J.; Wang, S. Removal of pharmaceuticals and personal care products (PPCPs) from wastewater: A review. *J. Environ. Manag.* **2016**, *182*, 620–640. [CrossRef] [PubMed]
22. Košutić, K.; Dolar, D.; Ašperger, D.; Kunst, B. Removal of antibiotics from a model wastewater by RO/NF membranes. *Sep. Purif. Technol.* **2007**, *53*, 244–249. [CrossRef]
23. Doğan, E.C. Investigation of ciprofloxacin removal from aqueous solution by nanofiltration process. *Glob. Nest J.* **2016**, *18*, 291–308.
24. Sun, S.P.; Hatton, T.A.; Chung, T.-S. Hyperbranched Polyethyleneimine Induced Cross-Linking of Polyamide–imide Nanofiltration Hollow Fiber Membranes for Effective Removal of Ciprofloxacin. *Environ. Sci. Technol.* **2011**, *45*, 4003–4009. [CrossRef]
25. Ahmadi, S.; Osagie, C.; Rahdar, S.; Khan, N.A.; Ahmed, S.; Hajini, H. Efficacy of persulfate-based advanced oxidation process (US/PS/Fe$_3$O$_4$) for ciprofloxacin removal from aqueous solutions. *Appl. Water Sci.* **2020**, *10*, 187. [CrossRef]
26. Mondal, S.K.; Saha, A.K.; Sinha, A. Removal of ciprofloxacin using modified advanced oxidation processes: Kinetics, pathways and process optimization. *J. Clean. Prod.* **2018**, *171*, 1203–1214. [CrossRef]
27. Barisci, S.; Turkay, O. Applications of response surface methodology (RSM) for the optimization of ciprofloxacin removal by electrocoagulation. In Proceedings of the IWA Balkan Young Water Professionals Conference, Thessaloniki, Greece, 10–12 May 2015; pp. 10–12.
28. Gupta, V.K.; Jain, R.; Varshney, S. Removal of Reactofix golden yellow 3 RFN from aqueous solution using wheat husk—An agricultural waste. *J. Hazard. Mater.* **2007**, *142*, 443–448. [CrossRef]
29. Ji, L.; Chen, W.; Duan, L.; Zhu, D. Mechanisms for strong adsorption of tetracycline to carbon nanotubes: A comparative study using activated carbon and graphite as adsorbents. *Environ. Sci. Technol.* **2009**, *43*, 2322–2327. [CrossRef]
30. Alnajrani, M.N.; Alsager, O.A. Removal of antibiotics from water by polymer of intrinsic microporosity: Isotherms, kinetics, thermodynamics, and adsorption mechanism. *Sci. Rep.* **2020**, *10*, 794. [CrossRef]
31. Aarab, N.; Hsini, A.; Essekri, A.; Laabd, M.; Lakhmiri, R.; Albourine, A. Removal of an emerging pharmaceutical pollutant (metronidazole) using PPY-PANi copolymer: Kinetics, equilibrium and DFT identification of adsorption mechanism. *Groundw. Sustain. Dev.* **2020**, *11*, 100416. [CrossRef]
32. Kong, Y.; Wang, L.; Ge, Y.; Su, H.; Li, Z. Lignin xanthate resin–bentonite clay composite as a highly effective and low-cost adsorbent for the removal of doxycycline hydrochloride antibiotic and mercury ions in water. *J. Hazard. Mater.* **2019**, *368*, 33–41. [CrossRef] [PubMed]

33. Kalhori, E.M.; Al-Musawi, T.; Ghahramani, E.; Kazemian, H.; Zarrabi, M. Enhancement of the adsorption capacity of the light-weight expanded clay aggregate surface for the metronidazole antibiotic by coating with MgO nanoparticles: Studies on the kinetic, isotherm, and effects of environmental parameters. *Chemosphere* **2017**, *175*, 8–20. [CrossRef]
34. Alshorifi, F.T.; Ali, S.L.; Salama, R.S. Promotional Synergistic Effect of Cs–Au NPs on the Performance of Cs–Au/MgFe$_2$O$_4$ Catalysts in Catalysis 3,4-Dihydropyrimidin-2(1H)-Ones and Degradation of RhB Dye. *J. Inorg. Organomet. Polym. Mater.* **2022**, 1–12. [CrossRef]
35. Bakry, A.M.; Alamier, W.M.; Salama, R.S.; El-Shall, M.S.; Awad, F.S. Remediation of water containing phosphate using ceria nanoparticles decorated partially reduced graphene oxide (CeO$_2$-PRGO) composite. *Surfaces Interfaces* **2022**, *31*, 102006. [CrossRef]
36. Ibrahim, A.A.; Salama, R.S.; El-Hakam, S.A.; Khder, A.S.; Ahmed, A.I. Synthesis of 12-tungestophosphoric acid supported on Zr/MCM-41 composite with excellent heterogeneous catalyst and promising adsorbent of methylene blue. *Colloids Surf. A Physicochem. Eng. Asp.* **2021**, *631*, 127753. [CrossRef]
37. Chen, W.-R.; Huang, C.-H. Adsorption and transformation of tetracycline antibiotics with aluminum oxide. *Chemosphere* **2010**, *79*, 779–785. [CrossRef]
38. Garg, V.K.; Amita, M.; Kumar, R.; Gupta, R.J.D. Basic dye (methylene blue) removal from simulated wastewater by adsorption using indian rosewood sawdust: A timber industry waste. *Dye. Pigment.* **2004**, *63*, 243–250. [CrossRef]
39. Ngah, W.W.; Hanafiah, M.M. Removal of heavy metal ions from wastewater by chemically modified plant wastes as adsorbents: A review. *Bioresour. Technol.* **2008**, *99*, 3935–3948. [CrossRef]
40. Khan, T.S.; Mubeen, U. Wheat straw: A pragmatic overview. *Curr. Res. J. Biol. Sci* **2012**, *4*, 673–675.
41. Li, G.; Zhu, W.; Zhang, C.; Zhang, S.; Liu, L.; Zhu, L.; Zhao, W. Effect of a magnetic field on the adsorptive removal of methylene blue onto wheat straw biochar. *Bioresour. Technol.* **2016**, *206*, 16–22. [CrossRef]
42. Oei, B.C.; Ibrahim, S.; Wang, S.; Ang, H.M.J.B.t. Surfactant modified barley straw for removal of acid and reactive dyes from aqueous solution. *Bioresour. Technol.* **2009**, *100*, 4292–4295. [CrossRef] [PubMed]
43. You, H.; Chen, J.; Yang, C.; Xu, L.J.C.; Physicochemical, S.A.; Aspects, E. Selective removal of cationic dye from aqueous solution by low-cost adsorbent using phytic acid modified wheat straw. *Colloids Surf. A Physicochem. Eng. Asp.* **2016**, *509*, 91–98. [CrossRef]
44. Kuang, Y.; Zhang, X.; Zhou, S. Adsorption of methylene blue in water onto activated carbon by surfactant modification. *Water* **2020**, *12*, 587. [CrossRef]
45. Irfan, M.; Asghar, U.; Nadeem, M.; Nelofer, R.; Syed, Q.; Shakir, H.A.; Qazi, J.I. Statistical optimization of saccharification of alkali pretreated wheat straw for bioethanol production. *Waste Bbiomass Valorization* **2016**, *7*, 1389–1396. [CrossRef]
46. Lin, Q.; Wang, K.; Gao, M.; Bai, Y.; Chen, L.; Ma, H. Effectively removal of cationic and anionic dyes by ph-sensitive amphoteric adsorbent derived from agricultural waste-wheat straw. *J. Taiwan Inst. Chem. Eng.* **2017**, *76*, 65–72. [CrossRef]
47. Avni, E.; Coughlin, R.W.; Solomon, P.R.; King, H.H. Mathematical modelling of lignin pyrolysis. *Fuel* **1985**, *64*, 1495–1501. [CrossRef]
48. Shah, J.; Jan, M.R.; Jamil, S.; Haq, A.U. Magnetic particles precipitated onto wheat husk for removal of methyl blue from aqueous solution. *Environ. Toxicol. Chem.* **2014**, *96*, 218–226. [CrossRef]
49. Han, X.; Chu, L.; Liu, S.; Chen, T.; Ding, C.; Yan, J.; Cui, L.; Quan, G. Removal of methylene blue from aqueous solution using porous biochar obtained by koh activation of peanut shell biochar. *BioResources* **2015**, *10*, 2836–2849. [CrossRef]
50. Zhu, Y.; Yi, B.; Yuan, Q.; Wu, Y.; Wang, M.; Yan, S. Removal of methylene blue from aqueous solution by cattle manure-derived low temperature biochar. *RSC advances* **2018**, *8*, 19917–19929. [CrossRef]
51. Liang, L.; Xi, F.; Tan, W.; Meng, X.; Hu, B.; Wang, X. Review of organic and inorganic pollutants removal by biochar and biochar-based composites. *Biochar* **2021**, *3*, 255–281. [CrossRef]
52. Guediri, A.; Bouguettoucha, A.; Chebli, D.; Chafai, N.; Amrane, A. Molecular dynamic simulation and DFT computational studies on the adsorption performances of methylene blue in aqueous solutions by orange peel-modified phosphoric acid. *J. Mol. Struct.* **2020**, *1202*, 127290. [CrossRef]
53. Patel, M.; Kumar, R.; Pittman, C.U., Jr.; Mohan, D. Ciprofloxacin and acetaminophen sorption onto banana peel biochars: Environmental and process parameter influences. *Environ. Res.* **2021**, *201*, 111218. [CrossRef] [PubMed]
54. Carabineiro, S.; Thavorn-Amornsri, T.; Pereira, M.; Serp, P.; Figueiredo, J. Comparison between activated carbon, carbon xerogel and carbon nanotubes for the adsorption of the antibiotic ciprofloxacin. *Catal. Today* **2012**, *186*, 29–34. [CrossRef]
55. Pooladi, H.; Foroutan, R.; Esmaeili, H. Synthesis of wheat bran sawdust/Fe$_3$O$_4$ composite for the removal of methylene blue and methyl violet. *Environ. Monit. Assess.* **2021**, *193*, 276. [CrossRef] [PubMed]
56. Hoslett, J.; Ghazal, H.; Mohamad, N.; Jouhara, H. Removal of methylene blue from aqueous solutions by biochar prepared from the pyrolysis of mixed municipal discarded material. *Sci. Total Environ.* **2020**, *714*, 136832. [CrossRef]
57. Dos Santos, K.J.L.; dos Santos, G.E.D.S.; de Sá, Í.M.G.L.; Ide, A.H.; da Silva Duarte, J.L.; de Carvalho, S.H.V.; Soletti, J.I.; Meili, L. *Wodyetia bifurcata* biochar for methylene blue removal from aqueous matrix. *Bioresour. Technol.* **2019**, *293*, 122093. [CrossRef]
58. Ji, B.; Wang, J.; Song, H.; Chen, W. Removal of methylene blue from aqueous solutions using biochar derived from a fallen leaf by slow pyrolysis: Behavior and mechanism. *J. Environ. Chem. Eng.* **2019**, *7*, 103036. [CrossRef]
59. Esmaeili, H.; Foroutan, R. Adsorptive Behavior of Methylene Blue onto Sawdust of Sour Lemon, Date Palm, and Eucalyptus as Agricultural Wastes. *J. Dispers. Sci. Technol.* **2018**, *40*, 990–999. [CrossRef]
60. Wang, Y.; Zhang, Y.; Li, S.; Zhong, W.; Wei, W. Enhanced methylene blue adsorption onto activated reed-derived biochar by tannic acid. *J. Mol. Liq.* **2018**, *268*, 658–666. [CrossRef]

61. Pathania, D.; Sharma, S.; Singh, P. Removal of methylene blue by adsorption onto activated carbon developed from *Ficus carica* bast. *Arab. J. Chem.* **2017**, *10*, S1445–S1451. [CrossRef]
62. Suganya, S.; Kumar, P.S.; Saravanan, A.; Rajan, P.S.; Ravikumar, C. Computation of adsorption parameters for the removal of dye from wastewater by microwave assisted sawdust: Theoretical and experimental analysis. *Environ. Toxicol. Pharmacol.* **2017**, *50*, 45–57.
63. Su, Y.; Zhao, B.; Xiao, W.; Han, R. Adsorption behavior of light green anionic dye using cationic surfactant-modified wheat straw in batch and column mode. *Environ. Sci. Pollut. Res.* **2013**, *20*, 5558–5568. [CrossRef] [PubMed]
64. Mohan, D.; Sarswat, A.; Ok, Y.S.; Pittman, C.U., Jr. Organic and inorganic contaminants removal from water with biochar, a renewable, low cost and sustainable adsorbent–A critical review. *Bioresour. Technol.* **2014**, *160*, 191–202. [CrossRef] [PubMed]
65. Yi, S.; Gao, B.; Sun, Y.; Wu, J.; Shi, X.; Wu, B.; Hu, X. Removal of levofloxacin from aqueous solution using rice-husk and wood-chip biochars. *Chemosphere* **2016**, *150*, 694–701. [CrossRef]
66. Gong, H.; Zhao, L.; Rui, X.; Hu, J.; Zhu, N. A review of pristine and modified biochar immobilizing typical heavy metals in soil: Applications and challenges. *J. Hazard. Mater.* **2022**, *432*, 128668. [CrossRef]
67. Zhen, M.; Tang, J.; Song, B.; Liu, X. Decontamination of Methylene Blue from Aqueous Solution by Rhamnolipid-modified Biochar. *BioResources* **2018**, *13*, 3061–3081. [CrossRef]

Article

CO Oxidation over Alumina-Supported Copper Catalysts

Guoyan Ma [1,2,3], Le Wang [1], Xiaorong Wang [4], Lu Li [2,3] and Hongfei Ma [5,*]

1. College of Chemistry and Chemical Engineering, Xi'an Shiyou University, Xi'an 710065, China
2. Key Laboratory of Auxiliary Chemistry and Technology for Chemical Industry, Ministry of Education, Shaanxi University of Science and Technology, Xi'an 710021, China
3. Shaanxi Collaborative Innovation Center of Industrial Auxiliary Chemistry and Technology, Shaanxi University of Science and Technology, Xi'an 710021, China
4. College of Chemistry and Chemical Engineering, Xianyang Normal University, Xianyang 712000, China
5. Department of Chemical Engineering, Norwegian University of Science and Technology, Sem Sælands vei 4, 7034 Trondheim, Norway
* Correspondence: hongfei.ma@ntnu.no

Citation: Ma, G.; Wang, L.; Wang, X.; Li, L.; Ma, H. CO Oxidation over Alumina-Supported Copper Catalysts. *Catalysts* **2022**, *12*, 1030. https://doi.org/10.3390/catal12091030

Academic Editors: Meng Li and Stuart H. Taylor

Received: 12 August 2022
Accepted: 6 September 2022
Published: 10 September 2022

Publisher's Note: MDPI stays neutral with regard to jurisdictional claims in published maps and institutional affiliations.

Copyright: © 2022 by the authors. Licensee MDPI, Basel, Switzerland. This article is an open access article distributed under the terms and conditions of the Creative Commons Attribution (CC BY) license (https://creativecommons.org/licenses/by/4.0/).

Abstract: CO oxidation, one of the most important chemical reactions, has been commonly studied in both academia and the industry. It is one good probe reaction in the fields of surface science and heterogeneous catalysis, by which we can gain a better understanding and knowledge of the reaction mechanism. Herein, we studied the oxidation state of the Cu species to seek insight into the role of the copper species in the reaction activity. The catalysts were characterized by XRD, N_2 adsorption-desorption, X-ray absorption spectroscopy, and temperature-programmed reduction. The obtained results suggested that adding of Fe into the Cu/Al_2O_3 catalyst can greatly shift the light-off curve of the CO conversion to a much lower temperature, which means the activity was significantly improved by the Fe promoter. From the transient and temperature-programmed reduction experiments, we conclude that oxygen vacancy plays an important role in influencing CO oxidation activity. Adding Fe into the Cu/Al_2O_3 catalyst can remove part of the oxygen from the Cu species and form more oxygen vacancy. These oxygen vacancy sites are the main active sites for CO oxidation reaction and follow a Mars-van Krevelen-type reaction mechanism.

Keywords: carbon monoxide; oxidation; copper; kinetic; reaction mechanism

1. Introduction

The catalytic oxidation of carbon monoxide (CO), a simple and typical heterogeneous catalytic reaction, is a key step in C1 chemistry and has been widely investigated for decades [1]. It has been considered the most studied probe reaction in heterogeneous catalysis, especially in the field of surface science, owing to its simple molecules [2–5]. With an in-depth understanding of this simple reaction, one can gain more and deeper fundamental new insights or knowledge of the reaction chemistry and mechanism [1,6]. Further, this knowledge is supposed to allow us to make progress in catalyst design and optimization. Therefore, CO oxidation, although seemingly a simple chemical reaction, is still widely under exploration [2,7]. It is not only useful as a model reaction system for fundamental studies for a better understanding of the reaction mechanism and the surface properties of the catalysts, but also imperative for some practical applications such as air cleaning, automotive emission control, and removal of CO impurities from H_2 for polymer electrolyte membrane fuel cells [8]. In addition, CO is a poisonous molecule for some catalytic chemical reactions [9–11], in which it can strongly be adsorbed on the surface of the working catalysts; it will block or inhibit the other reactants' adsorption and significantly hamper the catalytic performance. Those noble metals have long been used as the most efficient catalysts for CO oxidation with high activity and stability.

Many types of catalysts have been developed and proposed including noble metals, like Pt, Rh, Au, and Pd, either supported or non-supported catalysts [7,12]. However, owing to the high cost and limited availability of noble metals, the focus has been on transition metals and/or their oxides as the substitute for noble metal catalysts. Transition metals like Cu, Ru, and so on have been reported [13–20]. Among them, copper-based catalysts have been explored widely, and have been recognized as a possible substitute for noble metals for their high activity toward CO oxidation [13,21,22].

Much effort has been devoted to reaction mechanism studies in CO oxidation. The traditional Langmuir-Hinshelwood mechanism has been commonly reported [23–26], and the O_2 adsorption has been reported as the rate-determining step. Besides, Mars–van Krevelen reaction mechanisms were also reported [6,15,27,28], especially for the catalysts with O-vacancy, in which the catalyst participates in the reaction with the reactants. Although, CO oxidation has been studied for decades in both academia and industry and uses multiple techniques. It still has significant meaning to continuously place some more focus on this "simple and classical" chemical reaction.

In the present work, to gain better insights into the relationship between the oxidation state of Cu species and CO oxidation activity, we report the effect of adding Fe as the promoter to the Cu/γ-Al$_2$O$_3$ catalyst to boost the activity of CO oxidation. We found that the Cu/γ-Al$_2$O$_3$ catalyst shows a weak or poor CO oxidation activity in the tested temperature range. Meanwhile, the activity is greatly enhanced by adding Fe into the Cu/γ-Al$_2$O$_3$ catalyst. A full conversion can be obtained at 250 °C. The promoter of Fe can greatly increase the reduction of CuO. The discoveries in this work provide a systematic understanding of the redox dynamics of Cu species under reaction conditions, which has implications for a broad range of catalytic reactions beyond CO oxidation.

2. Results and Discussion

2.1. Catalyst Properties

The compositions of the prepared catalysts were analyzed by X-ray fluorescence spectroscopy (XRF). The final metal loadings can be obtained as nominal with preparation. The physical properties of the catalysts, such as specific surface area, pore volume, and pore size, are summarized in Table 1. From the table, we know that, when depositing the metals on the γ-Al$_2$O$_3$, the surface area, pore volume, and pore size decreased slightly compared with the support. The nitrogen adsorption/desorption isotherms of the fresh samples shown in Figure 1a can be categorized as type IV isotherms, the typical character of mesoporous materials [29,30]. The pore size distribution shown in Figure 1b also demonstrates the mesoporous properties of the prepared catalysts.

Figure 2 shows the XRD patterns of the fresh calcinated catalysts. The displayed diffraction peaks can be assigned to the phase of γ-Al$_2$O$_3$ [31,32], and no diffraction peaks of Fe and Cu species are observed. This indicates that both Fe and Cu are highly dispersed on the surface of γ-Al$_2$O$_3$. Besides, no mixed metal oxides can be detected on the XRD pattern. It was commonly reported that transition metal salts and oxides can be spontaneously highly dispersed on the surface of the oxide support (like γ-Al$_2$O$_3$ and TiO$_2$) and form a monolayer or even sub-monolayer [31–37]. This was proven to be a thermodynamic process during the impregnation process. This type of monolayer catalyst has been reported and applied in multiple catalytic systems, forming one imported supported catalyst. Owing to the high dispersion of the Cu species on the support, the influence of other dopants on the Cu species is supposed to be imperative in affecting either the chemical state of the Cu species or the catalytic performance of the total reaction.

Table 1. BET surface area, pore volume, and average pore size for the samples. Pore volume and diameter calculations were determined using the BJH model on desorption data.

Catalyst	Surface Area (m^2/g)	Pore Volume (cm^3/g)	Pore Size (nm)	Metal Loading (wt%)
Cu/γ-Al$_2$O$_3$	181	0.55	9.07	5
FeCu/γ-Al$_2$O$_3$	144	0.43	9.10	1/5 (Fe/Cu)

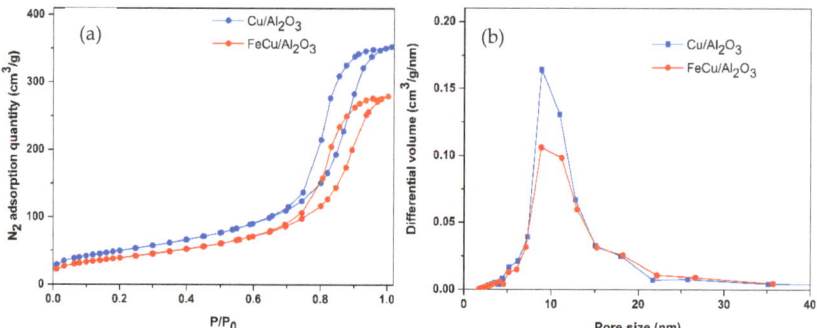

Figure 1. (a) N$_2$ adsorption/desorption isotherms and (b) pore size distribution of all of the catalysts (calculated by the BJH method).

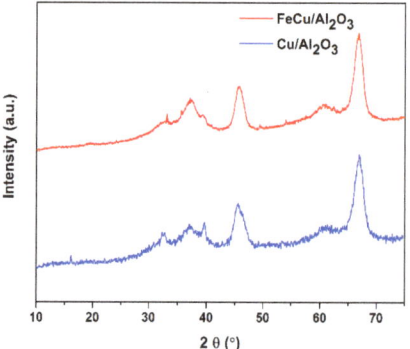

Figure 2. XRD patterns of the γ-Al$_2$O$_3$-supported catalysts.

2.2. Catalytic Evaluation of the CO Oxidation

In the following section, the CO oxidation is evaluated over the Cu- and FeCu-supported γ-Al$_2$O$_3$ catalysts. The light-off curve is the conversion–temperature plot of a catalytic reaction; it can directly be used for the criteria for the catalyst comparison and catalyst development. The light-off curve of the FeCu/Al$_2$O$_3$ and Cu/Al$_2$O$_3$ catalysts is shown in Figure 3. The conversion of CO oxidation over the two catalysts increases when the temperature increases. The activity of CO oxidation over the two catalysts is quite different. Cu/Al$_2$O$_3$ shows a rather low CO conversion, in the temperature range from 100 °C to 250 °C, and the conversion can be reached at about 60% at 250 °C. However, when the catalyst is promoted with Fe, the scenarios are different. The light-off curve is shifted to the lower temperature range. The activity of CO oxidation is significantly enhanced over the tested temperature range. A full conversion can be obtained at a temperature of 250 °C.

The stability of the FeCu/Al$_2$O$_3$ catalyst is also evaluated at a temperature of 250 °C, and the result is shown in Figure 4. The initial increasing phase was reported as an induction period of the catalyst. After the induction period, the conversion of CO is very stable over the 450 min test. No decreasing tendency is observed in the time-on-stream tests, and a longer reaction time can even be expected. This indicates that FeCu/Al$_2$O$_3$ is a good catalyst for CO oxidation.

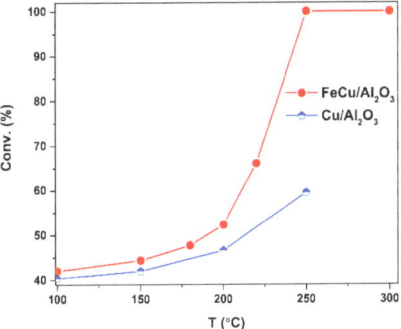

Figure 3. The temperature-dependent curve of the CO oxidation over the Al_2O_3-based catalysts. Reaction conditions: W_{cat} = 0.3 g, F_{tot} = 100 mL/min, P_{CO} = 0.02 bar, O_2/CO = 10/1, P_{tot} = 1 bar.

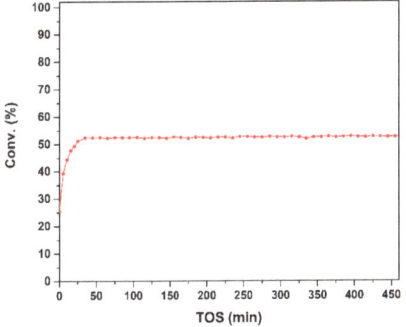

Figure 4. Stability test of the CO oxidation over the FeCu/Al_2O_3 catalyst. Reaction conditions: W_{cat} = 0.3 g, F_{tot} = 150 mL/min, P_{CO} = 0.02 bar, O_2/CO = 10/1, T = 250 °C, P_{tot} = 1 bar.

It has been commonly reported that CO oxidation follows the Mars-van Krevelen reaction mechanism over the Cu-based catalyst [13], in which Cu undergoes oxidation and reduction reactions via the oxygen vacancy. To gain a better understanding of the reaction, we also performed the transient experiment. The catalysts were first treated by the O_2 atmosphere so that the catalyst is in the highest oxidation state. Then, by introducing CO into the catalyst, the catalyst will be reduced with the production of CO_2, which can be traced by the online mass spectra. The normalized CO_2 formation signal over the catalysts is shown in Figure 5 for qualitative comparison. To make the comparison more reasonable and to eliminate the influence caused by the baseline of the mass spectra, the signal was normalized in the same time range. The relative difference between the two curves will be discussed. We can see that the production rate over the FeCu/Al_2O_3 catalyst is much faster than that over Cu/Al_2O_3, as the slope of the curve is much higher for the FeCu/Al_2O_3 catalyst, which is also confirmed by the activity test in Figure 3. The most likely reason is that adding Fe into the catalyst can enhance the reduction of CuO with CO to a lower Cu oxidation state and CO_2; therefore, the activity of CO oxidation is much higher on the FeCu/Al_2O_3 catalyst. Another parameter we should mention is the peak areas, which are related to the produced amount of CO_2 on the two pre-oxidized catalysts. We can know that more oxygen can be removed from the Cu/Al_2O_3 catalyst because more CO_2 is produced. As reported, the oxygen vacancy participates in the MvK type reaction cycle. Herein, we can know greater oxygen vacancy is produced on the FeCu/Al_2O_3 catalyst.

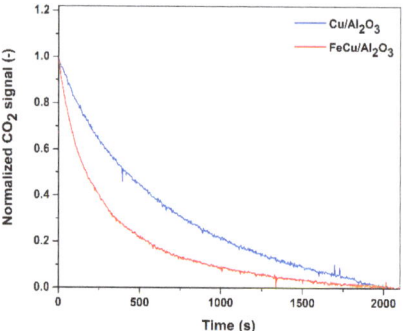

Figure 5. Transient CO reduction of the oxidized catalyst after treatment in O_2. Reaction conditions: W_{cat} = 0.3 g, F_{tot} = 100 mL/min, P_{CO} = 0.02 bar, T = 250 °C.

To further verify the discovery in the transient CO experiments, the CO temperature-programmed reduction (CO-TPR) was also performed to have an overview of the oxidation state and the reduction ability of the catalysts. The fresh catalyst was pre-treated in Ar at 100 °C to remove the adsorbed H_2O caused by the storage in the atmosphere. Then, the catalyst was reduced by introducing CO, and the temperature was increased to 350 °C and maintained for a certain period until a stable baseline was observed on the mass spectra, as shown in Figure 6. CO_2 is produced during the heating process; it indicates that oxygen is removed from the catalyst surface, and the catalyst is undergoing a reduction reaction with CO. This means the Cu or part of the Cu on the catalyst was in the oxidized state. While comparing the two catalysts, we can see that more CO_2 is produced on the Cu/Al_2O_3 catalyst than that on the $FeCu/Al_2O_3$ catalyst. This shows that more Cu is in the oxidized state on the Cu/Al_2O_3 catalyst, and more oxygen species are accessible. Adding Fe as the promoter to the Cu/Al_2O_3 catalyst can enhance the reduction of CuO during the synthesis process. Greater oxygen vacancy is formed on the $FeCu/Al_2O_3$ catalyst. Furthermore, this oxygen vacancy is supposed to influence the activity of CO oxidation. This can also be concluded from the X-ray adsorption near-edge spectroscopy (XANES), as shown in Figure 7. The white line intensity of the $FeCu/Al_2O_3$ catalyst is much lower than that of Cu/Al_2O_3, indicating that the oxidation state of Cu in Cu/Al_2O_3 is much higher than the $FeCu/Al_2O_3$ catalyst. This is consistent with the transient and CO-TPR experiments stating that more oxygen vacancy is formed on the $FeCu/Al_2O_3$ catalyst.

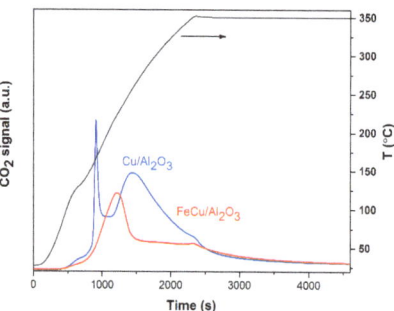

Figure 6. CO-TPR profile of the Cu/Al_2O_3 and $FeCu/Al_2O_3$ catalyst. Reaction conditions: W_{cat} = 0.3 g, $F_{CO/Ar}$ = 100 mL/min, ramping rate: 10 °C/min.

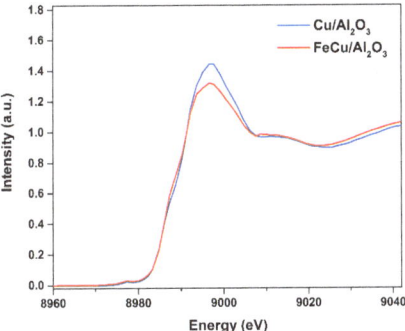

Figure 7. Normalized Cu K-edge XANES spectra of Cu/Al_2O_3 and $FeCu/Al_2O_3$ catalysts.

As mentioned above, CO oxidation follows a Mars van-Krevelen reaction mechanism. The catalyst undergoes reduction and oxidation via the oxygen vacancy, as summarized in the following equations.

$$CO + O_L \rightarrow CO_2 + \square_S \quad (1)$$

$$0.5O_2 + \square_S \rightarrow O_L \quad (2)$$

where O_L is surface lattice oxygen and $\square_S$ is the surface oxygen vacancy.

In this reaction mechanism, re-oxidation of the catalytic surface is usually faster than the step of withdrawal of oxygen from the copper oxide, so Equation 1 can be recognized as the rate-determining step [13,38]. From the CO-TPR, we know that, by adding Fe into the Cu/Al_2O_3 catalyst, the ability to withdraw oxygen from the catalyst surface becomes easier. It was reported that variations in copper valence during CO oxidation over CuO and Cu_2O cycled between 2 and 0 for CuO, but between 1 and 2 for Cu_2O [39]. The activity of CO oxidation over copper oxide species can be explained in terms of species transformation and changes in the amount of surface lattice oxygen. It seems the intermediate Cu oxidation state or non-stoichiometric metastable copper oxide species shows a good ability to transport surface lattice oxygen. Herein, it can be derived from the TPR that, in transient experiments, adding Fe into the Cu/Al_2O_3 catalyst increased the oxygen vacancy on the catalyst. Thus, the conversion of the $FeCu/Al_2O_3$ catalyst is much higher than that on Cu/Al_2O_3.

3. Materials and Methods

3.1. Catalyst Preparation

All of the catalysts were prepared by wetness impregnation methods. The precursor was $CuCl_2 \cdot 2H_2O$ (Sigma-Aldrich, St. Louis, MO, USA, ≥99%) and $Fe(NO_3)_3 \cdot 9H_2O$ (Sigma-Aldrich, St. Louis, MO, USA, ≥99%), which were impregnated on the γ-Al_2O_3 (Sasol Germany GmbH, Hamburg, Germany) and mixed well by stirring. The metal loadings for Cu and Fe are 5 wt% and 1 wt%, respectively. Then, the resultant mixture was placed into the oven and dried at 100 °C for 10 h, followed by calcination at a temperature of 500 °C for 2 h with a ramping rate of 5 °C/min. The obtained samples were represented as Cu/Al_2O_3 and $FeCu/Al_2O_3$ in the following context, and the fresh samples were directly used for characterizations and catalytic evaluation.

3.2. Catalyst Characterization

The specific surface areas of the two γ-Al_2O_3 were measured on a TriStar 3000 instrument at liquid nitrogen temperature using N_2 adsorption isotherms. BET and BJH analysis methods were used for the specific surface area and pore volume calculations. Samples were degassed under a vacuum condition at 200 °C overnight before measurements. XRD profiles were recorded with a Bruker D8 Davinci X-ray diffractometer (Bruker Nano GmbH, Berlin, Germany), using a Cu Kα1 (0.154 nm) wavelength.

The sample compositions were analyzed by X-ray fluorescence spectroscopy (XRF, Rigaku Supermini 200, Tokyo, Japan). The samples were dried and prepared in the form of pressed powder pellets.

3.3. X-Ray Absorption Spectroscopy Measurement

The X-ray absorption spectroscopy of the Cu K-edge was collected at BM 31 beamline station in ESRF (Swiss-Norwegian Beamline, European Synchrotron Radiation Facility, Grenoble, France) using the transmission mode with the use of a water-cooled flat Si [1 1 1] double-crystal monochromator. The Cu foil was used as the reference to conduct the energy calibrations. The spectra were normalized to the unity edge jump using Athena software (Demeter version 0.9.26, Bruce Ravel, BNL, NY, USA).

3.4. Catalytic Evaluation of CO Oxidation

The CO oxidation reaction was performed in a fixed bed reactor at 1 bar combined with an online mass spectrum (Omnistar GSD 3010, Asslar, Germany). The reactant gases were introduced into the reactor with specific mass flow controllers. Before the reaction, the catalysts were heated to the target temperature in Ar with a ramping rate of 5 °C/min. In one typical experiment, 0.3 g of catalyst was used at a total flow rate of 100 mL/min. The MS was used to perform the conversion calculation, in which Helium gas was used as the reference.

3.5. Carbon Monoxide Temperature-Programmed Reduction (CO-TPR)

CO-TPR was performed on the same setup with a fixed bed reactor, combined with an online MS recording the effluence gas. Before the TPR tests, the catalyst (0.3 g) was treated in Ar at 100 °C for 1 h to purge out the adsorbed water. Then, the samples were cooled down to room temperature in Ar. When a stable MS baseline was obtained, the samples were heated to 350 °C in 100 mL/min CO/Ar with a ramping rate of 10 °C/min. The final temperature was maintained until a stable MS baseline was obtained.

4. Conclusions

In summary, the γ-Al_2O_3 supported Cu with and without adding Fe as the promoter, and catalysts were prepared, characterized, and evaluated for CO oxidation reaction. Both Cu and Fe are highly dispersed on the surface of γ-Al_2O_3. The catalyst with Fe as the promoter shows much better activity than the neat Cu/Al_2O_3 catalyst over CO oxidation. Both the XAS results and transition experiments demonstrate that, by adding Fe inside the Cu/Al_2O_3 catalyst, the reduction of CuO_x is greatly enhanced, which benefits the CO oxidation reaction. Adding Fe into the base Cu/Al_2O_3 catalyst, part of the oxygen can be removed, leaving greater oxygen vacancy on the catalyst. We demonstrate the role of oxygen vacancy in influencing the activity of CO oxidation, which follows a Mars-van Krevelen-type reaction mechanism. From the transient experiment and temperature-programmed reduction, we know this oxygen vacancy will further contribute to the activity of CO oxidation, which makes the FeCu/Al_2O_3 catalyst highly active for CO oxidation. This work also demonstrates the relationship between the activity and the Cu oxidation state.

Author Contributions: Conceptualization, G.M. and H.M.; Methodology, G.M. and H.M.; Writing—Original Draft Preparation, G.M. and H.M.; Writing—Review and Editing, G.M., L.W., X.W., L.L., and H.M.; Funding Acquisition, G.M. All authors have read and agreed to the published version of the manuscript.

Funding: This research was funded by the Open Foundation of Key Laboratory of Auxiliary Chemistry and Technology for Chemical Industry, Ministry of Education, Shaanxi University of Science and Technology (No. KFKT2021-07); the Shaanxi Collaborative Innovation Center of Industrial Auxiliary Chemistry and Technology, Shaanxi University of Science and Technology (No. KFKT2021-07); and the Natural Science Basic Research Plan in Shaanxi Province of China (No. 2020JQ-765). And the APC was funded by the Norwegian University of Science and Technology.

Data Availability Statement: All relevant data are included in the paper.

Conflicts of Interest: The authors declare no conflict of interest.

References

1. Freund, H.-J.; Meijer, G.; Scheffler, M.; Schlögl, R.; Wolf, M. CO Oxidation as a Prototypical Reaction for Heterogeneous Processes. *Angew. Chem. Int. Ed.* **2011**, *50*, 10064–10094. [CrossRef] [PubMed]
2. Van Spronsen, M.A.; Frenken, J.W.M.; Groot, I.M.N. Surface science under reaction conditions: CO oxidation on Pt and Pd model catalysts. *Chem. Soc. Rev.* **2017**, *46*, 4347–4374. [CrossRef] [PubMed]
3. Stamenković, V.; Arenz, M.; Blizanac, B.; Mayrhofer, K.; Ross, P.; Marković, N. In situ CO oxidation on well characterized Pt$_3$Sn (hkl) surfaces: A selective review. *Surf. Sci.* **2005**, *576*, 145–157. [CrossRef]
4. Cui, H.; Liu, Z.; Jia, P. Pd-doped C3N monolayer: A promising low-temperature and high-activity single-atom catalyst for CO oxidation. *Appl. Surf. Sci.* **2020**, *537*, 147881. [CrossRef]
5. Meunier, F.C.; Cardenas, L.; Kaper, H.; Šmíd, B.; Vorokhta, M.; Grosjean, R.; Aubert, D.; Dembélé, K.; Lunkenbein, T. Synergy between metallic and oxidized Pt sites unravelled during room temperature CO oxidation on Pt/ceria. *Angew. Chem. Int. Ed.* **2021**, *60*, 3799–3805. [CrossRef]
6. Widmann, D.; Behm, R.J. Activation of Molecular Oxygen and the Nature of the Active Oxygen Species for CO Oxidation on Oxide Supported Au Catalysts. *Accounts Chem. Res.* **2014**, *47*, 740–749. [CrossRef]
7. Min, B.K.; Friend, C.M. Heterogeneous Gold-Based Catalysis for Green Chemistry: Low-Temperature CO Oxidation and Propene Oxidation. *Chem. Rev.* **2007**, *107*, 2709–2724. [CrossRef]
8. Lin, J.; Wang, X.; Zhang, T. Recent progress in CO oxidation over Pt-group-metal catalysts at low temperatures. *Chin. J. Catal.* **2016**, *37*, 1805–1813. [CrossRef]
9. Xu, S.; Chansai, S.; Xu, S.; Stere, C.E.; Jiao, Y.; Yang, S.; Hardacre, C.; Fan, X. CO poisoning of Ru catalysts in CO$_2$ hydrogenation under thermal and plasma conditions: A combined kinetic and diffuse reflectance infrared fourier transform spectroscopy–mass spectrometry study. *ACS Catal.* **2020**, *10*, 12828–12840. [CrossRef]
10. Knudsen, J.; Nilekar, A.U.; Vang, R.T.; Schnadt, J.; Kunkes, E.L.; Dumesic, J.A.; Mavrikakis, M.; Besenbacher, F. A Cu/Pt Near-Surface Alloy for Water-Gas Shift Catalysis. *J. Am. Chem. Soc.* **2007**, *129*, 6485–6490. [CrossRef]
11. Yang, X.; Wang, Y.; Wang, X.; Mei, B.; Luo, E.; Li, Y.; Meng, Q.; Jin, Z.; Jiang, Z.; Liu, C.; et al. CO-Tolerant PEMFC Anodes Enabled by Synergistic Catalysis between Iridium Single-Atom Sites and Nanoparticles. *Angew. Chem.* **2021**, *133*, 26381–26387. [CrossRef]
12. Mosrati, J.; Abdel-Mageed, A.M.; Vuong, T.H.; Grauke, R.; Bartling, S.; Rockstroh, N.; Atia, H.; Armbruster, U.; Wohlrab, S.; Rabeah, J.; et al. Tiny species with big impact: High activity of Cu single atoms on CeO$_2$–TiO$_2$ deciphered by operando spectroscopy. *ACS Catal.* **2021**, *11*, 10933–10949. [CrossRef]
13. Huang, T.-J.; Tsai, D.-H. CO oxidation behavior of copper and copper oxides. *Catal. Lett.* **2003**, *87*, 173–178. [CrossRef]
14. Falsig, H.; Hvolbæk, B.; Kristensen, I.S.; Jiang, T.; Bligaard, T.; Christensen, C.H.; Nørskov, J.K. Trends in the Catalytic CO Oxidation Activity of Nanoparticles. *Angew. Chem.* **2008**, *120*, 4913–4917. [CrossRef]
15. Jia, A.-P.; Hu, G.-S.; Meng, L.; Xie, Y.-L.; Lu, J.-Q.; Luo, M.-F. CO oxidation over CuO/Ce$_{1-x}$Cu$_x$O$_{2-\delta}$ and Ce$_{1-x}$Cu$_x$O$_{2-\delta}$ catalysts: Synergetic effects and kinetic study. *J. Catal.* **2012**, *289*, 199–209. [CrossRef]
16. Zhang, X.-m.; Tian, P.; Tu, W.; Zhang, Z.; Xu, J.; Han, Y.-F. Tuning the dynamic interfacial structure of copper–ceria catalysts by indium oxide during CO oxidation. *ACS Catal.* **2018**, *8*, 5261–5275. [CrossRef]
17. Liu, Z.-P.; Hu, P.; Alavi, A. Mechanism for the high reactivity of CO oxidation on a ruthenium–oxide. *J. Chem. Phys.* **2001**, *114*, 5956–5957. [CrossRef]
18. Huang, B.; Kobayashi, H.; Yamamoto, T.; Toriyama, T.; Matsumura, S.; Nishida, Y.; Sato, K.; Nagaoka, K.; Haneda, M.; Xie, W.; et al. A CO adsorption site change induced by copper substitution in a ruthenium catalyst for enhanced CO oxidation activity. *Angew. Chem. Int. Ed.* **2019**, *131*, 2252–2257. [CrossRef]
19. Kim, Y.; Over, H.; Krabbes, G.; Ertl, G. Identification of RuO$_2$ as the active phase in CO oxidation on oxygen-rich ruthenium surfaces. *Top. Catal.* **2000**, *14*, 95–100. [CrossRef]
20. Joo, S.H.; Park, J.Y.; Renzas, J.; Butcher, D.R.; Huang, W.; Somorjai, G.A. Size Effect of Ruthenium Nanoparticles in Catalytic Carbon Monoxide Oxidation. *Nano Lett.* **2010**, *10*, 2709–2713. [CrossRef]
21. Fedorov, A.; Saraev, A.; Kremneva, A.; Selivanova, A.; Vorokhta, M.; Šmíd, B.; Bulavchenko, O.; Yakovlev, V.; Kaichev, V. Kinetic and mechanistic study of CO oxidation over nanocomposite Cu-Fe-Al oxide catalysts. *ChemCatChem* **2020**, *12*, 4911–4921. [CrossRef]
22. Fedorov, A.V.; Tsapina, A.M.; Bulavchenko, O.A.; Saraev, A.A.; Odegova, G.V.; Ermakov, D.Y.; Zubavichus, Y.V.; Yakovlev, V.A.; Kaichev, V.V. Structure and chemistry of Cu–Fe–Al nanocomposite catalysts for CO oxidation. *Catal. Lett.* **2018**, *148*, 3715–3722. [CrossRef]
23. Engel, T.; Ertl, G. Surface residence times and reaction mechanism in the catalytic oxidation of CO on Pd(111). *Chem. Phys. Lett.* **1978**, *54*, 95–98. [CrossRef]
24. Cisternas, J.; Holmes, P.; Kevrekidis, I.G.; Li, X. CO oxidation on thin Pt crystals: Temperature slaving and the derivation of lumped models. *J. Chem. Phys.* **2003**, *118*, 3312–3328. [CrossRef]

25. Baxter, R.; Hu, P. Insight into why the Langmuir–Hinshelwood mechanism is generally preferred. *J. Chem. Phys.* **2002**, *116*, 4379–4381. [CrossRef]
26. Hopstaken, M.J.P.; Niemantsverdriet, J.W. Structure sensitivity in the CO oxidation on rhodium: Effect of adsorbate coverages on oxidation kinetics on Rh(100) and Rh(111). *J. Chem. Phys.* **2000**, *113*, 5457. [CrossRef]
27. Widmann, D.; Behm, R. Dynamic surface composition in a Mars-van Krevelen type reaction: CO oxidation on Au/TiO$_2$. *J. Catal.* **2018**, *357*, 263–273. [CrossRef]
28. Qi, L.; Yu, Q.; Dai, Y.; Tang, C.; Liu, L.; Zhang, H.; Gao, F.; Dong, L.; Chen, Y. Influence of cerium precursors on the structure and reducibility of mesoporous CuO-CeO$_2$ catalysts for CO oxidation. *Appl. Catal. B* **2012**, *119*, 308–320. [CrossRef]
29. Niu, J.; Liland, S.E.; Yang, J.; Rout, K.R.; Ran, J.; Chen, D. Effect of oxide additives on the hydrotalcite derived Ni catalysts for CO$_2$ reforming of methane. *Chem. Eng. J.* **2018**, *377*, 119763. [CrossRef]
30. Leofanti, G.; Padovan, M.; Tozzola, G.; Venturelli, B. Surface area and pore texture of catalysts. *Catal. Today* **1998**, *41*, 207–219. [CrossRef]
31. Ma, H.; Sollund, E.S.; Zhang, W.; Fenes, E.; Qi, Y.; Wang, Y.; Rout, K.R.; Fuglerud, T.; Piccinini, M.; Chen, D. Kinetic modeling of dynamic changing active sites in a Mars-van Krevelen type reaction: Ethylene oxychlorination on K-doped CuCl$_2$/Al$_2$O$_3$. *Chem. Eng. J.* **2021**, *407*, 128013. [CrossRef]
32. Ma, H.; Wang, Y.; Zhang, H.; Ma, G.; Zhang, W.; Qi, Y.; Fuglerud, T.; Jiang, Z.; Ding, W.; Chen, D. Facet-Induced Strong Metal Chloride-Support Interaction over CuCl$_2$/γ-Al$_2$O$_3$ Catalyst to Enhance Ethylene Oxychlorination Performance. *ACS Catal.* **2022**, *12*, 8027–8037. [CrossRef]
33. Ma, H.; Wang, Y.; Qi, Y.; Rout, K.R.; Chen, D. Critical Review of Catalysis for Ethylene Oxychlorination. *ACS Catal.* **2020**, *10*, 9299–9319. [CrossRef]
34. Xie, Y.-C.; Tang, Y.-Q. Spontaneous monolayer dispersion of oxides and salts onto surfaces of supports: Applications to heterogeneous catalysis. *Adv. Catal.* **1990**, *37*, 1–43. [CrossRef]
35. Ma, H.; Fenes, E.; Qi, Y.; Wang, Y.; Rout, K.R.; Fuglerud, T.; Chen, D. Understanding of K and Mg co-promoter effect in ethylene oxychlorination by operando UV–vis-NIR spectroscopy. *Catal. Today* **2020**, *369*, 227–234. [CrossRef]
36. Yang, G.; Haibo, Z.; Biying, Z. Monolayer dispersion of oxide additives on SnO$_2$ and their promoting effects on thermal stability of SnO$_2$ ultrafine particles. *J. Mater. Sci.* **2000**, *35*, 917–923. [CrossRef]
37. Yu, X.-F.; Wu, N.-Z.; Xie, Y.-C.; Tang, Y.-Q. A monolayer dispersion study of titania-supported copper oxide. *J. Mater. Chem.* **2000**, *10*, 1629–1634. [CrossRef]
38. Jernigan, G.G.; Somorjai, G.A. Carbon monoxide oxidation over three different oxidation states of copper: Metallic copper, copper (I) oxide, and copper (II) oxide-a surface science and kinetic study. *J. Catal.* **1994**, *147*, 567–577. [CrossRef]
39. Nagase, K.; Zheng, Y.; Kodama, Y.; Kakuta, J. Dynamic Study of the Oxidation State of Copper in the Course of Carbon Monoxide Oxidation over Powdered CuO and Cu$_2$O. *J. Catal.* **1999**, *187*, 123–130. [CrossRef]

Article

Sr₂TiO₄ Prepared Using Mechanochemical Activation: Influence of the Initial Compounds' Nature on Formation, Structural and Catalytic Properties in Oxidative Coupling of Methane

Svetlana Pavlova, Yulia Ivanova, Sergey Tsybulya, Yurii Chesalov, Anna Nartova, Evgenii Suprun and Lyubov Isupova *

Boreskov Institute of Catalysis, pr. Lavrentieva, 5, 630090 Novosibirsk, Russia
* Correspondence: isupova@catalysis.ru

Citation: Pavlova, S.; Ivanova, Y.; Tsybulya, S.; Chesalov, Y.; Nartova, A.; Suprun, E.; Isupova, L. Sr₂TiO₄ Prepared Using Mechanochemical Activation: Influence of the Initial Compounds' Nature on Formation, Structural and Catalytic Properties in Oxidative Coupling of Methane. *Catalysts* 2022, 12, 929. https://doi.org/10.3390/catal12090929

Academic Editors: Meng Li and Kevin J. Smith

Received: 28 June 2022
Accepted: 17 August 2022
Published: 23 August 2022

Publisher's Note: MDPI stays neutral with regard to jurisdictional claims in published maps and institutional affiliations.

Copyright: © 2022 by the authors. Licensee MDPI, Basel, Switzerland. This article is an open access article distributed under the terms and conditions of the Creative Commons Attribution (CC BY) license (https://creativecommons.org/licenses/by/4.0/).

Abstract: Methane oxidative coupling (OCM) is considered a potential direct route to produce C_2 hydrocarbons. Layered perovskite-like Sr_2TiO_4 is a promising OCM catalyst. Mechanochemical activation (MA) is known to be an environmentally friendly method for perovskite synthesis. Sr_2TiO_4 were synthesized using MA of the mixtures containing $SrCO_3$ or SrO and TiO_2 or $TiO(OH)_2$ and annealing at 900 and 1100 °C. XRD and FT-IRS showed that MA leads to the starting component disordering and formation of $SrTiO_3$ only for SrO being pronounced when using $TiO(OH)_2$. After annealing at 900 °C, Sr_2TiO_4 was mainly produced from the mixtures of $SrCO_3$ or SrO and $TiO(OH)_2$. The single-phase Sr_2TiO_4 was only obtained from MA products containing $SrCO_3$ after calcination at 1100 °C. The surface enrichment with Sr was observed by XPS for all samples annealed at 1100 °C depending on the MA product composition. The OCM activity of the samples correlated with the surface Sr concentration and the ratio of the surface oxygen amount in SrO and perovskite (O_o/O_p). The maximal CH_4 conversion and C_2 yield (25.6 and 15.5% at 900 °C, respectively), and the high long-term stability were observed for the sample obtained from ($SrCO_3$ + TiO_2), showing the specific surface morphology and optimal values of the surface Sr concentration and O_o/O_p ratio.

Keywords: layered strontium titanates; mechnochemical activation; methane oxidative coupling

1. Introduction

With large fossil resources of natural and shale gas, gas hydrates promote the development of processes for methane conversion into valuable chemicals. Among them, the direct catalytic conversion of methane to C_2 hydrocarbons by oxidative coupling (OCM) is considered a potential route for processing methane into useful products [1–3]. Researchers generally agree that a simplified heterogeneous–homogeneous mechanism of OCM occurs via the activation of methane to methyl radicals on the catalyst surface, the subsequent homogeneous coupling of two methyl radicals to ethane in the gas phase and oxidative dehydrogenation of ethane to ethylene. In parallel, nonselective (homogeneous and/or heterogeneous) oxidation can proceed to give COx. The reaction complexity hinders the development of efficient OCM catalysts [1–3]. On the whole, OCM activity of the catalysts is ensured by oxygen activation creating the active sites for generation of methyl radicals and high basicity for fast desorption of methyl radicals that could be tuned by catalyst composition [1–4]. Although many studies devoted to OCM have been performed and various catalytic materials have been investigated for this reaction, commercial implementation of OCM has not yet been reached, since the C_2 yield is relatively low [1,5–7]. Thus, design of the stable OCM catalysts with a high methane conversion rate and C_2 selectivity remains a relevant problem.

A number of previous studies revealed that some perovskite-type oxides containing alkaline, alkaline-earth and rare-earth elements exhibit certain activity and C_2 selectivity for the OCM reaction [1,8–19]. Perovskites have high thermal and chemical stability, while concentration of surface defects, active oxygen species and basic sites can be tailored by adjusting their chemical composition. Along with chemical composition, the surface and bulk properties could be affected by the perovskite crystalline structure. All these properties make perovskites promising catalysts for OCM. Among them, much attention has been paid to titanate and stannate perovskite catalysts based on regular ABO_3 [9–16] and layered Ruddlesden–Popper $A_{n+1}B_nO_{3n+1}$ [17–19] structures (A = Ca, Ba, Sr, B = Ti, Sn). The perovskite-like Ruddlesden–Popper-type oxides are comprised of alternating layers of ABO_3 perovskite and AO rock salt and exibit high oxygen mobility due to the peculiarities of its fine structure [10,20,21]. It was found in an earlier report by Yang et al. [17] that the layered Sr_2TiO_4 and Sr_2SnO_4 perovskites demonstrated better performance in OCM at 800 °C than the corresponding $SrTiO_3$ and $SnTiO_3$. Recently, it was shown that the fine crystal structure of different strontium stannates influences their OCM performance, which follows in the order of $Sr_2SnO_4 > Sr_3Sn_2O_7 > SrSnO_3$ [10]. The results of multiple methods reveal that the coordination environment of Sr cations is altered depending on perovskite type, making the formation of surface oxygen vacancies easier for Sr_2SnO_4 than for $Sr_3Sn_2O_7$ and $SrSnO_3$. This explains why Sr_2SnO_4 has the largest quantities of active surface oxygen and basic sites, as well as the best OCM performance.

The high OCM activity and selectivity of doped Sr_2TiO_4 (C_2 yield up to 25% and C_2 selectivity around 66% at 850–900 °C) were found in [18] and related to the segregation of SrO on the catalyst surface under the reaction conditions. SrO increases the basicity of the surface and lowers the amount of weakly adsorbed molecular oxygen species that can shift the reaction to the deep oxidation. The effect of Sr surface enrichment resulting from the surface reconstruction of the model $SrTiO_3$ on OCM activity was studied in [14]. CH_4 conversion, C_2 selectivity and the ratio of C_2H_4/C_2H_6 was found to increase at the Sr enrichment of the surface up to Sr/(Sr + Ti) of 0.66 and then levels off. Thus, the results evidence that the optimal surface concentration of Sr could promote the high OCM activity of the catalysts based on strontium titanates.

It is known that the presence of certain active oxygen species and basic sites determining the high catalytic performance of perovskite-type oxides in methane oxidative reactions are strongly influenced by preparation method and synthesis parameters, such as the nature of the raw chemicals, pH and temperature, etc. [1–3]. Along with the optimal surface properties, the effective OCM catalysts could have a low specific surface area (SSA) to prevent unselective homogeneous reactions [1–3]. Different methods can be used for synthesis of titanates, including: solid state reaction [11–13,15,17], spray and glycine-nitrate combustion [10,11], hydrothermal synthesis [15], coprecipitation [15,19], the polymer precursor method [16,22], the sol-gel method [19,23] and mechanochemical activation [18,19,24–26]. Fu et al. [11] investigated Ca, Sr, Ba titanates prepared by solid-state and spray combustion methods for the OCM reaction. They found that the samples prepared by the solid-state method, being of lower crystallinity, showed the better OCM performance due to the high surface concentration of alkaline earth metal cations, which is beneficial for C_2 formation. The study of $SrTiO_3$ samples synthesized using a solid-state reaction, molten salt and sol-precipitation hydrothermal treatment also shows that the surface atomic structure of perovskite determined by the morphology of crystalline nanoparticles depends on the synthesis method used [15]. The dependence of the texture and morphology of Sr_2TiO_4 prepared via the ultrasonic modified sol-gel method on the alkaline agent adjusting the pH was shown in [21]. It was shown in [19] that the phase composition of the OCM catalysts based on Sr_2TiO_4, their texture features (specific surface area, pore size and volume) and the distribution of active surface oxygen species depends on the synthesis method: co-precipitation, sol precipitation, citrate or mechanochemical activation (MA) methods with further calcinations at 1100 °C. The catalysts prepared by sol precipitation and MA were the most effective for OCM, giving a C_2 yield of about 12% at 800–900 °C. Doped Sr_2TiO_4 possessing high OCM activity and selectivity were also prepared using MA of the mixtures containing TiO_2, oxides and carbonates in [18].

Usually, the sintered complex oxides with a low SSA are prepared by a solid-state method which requires prolonged heating at high temperatures with the intermediate homogenization of the powder precursors yielding phases with nonuniform particle size distribution and variations in the stoichiometry [11–13,15,17]. So called "wet" methods, such as coprecipitation, the sol-gel method, spray combustion, etc., involve several steps which take many hours to achieve the formation of the target mixed oxide phase and require the use of various chemicals or special complex equipment. Mechanochemical activation (MA) of starting compounds through high energy milling is the alternative route to avoid these problems and it is an environmentally friendly method due to the absence of any wastes [25–30]. During the milling processes, the homogeneity of the mixture increases, the particle size decreases and the new contacts arise. A high-energy input into the reaction zone leads to localized heating and high pressure, resulting in disordering of the crystal structure and the generation of various types of defects in solids that increase their reactivity [29].

In the course of milling, the raw compounds can be transformed into precursors or directly into the target products, depending on their nature and synthesis conditions. The MA effect on the formation of strontium titanates was studied in the early report of Berbenni et al. [25]. The physical mixtures containing $SrCO_3$ and TiO_2 (rutile) (at a ratio of $SrCO_3:TiO_2$ = 1:1, 1:2, 3:2) were dry milled for 110–240 h, depending on their composition. $SrTiO_3$ and Sr_2TiO_4 were not formed during milling and were obtained only after the annealing of the activated mixtures at 800–850 °C for 12 h. MA was applied to obtain Ruddlesden–Popper titanates $Sr_2[Sr_{n-1}Ti_nO_{3n+1}]$ (n = 1–4) in [26]. Stoichiometric mixtures of SrO and TiO_2 (anatase) were activated in a planetary mill for 35–300 h. During MA, $SrTiO_3$ was formed after 35–150 h for all mixtures, except the two $SrO:TiO_2$ compositions. The kinetic study revealed that traces of Sr_2TiO_4 are observed at 70 h; increasing milling time up to 125 h results in the formation of the very low crystallinity Sr_2TiO_4 phase that is transformed into $SrTiO_3$ during further milling. After annealing of the corresponding MA products at 800–1200 °C, the single crystalline phases were only obtained for $SrTiO_3$, Sr_2TiO_4 and $Sr_3Ti_2O_7$. However, the study showed that MA of the initial mixtures leads to the substantial temperature decrease in the synthesis of all members of the layered $Sr_2[Sr_{n-1}Ti_nO_{3n+1}]$ series. It is noted in the paper that this is a result of the grains fracture and defects generated during grinding, which leads to a higher internal energy and reduces the thermal barrier for any subsequent reaction [26].

Thus, the analysis of the previous data shows that perovskite-type layered Sr_2TiO_4 are perspective catalysts for the OCM reaction due to peculiarities of its structure that provide for the large quantities of active surface oxygen and basic sites formed as a result of SrO segregation on the surface. Synthesis of Sr_2TiO_4 with mechnochemical activation is a prospective environmentally friendly method. However, systematic studies on the influence of the raw chemicals' nature on Sr_2TiO_4 synthesis using MA are absent in the literature. In this work, the interaction of raw compounds during MA in the high-energy planetary ball mill, the impact of MA product peculiarities on phase and surface composition, morphology and microstructure of Sr_2TiO_4 layered perovskite obtained after annealing and its catalytic activity in the OCM reaction are studied.

2. Results and Discussion

2.1. Study of MA Products by XRD and FT-IRS

The influence of the starting compounds nature on their interaction and phase composition of the samples after MA has been studied by XRD and FT-IR spectroscopy.

XRD data for the samples after mechanical activation are presented in Figure 1. The patterns of both MA-1 ($SrCO_3$ + TiO_2) and MA-2 ($SrCO_3$ + $TiO(OH)_2$) show the reflections of the starting $SrCO_3$ [PDF 05–418], in addition to the reflections of TiO_2 (rutile) [PDF 21-276] for MA-1. $TiO(OH)_2$ is an amorphous compound. The reflections of starting SrO [PDF 06-0520] and TiO_2 are presented in the pattern of MA-3 (SrO + TiO_2). For MA-4 (SrO + $TiO(OH)_2$), along with the reflections of SrO, a number of the wide peaks in the regions of

2θ~25–29° and 36–40° are observed that could evidence the presence of a minor admixture of strontium carbonate and hydroxide. In addition, the reflections of SrTiO$_3$ [PDF 35-0734] appear for MA-3 and MA-4. In the case of MA-4, their intensity is slightly higher, which could be due to some stronger interaction of SrO with TiO(OH)$_2$ compared to TiO$_2$.

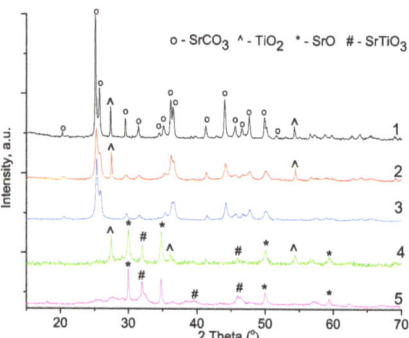

Figure 1. XRD patterns of the physical mixture (SrCO$_3$ + TiO$_2$) (1) and MA products. MA-1 (2), MA-2 (3), MA-3 (4), MA-4 (5).

The FT-IR spectroscopy data give additional information about the influence of the starting compounds' nature on their interaction during MA and the phase composition of MA products. The FT-IR spectra of the initial TiO$_2$, TiO(OH)$_2$, SrCO$_3$, SrO, and products of MA are presented in Figures 2–4. The wide absorption band (a.b.) at ~3470 cm^{-1} and a.b. at 1640 cm^{-1} in the spectrum of TiO$_2$ (Figure 2) could be accordingly assigned to the stretching and bending vibrations of physically absorbed H$_2$O or its hydroxyl groups [31,32]. The bands in the range of 1250–1050 cm^{-1} are related to the bending vibrations of Ti-OH, while the a.b. in the range of 950–350 cm^{-1} are the characteristic peaks of skeletal stretching and bending vibrations of Ti-O bonds in TiO$_2$ [32]. The spectrum of TiO(OH)$_2$ shows practically the same absorption bands as TiO$_2$ but the peaks related to H$_2$O and OH groups are noticeably higher in intensity [33,34]. Furthermore, two low-intensity a.b. at 1120 and 1050 cm^{-1} related to SO$_4^{2-}$ [35] are observed (Figure 5), which stems from the synthesis method of TiO(OH)$_2$ using TiOSO$_4$ hydrolysis.

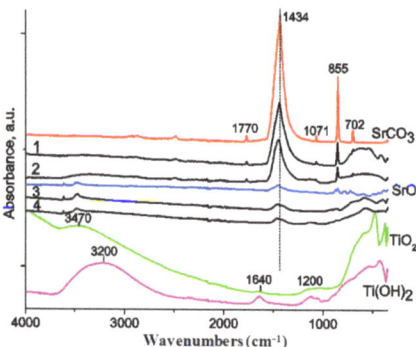

Figure 2. FT-IR spectra of the initial compounds and MA products after mechanical activation for 10 min. 1—MA-1, 2—MA-2, 3—MA-3, 4—MA-4.

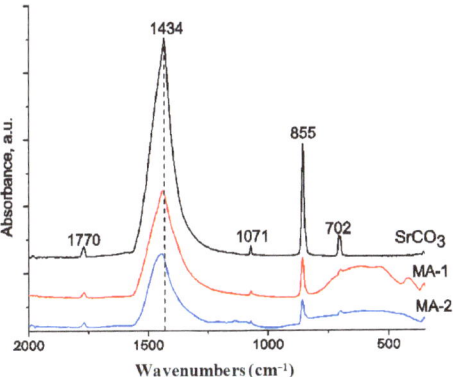

Figure 3. FT-IR spectra (fragment) of the initial SrCO₃, MA-1 and MA-2 after mechanical activation for 10 min.

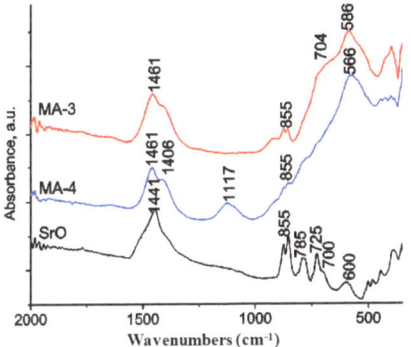

Figure 4. FT-IR spectra (fragment) of the initial SrO, MA-3 and MA-4 after mechanical activation for 10 min.

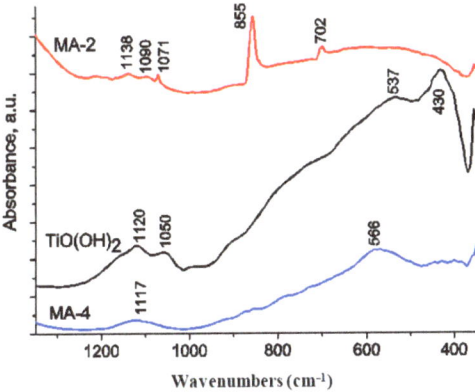

Figure 5. FT-IR spectra (fragment) of MA-2 and MA-4.

The spectrum of the initial SrCO₃ shows absorption peaks at 1770, 1434, 1071, 855, 704 and 702 cm^{-1} (Figures 2 and 3). The absorption band at 1770 cm^{-1} is assigned to the bond stretching vibration of C=O in CO_3^{2-} [36]. The intensive a.b. at 1434 cm^{-1} and the weak a.b. at 1071 cm^{-1} are related to asymmetric and symmetric stretching vibration of C-O band in carbonate anion, while the a.b. at 855, 704 and 699 cm^{-1} can be attributed to the bending out-of-plane and in-plane vibrations [36–38]. After activation in the spectra of the MA-1 (SrCO$_3$ + TiO$_2$) and MA-2 (SrCO$_3$+ TiO(OH)$_2$) samples (Figure 2), the bands of SrCO$_3$ and the wide a.b. at 850–350 cm^{-1} assigned to the characteristic vibrations of Ti-O are observed. For the MA-2 sample, the wide a.b. of the low intensity at 3400 cm^{-1} and 1200–1100 cm^{-1} attributed to vibrations of H$_2$O and OH groups in TiO(OH)$_2$ (Figure 2) are presented as well. Thus, in accordance with the XRD data, FT-IR spectra confirm the presence of the initial compounds in the MA products. However, the bands of SrCO$_3$, TiO$_2$ and TiO(OH) become wider and they are shifted compared with the starting compounds, which evidences the disordering of their crystal structure during MA. All alterations are more pronounced in the spectrum of MA-2, especially for the absorption bands corresponding to the vibrations of the H$_2$O and OH groups in TiO(OH)$_2$ at 3470 cm^{-1}, 1640 cm^{-1} and 1250–1050 cm^{-1} (Figure 2).

In the spectrum of the initial SrO a.b. at 3616 and 3472 cm^{-1} related to the stretching vibrations of OH groups, a.b. at 785 cm^{-1} and 725 cm^{-1} attributed to the bending vibrations of the Sr-OH bond and the bands at 1445 cm^{-1}, 850 cm^{-1} and 705 cm^{-1} assigned to SrCO$_3$ are observed (Figures 2 and 4). The high frequency of the bands corresponding to the OH group's stretching vibrations and their small width evidence the presence of strontium hydroxide [39,40]. Thus, the low intensity of the corresponding absorption bands implies that a small impurity of strontium carbonate and hydroxide presents in SrO, although the starting SrO was obtained shortly before the synthesis. This is explained by its high ability to hydration and carboxylation when exposed to atmospheric CO$_2$ and H$_2$O at room temperature and standard pressure [26]. The band at 600 cm^{-1} and the bands at 550–350 cm^{-1} correspond to the skeletal stretching and bending vibrations of Sr-O [39,40]. The spectra of the activated MA-3 (SrO + TiO$_2$) and MA-4 (SrO + TiO(OH)$_2$) samples are mainly similar (Figures 2 and 4). They show the bands at 3616 and 3472 cm^{-1} corresponding to the vibrations of OH groups, the bands at 1460 cm^{-1} and 900–705 cm^{-1} assigned to SrCO$_3$ and a pronounced a.b. at ~560–586 cm^{-1}, along with the bands in the range of 500–350 cm^{-1} that characterize the stretching and bending vibrations of Me (Sr, Ti)-O bonds. Furthermore, the bands shift and the substantial decrease in the intensity of a.b. characterizing the starting TiO$_2$ and TiO(OH)$_2$ (Figure 2) are observed, which could be a result of their disordering and interaction with SrO. Thus, the pronounced band at ~560–586 cm^{-1} assigned to the vibrations of TiO$_6$ octahedron [41,42] evidences the formation of SrTiO$_3$ in accordance with XRD data (Figure 1). The bands of SrCO$_3$ at 900–705 cm^{-1} are presented in the spectrum of MA-3, while they are absent in the case of MA-4 (Figure 4). This indicates a less effective interaction of SrO with TiO$_2$ compared to TiO(OH)$_2$ during MA, as the XRD data show (Figure 1).

Therefore, the XRD and FT-IRS data for MA products demonstrate that, during mechanical treatment, SrCO$_3$ possesses a lower reactivity compared with SrO, and only the disordering of the starting components is observed in the mixtures containing SrCO$_3$ (MA-1 and MA-2). Low SrCO$_3$ reactivity was also found in [25] when strontium titanates were not formed during prolonged milling of SrCO$_3$ + TiO$_2$ (rutile) mixture using a high-energy planetary mill at 400 rpm rotation. In the case of SrO-containing mixtures (MA-3 and MA-4), along with disordering of the starting components, formation of SrTiO$_3$ is observed in MA products. The formation of traces and the very low crystallinity Sr$_2$TiO$_4$ after milling of the 2SrO + TiO$_2$ mixture in the planetary mill at 200 rpm rotation speed for 70 and 125 h, correspondingly, were demonstrated by Hungrıa et al. [26]. In so doing, they did not observe the formation of intermediate SrTiO$_3$, in contrast with our results (Figures 1 and 4). Such a difference could be because a very-high power planetary ball mill at a 800 rpm rotation rate and acceleration of 40 g was used in our work. The high-energy input in the reaction zone results in the more effective dispersing and mixing of the components [43], as well as their disordering, which increases the number of contact sites and components reactivity, thus facilitating the appearance of SrTiO$_3$ in 10 min of milling.

XRD and FT-IRS data for MA products have also shown that the changes of the components structure and their interaction under milling are more pronounced in the mixtures containing TiO(OH)$_2$ (MA-2 and MA-4) compared to the ones with TiO$_2$. This difference could be attributed to such factors as the amorphous phase of TiO(OH)$_2$ and a large quantity of OH groups contained in it. It is well known that amorphization of the solid chemicals that can often occur in the course of mechanical treatment increases their reactivity [27–30]. In addition, the interaction during MA is more effective in the mixtures containing hydrated compounds due to the high reactivity of the OH group and the liberation of water increasing the efficiency of energy consumption in comparison with dry milling [29,44,45]. FT-IR spectra for MA-2 and MA-4 (Figure 2) showed a substantial decrease in the intensity of OH–Ti absorbance bands, which could be due to liberation of water from TiO(OH)$_2$ facilitating an acid–base reaction at the interface between acidic TiO(OH)$_2$ and basic SrCO$_3$ or, especially, SrO, which easily formed hydroxide.

2.2. Calcined Samples

2.2.1. Structural Properties

The data on qualitative and quantitative phase composition of MA products annealed at 900 and 1100 °C obtained by the analysis of XRD patterns are presented in Table 1 and Figures 6 and 7. The XRD patterns of MA-1 (SrCO$_3$ + TiO$_2$) and MA-2 (SrCO$_3$ + TiO(OH)$_2$) calcined at 900 °C (Figure 6) show that they comprise different strontium titanates and the initial SrCO$_3$, but the quantity of each phase in the samples varies significantly. Thus, MA-1 contains comparable amounts of Sr$_2$TiO$_4$ [39-1471], SrTiO$_3$ [35-0734] and SrCO$_3$, while for MA-2, the main phase is Sr$_2$TiO$_4$ (Table 1 and Figure 6). After annealing at 1100 °C, MA-1 and MA-2 are the single-phase Sr$_2$TiO$_4$ (Table 1 and Figure 7). There are no reflections of other phases (SrO or SrSO$_4$) in their XRD pattern but their presence as highly dispersed or surface compounds cannot be excluded. The phase composition of the MA-3 and MA-4 annealed at 900 °C differs considerably. The comparable quantity of SrTiO$_3$ and Sr$_2$TiO$_4$ is observed in the case of MA-3 (SrO + TiO$_2$), while MA-4 (SrO + TiO(OH)$_2$) comprises mainly Sr$_2$TiO$_4$. Their phase composition varies little after annealing at 1100 °C (Table 1 and Figure 7).

Table 1. Some characteristics of the samples after annealing of MA products at 900–1200 °C.

Sample/ T °C	Initial Compounds	Phase Composition/ Content,%	Lattice Parameters *, Å			Crystallite size, nm	Specific Surface Area, m^2/g
			a	b	c		
MA-1 900 1100	SrCO$_3$ + TiO$_2$	Sr$_2$TiO$_4$ (27.1) SrTiO$_3$ (26.4) Sr$_4$Ti$_3$O$_{10}$ (11) SrCO$_3$ (35.5) Sr$_2$TiO$_4$ (100)	3.8786	3.8786	12.593	120	3 1.2
MA-2 900 °C 1100 °C	SrCO$_3$ + TiO(OH)$_2$	Sr$_2$TiO$_4$ (92.5) Sr$_4$Ti$_3$O$_{10}$ SrTiO$_3$ SrCO$_3$ Sr$_2$TiO$_4$ (100)	3.8850 3.8756	3.8850 3.8756	12.580 12.561	120 120	1.7 1
MA-3 900 °C 1100 °C	SrO + TiO$_2$	Sr$_2$TiO$_4$ (40) SrTiO$_3$ (57) Sr$_4$Ti$_3$O$_{10}$ Sr$_2$TiO$_4$ (40) SrTiO$_3$ (56) Sr$_4$Ti$_3$O$_{10}$	3.8735 3.9049 3.8856	3.8735 3.8856	12.653 12.647	50	1.9 1.4
MA-4 900 °C 1100 °C	SrO + TiO(OH)$_2$	Sr$_2$TiO$_4$ (70) SrTiO$_3$ (14) Sr$_4$Ti$_3$O$_{10}$ (16) Sr$_2$TiO$_4$ (68) SrTiO$_3$ (14) Sr$_4$Ti$_3$O$_{10}$ (18)	3.8861 3.8862	3.8861 3.8862	12.549 12.564	50 120	1.6 1.3

* Sr$_2$TiO$_4$ [39-1471]: a = b = 3.8861, c = 12.5924.

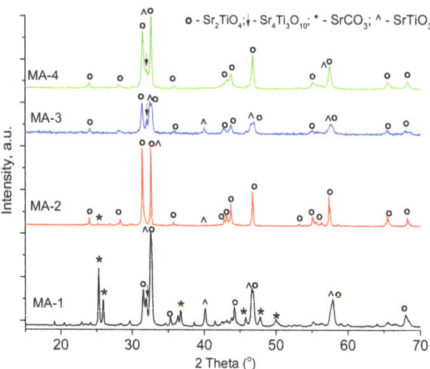

Figure 6. XRD patterns of the samples annealed at 900 °C.

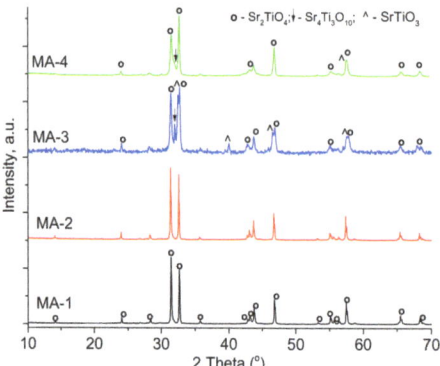

Figure 7. XRD patterns of the samples annealed at 1100 °C.

As an example, the results of Rietveld refinement for MA-2 (1100 °C) and MA-3 (1100 °C) are presented in Figures 7 and 8. The experimental pattern of MA-2 (1100 °C) and the theoretical one obtained using the known structural data for Sr_2TiO_4 [39-1471] (Figure 7) are qualitatively fit (the reliability factor Rwp = 20.12%), confirming the presence of layered Sr_2TiO_4 of tetragonal structure. The comparison of the experimental pattern for the multiphase MA-3 sample (1100 °C) and the theoretical pattern being superposition of the calculated curves for Sr_2TiO_4, $SrTiO_3$ and $Sr_4Ti_3O_{10}$ is illustrated in Figure 9.

The lattice parameters and the average size of Sr_2TiO_4 crystallites calculated for the samples contained mainly layered perovskite (Table 1). On the whole, the values of the Sr_2TiO_4 lattice parameters are similar and are closer to the ones known in the literature [21,25]. The narrow peaks in the XRD patterns of MA-1, MA-2 and MA-4 calcined at 1100 °C (Figure 7 and Table 1) evidence the presence of the large, well-crystallized particles with a mean size of about 120 nm. In the case of MA-3 (1100 °C), the broader diffraction peaks of perovskite stem from the smaller crystallites of a size of ~50 nm, which could be due to its multiphase composition.

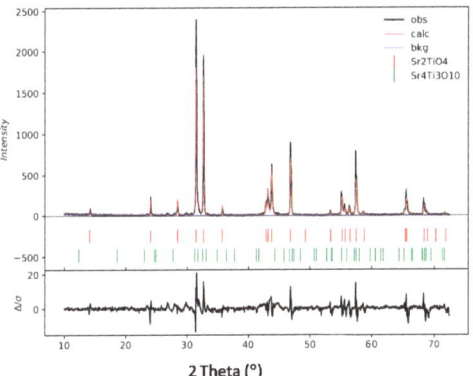

Figure 8. The experimental XRD pattern of MA-2 annealed at 1100 °C (black) and the theoretical one calculated using a model of crystal structure Sr_2TiO_4 [39-1471] (red). The bottom line is the difference plot between observed and calculated values.

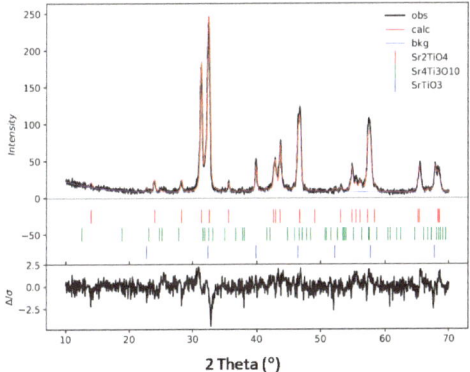

Figure 9. The experimental XRD pattern of MA-3 annealed at 1100 °C (black) and the theoretical one calculated using a model of crystal structure of strontium titanates (red).

Therefore, Sr_2TiO_4, along with other titanates, is formed in all samples after annealing at 900 °C because the interaction of the components in the activated mixtures mainly occurs at the temperatures up to 850–900 °C, was shown for mixtures containing SrO or $SrCO_3$ and TiO_2 [25,26]. However, the phase composition of the samples differs considerably in line with the results for mechanical activation of the corresponding mixtures. Thus, a more pronounced interaction of $SrCO_3$ or SrO with $TiO(OH)_2$ compared to TiO_2 during milling (MA-2, MA-4) leads to the formation mostly of Sr_2TiO_4 after annealing at 900 °C (Table 1 and Figure 6). The MA-1 sample obtained from the mixture of $SrCO_3$ and TiO_2 contains the initial $SrCO_3$, while only titanates are observed in the MA-3 (SrO + TiO_2) sample. Such a difference is due to the lower activity of $SrCO_3$ compared with SrO during activation. Indeed, $SrTiO_3$, being the intermediate in the formation of layered titanates [25,26], is already formed during activation of the mixtures with SrO (Figures 1 and 4).

2.2.2. Surface Composition

The surface composition of the samples calcined at 1100 °C was studied by XPS. The survey spectra of all samples and spectra of Ti $2p_{3/2}$, Sr $3d$, O $1s$ are presented in Figures 10 and 11. For all samples, a symmetric Ti $2p_{3/2}$ peak at 457.4–458.1 eV corresponding to the Ti^{4+} in perovskite is observed [46]. All spectra of Sr $3d$ show two doublets of peaks corresponding to Sr^{2+} $3d_{5/2}$ and $3d_{3/2}$ in oxide and perovskite (Figure 10). The peaks related to Sr^{2+} $3d_{5/2}$ are located at 132.2–132.5 eV and 132.7–133.0 eV for the Sr^{2+} in the oxide and perovskite phase, respectively [47–49]. The position of both peaks Sr^{2+} $3d_{5/2}$ (at 132.4 and 132.9 eV) is the same for the single-phase MA-1 and MA-2 annealed at 1100 °C. Some variation in the position of the peaks in the spectra of other samples could be due to their multiphase composition (Table 1), which determines the surface structure features and, as a result, a different charging effect in the position of XPS peaks. O $1s$ spectra show three peaks (Figure 10) corresponding to the oxygen species contained in perovskite (Op) at 529.1–529.4 eV, in oxide or carbonate (O_o) at 531.4–531.6 eV and in hydroxyl groups at 533.8–534 eV [46,47,50]. The low-intensity peak of S $2p$ is also observed for the MA-2 and MA-4 samples prepared using $TiO(OH)_2$ (Figures 10 and 11). The S $2p$ line is barely seen in the spectra of MA-2 and MA-4, which creates difficulties for quantitative analysis by XPS (Figure 10, S $2p$ line). The S/O atomic ratio is ~0.02, so oxygen from sulfate ions is about 8% of the total O $1s$ line intensity, which is close to the XPS error. The binding energy of oxygen from strontium sulfate is 531.8 eV [51], which overlaps with the 531.4 eV peak (Figure 11). Experimental spectra of O $1s$ regions do not allow the addition of extra components in deconvolution that are in agreement with sulfur quantity. At the same time, experimental O $1s$ spectra of MA-2 and MA-4 do not contradict the presence of strontium sulfate.

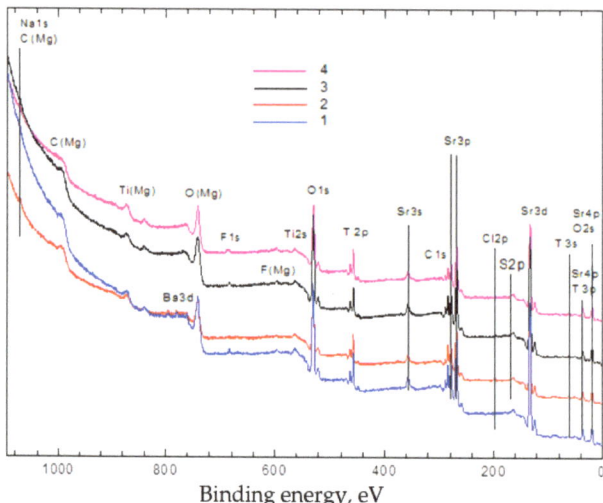

Figure 10. XPS survey spectra. 1—MA-1, 2—MA-2, 3—MA-3, 4—MA-4. F1s is from the set-up pumping system.

The surface composition and concentration of elements calculated using the XPS spectra of the samples are presented in Table 2. XPS analysis shows the presence of sulfur traces in MA-2 ($SrCO_3$ + $TiO(OH)_2$) and MA-4 (SrO + $TiO(OH)_2$) annealed at 1100 °C, which is due to its segregation from the initial $TiO(OH)_2$. The Sr/Ti ratio varies depending on the genesis of the samples. The enrichment of the surface layers with Sr compared with the bulk is observed for all samples, being more marked in the case of MA-1 ($SrCO_3$ + TiO_2) and MA-3 (SrO + TiO_2); the Sr/Ti ratio is equal to 2.5 and 2.9, correspondingly, at the stoichiometric value Sr/Ti = 2. The difference in the Sr/Ti value for MA-1 and MA-3 could result from the phase composition of the samples. Thus, MA-1 consists of the single-phase Sr_2TiO_4, while MA-3 is the multiphase system.

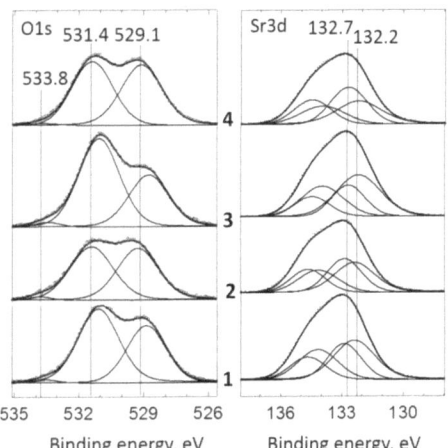

Figure 11. XPS spectra O *1s* and Sr *3d* of the samples calcined at 1100 °C: 1—MA-1, 2—MA-2, 3—MA-3 and 4—MA-4.

Table 2. Atomic ratio of elements in the surface layers and binding energy of oxygen in the different states for the samples calcined at 900–1100 °C obtained from XPS data.

Sample	Sr/Ti	O$_p$/Ti	O$_o$/Ti	O$_h$	O$_o$/O$_p$	E$_{Op}$	E$_{Oo}$	E$_{OOH}$
MA-1(1100)	2.5	3.2	4.6	0.086	1.4	529.8	531.1	533.5
MA-2 [1,2](1100)	2.1	3.1	3.3	0.1	1.1	529.2	531.4	533.8
MA-3(1100)	2.9	3.6	6.1	0.2	1.9	528.7	531.0	533.4
MA-4 [1,3](1100)	1.9	3.2	3.2	0.1	1.0	529.1	531.4	533.8

[1]—traces of sulfur; [2,3]—Sr/Ti~1.9 and 1.83 (Sr bulk content); O$_o$, O$_p$, O$_h$,—oxygen in Sr oxide, perovskite and hydroxyl groups, correspondingly.

(Table 1) that can lead to an easier enrichment of the surface with Sr. For MA-2 and MA-4, the smaller value of the Sr/Ti ratio (2.1 and 1.9, respectively) results from some lower Sr content in the samples, as shown by the chemical analysis (Table 2). The values of O$_p$/Ti equal to 3.1–3.6 show that all samples comprise SrTiO$_3$ as the layers in Sr$_2$TiO$_4$ or as the single phase. The high values of the O$_o$/O$_p$ ratio (1.4 and 1.9) for MA-1 and MA-3 correlate with the Sr enrichment of the surface, evidencing the segregation of SrO. In the case of MA-2 and MA-4, the strontium sulfate formation could be a reason for the smaller O$_o$/O$_p$ values of 1–1.1 (Table 2).

2.2.3. Textural and Morphological Properties

The specific surface area (SSA) of the calcined samples obtained using the BET method is in the range of 1–3 m^2/g (Table 1). The SSA of the samples annealed at 900 °C is higher for multiphase MA-1 and MA-3 samples, at which the highest SSA was shown by MA-1 (3 m^2/g), including a large amount of SrCO$_3$. By increasing the annealing temperature from 900 to 1100 °C, the value of SSA decreases for all samples. The most marked decline of SSA is observed for MA-1 and MA-2 prepared from the MA products containing SrCO$_3$, which could be a result of its decomposition.

The morphology and the elemental mapping of the samples calcined at 1100 °C were characterized by FE-SEM with EDX. The SEM images show that the form and the size of the particles in the samples depend on their genesis (Figure 12). The typical micrographs of the single-phase MA-1 sample synthesized from the SrCO$_3$ + TiO$_2$ mixture demonstrate the presence of large, well-crystallized particles with a size of ~500 nm and plates with a thickness of ~50 nm (Figure 12a,b). The micrographs of all other samples show well-crystallized, three-dimensional particles with a size in the range of 200–1000 nm

(Figure 12b–h). For the MA-1 sample, the ordered rows of the light spots of ~10 nm in size are clearly visible on the surface of the plates (Figure 12b), which could be related to the SrO nanospecies in accordance with previous HRTEM data for $Sr_2Ti_{0.9}Mg_{0.1}O_4$ [18]. In the case of MA-2 and MA-4 prepared from the mixtures containing $TiO(OH)_2$, slightly cambered, clear formations are observed. They could be attributed to strontium sulfate, which can be formed on the surface due to decomposition of the sulfate impurity from $TiO(OH)_2$, as XPS data show (Table 2). The elemental maps of Sr and Ti indicate their homogeneous distribution in MA-1 and MA-2 calcined at 1100 °C (Figure 13).

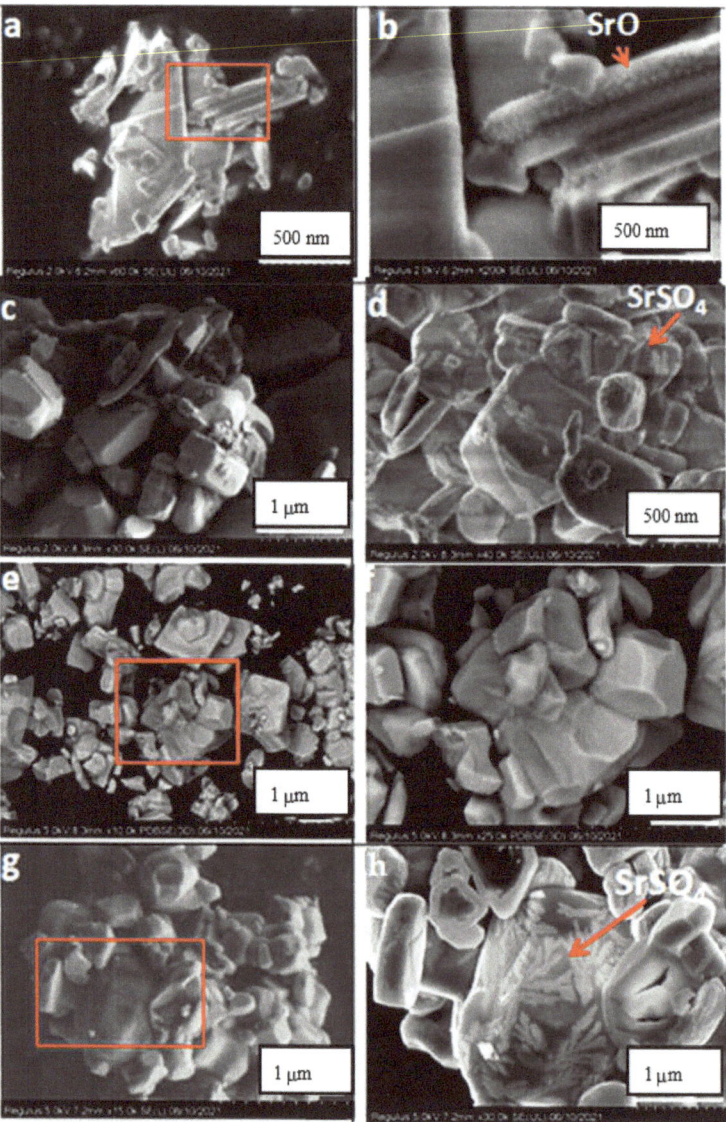

Figure 12. FE-SEM images at different magnifications for the samples calcined at 1100 °C: MA-1 (**a**,**b**), MA-2 (**c**,**d**), MA-3 (**e**,**f**), MA-4 (**g**,**h**).

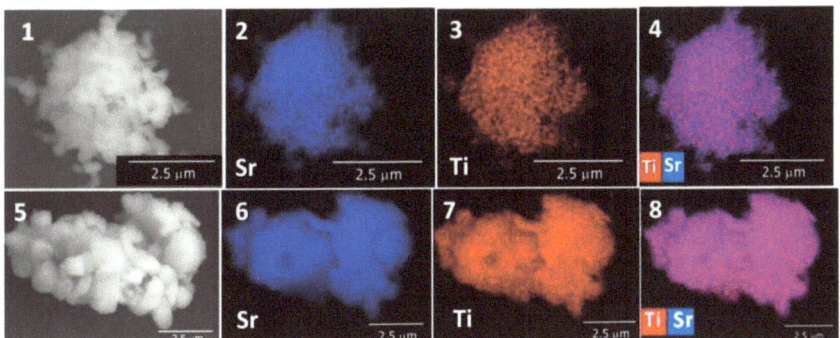

Figure 13. The elemental mapping images of Sr and Ti from the selected area of MA-1 (**1–4**) and MA-2 (**5-8**) annealed at 1100 °C.

2.2.4. Catalytic Activity

The activity characteristics of the samples annealed at 1100 °C in methane oxidative coupling at 800–900 °C are presented in Figure 14. For all samples, methane and oxygen conversion increases, with the temperature rise being maximal at 900 °C, but their values are significantly higher in the case of MA-1 and MA-3, reaching up to 24–25.6% and 82–92%, correspondingly. The activity of the MA-2 and MA-4 catalysts is considerably lower (CH_4 and O_2 conversion of 16–17% and 54–63% at 900 °C, respectively). For the yield of sum C_2 (ethane + ethylene) Y_{C2}, the same trend is observed. Thus, its value of ~15–15.5% is observed for MA-1 and MA-3, while Y_{C2} is only ~9% in the case of MA-2 and MA-4. For all the catalysts, excluding MA-2, the selectivity of C_2 hydrocarbons (S_{C2}) varies with the temperature within 50–63%, being maximal at 850 °C (Figure 14). In the case of MA-2, S_{C2} increases with the temperature rise, reaching ~55% at 900 °C. The selectivity of C_2H_4 grows with the temperature for all catalysts but it is highest In the case of MA-1 and MA-3, reaching up 40 and 37%, respectively. Compared to MA-1 and MA-3, the C_2H_4 selectivity for MA-2 and MA-4 is noticeably lower at 800–850 °C and is close at 900 °C. The ratio of CO/CO_2 selectivities for MA-1 and MA-3 changes with the temperature rise conversely to C_2H_4 selectivity, while it tends to the increase in line with S_{C2H4} for MA-2 and MA-4. Thus, on the whole, the MA-1 and MA-3 samples obtained from MA products of $SrCO_3$ + TiO_2 and SrO + TiO_2 mixtures, correspondingly, are substantially active compared with MA-2 and MA-4 prepared from the mixtures containing $TiO(OH)_2$. This difference could be attributed to the presence of sulfate traces in the surface layers of two latter catalysts (Table 2), blocking methane activation centers and thus reducing their activity. On the whole, the performance of the most active catalysts, MA-1 and MA-3, is comparable with the literature results (Table 3). Note that the analysis and comparison of the catalytic testing data is problematic due to different reaction conditions used. Thus, the most data were obtained at a GHSV of 10,000–18,000 h^{-1}, while our experiments were conducted at a higher GHSV of 75,000 h^{-1}.

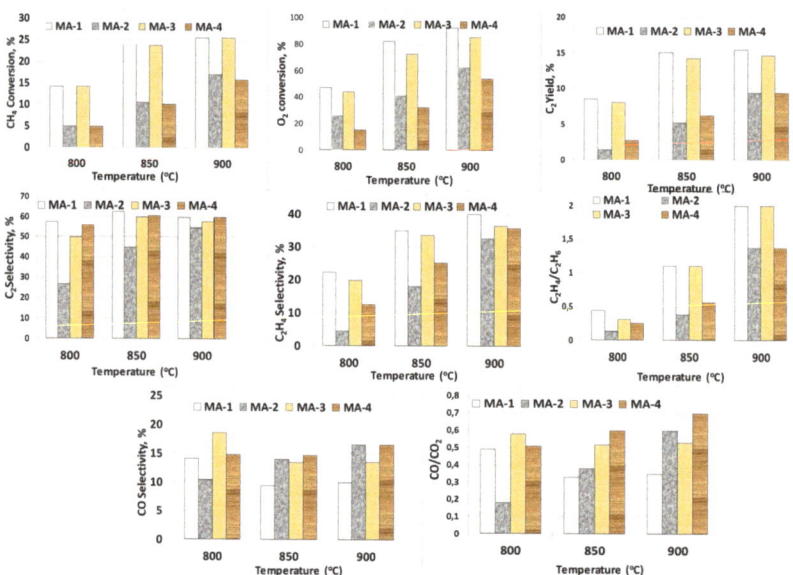

Figure 14. Methane and oxygen conversion, C2 and CO_x selectivity, C2 yield in OCM for the samples annealed at 1100 °C. The reaction mixture—$CH_4:O_2:N_2$ = 46:11.5:42.5% vol., $CH_4:O_2$ = 4, GHSV = 75,000 h^{-1}, 800–900 °C.

Table 3. Some recent studies on the OCM performance of different titanates with a perovskite structure.

Catalyst	Reaction Conditions	Methane Conversion, %	C2 selectivity(S)/ Yield(Y),%	Ref.
$CaTiO_3$ $SrTiO_3$ $BaTiO_3$	700 °C, CH4/O2 = 3, GHSV = 10,000 h^{-1}	13 24 19	2/ 12/ 36/	[16]
$SrTiO_3$ $SrTi_{0.8}Sn_{0.2}O_3$ $SrTi_{0.8}Nd_{0.2}O_3$	800 °C, CH4/O2 = 3, GHSV = 10,000 h^{-1}	32.5 30.8 30.5	48.9/ 52.1/ 54.4/	[52]
$SrZrO_3$	775 °C, CH4/O2 = 3, GHSV = 10,000 h^{-1}	30.7	45.4/	[8]
Sr_2TiO4	850 °C, CH4/O2 = 4, GHSV = 75,000 h^{-1}	19.8	59.6/11.8	[19]
$SrTiO_3$ Sr_2TiO_4 $Sr_2Ti_{0.9}Mg_{0.1}O4$	850 °C, CH4/O2 = 4, GHSV = 75,000 h^{-1}	-	65.9/12.8 68.5/17.3 71.2/18.2	[18]
Sr_2TiO4 (MA-1) Sr_2TiO4 (MA-3)	850 °C, CH4/O2 = 4, GHSV = 75,000 h^{-1}	24.1	62.7/15.1	This work

OCM is mainly considered to be a heterogeneous–homogeneous process including activation of methane to methyl radicals on the catalyst surface; their subsequent coupling to ethane in the gas phase is then converted into ethylene or CO_x, depending on the peculiarities of the catalyst [1–4]. It is generally implied that formation of methyl radicals by abstraction of hydrogen and oxidative dehydrogenation of ethane to ethylene occur over the active sites if the oxygen species is available on the oxide catalyst surface. Meanwhile, the reagent conversion and product selectivity are affected by the type and concentration of the oxygen species present on the surface. XPS results for all catalysts studied revealed two main types of surface oxygen, corresponding to oxygen in SrO and perovskite $SrTiO_3$

(Table 2). The MA-1 and MA-3 samples demonstrating the most effective performance in OCM show the high ratio of these oxygen forms (O_o/O_p = 1.44 and 1.87) and surface Sr enrichment (Sr/Ti = 2.47 and 2.86, correspondingly). In so doing, the MA-1 sample exhibits some higher oxygen conversion, C_2 selectivity and yield (Figure 14). The presence of Sr cations in the surface layers of the Sr-Ti perovskites are considered to enhance methane conversion and promote C_2 selectivity, especially C_2H_4 [11,14,18]. Bai et al. investigated the effect of the different surface compositions caused by the surface reconstruction of $SrTiO_3$ on its OCM activity [14]. They revealed that in CH_4 conversion, S_{C2} and S_{C2H4} depend on the Sr surface concentration, and they reach maximal values at a certain Sr concentration. It may be assumed that the higher activity of MA-1 compared with MA-3 could be due to an optimal surface concentration of SrO (Sr/Ti = 2.47) and an O_o/O_p ratio equal to 1.44. This difference can result from the different structural and morphological peculiarities of the samples. MA-1 is the single-phase catalyst, while MA-3 is the multiphase one comprising a comparable quantity of $SrTiO_3$ and Sr_2TiO_4. Furthermore, the presence of the plate-like particles with ordered SrO nanospecies on the surface (Figure 12a,b) could contribute to the optimal activity of the surface oxygen and thus to the effective performance of the MA-1 catalyst in OCM. The activity of the multiphase MA-3 catalyst can be attributed to both the surface oxygen of SrO and, possibly, to more weakly bound active oxygen species arising out of the interface boundaries. The latter could be involved in CH_4 partial oxidation that leads to some higher CO selectivity and lower C_2 selectivity for MA-3 compared with the MA-4 catalyst (Figure 14).

The low activity of MA-2 and MA-4 synthesized from mixtures containing $TiO(OH)_2$ is correlated with the low values of the surface Sr concentration (Sr/Ti ~ 2) and O_o/O_p ~ 1 ratio (Table 2). This could be caused by blocking the active oxygen sites with strontium sulfate, which is observed on the surface of the catalysts, as SEM and XPS data shows (Figure 12d,k and Table 2). The methane conversion and sum C_2 yield for both MA-2 and MA-4 are close, while oxygen conversion and C_2 selectivity clearly differs, especially at temperatures below 900 °C: the lower oxygen conversion and the higher C_2 selectivity are observed for MA-4 compared with MA-2 (Figure 14). Such a difference at the close O_o/O_p ratio could be due to the lower surface Sr concentration in the case of MA-4, along with blocking the oxygen active sites to some larger extent compared to MA-2. Thus, the lower CO/CO_2 ratio for MA-2 could suggest the presence of a larger amount of surface oxygen species being active in the deep oxidation of CH_4 to CO_2.

The most active catalysts (MA-1 and MA-3) were tested for 10 h in OCM at 850 °C. The time dependence of the methane conversion and C_2 yield (Figure 15) shows the high long-term stability of the catalyst performance under the highly concentrated reaction mixture and short contact time.

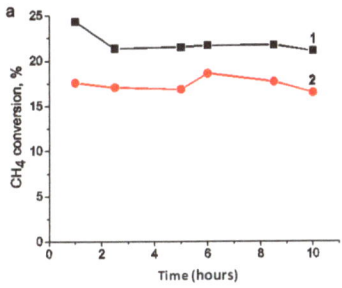 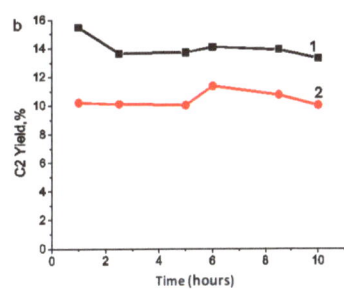

Figure 15. Time dependence of methane conversion (**a**) and C_2 yield (**b**) in OCM for the MA-1 (1) and MA-3 (2) annealed at 1100 °C. The reaction mixture—$CH_4:O_2:N_2$ = 46:11.5:42.5% vol., $CH_4:O_2$ = 4, GHSV = 75,000 h^{-1}, 800–900 °C.

3. Experimental

3.1. Sr_2TiO_4 Preparation

$SrCO_3$, SrO, TiO_2 (rutile) and $TiO(OH)_2$ prepared by hydrolysis of $TiOSO_4$ were used to prepare the Sr_2TiO_4 samples. SrO was obtained by calcination of $Sr(NO_3)_2$ at 900 °C just before the synthesis. To provide a target stoichiometry of samples, corresponding amounts of starting compounds were taken on the basis of their thermal analysis.

The stoichiometric mixtures of the starting chemicals were mixed and then activated in a high-power planetary APF-5 ball mill with two steel drums (25 cm^3 volume) at a 800 rpm rotation rate and an acceleration of 40 g. Mechanochemical activation (MA) of powders was conducted for 10 min under the following conditions: air atmosphere, zirconium balls of 5 mm diameter and a powder-to-ball mass ratio of 1:10. Before each synthesis, a preliminary treatment of drums and balls with the corresponding mixture was performed to cover the surface of the drums and balls by a layer of the initial mixture to minimize contamination of the samples with Fe and Zr due to their rubbing during MA.

Activated mixtures were pressed in tablets and annealed at 900 and 1100 °C.

The composition of the starting mixtures, some characteristics of the samples and their abbreviations are presented in Table 1.

3.2. Catalysts Characterization

Thermal analysis of the starting compounds was carried out with a Q-1500D thermo-analyzer. The phase composition of the samples after MA and annealing were examined by using powder X-ray diffraction (XRD) and IR spectroscopy. XRD patterns were recorded in an X'TRA (Thermo ARL) diffractometer with Cu Kα (λ = 1.5418 Å) radiation in the 2θ angle range 10−70° with a step of 0.05° and an exposure time of 5 s at each step, and graphically processed with the Fityk program. Qualitative phase analysis was carried out by using PDF-2–ICDD files and the ICSD/retrieve database. The quantitative phase analysis was performed through the full pattern simulation method using PCW 2.4 software (http://powdercell-forwindows.software.informer.com/2.4, access date 16 February 2022). The callculated XRD patterns were obtained using the crystal structure database ICSD. To identify the phase structure of the samples, the lattice parameters of the prepared catalysts were obtained by the XRD Rietveld refinement method. FT-IR spectra were recorded in the range of 200–4000 cm^{-1} using a BOMEM MB-102 FT-IR spectrometer.

The specific surface area (S_{BET}, m^2/g) was determined by a routine BET procedure using the Ar thermal desorption data. The morphology and elemental mapping of the samples annealed at 1100 °C were studied using a Field Emission Scanning Electron Microscope (FE-SEM) Hitachi Regulus SU8230 equipped with an X-ray microanalysis.

The XPS experiments were performed with a SPECS (Germany) spectrometer equipped with a hemispherical PHOIBOS-150-MCD-9 analyzer. Non-monochromatic MgKα radiation (hν = 1253.6 eV) at 200W was used as the primary excitation. The spectrometer was calibrated using the Au4f$_{7/2}$ (84.0 eV), Ag3d$_{5/2}$ (368.3 eV) and Cu2p$_{3/2}$ (932.7 eV) peaks from metallic gold and copper foils [20]. The binding energies of detected peaks were calibrated by the position of the C1s peak (BE = 284.5 eV). The binding energy values and the areas of XPS peaks were determined after Shirley background subtraction and analysis of the line shapes. The ratios of the surface atomic concentrations of elements were calculated from the integral photoelectron peak intensities corrected by the corresponding relative atomic sensitivity factors based on the Scofield's photo-ionization cross sections and transmission function of the analyzer.

3.3. Activity Tests

The catalytic activity in the OCM reaction was studied in a fixed-bed quartz tube reactor (5 mm inner diameter) at 750–900 °C and ambient pressure. The catalyst (0.2 mL) of 25–50 mesh was diluted with the quartz (1:3) to prevent a temperature gradient in the catalyst bed. The temperature of the catalyst was measured by a chromel–alumel thermocouple placed in the quartz well which was located at the middle of the catalyst bed.

Methane was mixed with air so that the reaction mixture was $CH_4:O_2:N_2 = 46:11.5:42.5\%$ vol., $CH_4:O_2 = 4$. The total feed was 15 L/h corresponding to a gas hourly space velocity (GHSV) of 75,000 h^{-1}. Reactant and product concentrations were analyzed by on-line gas chromatography with Porapack Q (i.d. = 3 mm, l = 3 m) and CaX (i.d. = 3 mm, l = 2 m) columns using a thermal conductivity detector. The reaction products were ethane, ethylene, water, CO, hydrogen and CO_2. Water was removed from the probe with an SiO_2 trap. A blank run with the inert SiO_2 particles (0.25–0.5 mesh) showed no conversion in the reaction conditions.

The methane conversion (X_{CH4}), C_2 selectivity (S_{C2}) and C_2 yield (Y_{C2}) in this study were calculated using the standard normalization method [18], defined as the following equations:

$$X_i = \left(1 - \frac{C_i}{C_i^0} \Delta V\right), i = CH_4, O_2,$$

$$S_j = \frac{2C_j}{2C_{C_2H_6} + 2C_{C_2H_4} + 2C_{CO} + 2C_{CO_2}}, j = C_2H_6, C_2H_4,$$

$$Y_{C_j} = X_{CH_4} \times S_{C_j} \times 100\%,$$

$$S_k = \frac{2C_k}{2C_{C_2H_6} + 2C_{C_2H_4} + 2C_{CO} + 2C_{CO_2}}, k = CO, CO_2,$$

$$C - balance = \frac{2C_{C_2H_6} + 2C_{C_2H_4} + 2C_{CO} + 2C_{CO_2}}{C_{CH_4}^0}, \Delta V = \frac{C_{N_2}^0}{C_{N_2}},$$

where ΔV is the molar flow rate change, $C_{N_2}^0$, C_{N_2} is the inlet and outlet concentration of N_2, C_i is a component mole fraction. The average carbon balance was not less than 98% in all the tests.

4. Conclusions

The samples of Sr_2TiO_4 were synthesized using mechanochemical activation (MA) of the four mixtures: ($SrCO_3 + TiO_2$), ($SrCO_3 + TiO(OH)_2$), ($SrO + TiO_2$) and ($SrO + TiO(OH)_2$) in a high-energy planetary ball mill, with subsequent annealing at 900 and 1100 °C. XRD and FT-IRS data show that MA for 10 min led to the disordering of the starting component structure in all mixtures and the formation of $SrTiO_3$ only in the case of SrO. These effects were more pronounced in the mixtures with amorphous $TiO(OH)_2$ containing a lot of OH groups.

The XRD Rietveld refinement method showed that, after annealing of the MA products at 900 °C, Sr_2TiO_4 was contained in all samples but its amount was considerably different in line with the peculiarities of MA products. Sr_2TiO_4 was mainly formed in the case of $SrCO_3$ or SrO and $TiO(OH)_2$, in contrast to the mixtures with TiO_2. The single-phase Sr_2TiO_4 is obtained after annealing at 1100 °C only from MA products containing $SrCO_3$.

FE-SEM data demonstrated that the single-phase sample synthesized from the $SrCO_3+TiO_2$ comprised of plates with a thickness of ~50 nm and large, well-crystallized particles with a size of ~500 nm, while all other samples contained only three-dimensional particles of 200–1000 nm. In addition, the presence of the SrO nanospecies of ~10 nm on the plate surface was assumed, while strontium sulfate was probably formed on the surface of the samples prepared using $TiO(OH)_2$ comprised of sulfate impurities.

The surface enrichment with Sr was observed by XPS data in all samples, being more marked in the case of the ones obtained from ($SrCO_3 + TiO_2$) and ($SrO + TiO_2$): the Sr/Ti ratio was equal to 2.47 and 2.86, correspondingly, at the stoichiometric value Sr/Ti = 2. The testing of the samples in OCM revealed that the CH_4 conversion and C_2 yield correlated with the surface Sr concentration and the ratio of oxygen amount in SrO and perovskite (O_o/O_p). The low activity of the catalysts obtained using $TiO(OH)_2$ could be related to the sulfate formation on the surface after annealing at 1100 °C due to blocking the active sites. The maximal CH_4 conversion and C_2 yield (25.6 and 15.5% at 900 °C, respectively) were observed for the sample obtained from ($SrCO_3 + TiO_2$), showing the specific surface morphology and optimal values of the surface Sr concentration (Sr/Ti = 2, 47) and a ratio of $O_o/O_p = 1.44$. The most active catalysts (MA-1 and MA-3) showed a high long-term stability performance under the highly concentrated reaction mixture and short contact time.

Author Contributions: S.P.—methodology, conceptualization; Y.I.—validation, investigation; S.T.—investigation; Y.C.—investigation, A.N.—investigation, E.S.—investigation; L.I.—supervision. All authors have read and agreed to the published version of the manuscript.

Funding: This work was supported by the budget project AAAA-A21-121011490008-3 from the Boreskov Institute of Catalysis.

Data Availability Statement: Informed consent was obtained from all subjects involved in the study.

Conflicts of Interest: The authors declare no conflict of interest.

References

1. Gambo, Y.; Jalil, A.A.; Triwahyono, S.; Abdulrasheed, A.A. Recent advances and future prospect in catalysts for oxidative coupling of methane to ethylene: A review. *J. Ind. Eng. Chem.* **2018**, *59*, 218–229. [CrossRef]
2. Galadima, A.; Muraza, O. Revisiting the oxidative coupling of methane to ethylene in the golden period of shale gas: A review. *J. Ind. Eng. Chem.* **2016**, *37*, 1–13. [CrossRef]
3. Schwach, P.; Pan, X.; Bao, X. Direct Conversion of Methane to Value-Added Chemicals over Heterogeneous Catalysts: Challenges and Prospects. *Chem. Rev.* **2017**, *117*, 8497–8520. [CrossRef] [PubMed]
4. Alexiadis, V.I.; Chaar, M.; van Veen, A.; Muhler, M.; Thybaut, J.W.; Marin, G.B. Quantitative screening of an extended oxidative coupling of methane catalyst library. *Appl. Catal. B Environ.* **2016**, *199*, 252–259. [CrossRef]
5. Kondratenko, E.V.; Schlüter, M.; Baerns, M.; Linke, D.; Holena, M. Developing catalytic materials for the oxidative coupling of methane through statistical analysis of literature data. *Catal. Sci. Technol.* **2015**, *5*, 1668–1677. [CrossRef]
6. Tang, L.; Yamaguchi, D.; Wong, L.; Burke, N.; Chiang, K. The promoting effect of ceria on Li/MgO catalysts for the oxidative coupling of methane. *Catal. Today* **2011**, *178*, 172–180. [CrossRef]
7. Thum, L.; Rudolph, M.; Schomäcker, R.; Wang, Y.; Tarasov, A.; Trunschke, A.; Schlögl, R. Activation in Oxidative Coupling of Methane on Calcium Oxide. *J. Phys. Chem.* **2018**, *123*, 8018–8026. [CrossRef]
8. Sim, Y.; Kwon, D.; An, S.; Ha, J.-M.; Oh, T.-S.; Jung, J.C. Catalytic behavior of ABO3 perovskites in the oxidative coupling of methane. *Mol. Catal.* **2020**, *489*, 110925. [CrossRef]
9. Xu, J.; Xi, R.; Zhang, Z.; Zhang, Y.; Xu, X.; Fang, X.; Wang, X. Promoting the surface active sites of defect BaSnO3 perovskite with BaBr2 for the oxidative coupling of methane. *Catal. Today* **2021**, *374*, 29–37. [CrossRef]
10. Xu, J.; Xi, R.; Xiao, Q.; Xu, X.; Liu, L.; Li, S.; Gong, Y.; Zhang, Z.; Fang, X.; Wang, X. Design of strontium stannate perovskites with different fine structures for the oxidative coupling of methane (OCM): Interpreting the functions of surface oxygen anions, basic sites and the structure–reactivity relationship. *J. Catal.* **2022**, *408*, 465–477. [CrossRef]
11. Ding, W.; Chen, Y.; Fu, X. Influence of surface composition of perovskite-type complex oxides on methane oxidative coupling. *Appl. Catal. A Gen.* **1993**, *104*, 61–75. [CrossRef]
12. Yu, C.Y.; Li, W.Z.; Martin, G.A.; Mirodatos, C. Studies of CaTiO3 based catalysts for the oxidative coupling of methane. *Appl. Catal. A.* **1997**, *158*, 201–214. [CrossRef]
13. Li, X.H.; Fujimoto, K. Low Temperature Oxidative Coupling of Methane by Perovskite Oxide. *Chem. Lett.* **1994**, *23*, 1581–1584. [CrossRef]
14. Bai, L.; Polo-Garzon, F.; Bao, Z.H.; Luo, S.; Moskowitz, B.M.; Tian, H.J.; Wu, Z.L. Impact of Surface Composition of SrTiO3 Catalysts for Oxidative Coupling of Methane. *ChemCatChem* **2019**, *11*, 2107–2117. [CrossRef]
15. Rabuffetti, F.A.; Stair, P.C.; Poeppelmeier, K.R. Synthesis-Dependent Surface Acidity and Structure of SrTiO3 nanoparticles solid-state reaction, molten salt, and sol-precipitation–hydrothermal treatment. *J. Phys. Chem.* **2010**, *114*, 11056–11067. [CrossRef]
16. Lim, S.; Choi, J.-W.; Jin Suh, D.; Lee, U.; Song, K.H.; Ha, J.-M. Low-temperature oxidative coupling of methane using alkaline earth metal oxide-supported perovskites. *Catal. Today* **2020**, *352*, 127–133. [CrossRef]
17. Yang, W.M.; Yan, Q.J.; Fu, X.C. Oxidative coupling of methane over Sr−Ti, Sr−Sn perovskites and corresponding layered perovskites. *React. Kinet. Catal. Lett.* **1995**, *54*, 21–27. [CrossRef]
18. Ivanov, D.V.; Isupova, L.A.; Gerasimov, E.Y.; Dovlitova, L.S.; Glazneva, T.S.; Prosvirin, I.P. Oxidative methane coupling over Mg, Al, Ca, Ba, Pb-promoted SrTiO3 and Sr2TiO4: Influence of surface composition and microstructure. *Appl. Catal. A Gen.* **2014**, *485*, 10–19. [CrossRef]
19. Ivanova, Y.A.; Sutormina, E.F.; Rudina, N.A.; Nartova, A.V.; Isupova, L.A. Effect of preparation route on Sr2TiO4 catalyst for the oxidative coupling of methane. *Catal. Commun.* **2018**, *117*, 43–48. [CrossRef]
20. Ruddlesden, S.N.; Popper, P. New compounds of the K2NIF4 type. *Acta Crystallogr.* **1957**, *10*, 538–539. [CrossRef]
21. Lee, K.H.; Kim, S.W.; Ohta, H.; Koumoto, K. Ruddlesden-Popper phases as thermoelectric oxides: Nb-doped SrO (SrTiO3)$_n$ (n = 1,2). *J. Appl. Phys.* **2006**, *100*, 063717. [CrossRef]

22. Sorkh-Kaman-Zadeh, A.; Dashtbozorg, A. Facile chemical synthesis of nanosize structure of Sr_2TiO_4 for degradation of toxic dyes from aqueous solution. *J. Mol. Liq.* **2016**, *223*, 921–925. [CrossRef]
23. Kwak, B.S.; Do, J.Y.; Park, N.K.; Kang, M. Surface modification of layered perovskite Sr_2TiO_4 for improved CO_2 photoreduction with H_2O to CH_4. *Sci. Rep.* **2017**, *7*, 16370. [CrossRef] [PubMed]
24. Kesić, Ž.; Lukić, I.; Zdujić, M.; Jovalekić, Č.; Veljković, V.; Skala, D. Assessment of $CaTiO_3$, $CaMnO_3$, $CaZrO_3$ and $Ca_2Fe_2O_5$ perovskites as heterogeneous base catalysts for biodiesel synthesis. *Fuel Processing Technol.* **2016**, *143*, 162–168. [CrossRef]
25. Berbenni, V.; Marini, A.; Bruni, G. Effect of mechanical activation on the preparation of $SrTiO_3$ and Sr2TiO4 ceramics from the solid state system $SrCO_3$–TiO_2. *J. Alloy. Compd.* **2001**, *329*, 230–238. [CrossRef]
26. Hungría, T.; Hungría, A.-B.; Castro, A. Mechanosynthesis and mechanical activation processes to the preparation of the $Sr_2[Sr_{n-1}inO_{3n+1}]$ Ruddlesden–Popper family. *J. Solid State Chem.* **2004**, *177*, 1559–1566. [CrossRef]
27. Fiss, B.G.; Richard, A.J.; Douglas, G.; Kojic, M.; Friščić, T.; Moores, A. Mechanochemical methods for the transfer of electrons and exchange of ions: Inorganic reactivity from nanoparticles to organometallics. *Chem. Soc. Rev.* **2021**, *50*, 8279–8318. [CrossRef]
28. Sepelak, V.; Duvel, A.; Wilkening, M.; Beckerb, K.D.; Heitjans, P. Mechanochemical reactions and syntheses of oxide. *Chem. Soc. Rev.* **2013**, *42*, 7507–7520. [CrossRef]
29. Balaz, P.; Achimovicova, M.; Balaz, M.; Billik, P.; Cherkezova-Zheleva, Z.; Criado, J.M.; Delogu, F.; Dutkova, E.; Gaffet, E.; Gotor, F.J.; et al. Hallmarks of mechanochemistry: From nanoparticles to technology. *Chem. Soc. Rev.* **2013**, *42*, 7571–7639. [CrossRef]
30. Amrute, A.P.; De Bellis, J.; Felderhoff, M.; Schüth, F. Mechanochemical Synthesis of Catalytic Materials. *Chem. Eur. J.* **2021**, *27*, 6819–6847. [CrossRef]
31. Ba-Abbad, M.M.; Kadhum, A.A.H.; Mohamad, A.B.; Takriff, M.S.; Sopian, K. Synthesis and catalytic activity of TiO_2 nanoparticles for photochemical oxidation of concentrated chlorophenols under direct solar radiation. *Int. J. Electrochem. Sci.* **2012**, *7*, 4871–4888.
32. Chen, H.; Chen, D.; Bai, L.; Shu, K. Hydrothermal Synthesis and Electrochemical Properties of TiO_2 Nanotubes as an Anode Material for Lithium Ion Batteries. *J. Electrochem. Sci.* **2018**, *13*, 2118–2125. [CrossRef]
33. Yao, H.; Toan, S.; Huang, L.; Fan, M.; Wang, Y.; Russell, A.G.; Luo, G.; Fe, W. $TiO(OH)_2$—Highly effective catalysts for optimizing CO_2 desorption kinetics reducing CO_2 capture cost: A new pathway. *Sci. Rep.* **2017**, *7*, 2943. [CrossRef] [PubMed]
34. Dutcher, B.; Fan, M.; Leonard, B. Use of multifunctional nanoporous $TiO(OH)_2$ for catalytic $NaHCO_3$ decomposition-eventually for $Na_2CO_3/NaHCO_3$ based CO_2 separation technology. *Sep. Purif. Technol.* **2011**, *80*, 364–374. [CrossRef]
35. Silva, G.M.; Wypych, F. A novel and facile synthesis route for obtaining highly urity free layered hydroxide sulfates: Gordaite and osakaite. *Inorg. Chem. Commun.* **2022**, *143*, 109723. [CrossRef]
36. Asgari-Fard, Z.; Sabet, M.; Salavati-Niasari, M. Synthesis and Characterization of Strontium Carbonate Nanostructures via Simple Hydrothermal Method. *High Temp. Mater. Proc.* **2016**, *35*, 215–220. [CrossRef]
37. Lu, P.; Hu, X.; Li, Y.; Zhang, M.; Liu, X.; He, Y.; Dong, F.; Fub, M.; Zhang, Z. One-step preparation of a novel $SrCO_3/g$-C_3N_4 nano-composite and its application in selective adsorption of crystal violet. *RSC Adv.* **2018**, *8*, 6315–6325. [CrossRef]
38. Márquez-Herrera, A.; Ovando-Medina, V.M.; Castillo-Reyes, B.E.; Zapata-Torres, M.; Meléndez-Lira, M.; González-Castañeda, J. Facile Synthesis of $SrCO_3$-$Sr(OH)_2$/PPy Nanocomposite with Enhanced Photocatalytic Activity under Visible Light. *Materials* **2016**, *9*, 30. [CrossRef]
39. Granados-Correa, F.; Bonifacio-Martínez, J. Combustion synthesis process for the rapid preparation of high-purity SrO powders. *Mater. Sci. -Pol.* **2014**, *32*, 682–687. [CrossRef]
40. Tabah, B.; Nagvenkar, A.P.; Perkas, N.; Gedanken, A. Solar-Heated Sustainable Biodiesel Production from Waste Cooking Oil Using a Sonochemically Deposited SrO Catalyst on Microporous Activated Carbon. *Energy Fuels.* **2017**, *31*, 6228–6239. [CrossRef]
41. Xie, T.; Wang, Y.; Liu, C.; Xu, L. New Insights into Sensitization Mechanism of the Doped Ce (IV) into Strontium Titanate. *Materials* **2018**, *11*, 646. [CrossRef]
42. Grabowska, E.; Marchelek, M.; Klimczuk, T.; Lisowski, W.; Zaleska-Medynska, A. $TiO_2/SrTiO_3$ and $SrTiO_3$ microspheres decorated with Rh, Ru or Pt nanoparticles: Highly UV–vis responsible photoactivity and mechanism. *J. Catal.* **2017**, *350*, 159–173. [CrossRef]
43. Senna, M. A straight way toward phase pure complex oxides. *J. Eur. Cer. Soc.* **2005**, *25*, 1977–1984. [CrossRef]
44. Senna, M. Grinding of mixture under mild condition for mechanochemical complexation. *Int. J. Miner. Processing* **1996**, *187*, 44–45.
45. Avvakumov, E.G.; Karakchiev, L.G. Prospects for soft mechanochemical synthesis. *Chem. Sustain. Dev.* **2014**, *22*, 359–369.
46. Sun, X.; Mi, Y.; Jiao, F.; Xu, X. Activating Layered Perovskite Compound Sr_2TiO_4 via La/N Codoping for Visible Light Photocatalytic Water Splitting. *ACS Catal.* **2018**, *8*, 3209–3221. [CrossRef]
47. Vasquez, R.P. X-ray photoelectron spectroscopy study of Sr and Ba compounds. *J. Electron Spectrosc. Relat. Phenom.* **1991**, *56*, 217. [CrossRef]
48. Pilleux, M.E.; Grahmann, C.R.; Fuenzalida, V.M. Hydrothermal Strontium Titanate Films on Titanium: An XPS and AES Depth-Profiling Study. *J. Am. Ceram. Soc.* **1994**, *77*, 1601–1604. [CrossRef]
49. Wang, K.; Ji, S.; Shi, X.; Tang, J. Autothermal oxidative coupling of methane on the $SrCO_3/Sm_2O_3$ catalysts. *Catal. Commun.* **2009**, *10*, 807–810. [CrossRef]
50. Zhang, W.; Sun, X.; Feifei, B.; He, H. Sol-gel Preparation of Photocatalytic Porous Strontium Titanate using PEG4000 as the Template. *Nanosci. Nanotechnol. -Asia* **2012**, *2*, 183–189. [CrossRef]
51. Vasquez, R.P. $SrSO_4$ by XPS. *Surf. Sci. Spectra* **1992**, *1*, 117. [CrossRef]

52. Lim, S.; Choi, J.W.; Suh, D.J.; Song, K.H.; Ham, H.C.; Ha, J.M. Combined experimental and density functional theory (DFT) studies on the catalyst design for the oxidative coupling of methane. *J. Catal.* **2019**, *375*, 478–492. [CrossRef]

Article

Tungsten Trioxide and Its TiO$_2$ Mixed Composites for the Photocatalytic Degradation of NO$_x$ and Bacteria (*Escherichia coli*) Inactivation

Ermelinda Falletta [1], Claudia Letizia Bianchi [1], Franca Morazzoni [2,*], Alessandra Polissi [3], Flavia Di Vincenzo [3] and Ignazio Renato Bellobono [4]

1. Dipartimento di Chimica, Università degli Studi di Milano, Via Camillo Golgi 19, 20133 Milano, Italy; ermelinda.falletta@unimi.it (E.F.); claudia.bianchi@unimi.it (C.L.B.)
2. Dipartimento di Scienza dei Materiali, Università di Milano-Bicocca, Via Roberto Cozzi, 55, 20125 Milano, Italy
3. Dipartimento di Scienze Farmacologiche e Biomolecolari, Università degli Studi di Milano, Via Giuseppe Balzaretti, 9, 20133 Milan, Italy; alessandra.polissi@unimi.it (A.P.); flavia.divincenzo@unimi.it (F.D.V.)
4. Dipartimento di Fisica, Università degli Studi di Milano, Via Giovanni Celoria, 16, 20133 Milano, Italy; i.r.bellobono@alice.it
* Correspondence: franca.morazzoni@unimib.it

Citation: Falletta, E.; Bianchi, C.L.; Morazzoni, F.; Polissi, A.; Di Vincenzo, F.; Bellobono, I.R. Tungsten Trioxide and Its TiO$_2$ Mixed Composites for the Photocatalytic Degradation of NO$_x$ and Bacteria (*Escherichia coli*) Inactivation. *Catalysts* **2022**, *12*, 822. https://doi.org/10.3390/catal12080822

Academic Editor: Meng Li

Received: 5 July 2022
Accepted: 22 July 2022
Published: 26 July 2022

Publisher's Note: MDPI stays neutral with regard to jurisdictional claims in published maps and institutional affiliations.

Copyright: © 2022 by the authors. Licensee MDPI, Basel, Switzerland. This article is an open access article distributed under the terms and conditions of the Creative Commons Attribution (CC BY) license (https:// creativecommons.org/licenses/by/ 4.0/).

Abstract: The increased air pollution and its impact on the environment and human health in several countries have caused global concerns. Nitrogen oxides (NO$_2$ and NO) are principally emitted from industrial activities that strongly contribute to poor air quality. Among bacteria emanated from the fecal droppings of livestock, wildlife, and humans, *Escherichia coli* is the most abundant, and is often associated with the health risk of water. TiO$_2$/WO$_3$ heterostructures represent emerging systems for photocatalytic environmental remediation. However, the results reported in the literature are conflicting, depending on several parameters. In this work, WO$_3$ and a series of TiO$_2$/WO$_3$ composites were properly synthesized by an easy and fast method, abundantly characterized by several techniques, and used for NO$_x$ degradation and *E. coli* inactivation under visible light irradiation. We demonstrated that the photoactivity of TiO$_2$/WO$_3$ composites towards NO$_2$ degradation under visible light is strongly related to the WO$_3$ content. The best performance was obtained by a WO$_3$ load of 20% that guarantees limited e$^-$/h$^+$ recombination. On the contrary, we showed that *E. coli* could not be degraded under visible irradiation of the TiO$_2$/WO$_3$ composites.

Keywords: photocatalysts; visible light; tungsten trioxide; composites

1. Introduction

Due to growing industrialization, urban environments have faced chronic air pollution issues in the last decades. Exhaust gases and burning fuels from factories represent the primary sources of air pollutants on a global scale, causing a significant impact on human health, animal and plant life, and climate [1]. Although natural sources responsible for air pollutants production, such as broad forest fires, volcanic eruptions, and soil erosion, can play a role in air pollution, the emissions resulting from human activities, such as motor vehicle exhaust, combustion of fossil fuels, and industrial processes, are the most active and concerning cause of air quality decline [2].

Nitrogen monoxide (NO) and nitrogen dioxide (NO$_2$), known as nitrogen oxides (NO$_x$), are relevant pollutants whose emissions are directly related to human health problems [3], as they affect respiratory and immune systems [4], to the production of tropospheric ozone, acid rains, and in general to global air pollution.

Over the years, different techniques have been developed for NO$_x$ abatement. Among the traditional techniques, selective catalytic reduction (SCR) with ammonia in the presence of oxygen is the most used, mainly applied to reduce NO_x emission from combustion processes [5], as well as absorption, adsorption, or electrical discharge processes [6]. However, all these methods are characterized by several limitations and disadvantages that make actual application hard. Moreover, the growing environmental constraints invoke restrictions regarding NO_x emission, requiring more efficient techniques for NO_x abatement.

In addition, awareness about the importance of supplying adequate drinking water has recently increased. In 2012, the United Nations estimated that nearly 11% of the world's population did not have access to improved drinking water sources. African water resources indeed contain high levels of microbial pathogens, including bacteria, viruses, and protozoa, as well as chemical contaminants. *Escherichia coli* and related bacteria constitute approximately 0.1% of gut flora, and fecal–oral transmission is the primary route through which pathogenic strains of the species cause disease. For that reason, new disinfection technologies are currently in development to fulfill the WHO Guidelines for drinking-water quality (World Health Organization, 2008). The traditional disinfection methods lead to chloro-organic disinfection by-products (DBPs) with carcinogenic and mutagenic effects.

In both study cases, using a TiO_2 semiconductor as a catalyst under UV or visible irradiation seems the most promising method.

Titania (TiO_2) has been considered the most efficient photocatalyst for a wide variety of applications, such as pollution abatement [7,8], water and air purification [9], antimicrobial applications [10], and energy conversion [11]. However, TiO_2 in its photoactive anatase phase has a wide band gap of 3.2 eV, limiting the photoactivity of the semiconductor only under UV irradiation [12]. Moreover, because of its suspected carcinogenic nature [13], researchers are willing to replace TiO_2 with new low-cost and visible-light-active smart materials.

Though many studies have focused on using TiO_2 [14–18], WO_3 and its composites have been poorly investigated to date [19–22].

WO_3 is a cheap, physiochemically stable, and mechanically robust semiconductor with a narrow band gap energy (2.4–2.8 eV), making it a visible-light-responsive photocatalyst for different applications [23–30]. Therefore, WO_3 represents a suitable choice for photocatalytic degradation under visible light irradiation.

As described in Figure 1, because the VB (valence band) edge potential of WO_3 is lower than that of TiO_2, upon photon absorption, electrons can be transferred from the conduction band of TiO_2 to WO_3, whereas photogenerated holes move in the opposite direction from electrons.

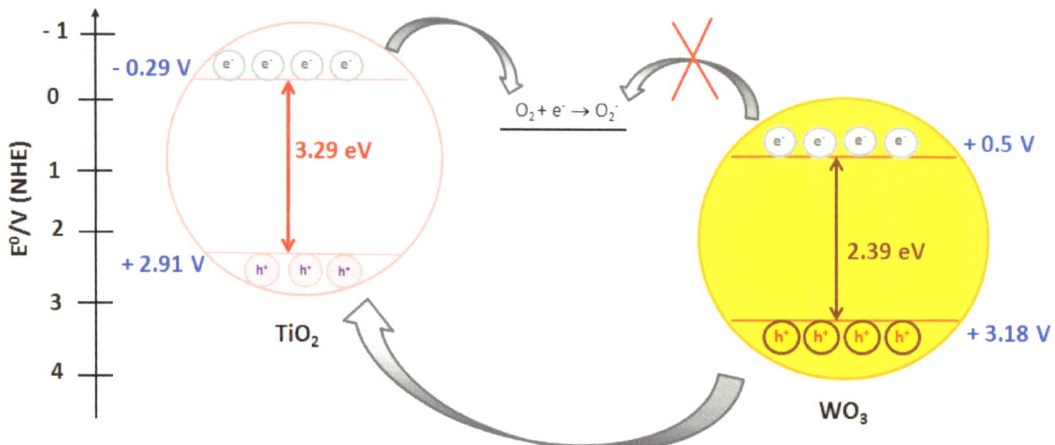

Figure 1. Photocatalytic mechanism of WO_3−loaded TiO_2 under light irradiation.

The transfer of photogenerated carriers is accompanied by consecutive W^{6+} reduction into W^{5+} by capturing photogenerated electrons at trapping sites in WO_3. In addition, W^{5+} ions on the surface of WO_3 are reoxidized into W^{6+}. However, the reduction potential value for the photogenerated electrons in the conduction band is not high enough for the single electron reduction in O_2 (Figure 1) [30]. The holes accumulated in the TiO_2 VB take part in the oxidation process to make OH^- or $OH^·$ hydroxyl radical reactive species. These processes in WO_3/TiO_2 heterostructures restrain the recombination of electron–hole pairs significantly.

Different approaches have been developed to produce highly active TiO_2/WO_3 composites aiming to optimize the WO_3 content [31]. It has been demonstrated that the influence of WO_3 on TiO_2 photoactivity depends on several factors, such as crystal phase, electrons accumulation ability of WO_3, type of pollutants, and degradation pathways involved [31].

Yang et al. investigated the role of amorphous WO_x species, demonstrating that they are more active than the crystalline ones toward methylene blue degradation [32].

Other experiments by Žerjav et al. explained the correlation between the photocatalytic performance of TiO_2/WO_3 and their shallow and deep electron trapping states [33].

However, concerning WO_3 and its mixed oxides, the obtained results are conflicting because in the same cases, the presence of WO_3 seems to positively affect the photoactivity and the performances of TiO_2. In other cases, the results worsen [19,31,32,34].

Regarding NO_x degradation, Luévano-Hipólito et al. demonstrated that WO_3 with a polyhedral shape leads to 50% NO oxidation to NO_2 [19]. On the other hand, Yu and coworkers noticed for the first time the photo-transformation of NO_2 into NO in the presence of N_2 on the surface of a WO_3 photocatalyst under UV/visible light irradiation [20].

Recently, Mendoza et al. proposed TiO_2/WO_3 composites as efficient materials for NO_x abatement under visible light, leading to 90% of photodegradation in 1 h [22], whereas Paula and coworkers observed the decay of the photocatalytic activity of TiO_2/WO_3 heterostructures as a function of the W(VI) content [31].

Jawwad A. Darr et al. reported the easy disinfection of water by TiO_2/WO_3 mixed composites, which induce bacterial inactivation after 30 min of photo-irradiation [35].

In order to clarify the behavior of WO_3 and TiO_2/WO_3 heterostructures in the photocatalytic degradation of NO_x under visible light irradiation, in this work, WO_3 and a series of TiO_2/WO_3 composites were synthesized by a fast and cost-effective chemical procedure and tested for the photodegradation of NO_x and the inactivation of *E. coli* under visible light irradiation.

The role of the calcination temperature in the TiO_2/WO_3 preparation has been investigated and critically discussed, as well as the effect of the WO_3 loading in the final composites. Differently from the recent literature, the results proved that high calcination temperatures could cause complete or partial WO_3 sublimation with adverse effects on the activity of the TiO_2/WO_3 heterostructures.

Finally, while the synthesized catalysts were active in NO_2 photodegradation, they were inert to the antibacterial activity under visible light irradiation, in line with the scientific literature [36].

2. Results

2.1. Materials Characterization

Figure 2 shows the XRD patterns of all the synthesized materials.

TiO_2 exhibits the characteristics of diffraction peaks of anatase, as confirmed by the peaks at 25.3°, 37.7°, 48.0°, 53.8°, and 55.0°, with (101), (004), (200), (105), and (211) diffraction planes, respectively. The XRD pattern of WO_3 shows a crystalline phase characterized by diffraction peaks at 23.1°, 23.6°, 24.4°, and 34.2°, corresponding to the (002), (020), (200), and (202) crystal planes of monoclinic phase.

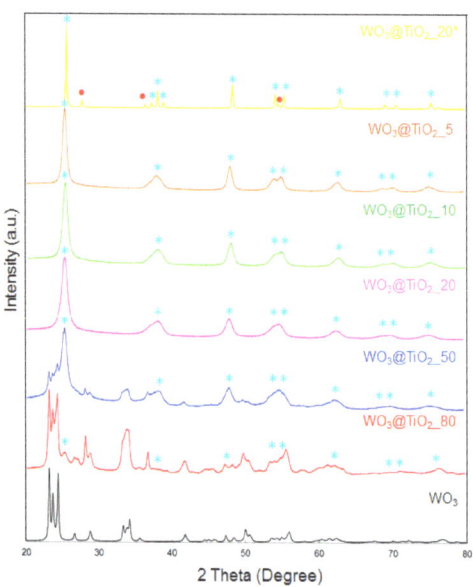

Figure 2. XRD patterns of the samples of Table 1 (* peaks of TiO_2 anatase, •peaks of TiO_2 rutile).

Table 1. Energy of band gap (eV), specific surface area, CBET, Vm, and mean pore diameter of the WO_3/TiO_2 composites series calcined at 400 °C. * Surface area by BET equation (2-parameters), ** mean pore diameter by BJH model from isotherm desorption branch ($0.3 < p/p0 < 0.95$).

Sample	Band Gap (eV)	* Specific Surface Area (m²/g)	CBET	Vm (cm³/g)	** Mean Pore Diameter (nm)
WO_3	2.39	4.00	75.75	0.94	21.17
WO_3@TiO_2_80	2.63	42.78	123.22	9.92	8.6
WO_3@TiO_2_50	3.05	110.65	98.11	25.87	6.4
WO_3@TiO_2_20	3.14	179.78	75.50	43.26	6.0
WO_3@TiO_2_10	3.26	139.47	94.7	33.15	6.8
WO_3@TiO_2_5	3.20	111.08	112.87	31.29	9.1
TiO_2	3.29	318.00	84.48	75.50	4.70

As expected, in the WO_3@TiO_2 composites, the intensity of the diffraction peaks of WO_3 declines by decreasing the percentage of WO_3; on the contrary, anatase peaks intensity increases or appears with the higher concentration of TiO_2 in each composite.

Based on its XRD pattern, WO_3@TiO_2_20* exhibits higher crystallinity degree if compared to the others and the appearance of new diffraction peaks can be observed.

As reported in the literature [37], it is directly correlated to the high temperature (600 °C) used for the calcination of this material. The degree of crystallinity increases with the temperature, and in the case of TiO_2-based compounds, at 600 °C the phase conversion from anatase to rutile starts.

According to the literature [22,34], when the WO_3 content in the composite materials is 20% or lower, the diffraction peaks of this semiconductor are undetected. Some authors justify this result with the presence of highly dispersed WO_3 small particles in TiO_2/WO_3 composites, which makes it hard to detect them by this technique [22,34]. In addition, Yang et al. demonstrated that when the loading amount of WO_3 was below 3 mol%, it exists in highly dispersed amorphous species that do not respond to XRPD. However, accurate quantification of WO$_3$ loading on TiO$_2$ after calcination is necessary to verify unequivocally the WO$_3$/TiO$_2$ composite formation rather than a superficial W doping on the TiO$_2$ surface. In the present work, this was easily carried out by the reaction yield calculation (Equation (2)) and by EDS analysis (Table S1) for two composites with a nominal WO$_3$ load of 20% (WO$_3$@TiO$_2$_20 and WO$_3$@TiO$_2$_20*) calcined at two different temperatures (400 °C and 600 °C). From the results of the reaction yield, the WO$_3$@TiO$_2$_20* composite calcined at 600 °C exhibits a mass loss of approximately 40%, unlike the same sample calcined at 400 °C, obtained with a yield of 94.8%. Since WO$_3$ is a low-temperature sublimation material [38], it cannot be excluded that by increasing the calcination temperature, a complete or partial WO$_3$ sublimation can occur, as also confirmed by the EDS results (Table S1), where the measured percentage of WO$_3$ in the final composite is lower than 3%, whereas the material calcined at 400 °C shows a 27% of WO$_3$.

Figure 3 displays the FT-IR for all the synthesized composites calcined at 400 °C.

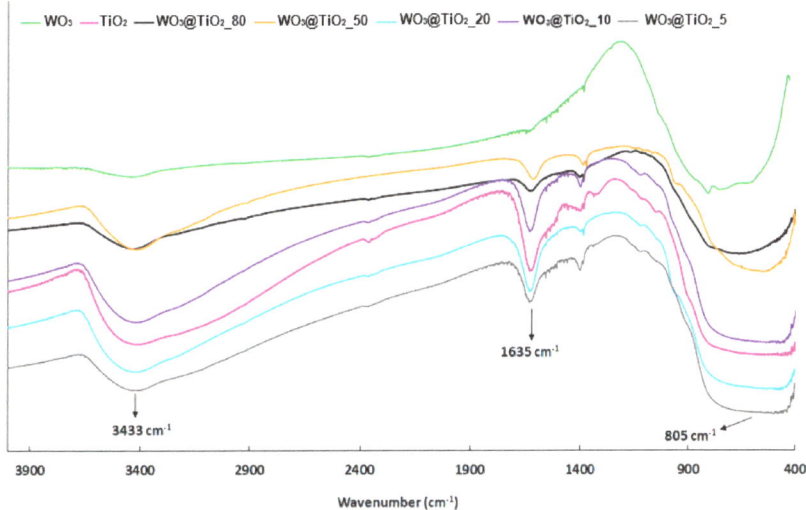

Figure 3. FT−IR spectra of the samples calcined at 400 °C.

The FT-IR spectrum of WO$_3$ exhibits characteristic vibration bands, such as those at approximately 3433 cm^{-1} and 1635 cm^{-1}, that can be associated with the symmetric stretching vibrations of WO$_3$ and intercalated water molecules and the deformation vibrations of H–O–H bonds of the adsorbed water molecules, respectively, and the signal at 805 cm^{-1}, attributed to O–W–O stretching modes of WO$_3$ [39].

On the other hand, the FT-IR spectrum of TiO$_2$ nanoparticles is characterized by several peaks. The OH stretching mode of the hydroxyl groups is responsible for the broad band observed in the range of 3600–3000 cm^{-1}, indicating the presence of moisture in the sample. The band at approximately 1605 cm^{-1} is due to the OH bending vibrations of the absorbed water molecules. Finally, the broadband between 1000 and 500 cm^{-1} can be related to the Ti–O stretching and Ti–O–Ti bridging stretching modes [40].

As expected, in the FT-IR spectra of the composites, the WO$_3$ characteristic bands are covered by the more intense ones of TiO$_2$.

The optical properties of the synthesized WO$_3$/TiO$_2$ series calcined at 400 °C, as well as of single-phase semiconductors, were investigated by UV–Vis scanning spectrophotometry (Figure 4).

The main absorption edges of the samples are all around 400 nm, attributing to the excitation of electrons from the valence band to the conduction band. As reported in the literature [41], the empty orbitals of W^{6+} (W 5d) are closed to the Ti 3d orbitals of the conduction band. Therefore, the $O_{2-} \rightarrow W^{6+}$ charge transfer transitions are overlapped with the $O_2 \rightarrow Ti^{4+}$ charge transfer transitions. Increasing the WO_3 loading, the absorption edge of photocatalysts red-shifts. If this is only slightly noticeable up to 10%, increasing the WO_3 percentage to 50% and above, the effect is much more evident, and the absorption edges of these materials are shifted at higher wavelengths.

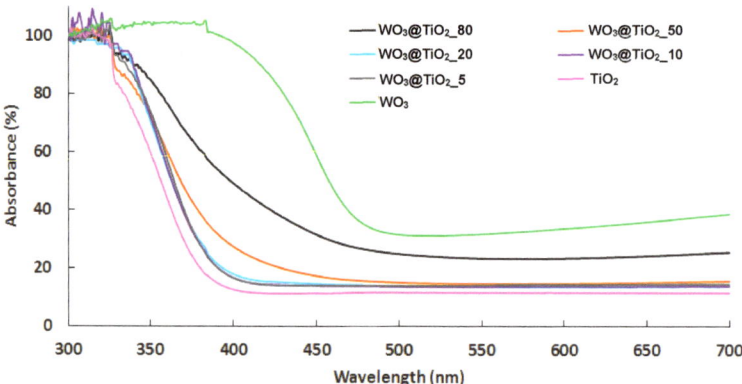

Figure 4. UV–Vis absorption spectra of the WO_3/TiO_2 composites series calcined at 400 °C.

The band gaps of the materials, estimated by the Kubelka–Munk function, are summarized in Table 1, and the Tauc plots are reported in Figure S2.

Samples Eg decreases, increasing the tungsten content, due to the formation of defective energy levels within the forbidden band gap of WO_3. On the other hand, by increasing the TiO_2 content, the total band gap of the photocatalyst decreases [42].

According to the shape of nitrogen adsorption–desorption isotherms reported in Figure 5 and the IUPAC classification [43], all the photocatalysts calcined at 400 °C can be classified as mesoporous materials of type IV, as confirmed by the values of C_{BET} in Table 1, containing other quantitative data.

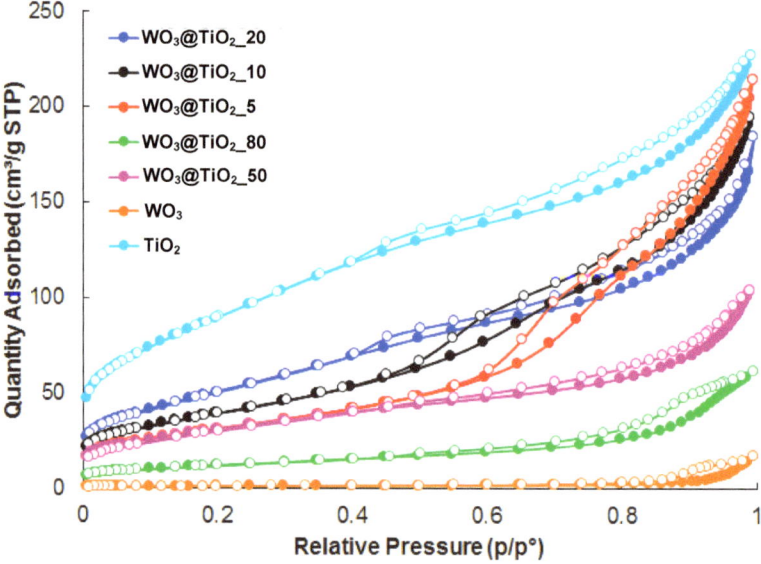

Figure 5. Nitrogen adsorption–desorption isotherms for the synthesized samples.

Figure 6 displays the SEM images and the elemental mapping of Ti and W for the synthesized composites, whereas the EDX spectra are reported in the (Supplementary Information Table S1).

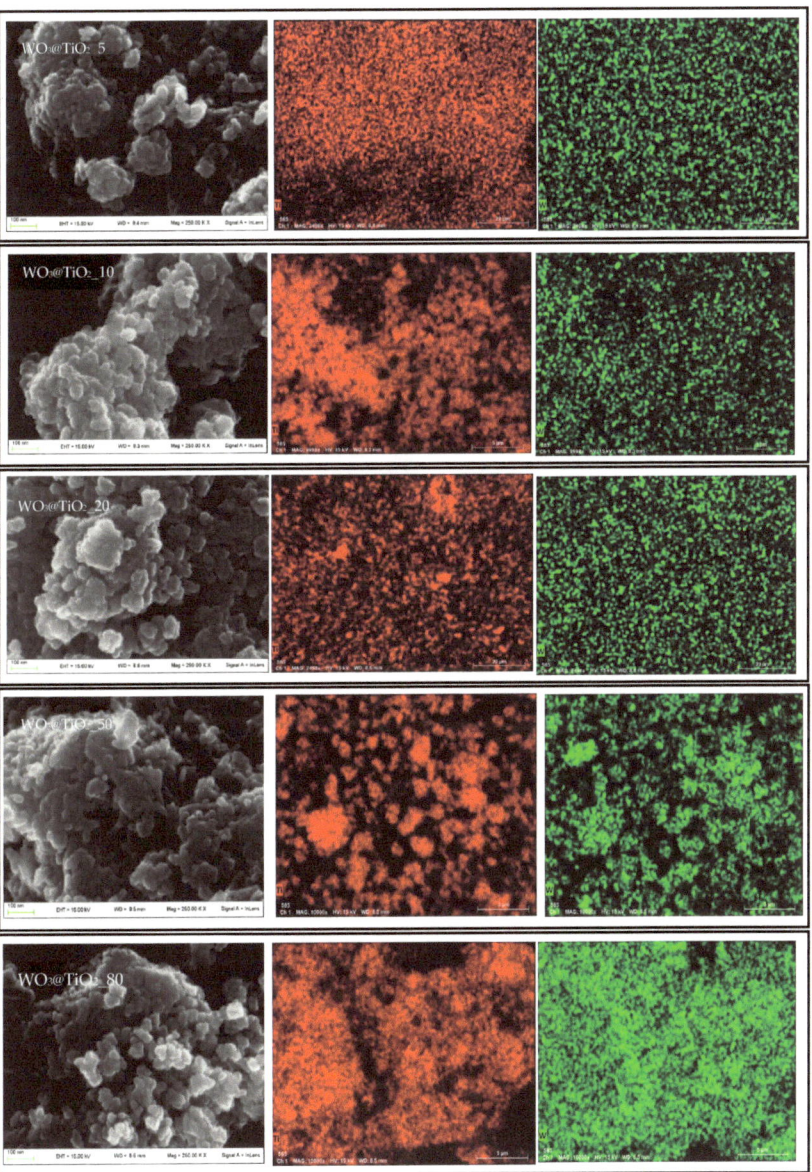

Figure 6. SEM images and elemental mapping of the synthesized materials. (**left**) sample analyzed, ((**middle**), red-colored) titanium map, ((**right**), green-colored) tungsten map.

The elemental maps of W and Ti by X-ray energy dispersion (EDX) in the composites demonstrate that TiO_2 and WO_3 are well dispersed in each material. All the $WO_3@TiO_2$ show a globular-like morphology with particle sizes ranging from 60 to 5 nm. The particles are aggregated by sharing corners or edges that probably involve the formation of Ti–O–W bonds [44]. The same information was obtained by TEM investigations (Figure 7) showing nanoparticles of 12–35 nm that gradually aggregate with the WO$_3$ load, reaching up to 60 nm in size.

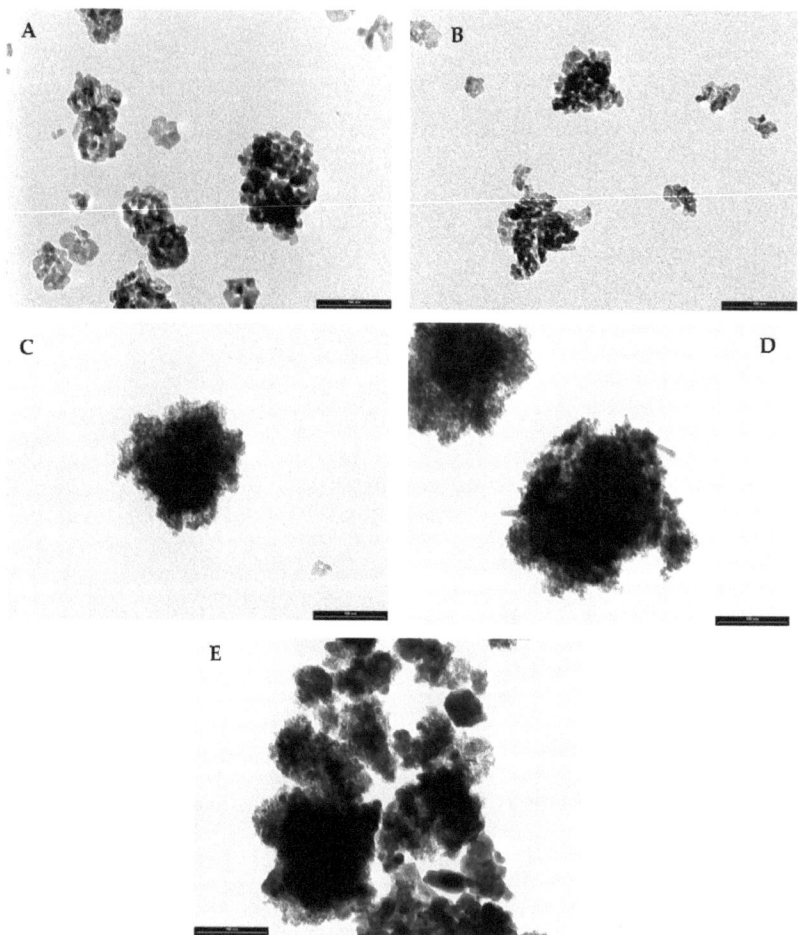

Figure 7. TEM images of WO$_3$@TiO$_2$_5 (**A**), WO$_3$@TiO$_2$_10 (**B**), WO$_3$@TiO$_2$_20 (**C**), WO$_3$@TiO$_2$_50 (**D**), WO$_3$@TiO$_2$_80 (**E**).

2.2. Photocatalytic and Biological Activities of Catalysts

2.2.1. NO$_x$ Photocatalytic Degradation

The photocatalytic activity of the TiO$_2$/WO$_3$ composites, as well as single-phase photocatalysts, referred to both NO$_x$ (NO + NO$_2$) and NO$_2$ conversion under visible light irradiation (Figure 8).

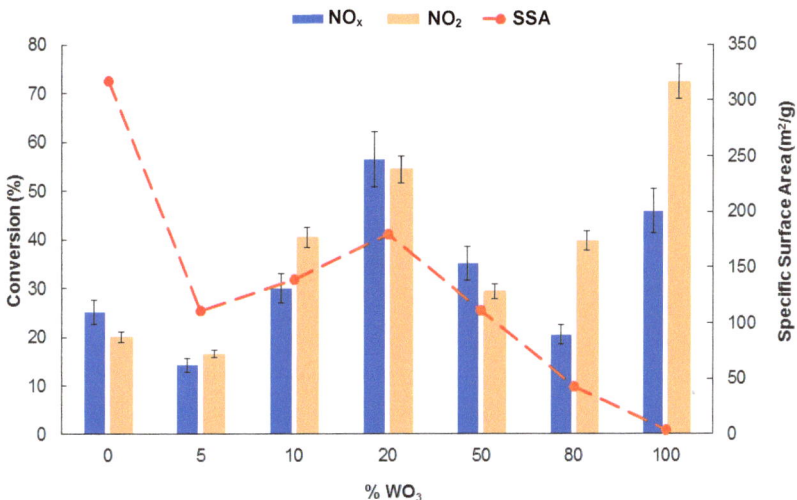

Figure 8. Dependence of NO_x and NO_2 photodegradation on WO_3 content and specific surface area (SSA) of the materials.

According to the literature [22,31,45], for this type of material the photocatalytic degradation of NO_x consists of a photo-oxidation process, where both NO and NO_2 species are first adsorbed on the surface of the heterostructures and then converted into the corresponding oxidation product (NO_3^-) under light irradiation. Although the formation of the oxidation products was well documented in the literature [46,47], an analytical confirmation was not performed in the present work due to the low NO_x concentrations used during the test.

For all the experiments, the initial concentration of NO_x was approximately 500 ppb. The NO_x photodegradation results are summarized in Figure 8.

As expected, despite its extraordinarily high surface area, TiO_2 shows poor activity towards NO_x conversion under visible light irradiation, leading to NO_x and NO_2 degradation in 25% and 20%, respectively. On the other hand, according to the band gap value, regardless of its low surface area, pristine WO_3 exhibits a good photoactivity towards NO_2 photodegradation (72%) in 3 h, whereas the NO_x abatement is only 46%. In fact, as reported in the scientific literature [19–22], WO_3 can remarkably reduce NO_2 into NO in the presence of N_2. This is confirmed by the results reported in Figure S3, showing for the WO_3 sample an increase in the NO concentration during the reaction. This makes pristine WO_3 not efficient in the NO_x abatement, because, as it is known, in air NO is immediately reoxidized to NO_2. In this regard, from the pioneering investigations of Yu et al., carried out under nitrogen atmosphere and UV irradiation, a 20% conversion of NO_2 into NO can be inferred [20]. The present results demonstrate that even under visible light irradiation, the percentage of NO formation from NO_2 is of the same order (24%), calculated by Equation (1):

$$\text{NO produced } (\%) = \frac{[NOt] - [NOi]}{[NOt]} \quad (1)$$

where [NOt] is the NO concentration at the end of the reaction (after 3 h of light irradiation), [NOi] is the NO concentration before light irradiation.

If compared to the single-phase photocatalysts (TiO_2 and WO_3), the photoactivity of WO_3/TiO_2 heterostructures strongly depends on their composition. More in detail, the activity of the catalysts gradually increases with the WO_3 load, reaching the highest photodegradation efficiency by $WO_3@TiO_2\_20$ (54.4% NO_x conversion and 56.4% NO_2

abatement), whereas it decreases for a percentage of $WO_3 > 20$. These results are in line with the pioneering investigations of Balayeva et al., who tested the photocatalytic activity of TiO_2/WO_3 composites towards NO degradation under UV irradiation, obtaining a ca. 35% of conversion for heterostructures characterized by a 1% and 2.5% of WO_3 load [48].

For a very high amount of WO_3 ($WO_3@TiO_2\_80$), these latter composites maintain the photoactive capability of pristine WO_3, converting NO_2 to NO. The different photoactivity of the materials may be due to a combination of factors.

First of all, it can be assumed that for a WO_3 load < 20%, TiO_2 and WO_3 only play their own photocatalytic role, and coupled photocatalysts are not formed. In this case, the low activity of TiO_2 prevails because it is the major component. In contrast, for a large amount of WO_3, the fast e^-/h^+ recombination of the WO_3 component predominates.

On the contrary, the absence of WO_3 peaks in the XRD spectra of the $WO_3@TiO_2\_20$ sample suggests that its increased photoactivity is not related to the formation of crystalline tungsten oxide but is probably due to the presence of WO_3 centers on the surface of TiO_2 acting as electrons/holes separators [49]. When the test was carried out using $WO_3@TiO_2\_20^*$, in order to observe the effect of calcination temperature on the photoactivity of the material, the percentage of NO_x and NO_2 degradation dropped to ca. 20%, confirming that the thermal treatment acts by reducing the WO_3 content in the $WO_3@TiO_2$ heterostructure and as a consequence of its activity.

It is known that the quantum efficiency of photocatalytic reactions carried out by heterogeneous photocatalysts depends on the competition between the recombination of photogenerated electrons and holes and the transfer of both electrons and holes at the interface of the material. Extending the electrons and holes recombination time and increasing the transfer rate of electrons at the interface enhance the quantum efficiency positively. As reported in the literature [31], the formation of TiO_2/WO_3 heterostructures leads to enhanced charge carrier lifetimes, due to the transfer of photogenerated electrons in the TiO_2 to WO_3 CB, and at the same time to the entrapment of the photogenerated holes within the TiO_2 particle. Both these phenomena make charge separation more efficient.

Finally, the effect related to the different surface area values cannot be ignored. As for the photocatalytic activity, the surface area values also seem to be correlated to the WO_3 load and the most active catalyst ($WO_3@TiO_2\_20$) is also the one with the highest surface area (Figure 8, Table 1). At first glance, the results obtained by the $WO_3@TiO_2\_20$ photocatalyst seem to contrast with those of Mendoza et al. [22], who report very high NO conversion values under visible light irradiation in similar conditions. The different photocatalytic activity of the $WO_3@TiO_2\_20$ sample compared to those reported by Mendoza and coworkers can be reasonably attributed to the real WO_3 content in the synthesized composites that it is not specified in the work of the author [22].

2.2.2. E. coli Photoinactivation

The antibacterial activity of $WO_3@TiO_2$ composites, as well as single-phase photocatalysts, was assessed by determining the percentage of *E. coli* cell survival following exposure to visible light. According to standard methods (ASTM E2149, 2001), values of survival $\leq 90\%$ indicate the antibacterial activity of a given photocatalytic film.

As shown in Figure 9, none of the tested samples displays antibacterial activity under visible light irradiation, in agreement with what was reported in the literature for the WO_3/TiO_2 catalyst [36].

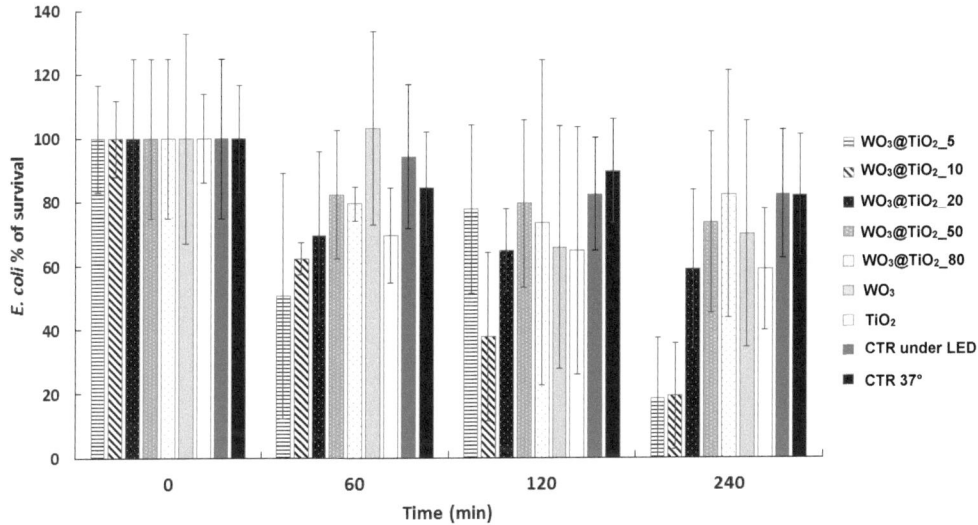

Figure 9. Percentage of *E. coli* cell survival after exposure to visible light in the presence of the synthesized heterostructures.

Different catalysts (e.g., sulfur-doped carbon quantum dots loaded hollow tubular g-C$_3$N$_4$) give degradation of *E. coli* cells instead, under visible light [50].

Based on the morphology of nanoparticles (Figure 7), we speculate that the nanoparticle aggregation of TiO$_2$/WO$_3$ hinders a suitable surficial interaction with the bacteria and the catalyst cytotoxicity.

3. Materials and Methods

3.1. Chemicals

Tungstic acid (H$_2$WO$_4$, 99% Merck), AMT 100 TiO$_2$ (Tayca Corporation, WP0097, Osaka, Japan), ammonium hydroxide solution (ACS reagent, 28.0–30.0% Merck & Co., St. Louis, MO, USA), hydrochloric acid (HCl 36%, Suprapur®, Supelco, Belfont, PA, USA) were used as received.

3.2. Synthesis of TiO$_2$/WO$_3$ Series

To synthesize 1 g of TiO$_2$/WO$_3$ composite, a proper amount of TiO$_2$ was dispersed in 25 mL of 2 M ammonium hydroxide solution under constant stirring (solution A). The WO$_3$ precursor solution was prepared by dissolving a proper amount of tungstic acid H$_2$WO$_4$ in 25 mL of 2 M ammonium hydroxide solution under constant stirring (solution B). The quantity of TiO$_2$ and H$_2$WO$_4$ used is reported in Table 2. The two solutions were stirred for 30 min at room temperature, then solution B was added to solution A and the stirring was continued for another 2 h at room temperature. Then, the solvent was evaporated, heating the mixture at 110 °C. The white-yellow powder was treated with 0.5 M hydrochloric acid solution and dried again. The final powder was washed with deionized water abundantly, dried at 100 °C overnight, and calcinated in the air at 400 °C for 2 h (heating rate 8 °C·min^{-1}). The synthesized samples containing different TiO$_2$/WO$_3$ w/w ratios (95:5, 90:10, 80:20, 20:80) were properly characterized and tested for NO$_x$ photodegradation under visible light irradiation.

Table 2. Labels of the WO_3/TiO_2 composites, WO_3/TiO_2 w/w ratio, calcination temperature, and reaction yield.

Label	TiO_2 (g)	H_2WO_4 (g)	WO_3/TiO_2 (w/w Ratio)	Calcination Temperature (°C)	Yield (%)
WO_3	0.00	1.08	100:0	400	95.2
$WO_3@TiO_2\_80$	0.20	0.86	80:20	400	95.7
$WO_3@TiO_2\_50$	0.50	0.55	50:50	400	95.1
$WO_3@TiO_2\_20$	0.80	0.22	20:80	400	94.8
$WO_3@TiO_2\_10$	0.90	0.11	10:90	400	96.2
$WO_3@TiO_2\_5$	0.95	0.05	5:95	400	95.7
$WO_3@TiO_2\_20*$	0.80	0.22	20:80	600	60.9
TiO_2	1.00	0.00	0:100	400	98.5

WO_3 was synthesized by the same procedure using a TiO_2-free solution A.

An aliquot of the TiO_2/WO_3 composite with an 80:20 w/w ratio was calcined at 600 °C.

Table 2 reports a list of synthesized composites with the corresponding label, WO_3/TiO_2 w/w ratio, calcination temperature, and reaction yield, calculated by Equation (2):

$$\text{Yield (\%)} = \frac{\frac{g\ (H_2WO_4)\ \times\ molar\ mass(WO_3)}{molar\ mass\ (H_2WO_4)} + g\ TiO_2}{g\ final\ product} \quad (2)$$

3.3. Characterization Methods

X-Ray Diffraction (XRD) measurements investigated the crystalline structure on a PW3830/3020 X'Pert diffractometer (PANalytical, Almelo, The Netherlands) working Bragg–Brentano, using the Cu Kα1 radiation (k = 1.5406 Å).

FT-IR spectra were recorded in the range of 400–4000 cm^{-1} with a resolution of 0.5 cm^{-1} by anPerkin-Elmer spectrometer (Perkin Elmer, Waltham, MA, USA) dispersing a few milligrams of each material in anhydrous KBr. The morphology of the catalysts was inspected employing high-resolution electron transmission microscopy (HR-TEM), using a JEOL 3010-UHR instrument (Musashino Akishima, Japan; acceleration potential: 300 kV; LaB6 filament), and by scanning electron microscopy (SEM), using a Zeiss LEO 1525 field emission microscope (Jena, Germany). The samples were "dry" dispersed on lacey carbon Cu grids for TEM analyses, whereas SEM analyses were carried out without any pre-treatment of the samples.

For the band gap determinations, diffuse reflectance spectra of the powders were collected on a UV–Vis diffuse reflectance spectra using a scanning spectrophotometer PerkinElmer, Lambda 35 (Perkin Elmer, Waltham, MA, USA), which was equipped with a diffuse reflectance accessory. A thin film of each sample was placed in the sample holder on an integrated sphere for the reflectance measurements. A KBr pill was used as the reference material. Data were elaborated using the Kubelka–Munk function (Equation (3)), which expresses the adsorbance as a function of reflectance (F(R)) [51]:

$$F(R) = (1 - R)^2/2R \quad (3)$$

where R = reflectance of the powder.

The band gap values were determined by performing the first derivative of the Kubelka–Munk function (Equation (4)):

$$dF(R)/d\lambda \quad (4)$$

where λ = wavelength of the incident radiation. The energy of the radiation at which the first derivative dF(R)/d λ shows the maximum was taken to estimate the band gap values.

Specific surface area and porosity distribution were determined by processing N_2 adsorption–desorption isotherms at 77 K (Coulter SA3100 instrument, Beckman Life Sciences, Los Angeles, CA, USA) with Brunauer–Emmett–Teller (BET) and Barrett–Joyner–Halenda analyses. Before the analysis, samples were heat-treated (T = 150 °C, 4 h, N_2) to remove adsorbed foreign species.

3.4. NO_x Photodegradation Tests

A photocatalytic film of each sample was deposited by drop-casting on glass supports as follows: a suspension of 0.050 ± 0.001 g of photocatalyst in 5 mL of isopropanol was deposited on a glass plate (230 × 19 mm). Once the solvent was evaporated, the photocatalyst was placed inside a 20 L Pyrex glass cylindrical batch reactor for the photocatalytic tests. The photocatalytic tests were performed by a mixture of NO and NO_2 in air. The starting inlet gas contains only NO_2, but the chemical equilibrium between NO and NO_2 is established as it is exposed to air. An LED lamp (350 mA, 9–48 V DC, 16.8 W) with emissions in the 400–700 nm range was used as the light source. The luminous intensity (lux) was measured using an illumination meter (Delta Ohm photo/radiometer HD 2102.2) and was 2900 lx to estimate the light intensity. It was then converted to the irradiance unit (in mW/cm^2) [19], obtaining a light intensity of 3.24 mW/cm^2.

The NO_x initial concentration was 500 ± 50 ppb. A chemiluminescence analyzer measured the NO_x concentration after 30, 60, and 180 min of exposure to light irradiation (ENVEA AC32e).

3.5. Antibacterial Assay

Cultures of *E. coli* MG1655 [52] were grown at 37 °C in Luria–Bertani (LB) medium (10 g/L tryptone, 5 g/L yeast extract, 10 g/L NaCl) or LB-agar medium (LB medium with 10 g/L agar). Bacteria (200 mL) were collected by 10 min centrifugation at 5000 rpm, washed in PBS 1X (Merck & Co., St. Louis, MO, USA), and resuspended in the same volume of PBS 1X. Stationary phase cultures of *E. coli* were diluted up to optical density at 600 nm (OD600) of 0.05 and then grown aerated up to 0.6.

Films of the synthesized heterostructures were prepared as follows. A total of 100 mg of each material was dispersed in 8 mL of isopropanol and deposited on a Petri dish (90 mm in diameter) by drop-casting and air-dried.

10 mL of bacterial cells were added to Petri dishes. The plates were irradiated with visible light (2900 lux obtained with an LED lamp) and removed from light at different time points (60, 120, and 240 min). As controls, bacterial cells were deposited onto empty Petri dishes and irradiated (CTR under LED) or incubated at 37 °C under dark (CTR). Viable bacteria expressed as CFU/mL (colony-forming unit /mL) were enumerated at t = 0, 60, 120, and 240 min by plating suitable dilutions onto LB-agar plates following incubation at 37 °C for 18 h.

The percentage of bacterial survival is expressed as follows:
(average of viable bacteria at a given time/average of viable bacteria CTR 37° at t = 0 min) × 100.

The average of viable bacteria is calculated from at least three independent experiments. Values of survival ≤90% indicate the antibacterial activity of a given photocatalytic film.

4. Conclusions

In this study, we investigated the photoactivity of WO_3 and TiO_2/WO_3 composites towards NO_2 degradation under visible light, in order to clarify the numerous conflicting data reported in the literature so far. It was demonstrated that the photoactivity of TiO_2/WO_3 heterostructures are strongly related to their composition. For $WO_3@TiO_2$ materials characterized by low tungsten trioxide content (<20%), TiO_2 and WO_3 are present as separate phases, each playing their own photocatalytic role, whereas coupled photocatalysts are not formed.

The composite with a WO_3 load of 20% was the most efficient photocatalyst, extending the electrons and holes recombination time and promoting the transfer rate of electrons at the interface. The high activity of the material can be explained with its high surface area value and with the presence of WO_3 centers on the surface of TiO_2 acting as electrons/holes separators. However, if the WO_3 load is higher than 20%, a fast e^-/h^+ recombination can occur and the ability of tungsten trioxide to reduce NO_2 to NO prevails over the composites' capability to photo-oxidize NO_2 to NO_3^-. Moreover, the photodegradation activity of the heterostructures can be attributed to the oxidizing effect of holes. Moreover, it was demonstrated that high-temperature calcination leads to a partial sublimation of the WO_3 component that causes a decrease in heterostructure activity. As for the lack of the bacteria degradation, we tentatively suggest that the aggregation of nanoparticles hinders an efficient surface contact between bacteria and catalyst.

Supplementary Materials: The following supporting information can be downloaded at: https://www.mdpi.com/article/10.3390/catal12080822/s1: Table S1: EDS analysis, Figure S1: EDX spectrum of (A) $WO_3@TiO_2\_20*$ and (B) $WO_3@TiO_2\_20$, Figure S2: Tauc plot of TiO_2 (A), $WO_3@TiO_2\_5$ (B), $WO_3@TiO_2\_10$ (C), $WO_3@TiO_2\_20$ (D), $WO_3@TiO_2\_50$ (E), $WO_3@TiO_2\_80$ (F), WO_3 (G), Figure S3: Dependence of WO_3 content (%) in the composites versus and NO production.

Author Contributions: Conceptualization, C.L.B., F.M. and A.P.; methodology, C.L.B. and A.P.; investigation, E.F. and F.D.V.; data curation, C.L.B., F.M. and A.P.; writing—original draft preparation, E.F., F.M., C.L.B. and A.P.; writing—review and editing, E.F.; supervision, F.M.; funding acquisition, I.R.B. All authors have read and agreed to the published version of the manuscript.

Funding: This research was funded by Fondazione di Comuntà Milano, Fondo Ignazio Renato Bellobono Letizia Stefanelli.

Data Availability Statement: The data that support the plots within this paper are available from the corresponding author on reasonable request.

Conflicts of Interest: The authors declare no conflict of interest.

References

1. Colls, J. *Air Pollution*, 2nd ed.; CRC Press: London, UK, 2002; pp. 1–360. [CrossRef]
2. Chaloulakou, A.; Mavroidis, I.; Gavriil, I. Compliance with the annual NO_2 air quality standard in Athens. Required NO_x levels and expected health implications. *Atmos. Environ.* **2008**, *42*, 454–465. [CrossRef]
3. Roy, S.; Hegde, M.S.; Madras, G. Catalysis for NO_x abatement. *Appl. Energy* **2009**, *86*, 2283–2297. [CrossRef]
4. Ângelo, J.; Andrade, L.; Madeira, L.M.; Mendes, A. An overview of photocatalysis phenomena applied to NO_x abatement. *J. Environ. Manag.* **2013**, *129*, 522–539. [CrossRef] [PubMed]
5. Brüggemann, T.C.; Keil, F.J. Theoretical investigation of the mechanism of the selective catalytic reduction of nitric oxide with ammonia on H-form zeolites. *J. Phys. Chem. C* **2008**, *112*, 17378–17387. [CrossRef]
6. Mok, Y.S.; Lee, H. Removal of sulfur dioxide and nitrogen oxides by using ozone injection and absorption–reduction technique. *Fuel Process. Technol.* **2006**, *87*, 591–597. [CrossRef]
7. German, R.M.; Suri, P.; Park, S.J. Review: Liquid phase sintering. *J. Mater. Sci.* **2009**, *44*, 1–39. [CrossRef]
8. Zeng, Y.; Haw, K.-G.; Wang, Y.; Zhang, S.; Wang, Z.; Zhong, Q.; Kawi, S. Recent Progress of CeO_2-TiO_2 Based Catalysts for Selective Catalytic Reduction of NO_x by NH_3. *ChemCatChem* **2021**, *13*, 491–505. [CrossRef]
9. Nakata, K.; Ochiai, T.; Murakami, T.; Fujishima, A. Photoenergy conversion with TiO_2 photocatalysis: New materials and recent applications. *Electrochim. Acta* **2012**, *84*, 103–111. [CrossRef]
10. Negishi, N.; Sugasawa, M.; Miyazaki, Y.; Hirami, Y.; Koura, S. Effect of dissolved silica on photocatalytic water purification with a TiO_2 ceramic catalyst. *Water Res.* **2019**, *150*, 40–46. [CrossRef]
11. Ge, M.; Caia, J.; Iocozzi, J.; Cao, C.; Huang, J.; Zhang, X.; Shen, J.; Wang, S.; Zhang, S.; Zhang, K.-Q.; et al. A review of TiO_2 nanostructured catalysts for sustainable H_2 generation. *Int. J. Hydrog. Energy* **2017**, *42*, 8418–8449. [CrossRef]
12. Hamrouni, A.; Azzouzi, H.; Rayes, A.; Palmisano, L.; Ceccato, R.; Parrino, F. Enhanced Solar Light Photocatalytic Activity of Ag Doped $TiO_2-Ag_3PO_4$ Composites. *Nanomaterials* **2020**, *10*, 795. [CrossRef]
13. Braakhuis, H.M.; Gosens, I.; Heringa, M.B.; Oomen, A.G.; Vandebriel, R.J.; Groenewold, M.; Cassee, F.R. Mechanism of Action of TiO_2: Recommendations to Reduce Uncertainties Related to Carcinogenic Potential. *Annu. Rev. Pharmacol. Toxicol.* **2021**, *61*, 203–223. [CrossRef]
14. Luna, M.; Gatica, J.M.; Vidal, H.; Mosquera, M.J. One-pot synthesis of $Au/N-TiO_2$ photocatalysts for environmental applications: Enhancement of dyes and NO_x photodegradation. *Powder Technol.* **2019**, *355*, 793–807. [CrossRef]

15. Xu, M.; Wang, Y.; Geng, J.; Jing, D. Photodecomposition of NO_x on Ag/TiO_2 composite catalysts in a gas phase reactor. *Chem. Eng. J.* **2017**, *307*, 181–188. [CrossRef]
16. Bianchi, C.L.; Pirola, C.; Galli, F.; Cerrato, G.; Morandi, S.; Capucci, V. Pigmentary TiO_2: A challenge for its use as photocatalyst in NO_x air purification. *Chem. Eng. J.* **2015**, *261*, 76–82. [CrossRef]
17. Cerrato, G.; Galli, F.; Boffito, D.C.; Operti, L.; Bianchi, C.L. Correlation preparation parameters/activity for microTiO_2 decorated with SilverNPs for NO_x photodegradation under LED light. *Appl. Catal. B Environ.* **2019**, *253*, 218–225. [CrossRef]
18. Bianchi, C.L.; Cerrato, G.; Pirola, C.; Galli, F.; Capucci, V. Photocatalytic porcelain grés large slabs digitally coated with AgNPs-TiO_2. *Environ. Sci. Pollut. Res.* **2019**, *26*, 36117–36123. [CrossRef]
19. Luévano-Hipólito, E.; Martínez-de la Cruz, A.; Yu, Q.L.; Brouwers, H.J.H. Precipitation synthesis of WO^3 for NO_x removal using PEG as template. *Ceram. Int.* **2014**, *40*, 12123–12128. [CrossRef]
20. Yu, J.C.-C.; Lasek, J.; Nguyen, V.-H.; Yu, Y.-H.; Wu, J.C.S. Visualizing reaction pathway for the phototransformation of NO_2 and N_2 into NO over WO_3 photocatalyst. *Res. Chem. Intermed.* **2017**, *43*, 7159–7169. [CrossRef]
21. Kowalkińska, M.; Borzyszkowska, A.F.; Grzegórska, A.; Karczewski, J.; Głuchowski, P.; Łapiński, M.; Sawczak, M.; Zielińska-Jurek, A. Pilot-Scale Studies of WO_3/S-Doped g-C_3N_4 Heterojunction toward Photocatalytic NO_x Removal. *Materials* **2022**, *15*, 633. [CrossRef]
22. Mendoza, J.A.; Lee, D.H.; Kang, J.-H. Photocatalytic removal of gaseous nitrogen oxides using WO_3/TiO_2 particles under visible light irradiation: Effect of surface modification. *Chemosphere* **2017**, *182*, 539–546. [CrossRef]
23. Tahir, M.B.; Nabi, G.; Rafique, M.; Khalid, N.R. Nanostructured-based WO_3 photocatalysts: Recent development, activity enhancement, perspectives and applications for wastewater treatment. *Int. J. Environ. Sci. Technol.* **2017**, *14*, 2519–2542. [CrossRef]
24. Wang, J.; Chen, Z.; Zhai, G.; Men, Y. Boosting photocatalytic activity of WO_3 nanorods with tailored surface oxygen vacancies for selective alcohol oxidations. *Appl. Surf. Sci.* **2018**, *462*, 760–771. [CrossRef]
25. Li, L.; Xiao, S.; Li, R.; Cao, Y.; Chen, Y.; Li, Z.; Li, G.; Li, H. Nanotube array-like WO_3 photoanode with dual-layer oxygen-evolution cocatalysts for photoelectrocatalytic overall water splitting. *ACS Appl. Energy Mater.* **2018**, *1*, 6871–6880. [CrossRef]
26. Aslam, M.; Ismail, I.M.; Chandrasekaran, S.; Hameed, A. Morphology controlled bulk synthesis of disc-shaped WO_3 powder and evaluation of its photocatalytic activity for the degradation of phenols. *J. Hazard. Mater.* **2014**, *276*, 120–128. [CrossRef]
27. Chu, W.; Rao, Y.F. Photocatalytic oxidation of monuron in the suspension of WO_3 under UV–visible light. *Chemosphere* **2012**, *86*, 1079–1086. [CrossRef]
28. Kim, D.-S.; Yang, J.-H.; Balaji, S.; Cho, H.-J.; Kim, M.-K.; Kang, D.-U.; Djaoued, Y.; Kwon, Y.-U. Hydrothermal synthesis of anatase nanocrystals with lattice and surface doping tungsten species. *CrystEngComm* **2009**, *11*, 1621. [CrossRef]
29. Riboni, F.; Bettini, L.G.; Bahnemann, D.W.; Selli, E. WO_3–TiO_2 vs. TiO_2 photocatalysts: Effect of the W precursor and amount on the photocatalytic activity of mixed oxides. *Catal. Today* **2013**, *209*, 28–34. [CrossRef]
30. Dozzi, M.V.; Marzorati, S.; Longhi, M.; Coduri, M.; Artiglia, L.; Selli, E. Photocatalytic activity of TiO_2-WO_3 mixed oxides in relation to electron transfer efficiency. *Appl. Catal. B Environ.* **2016**, *186*, 157–165. [CrossRef]
31. Paula, L.F.; Hofer, M.; Lacerda, V.P.B.; Bahnemann, D.W.; Patrocinio, A.O.T. Unraveling the photocatalytic properties of TiO_2/WO_3 mixed oxides. *Photochem. Photobiol. Sci.* **2019**, *18*, 2469–2483. [CrossRef]
32. Yang, L.; Si, Z.; Weng, D.; Yao, Y. Synthesis, characterization and photocatalytic activity of porousWO_3/TiO_2 hollow microspheres. *Appl. Surf. Sci.* **2014**, *313*, 470–478. [CrossRef]
33. Žerjav, G.; Arshad, M.S.; Djinović, P.; Zavašnik, J.; Pintar, A. Electron trapping energy states of TiO_2–WO_3 composites and their influence on photocatalytic degradation of bisphenol. *Appl. Catal. B Environ.* **2017**, *209*, 273–284. [CrossRef]
34. Shifu, C.; Lei, C.; Shen, G.; Gengyu, C. The preparation of coupled WO_3/TiO_2 photocatalyst by ball milling. *Powder Technol.* **2005**, *160*, 198–202. [CrossRef]
35. Makwana, N.M.; Hazael, R.; McMillan, P.F.; Darr, J.A. Photocatalytic water disinfection by simple and low-cost monolithic and heterojunction ceramic wafers. *Photochem. Photobiol. Sci.* **2015**, *14*, 1190–1196. [CrossRef] [PubMed]
36. Dhanalekshmi, K.I.M.; Umapathy, J.; Magesan, P.; Zhang, X. Biomaterial (Garlic and Chitosan)-Doped WO_3-TiO_2 Hybrid Nanocomposites: Their Solar Light Photocatalytic and Antibacterial Activities. *ACS Omega* **2020**, *5*, 31673–31683. [CrossRef] [PubMed]
37. Manmohan, L.; Praveen, S.; Chhotu, R. Calcination temperature effect on titanium oxide (TiO_2) nanoparticles synthesis. *Optik* **2021**, *241*, 166934. [CrossRef]
38. El-Yazeed, W.S.A.; Ahmed, A.I. Photocatalytic activity of mesoporous WO_3/TiO_2 nanocomposites for the photodegradation of methylene blue. *Inorg. Chem. Commun.* **2019**, *105*, 102–111. [CrossRef]
39. Boruah, P.J.; Khanikar, R.R.; Bailung, H. Synthesis and Characterization of Oxygen Vacancy Induced Narrow Bandgap Tungsten Oxide (WO_{3-x}) Nanoparticles by Plasma Discharge in Liquid and Its Photocatalytic Activity. *Plasma Chem. Plasma Process* **2020**, *40*, 1019–1036. [CrossRef]
40. Kathiravan, A.; Renganathan, R. Photosensitization of colloidal TiO_2 nanoparticles with phycocyanin pigment. *J. Colloid Interface Sci.* **2009**, *335*, 196–202. [CrossRef]
41. Gutiérrez-Alejandre, A.; Castillo, P.; Ramírez, J.; Ramis, G.; Busca, G. Redox and acid reactivity of wolframyl centers on oxide carriers: Brønsted, Lewis and redox sites. *Appl. Catal. A* **2001**, *216*, 181–194. [CrossRef]
42. Lv, K.; Li, J.; Qing, X.; Li, W.; Chen, Q. Synthesis and photodegradation application of WO_3/TiO_2 hollow spheres. *J. Hazard. Mater.* **2011**, *189*, 329–335. [CrossRef]

43. Nguyen, T.T.; Nam, S.N.; Son, J.; Oh, J. Tungsten Trioxide (WO$_3$)-assisted Photocatalytic Degradation of Amoxicillin by Simulated Solar Irradiation. *Sci. Rep.* **2019**, *9*, 9349. [CrossRef]
44. Dirany, N.; Arab, M.; Madigou, V.; Leroux, C.; Gavarri, J.R. A facile one step route to synthesize WO$_3$ nanoplatelets for CO oxidation and photodegradation of RhB: Microstructural, optical and electrical studies. *RSC Adv.* **2016**, *6*, 6961–69626. [CrossRef]
45. Lasek, J.; Yu, Y.-H.; Wu, Y.C.S. Removal of NO$_x$ by photocatalytic processes. *J. Photochem. Photobiol. C Photochem. Rev.* **2013**, *14*, 29–52. [CrossRef]
46. Dalton, J.S.; Janes, P.A.; Jones, N.G.; Nicholson, J.A.; Hallam, K.R.; Alle, G.C. Photocatalytic oxidation of NO$_x$ gases using TiO$_2$: A surface spectroscopic approach. *Environ. Pollut.* **2002**, *120*, 415–422. [CrossRef]
47. Wu, J.C.S.; Cheng, Y.-T. In situ FTIR study of photocatalytic NO reaction on photocatalysts under UV irradiation. *J. Catal.* **2006**, *237*, 393–404. [CrossRef]
48. Balayeva, N.O.; Fleisch, M.; Bahnemann, D.W. Surface-grafted WO$_3$/TiO$_2$ photocatalysts: Enhanced visible-light activity towards indoor air purification. *Catal. Today* **2018**, *313*, 63–71. [CrossRef]
49. Rampaul, A.; Parkin, I.P.; O'Neill, S.A.; DeSouza, J.; Mills, A.; Elliott, N. Titania and tungsten doped titania thin films on glass; active photocatalysts. *Polyhedron* **2003**, *22*, 35–44. [CrossRef]
50. Wang, W.; Zeng, Z.; Zeng, G.; Zhang, C.; Xiao, R.; Zhou, C.; Xiong, W.; Yang, Y.; Lei, L.; Liu, Y.; et al. Sulfur doped carbon quantum dots loaded hollow tubular g-C$_3$N$_4$ as novel photocatalyst for destruction of *Escherichia coli* and tetracycline degradation under visible light. *Chem. Eng. J.* **2019**, *378*, 122132. [CrossRef]
51. Yang, Y.A.; Ma, Y.; Yao, J.N.; Loo, B.H. Simulation of the sublimation process in the preparation of photochromic WO$_3$ film by laser microprobe mass spectrometry. *J. Non-Cryst. Solids* **2000**, *272*, 71–74. [CrossRef]
52. Blattner, F.R.; Plunkett, G.; Bloch, C.A.; Perna, N.T.; Burland, V.; Riley, M.; Collado-Vides, J.; Glasner, J.D.; Rode, C.K.; Shao, Y. The complete genome sequence of *Escherichia coli* K-12. *Science* **1997**, *277*, 1453–1462. [CrossRef] [PubMed]

Article

Strong Pyro-Electro-Chemical Coupling of Elbaite/H$_2$O$_2$ System for Pyrocatalysis Dye Wastewater

Fei Chen [1], Jiesen Guo [1], Dezhong Meng [1,2,*], Yuetong Wu [1], Ruijin Sun [1] and Changchun Zhao [1,*]

[1] School of Science, China University of Geosciences (Beijing), Beijing 100083, China; cf15612189105@163.com (F.C.); Jason950616@163.com (J.G.); wuyuetong2020@outlook.com (Y.W.); sunruijin@cugb.edu.cn (R.S.)

[2] Zhengzhou Institute, China University of Geosciences (Beijing), Zhengzhou 451283, China

* Correspondence: meng@cugb.edu.cn (D.M.); 2011010020@cugb.edu.cn (C.Z.); Tel.: +86-10-8232-1062 (D.M.); +86-10-8232-3426 (C.Z.)

Citation: Chen, F.; Guo, J.; Meng, D.; Wu, Y.; Sun, R.; Zhao, C. Strong Pyro-Electro-Chemical Coupling of Elbaite/H$_2$O$_2$ System for Pyrocatalysis Dye Wastewater. *Catalysts* **2021**, *11*, 1370. https://doi.org/10.3390/catal11111370

Academic Editor: Li Meng

Received: 28 October 2021
Accepted: 11 November 2021
Published: 13 November 2021

Publisher's Note: MDPI stays neutral with regard to jurisdictional claims in published maps and institutional affiliations.

Copyright: © 2021 by the authors. Licensee MDPI, Basel, Switzerland. This article is an open access article distributed under the terms and conditions of the Creative Commons Attribution (CC BY) license (https://creativecommons.org/licenses/by/4.0/).

Abstract: Elbaite is a natural silicate mineral with a spontaneous electric field. In the current study, it was selected as a pyroelectric catalyst to promote hydrogen peroxide (H$_2$O$_2$) for dye decomposition due to its pyro-electro-chemical coupling. The behaviors and efficiency of the elbaite/H$_2$O$_2$ system in rhodamine B (RhB) degradation were systematically investigated. The results indicate that the optimal effective degradability of RhB reaches 100.0% at 4.0 g/L elbaite, 7.0 mL/L H$_2$O$_2$, and pH = 2.0 in the elbaite/H$_2$O$_2$ system. The elbaite/H$_2$O$_2$ system exhibits high recyclability and stability after recycling three times, reaching 94.5% of the degradation rate. The mechanisms of RhB degradation clarified that the hydroxyl radical (·OH) is the main active specie involved in catalytic degradation in the elbaite/H$_2$O$_2$ system. Moreover, not only does elbaite act as a pyroelectric catalyst to activate H$_2$O$_2$ in order to generate the primary ·OH for subsequent advanced oxidation reactions, but it also has the role of a dye sorbent. The elbaite/H$_2$O$_2$ system shows excellent application potential for the degradation of RhB.

Keywords: tourmaline; elbaite/H$_2$O$_2$ system; pyroelectric catalyst; dye wastewater; RhB degradation

1. Introduction

The development of synthetic dyes has displayed notable progression since the late 19th century [1]. Dyes without proper treatment are the main source of water pollution, which contains large amounts of polyphenylene ring substituents, showing biotoxicity and chromaticity [2–4]. Moreover, polluted water is gradually developing a resistance to photolysis, antioxidants, and biology [5,6]. Hence, it is an urgent problem for scientists to improve the removal technology of these harmful dyes from wastewater [4]. In order to solve this problem, traditional and modern techniques are raised as common methods to treat dye wastewater. These technologies mainly include coagulation [7], air stripping [8], incineration [9,10], filtration through a membrane [11], adsorption on stimulated carbon [12], electrochemical oxidation [13], wet oxidation [14], biological oxidation [15], and chemical oxidation techniques [16,17]. Among these technologies, chemical oxidation techniques are the most widely used due to the degradation of dye molecules. As one kind of chemical oxidation technique, advanced oxidation process (AOPs) has drawn an increasing amount of attention in the degradation of dyes, specifically, photocatalytic degradation [18–23], Fenton oxidation [24], Fenton-like oxidation [25], and activation of persulfate/peroxymonosulfate (PS/PMS) [26,27]. The reactive species in AOPs are usually the hydroxyl radical (·OH) [28] and hydrated electron (eaq−) [29]. However, there might still be some problems with the application of these technologies. For instance, the possibility of releasing residual-free radical oxidants such as ·OH and producing toxic degradation products [28]. Therefore, it is necessary to develop an efficient and environmentally friendly catalytic treatment technology for dye wastewater. It is well accepted that pyroelectric materials can generate polarized charges under external thermal excitation. Furthermore, the polarized charge can promote the emergence of powerful oxidant active substances such as ·OH. As a result, heat energy and electrical energy can be used for the degradation of dye wastewater due to the pyro-electric-chemical coupling effect [30,31]. To date, pyroelectric material, $Ba_{0.7}Sr_{0.3}TiO_3$@Ag nanoparticles [32], $NaNbO_3$ nanoparticles [33], $BaTiO_3$ nanofibers [34], ZnO nanorods [35], and $Pb(Zr_{0.52}Ti_{0.48})O_3$ polarized ceramic [36] were found to be effective for the degradation of dyes. Tourmaline is one of the typical natural pyroelectric materials [37]. There are currently several studies on the degradation of dyes by tourmaline [38–42]. Schorl, as an iron-rich tourmaline, can be used as an effective catalyst to enhance the efficiency of reactive free radicals and has been shown to assist in the removal of many organic pollutants, such as fosfomycin and tetracycline in schorl/H_2O_2 systems [39,40]. The heterogeneous Fenton reaction formed by the schorl/H_2O_2 system mainly uses the presence of iron in schorl, but the iron content in schorl is generally not higher than 10%, which has certain limitations in practical application [41,42]. Furthermore, it is reported that tourmaline can be regarded as an excellent adsorbent candidate [38]. The adsorption of tourmaline results from the diffusion of dye molecules into the pores of tourmaline. Although drawbacks remain, it is considered that not only can tourmaline be regarded as a candidate for physical adsorption, but it also has the potential as a catalyst for advanced oxidation technologies.

Elbaite is a typical representative of the tourmaline family. Meanwhile, it is also one of the most promising pyroelectric materials, and its high pyroelectric coefficient can be up to 0.46 $nC/cm^2/K$ [43]. In this study, a new method was proposed in which elbaite was selected to catalyze H_2O_2 for the decomposition of rhodamine B (RhB). The pyro-electro-chemical coupling effect is found in the elbaite/H_2O_2 system, and the decomposition mechanism is well understood. The elbaite/H_2O_2 system has proved to be a potential candidate to deal with dyes.

2. Results and Discussion

2.1. Characterization of Elbaite

The XRD patterns of elbaite are shown in Figure 1a. In these patterns, all reflection peaks can be indexed to previous studies (JCPDS card number PDF#71–716, space group: R3m). No characteristic peaks of other impurities are detected in the pattern. Figure 1b demonstrates the morphology of elbaite with micron-sized particles, generally less than 10 μm. In addition, the EDX spectrums in Figure 1c,d demonstrate the existence of O, Si, Al, Na, F, Ca, and C elements in the sample. Here, C comes from the sample carrier during the SEM sample preparation process. Furthermore, based on the analysis of chemical composition in Table 1, the elbaite applied to this study mainly contains the elements O, Si, Al, S, Mn, Ca, K, Lu, Li, Pb, and Fe, which are virtually consistent with the chemical composition of Figure 1c,d. Since the content of Fe in elbaite is very rare, the effect of the Fenton reaction between elbaite and H_2O_2 can be excluded.

Table 1. Chemical composition analyses of elbaite.

Component	SiO_2	Al_2O_3	SO_3	MnO	CaO	K_2O	Lu_2O_3	Li	PbO	Fe	Total
wt%	51.80	40.50	3.16	1.83	1.68	0.31	0.25	0.23	0.10	0.07	99.93

Note: Li and Fe elements are detected by ICP-MS. Other elements are detected by XRF and expressed in oxide.

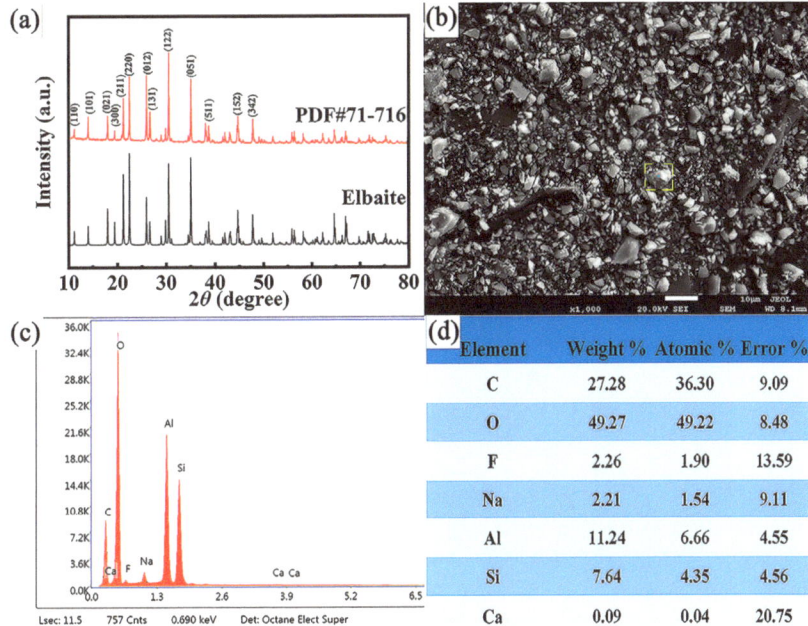

Figure 1. Characterization of elbaite: (**a**) the XRD pattern; (**b**) the SEM image; (**c**) the EDX analysis; and (**d**) the elements content.

2.2. Degradation Factors for Contaminant RhB

The elbaite's pyroelectrically driven activity is determined based on the decomposition of RhB dye. Figure 2a shows the change in concentration of RhB solution without any catalyst under acidic conditions of pH = 2.0. The results indicate that no change in RhB dye solution occurs during the temperature change. The RhB degradation performances of the elbaite-alone, H_2O_2-alone and elbaite/H_2O_2 systems were compared. Figure 2b indicates that 16.9% RhB is removed in the elbaite-alone system after 10 min of heating. As a consequence, it can be concluded that elbaite alone also affected the RhB dye solution. When the temperature increased, a polarized electric field was generated on the surface of the elbaite. The dye molecules in the solution are adsorbed by tourmaline elbaite, and this reduces the concentration of the dye solution. Figure 2c shows the N_2 adsorption–desorption isotherm of elbaite powder. According to the adsorption isotherm, the calculated values of specific surface area, pore volume, and pore width are 2.48 m^2/g, 2.2 × 10^{-2} cm^3/g, and 3.847 nm, respectively. The N_2 adsorption–desorption isotherms indicate that elbaite has a good adsorption capacity compared to other adsorbents [38]. Moreover, there is an obvious red adsorbate on the surface of elbaite after the reaction shown in the inset of Figure 2c. It is reported that the major surface functional group ·OH in tourmaline is responsible for the adsorption of the dye molecule [38]. The increase in the concentration of the RhB solution at 20 and 30 min of heating may come from the slightly evaporated water as the reaction is carried out. Compared with the elbaite alone system, the degradation rate of RhB is 24.8% in the H_2O_2 only system, indicating that RhB can directly react with H_2O_2 without the catalyst. However, the degradation rate of RhB in the elbaite/H_2O_2 system is much higher than in the other two systems. The RhB degradation percentage is 100% at 30 min of heating, with a clear and transparent solution. To investigate the best conditions of the reaction, the impact of different elbaite and H_2O_2 doses on the reaction was investigated.

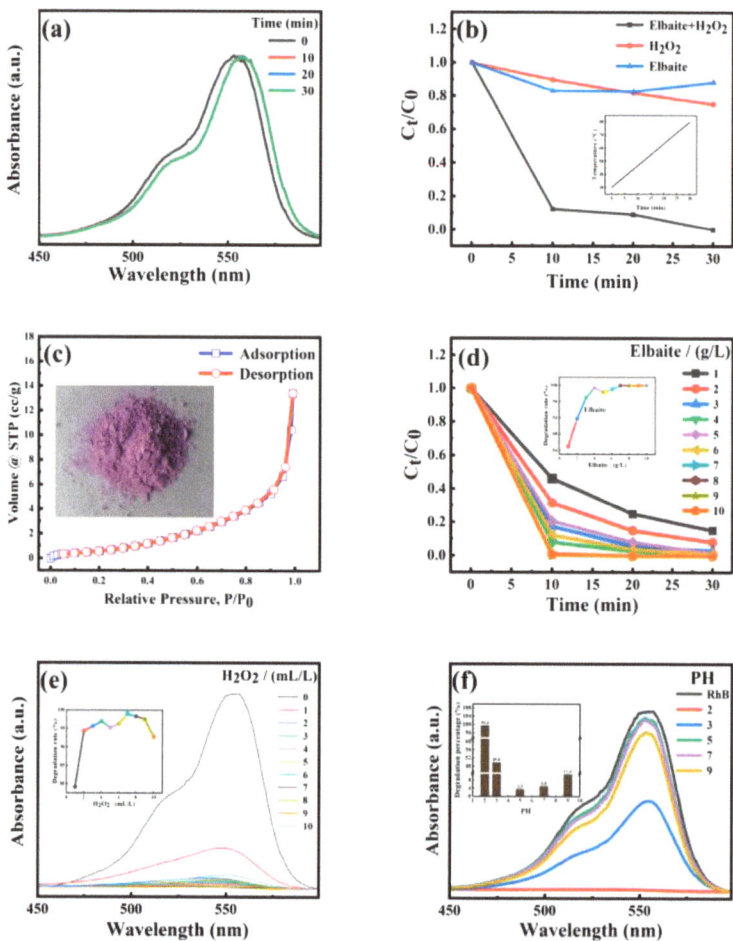

Figure 2. (**a**) Pyrocatalytic UV absorption spectrum of RhB without catalyst. (**b**) RhB removal rates in three different systems. (**c**) N2 adsorption–desorption isotherm of elbaite powder and the picture of elbaite powder after adsorption reaction. (**d**) Effect of elbaite dose on RhB degradation in the elbaite/H_2O_2 system, and the degradation rate of different elbaite doses. (**e**) Effect of H_2O_2 dose on RhB degradation in the elbaite/H_2O_2 system, and the degradation rate of different H_2O_2 doses. (**f**) Effect of initial pH on the degradation of RhB in the elbaite/H_2O_2 system, and the degradation rate at different pH. Experimental conditions: RhB = 100 mg/L, elbaite dose = 1.0–10.0 g/L, H_2O_2 dose = 1.0–10.0 mL/L, initial pH = 2.0, initial pH = 2.0, 3.0, 5.0, 7.0, 9.0, T = 30 °C–80 °C, time = 30 min.

As shown in Figure 2d, the dose of H_2O_2 is 7.0 mL/L, and the degradation efficiency of RhB increases significantly when the elbaite dose changes from 1.0 g/L to 4.0 g/L, reaching 99.4% at 4.0 g/L. With a further increase in the dose of elbaite, the degradation efficiency sinks slightly and then increases gradually to 100% at 7.0 g/L. This is due to the excessive addition of elbaite. On the one hand, the collision frequency between the elbaite particles may enhance the contact between positive and negative charges on the surface of the elbaite. On the other hand, the excessive addition of elbaite leads to a decrease in the dose of charge that can produce the active species. That is, excess elbaite does not lead to higher degradation efficiency. However, as the dose of elbaite continues to increase, the amount of charge generated increases, eliminating these effects. Therefore, the optimum dose of elbaite was found to be 4.0 g/L. Based on the above results, the dose of elbaite was designated at 4.0 g/L for further study.

The effect of H_2O_2 dose was examined in the elbaite/H_2O_2 system (given in Figure 2e). Firstly, the dose of elbaite was designated as 4.0 g/L. Then, with the increase in H_2O_2 dose from 1.0 mL/L to 7.0 mL/L, the degradation rate increased from 79.1% to 99.1%. After that, when the H_2O_2 dose was increased to 10.0 mL/L, the degradation rate decreased slightly, reaching 92.9%. The results also proved that the degradation rate changed significantly with the dose of H_2O_2. Hence, ·OH is the predicted active substance in the solution, and the direction of change can be explained as follows. With the increase in H_2O_2 dose, H_2O_2 will constantly generate ·OH to attack the RhB structure. However, H_2O_2 is also a scavenger of ·OH. Excess H_2O_2 consumes ·OH before these can react with organic pollutants, which can be explained by reaction Equations (1) and (2) [44,45]. As a result, the optimal efficiency for H_2O_2 is 7.0 mL/L. Thus, 7.0 mL/L H_2O_2 was chosen for further study.

$$H_2O_2 + \cdot OH \rightarrow H_2O + \cdot HO_2 \quad (1)$$

$$HO_2 + \cdot OH \rightarrow H_2O + O_2 \quad (2)$$

In order to study the influence of pH on RhB degradation performance, the influence of the initial pH of the RhB solution on the degradation rate under acidic, neutral, and alkaline conditions was studied in the elbaite/H_2O_2 system, and the results are illustrated in Figure 2f. It can be seen that the highest degradation rate is up to 99.4% at pH = 2.0, while the degradation rate is detected to be only 3.5% at pH = 5.0, 5.0% at pH = 7.0, and 11.4% at pH = 9.0. Clearly, the elbaite/H_2O_2 system shows an excellent degradation rate under a strong acid environment (pH = 2.0) compared to the neutral and alkaline conditions. It is speculated that the pyroelectricity of elbaite in acidic conditions is stronger than that in neutral and alkaline conditions, which leads to higher efficiency. Furthermore, it was found that the amount of hydroxyl decreased first and then increased with the increase in pH. The result is shown in Figure S1. Therefore, the elbaite/H_2O_2 system is pH dependent for the degradation of RhB solution and shows high removal efficiency in low pH conditions.

2.3. Detection of Hydroxyl Radicals

To further determine into the mechanism of the elbaite/H_2O_2 system, the surface radical-trapping experiments proceeded. As shown in Figure 3a, the addition of TBA (a quencher of ·OH) [46] significantly inhibited the decomposition of RhB in the elbaite/H_2O_2 system, and the inhibition degree increased with an increase in TBA content. Without the addition of the scavenger, the degradation rate of RhB dye was 89.1% after 10 min of reaction, where the temperature was 46.7 °C. However, the degradation rate was only 13% after the addition of 50 mM TBA. Therefore, it is presumed that ·OH is the dominant radical in the elbaite/H_2O_2 system, because the scavenger TBA can greatly inhibit the degradation of RhB dye.

To monitor the intermediate products of the pyrocatalysis process, we further demonstrated the generation of ·OH radicals in the elbaite/H_2O_2 system by using TA as a probe molecule. TA reacts with the ·OH to produce 2-hydroxyterephthalic acid (·OH capture), a substance with strong fluorescent properties. It emits a unique fluorescence signal under a PL excitation wavelength of 315 nm with a peak around 425 nm [47]. In turn, the intensity of the fluorescence signal of 2-hydroxyterephthalic acid is proportional to the number of ·OH produced in water [48]. As shown in Figure 3b, with the increase in pyrocatalysis time, the fluorescence intensity of 2-Hydroxyterephthalic acid at 425 nm increased, indicating the presence of ·OH.

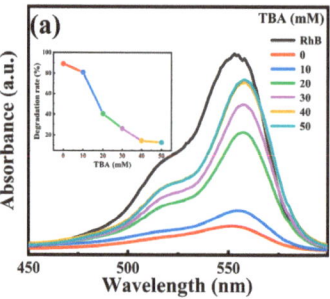 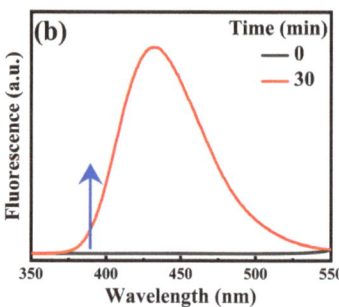

Figure 3. (a) The effect of different doses of TBA on the degradation of RhB solution by the elbaite/H_2O_2 system. (b) The ·OH capture spectra of suspension containing elbaite and TA were studied at a 315 nm photoluminescence excitation wavelength. Experimental conditions: RhB = 100 mg/L, elbaite dose = 4.0 g/L, H_2O_2 dose = 7.0 mL/L, initial pH = 2.0, T = 30 °C–80 °C, time = 10 min, 30 min.

2.4. Degradation Factors for Contaminant RhB

Reusability and stability are significant and necessary for the practical application of the elbaite/H_2O_2 system. Thus, the recycling utilization of elbaite catalysts was investigated. As shown in Figure 4a, the degradation rate reduced from 100% to 94.5% as the number of cycles increased to three. After three repeated experiments, the degradation rate decreased slightly due to the small loss of elbaite powder during centrifugal filtration. The degradation rate of the third cycle was more than 90% of that of the first cycle, indicating that elbaite has good recycling performance. The FTIR spectra of bare elbaite particles and the elbaite particles after reaction with RhB are shown in Figure 4b. The structure of elbaite in the elbaite/H_2O_2 system is essentially the same before and after the reaction. The bands at 781 cm^{-1} and 1100 cm^{-1} are caused by the stretching vibration of the Si-O-Si bond. The O-Si-O bond and the B-O bond cause stretching vibration at 969 cm^{-1} and 1290 cm^{-1}, respectively. Furthermore, the bands at 3590 cm^{-1} and 1356 cm^{-1} are the results of the stretching vibration of the O-H bond [49]. It is shown that the reaction process of the elbaite/H_2O_2 system does not cause any change to elbaite, further demonstrating its structural stability. Thus, the excellent recycling performance of elbaite is proved once again.

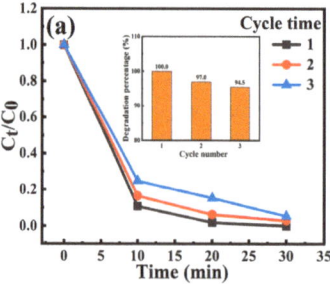 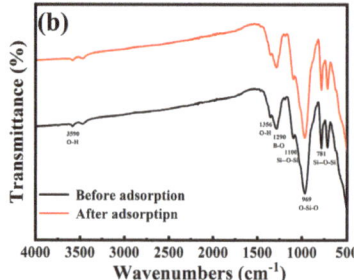

Figure 4. (a) Recycling utilization tests with elbaite. (b) FTIR spectra of powdered elbaite before reaction and after reaction of the elbaite/H_2O_2 system. Experimental conditions: RhB = 100 mg/L, elbaite dose = 4.0 g/L, H_2O_2 dose = 7.0 mL/L, initial pH = 2.0, T = 30 °C–80 °C, time = 30 min.

2.5. Degradation Factors for Contaminant RhB

The pyroelectricity of elbaite played a major role in the elbaite/H_2O_2 system. The intensity of polarization formed in pyroelectric materials changes as the temperature (T) changes and, therefore, is defined as the pyroelectric coefficient [50], and the reaction equation is expressed as Equation (3) as follows:

$$P = \frac{\partial P_s}{\partial T} \tag{3}$$

where ∂P_S is the vector of polarization intensity change in the crystal, and ∂T is the small temperature change that occurs uniformly throughout the crystal. The P value represents the strength of the pyro-driven catalytic performance, and the higher the P value, the better the pyro-driven catalytic performance. The polarization intensity (P_s) and pyroelectric coefficient (P) of the selected elbaite at different temperatures were measured by a ferroelectric analyzer. As shown in Figure 5a, the polarization intensity of elbaite changed from 1.30 nC/cm^2 at 30 °C to 7.79 nC/cm^2 at 80 °C. Furthermore, the pyroelectric coefficient of elbaite can reach around 0.20 nC/cm^2/K at 80 °C, according to Figure 5b. The results indicate that the selected elbaite has a relatively high pyroelectric coefficient.

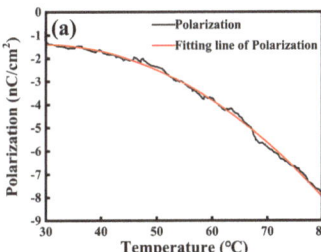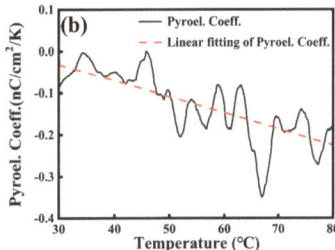

Figure 5. (**a**) Polarization (Ps) and (**b**) pyroelectric coefficient (P) of elbaite.

Figure 6 shows the schematic diagram of pyrocatalysis in the elbaite/H$_2$O$_2$ system. Elbaite contain a high pyroelectric coefficient P of 0.2 nC/cm^2/K. In the case of a thermodynamic equilibrium state ($\Delta T = 0$ °C), the surface of elbaite does not have a polarized charge. When the elbaite particles are heated, the temperature of elbaite particles changes, the total electric dipole moment vector sum decreases, and the polarization density decreases, causing a large amount of polarized charge to accumulate on the surface of elbaite particles, which can be expressed as Equation (4). At the same time, the negative charges react with H$_2$O$_2$ to produce ·OH- and ·OH, as shown in Equation (5). Not only that, but the positive charge (h$^+$) accumulated on the positive charge side of elbaite also reacts with ·OH$^-$ to produce ·OH, as shown in Equation (6). Furthermore, ·OH in the solution causes the dye to decompose through redox reaction so as to achieve the purpose of degradation, as shown in Equation (7) [51,52]. As a result, the pyro-electro-chemical coupling of the elbaite/H$_2$O$_2$ system contributes to the decomposition of RhB.

$$\text{Elbaite} \xrightarrow{\Delta T} \text{Elbaite} + h^+ + q^- \tag{4}$$

$$H_2O_2 + q^- \rightarrow OH^- + \cdot OH \tag{5}$$

$$h^+ + OH^- \rightarrow \cdot OH \tag{6}$$

$$OH + \text{Dye} \rightarrow \text{Dye decomposition} \tag{7}$$

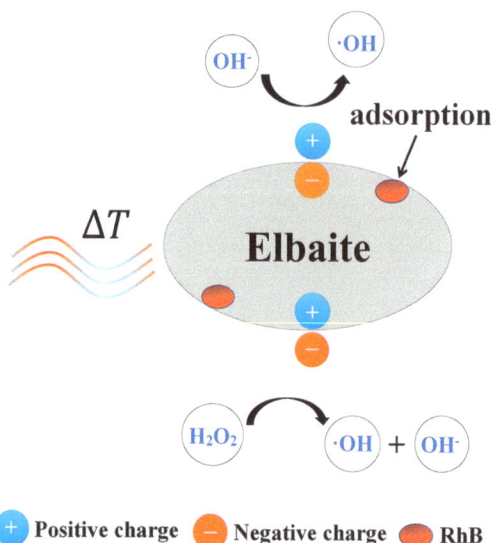

Figure 6. Devised mechanism for RhB degradation in the elbaite/H_2O_2 system.

3. Materials and Methods

RhB, TA, and tert-butyl alcohol (TBA) were purchased from Aladdin. Deionized water was used for the preparation of RhB solution. The sodium hydroxide (NaOH) and sulfuric acid (H_2SO_4) that were used in the experiment were purchased from Sinopharm Chemical Reagent Co. H_2O_2 (30%, v/v) was obtained from Nanjing Chemical Reagent Co., Ltd. Elbaite was purchased from Brazil with high purity, and it was ground into micron-sized particles using an intelligent high-throughput tissue grinder. Then, it was washed five times with deionized water, dried at about 80 °C, and stored in a dry container before use. The reagents used in this study are all analytical grade reagents and can be used directly without further purification.

The phases of elbaite were characterized by PANalytical X'pert PRO powder X-ray diffractometer (Cu Kα radiation, λ = 1.5406 Å) with 2θ ranging from 10° to 80° at step width of 0.01°. The morphology of the elbaite powder sample was characterized by using a scanning electron microscope (SEM, The Netherlands). The chemical composition of the elbaite samples was analyzed by energy-dispersive X-ray spectroscopy (EDS, Japan), iCAP Q inductively coupled plasma mass spectrometry (ICP-MS, USA), and EDX-7000 X-ray fluorescence spectrometry (XRF, Japan). The N_2 adsorption isotherm was measured at 77.3 K to determine the specific surface area (S_{BET}) via the Brunauer–Emmett–Teller (BET, autosorb iQ, USA) method, while the total pore volumes (Vtotal) and pore width were directly obtained from the isotherm at P/P_0 = 0.99. The pyroelectricity of the sample was measured using a Ferro-electric Analyzer (TF-3000E, aix-ACCT, Germany) at different temperature ranges, where the elbaite sample was cut into small slices (perpendicular to c-axis) and polished. Furthermore, the size of each slice was approximately 6 mm × 6 mm × 1.0 mm. The upper and lower surfaces of each piece were coated with silver glue. A Fourier transform infrared spectroscopy (FTIR, spectrometer Nicolet Avatar 370DTGS, Thermo, USA) covering a wave number between 500 cm^{-1} and 4000 cm^{-1} was used to identify the functional groups on the bare elbaite particles and the elbaite particles after reaction with RhB.

The pyrocatalytic activities of the elbaite/H_2O_2 system were diagnosed through the degradation situation of RhB aqueous solution (100 mg/L). The degradation experiments were implemented in a thermostatic water bath. In the degradation experiments, elbaite was immersed in 100 mL of RhB aqueous solution and stirred for ∼1 h to establish an adsorption–desorption equilibrium between the elbaite and the RhB solution. Then, the H_2O_2 aqueous solution was added into the mixture solution, where the pH was adjusted to 2 by using H_2SO_4 and NaOH. The solution was heated from 30 °C to 80 °C at a constant rate in a water bath within 30 min to make full use of pyroelectric properties. Then, 5 mL of RhB aqueous solution was taken out and centrifuged at 11,000 rpm for 2 min in a TG20-WS table top high-speed centrifuge (China) to remove impurities. Accordingly, the concentration of RhB aqueous solution was measured using a UV–Vis spectrophotometer (T6, Beijing Puxi General Instrument Co. Ltd.). For comparison, adsorption capacity of elbaite was determined by an adsorption experiment. Elbaite was added to RhB aqueous solution and stirred for ~1 h without H_2O_2, and the experiment conditions were the same as for the elbaite/H_2O_2 system. In addition, H_2O_2 was added to RhB aqueous solution without elbaite to determine its effect on the solution, and the experiment conditions were the same as those used for the elbaite/H_2O_2 system. RhB concentration was detected during the reaction, and the change in RhB concentration can reflect the adsorption capacity of elbaite. Here, the degradation rate (DR) was calculated using the following equation:

$$DR\ (\%) = 1 - C_t/C_0 \times 100\% \tag{8}$$

where C_0 and C_t denote the premier and residual RhB concentration at t min, respectively, and DR is the final degradation rate. Subsequently, the influence of various factors on RhB degradation efficiency was investigated. Firstly, the optimum dose of elbaite (1.0–10.0 g/L) was researched during the reaction progress. Secondly, we explored the optimum dose of H_2O_2 (1.0–10.0 mL/L). Afterwards, the initial pH of RhB solution was designed to 2.0, 3.0, 5.0, 7.0, and 9.0 with 0.5 mol/L H_2SO_4 or NaOH aqueous solution to reveal the pH dependence of the elbaite/H_2O_2 system. Finally, the recyclability of the elbaite/H_2O_2 system was tested. During the recycling experiment, the elbaite particles were collected by centrifugation and washed by deionized water and ethanol five times. Afterwards, the collected elbaite was dried to a constant weight in an oven at 80 °C. The experiment was repeated for the second and third cycles.

In order to verify the active substances produced in the pyrocatalysis process, the TBA trapping experiment was performed [46]. During the pyrocatalysis process, TBA was added into the optimal elbaite/H_2O_2 system with the content ranging from 10–50 mM. RhB aqueous solution concentration changes with time, thus confirming the role of ·OH in the pyrocatalysis process. In order to further determine the presence of ·OH, the photoluminescence (PL) technique of terephthalic acid (TA) as probe molecule was used to achieve more accurate detection of ·OH radical induced pyroelectrically [48]. TA reacts with ·OH to produce 2-hydroxyl terephthalic acid, which presents a highly fluorescent state with a peak center at 425 nm, and its PL intensity is positively correlated with the amount of ·OH produced in water [47,53]. Definitively, 2 mM NaOH and 0.5 mM TA were mixed into the 100mL aqueous solution, and then the solution was transferred to the RhB aqueous solution. Thereafter, elbaite was added to RhB aqueous solution and stirred for 1 h to establish the adsorption–desorption equilibrium between elbaite and solution. The result is shown in Figure S2. Finally, 0.7 mL H_2O_2 was added, and the solution was transferred to the same experimental procedure. At the excitation wavelength of 315 nm, the PL intensity of the solution was measured at different reaction times.

4. Conclusions

An elbaite/H_2O_2 pyro-electro-chemical coupling catalytic system was developed and successfully applied to the degradation of RhB aqueous solution. Extensive degradation experiments showed that the optimized degradation efficiency of RhB reached 100.0% at 4.0 g/L elbaite, 7.0 mL/L H_2O_2, and pH = 2.0 in the elbaite/H_2O_2 system. Recycling utilization showed that the third removal rate was 94.5%, which demonstrates the economic usefulness of the silicate mineral elbaite. Furthermore, it is shown that ·OH was the main active specie involved in catalytic degradation in the elbaite/H_2O_2 pyro-electro-chemical coupling catalytic system. Elbaite plays the role of a dye sorbent and a pyroelectric catalyst to activate H_2O_2 in order to generate the primary ·OH for subsequent advanced redox reactions. Therefore, the effective application of the elbaite/H_2O_2 pyroelectric catalytic system has a high potential for the purification of dye wastewater.

Supplementary Materials: The following are available online at https://www.mdpi.com/article/10.3390/catal11111370/s1, Figure S1. The ·OH capture (TA) spectra of suspension at different pH. Figure S2. Pyrocatalytic UV absorption spectrum of RhB after adsorption-desorption.

Author Contributions: D.M. and C.Z. conceived and designed the experiments; F.C. and J.G. performed the pyro-electro-chemical catalytic experiments and measured the strong oxidation actives; F.C. and J.G. described the mechanism of the elbaite/H_2O_2 system; R.S. and Y.W. conducted the XRD, SEM, ICP-MS, XRF, and BET experiments and performed recycling; F.C. wrote the original draft. All authors have read and agreed to the published version of the manuscript.

Funding: This research was funded by the Natural Science Foundation of Beijing Municipality (3214052), Fundamental Research Funds for the Central Universities (2652019109), and National Natural Science Foundation of China (51472224).

Data Availability Statement: The data presented in this study are available on request from the corresponding authors.

Conflicts of Interest: The authors declare no conflict of interest.

References

1. Ahmad, A.; Mohd-Setapar, S.H.; Chuong, C.S.; Khatoon, A.; Wani, W.A.; Kumar, R.; Rafatullah, M. Recent advances in new generation dye removal technologies: Novel search for approaches to reprocess wastewater. *RSC Adv.* **2015**, *5*, 30801–30818. [CrossRef]
2. Lin, H.; Wu, Z.; Jia, Y.; Li, W.; Zheng, R.K.; Luo, H. Piezoelectrically induced mechano-catalytic effect for degradation of dye wastewater through vibrating Pb ($Zr_{0.52}Ti_{0.48}$) O_3 fibers. *Appl. Phys. Lett.* **2014**, *104*, 162907. [CrossRef]
3. Krawczyk, K.; Wacławek, S.; Kudlek, E.; Silvestri, D.; Kukulski, T.; Grübel, K.; Černík, M.; Padil, V.V.T. UV-catalyzed persulfate oxidation of an anthraquinone based dye. *Catalysts* **2020**, *10*, 456. [CrossRef]
4. Forgacs, E.; Cserhati, T.; Oros, G. Removal of synthetic dyes from wastewaters: A review. *Environ. Int.* **2004**, *30*, 953–971. [CrossRef]
5. Xu, Y.; Li, X.; Cheng, X.; Sun, D.; Wang, X. Degradation of cationic red GTL by catalytic wet air oxidation over Mo–Zn–Al–O catalyst under room temperature and atmospheric pressure. *Environ. Sci. Technol.* **2012**, *46*, 2856–2863. [CrossRef] [PubMed]
6. Mat Yasin, N.M.F.; Hossain, M.S.; HPS, A.K.; Zulkifli, M.; Al-Gheethi, A.; Asis, A.J.; Yahaya, A.N. Treatment of palm oil refinery effluent using tannin as a polymeric coagulant: Isotherm, kinetics, and thermodynamics analyses. *Polymers* **2020**, *12*, 2353. [CrossRef] [PubMed]
7. Anjaneyulu, Y.; Chary, N.S.; Raj, D.S.S. Decolourization of industrial efflfluents–Available methods and emerging technologies—A review. *Rev. Environ. Sci. Bio/Technol.* **2005**, *4*, 245–273. [CrossRef]
8. Sutherland, J.; Adams, C.; Kekobad, J. Treatment of MTBE by air stripping, carbon adsorption, and advanced oxidation: Technical and economic comparison for five groundwaters. *Water Res.* **2004**, *38*, 193–205. [CrossRef]
9. Chiang, K.Y.; Wang, K.S.; Lin, F.L.; Chu, W.T. Chloride effects on the speciation and partitioning of heavy metal during the municipal solid waste incineration process. *Sci. Total Environ.* **1997**, *203*, 129–140. [CrossRef]
10. Cimini, S.; Prisciandaro, M.; Barba, D. Simulation of a waste incineration process with flue-gas cleaning and heat recovery sections using Aspen Plus. *Waste Manag.* **2005**, *25*, 171–175. [CrossRef]
11. Zagklis, D.P.; Vavouraki, A.I.; Kornaros, M.E.; Paraskeva, C.A. Purification of olive mill wastewater phenols through membrane filtration and resin adsorption/desorption. *J. Hazard. Mater.* **2015**, *285*, 69–76. [CrossRef]
12. Boucher, F.R.; Lee, G.F. Adsorption of lindane and dieldrin pesticides on unconsolidated aquifer sands. *Environ. Sci. Technol.* **1972**, *6*, 538–543. [CrossRef]
13. Martinez-Huitle, C.A.; Ferro, S. Electrochemical oxidation of organic pollutants for the wastewater treatment: Direct and indirect processes. *Chem. Soc. Rev.* **2006**, *35*, 1324–1340. [CrossRef]
14. Lin, S.H.; Chuang, T.S. Combined treatment of phenolic wastewater by wet air oxidation and activated sludge. *Toxicol. Environ. Chem.* **1994**, *44*, 243–258. [CrossRef]
15. Scott, J.P.; Ollis, D.F. Integration of chemical and biological oxidation processes for water treatment: Review and recommendations. *Environ. Prog.* **1995**, *14*, 88–103. [CrossRef]
16. Wang, Y.-T. Effect of chemical oxidation on anaerobic biodegradation of model phenolic compounds. *Water Environ. Res.* **1992**, *64*, 268–273. [CrossRef]
17. Munter, R. Advanced oxidation processes–current status and prospects. *Proc. Est. Acad. Sci. Chem.* **2001**, *50*, 59–80. [CrossRef]
18. Augugliaro, V.; Bellardita, M.; Loddo, V.; Palmisano, G.; Palmisano, L.; Yurdakal, S. Overview on oxidation mechanisms of organic compounds by TiO_2 in heterogeneous photocatalysis. *J. Photoch. Photobio. C Photochem. Rev.* **2012**, *13*, 224–245. [CrossRef]

19. Yurdakal, S.; Palmisano, G.; Loddo, V.; Augugliaro, V.; Palmisano, L. Nanostructured rutile TiO_2 for selective photocatalytic oxidation of aromatic alcohols to aldehydes in water. *J. Am. Chem. Soc.* **2008**, *130*, 1568–1569. [CrossRef]
20. Augugliaro, V.; Prevot, A.B.; Vázquez, J.C.; Garcıa-López, E.; Irico, A.; Loddo, V.; Malato Rodrıguez, S.; Marcı, G.; Palmisano, L.; Pramauro, E. Photocatalytic oxidation of acetonitrile in aqueous suspension of titanium dioxide irradiated by sunlight. *Adv. Environ. Res.* **2004**, *8*, 329–335. [CrossRef]
21. Marin, M.L.; Santos-Juanes, L.; Arques, A.; Amat, A.M.; Miranda, M.A. Organic photocatalysts for the oxidation of pollutants and model compounds. *Chem. Rev.* **2012**, *112*, 1710–1750. [CrossRef]
22. Martinez-Haya, R.; Luna, M.M.; Hijarro, A.; Martinez-Valero, E.; Miranda, M.A.; Marin, M.L. Photocatalytic degradation of phenolic pollutants using N-methylquinolinium and 9-mesityl-10-methylacridinium salts. *Catal. Today* **2019**, *328*, 243–251. [CrossRef]
23. Martinez-Haya, R.; Gomis, J.; Arques, A.; Marin, M.L.; Amat, A.M.; Miranda, M.A. Time-resolved kinetic assessment of the role of singlet and triplet excited states in the photocatalytic treatment of pollutants at different concentrations. *Appl. Catal. B-Environ.* **2017**, *203*, 381–388. [CrossRef]
24. Zhu, Y.; Zhu, R.; Xi, Y.; Zhu, J.; Zhu, G.; He, H. Strategies for enhancing the heterogeneous Fenton catalytic reactivity: A review. *Appl. Catal. B-Environ.* **2019**, *255*, 117739. [CrossRef]
25. Wang, A.; Wang, H.; Deng, H.; Wang, S.; Shi, W.; Yi, Z.; Yan, K. Controllable synthesis of mesoporous manganese oxide microsphere efficient for photo-Fenton-like removal of fluoroquinolone antibiotics. *Appl. Catal. B-Environ.* **2019**, *248*, 298–308. [CrossRef]
26. Wang, J.; Wang, S. Activation of persulfate (PS) and peroxymonosulfate (PMS) and application for the degradation of emerging contaminants. *Chem. Eng. J.* **2018**, *334*, 1502–1517. [CrossRef]
27. Jiang, L.; Zhang, Y.; Zhou, M.; Liang, L.; Li, K. Oxidation of Rhodamine B by persulfate activated with porous carbon aerogel through a non-radical mechanism. *J. Hazard. Mater.* **2018**, *358*, 53–61. [CrossRef] [PubMed]
28. Ribeiro, A.R.; Nunes, O.C.; Pereira, M.F.; Silva, A.M. An overview on the advanced oxidation processes applied for the treatment of water pollutants defined in the recently launched Directive 2013/39/EU. *Environ. Int.* **2015**, *75*, 33–51. [CrossRef]
29. Mezyk, S.P.; Neubauer, T.J.; Cooper, W.J.; Peller, J.R. Free-radical-induced oxidative and reductive degradation of sulfa drugs in water: Absolute kinetics and efficiencies of hydroxyl radical and hydrated electron reactions. *J. Phys. Chem. A* **2007**, *111*, 9019–9024. [CrossRef] [PubMed]
30. Sebald, G.; Guyomar, D.; Agbossou, A. On thermoelectric and pyroelectric energy harvesting. *Smart Mater. Struct.* **2009**, *18*, 125006. [CrossRef]
31. Lang, S.B.; Muensit, S. Review of some lesser-known applications of piezoelectric and pyroelectric polymers. *Appl. Phys. A* **2006**, *85*, 125–134. [CrossRef]
32. Xu, X.; Chen, S.; Wu, Z.; Jia, Y.; Xiao, L.; Liu, Y. Strong pyro-electro-chemical coupling of $Ba_{0.7}Sr_{0.3}TiO_3$@Ag pyroelectric nanoparticles for room-temperature pyrocatalysis. *Nano Energ.* **2018**, *50*, 581–588. [CrossRef]
33. You, H.; Wu, Z.; Wang, L.; Jia, Y.; Li, S.; Zou, J. Highly efficient pyrocatalysis of pyroelectric $NaNbO_3$ shape-controllable nanoparticles for room-temperature dye decomposition. *Chemosphere* **2018**, *199*, 531–537. [CrossRef] [PubMed]
34. Xia, Y.; Jia, Y.; Qian, W.; Xu, X.; Wu, Z.; Han, Z.; Hong, Y.; You, H.; Ismail, M.; Bai, G. Pyroelectrically induced pyro-electro-chemical catalytic activity of $BaTiO_3$ nanofibers under room-temperature cold–hot cycle excitations. *Metals* **2017**, *7*, 122. [CrossRef]
35. Qian, W.; Wu, Z.; Jia, Y.; Hong, Y.; Xu, X.; You, H.; Zheng, Y.; Xia, Y. Thermo-electro-chemical coupling for room temperature thermocatalysis in pyroelectric ZnO nanorods. *Electrochem. Commun.* **2017**, *81*, 124–127. [CrossRef]
36. Ma, J.; Chen, L.; Wu, Z.; Chen, J.; Jia, Y.; Hu, Y. Pyroelectric Pb $(Zr_{0.52}Ti_{0.48})$ O_3 polarized ceramic with strong pyro-driven catalysis for dye wastewater decomposition. *Ceram. Int.* **2019**, *45*, 11934–11938. [CrossRef]
37. Zhou, G.; Liu, H.; Chen, K.; Gai, X.; Zhao, C.; Liao, L.; Shan, Y. The origin of pyroelectricity in tourmaline at varying temperature. *J. Alloys Compd.* **2018**, *744*, 328–336. [CrossRef]
38. Liu, N.; Wang, H.; Weng, C.H.; Hwang, C.C. Adsorption characteristics of Direct Red 23 azo dye onto powdered tourmaline. *Arab. J. Chem.* **2018**, *11*, 1281–1291. [CrossRef]
39. Shi, J.; Yin, D.; Xu, Z.; Song, D.; Cao, F. Fosfomycin removal and phosphorus recovery in a schorl/H_2O_2 system. *RSC Adv.* **2016**, *6*, 68185–68192. [CrossRef]
40. Zhang, Y.; Shi, J.; Xu, Z.; Chen, Y.; Song, D. Degradation of tetracycline in a schorl/H_2O_2 system: Proposed mechanism and intermediates. *Chemosphere* **2018**, *202*, 661–668. [CrossRef] [PubMed]
41. Tokumura, M.; Znad, H.T.; Kawase, Y. Modeling of an external light irradiation slurry photoreactor: UV light or sunlight-photoassisted Fenton discoloration of azo-dye Orange II with natural mineral tourmaline powder. *Chem. Eng. Sci.* **2006**, *61*, 6361–6371. [CrossRef]
42. Li, J.; Wang, C.; Wang, D.; Zhou, Z.; Sun, H.; Zhai, S. A novel technology for remediation of PBDEs contaminated soils using tourmaline-catalyzed Fenton-like oxidation combined with P. chrysosporium. *Chem. Eng. J.* **2016**, *296*, 319–328. [CrossRef]
43. Zhang, X.H.; Wu, R.H. Mechanism and experimental Research on improvement in conductive performances of ZnO coated tourmaline powder. *Adv. Mat. Res.* **2013**, *750*, 2108–2112. [CrossRef]
44. Zhang, H.; Wu, J.; Wang, Z.; Zhang, D. Electrochemical oxidation of crystal violet in the presence of hydrogen peroxide. *J. Chem. Technol. Biotechnol.* **2010**, *85*, 1436–1444. [CrossRef]

45. Li, Y.; Lu, Y.; Zhu, X. Photo-Fenton discoloration of the azo dye X-3B over pillared bentonites containing iron. *J. Hazard. Mater.* **2006**, *132*, 196–201. [CrossRef]
46. Xu, T.; Cai, Y.; O'Shea, K.E. Adsorption and photocatalyzed oxidation of methylated arsenic species in TiO_2 suspensions. *Environ. Sci. Technol.* **2007**, *41*, 5471–5477. [CrossRef]
47. Yu, J.; Wang, W.; Cheng, B.; Su, B.L. Enhancement of photocatalytic activity of mesoporous TiO_2 powders by hydrothermal surface fluorination treatment. *J. Mater. Chem. C* **2009**, *113*, 6743–6750. [CrossRef]
48. Xu, X.; Wu, Z.; Xiao, L.; Jia, Y.; Ma, J.; Wang, F.; Huang, H. Strong piezo-electro-chemical effect of piezoelectric $BaTiO_3$ nanofibers for vibration-catalysis. *J. Alloys Compd.* **2018**, *762*, 915–921. [CrossRef]
49. Li, G.; Chen, D.; Zhao, W.; Zhang, X. Efficient adsorption behavior of phosphate on La-modified tourmaline. *J. Environ. Chem. Eng.* **2015**, *3*, 515–522. [CrossRef]
50. Damjanovic, D. Ferroelectric, dielectric and piezoelectric properties of ferroelectric thin films and ceramics. *Rep. Prog. Phys.* **1998**, *61*, 1267. [CrossRef]
51. Zhan, H.; Tian, H. Photocatalytic degradation of acid azo dyes in aqueous TiO_2 suspension I, The effect of substituents. *Dye. Pigment.* **1998**, *37*, 231–239. [CrossRef]
52. Houas, A.; Lachheb, H.; Ksibi, M.; Elaloui, E.; Guillard, C.; Herrmann, J.M. Photocatalytic degradation pathway of methylene blue in water. *Appl. Catal. B* **2001**, *31*, 145–157. [CrossRef]
53. Xiao, Q.; Si, Z.; Zhang, J.; Xiao, C.; Tan, X. Photoinduced hydroxyl radical and photocatalytic activity of samarium-doped TiO_2 nanocrystalline. *J. Hazard. Mater.* **2008**, *150*, 62–67. [CrossRef] [PubMed]

MDPI AG
Grosspeteranlage 5
4052 Basel
Switzerland
Tel.: +41 61 683 77 34

Catalysts Editorial Office
E-mail: catalysts@mdpi.com
www.mdpi.com/journal/catalysts

Disclaimer/Publisher's Note: The title and front matter of this reprint are at the discretion of the Guest Editor. The publisher is not responsible for their content or any associated concerns. The statements, opinions and data contained in all individual articles are solely those of the individual Editor and contributors and not of MDPI. MDPI disclaims responsibility for any injury to people or property resulting from any ideas, methods, instructions or products referred to in the content.

www.ingramcontent.com/pod-product-compliance
Lightning Source LLC
LaVergne TN
LVHW072321090526
838202LV00019B/2324